AF615409

THE BASAL GANGLIA

Structure and Function

ADVANCES IN BEHAVIORAL BIOLOGY

Editorial Board:

Jan Bures *Institute of Physiology, Prague, Czechoslovakia*
Irwin Kopin *National Institute of Mental Health, Bethesda, Maryland*
Bruce McEwen *Rockefeller University, New York, New York*
James McGaugh *University of California, Irvine, California*
Karl Pribram *Stanford University School of Medicine, Stanford, California*
Jay Rosenblatt *Rutgers University, Newark, New Jersey*
Lawrence Weiskrantz *University of Oxford, Oxford, England*

Recent Volumes in this Series

Volume 17 ENVIRONMENTS AS THERAPY FOR BRAIN DYSFUNCTION
Edited by Roger N. Walsh and William T. Greenough

Volume 18 NEURAL CONTROL OF LOCOMOTION
Edited by Richard M. Herman, Sten Grillner, Paul S. G. Stein, and Douglas G. Stuart

Volume 19 THE BIOLOGY OF THE SCHIZOPHRENIC PROCESS
Edited by Stewart Wolf and Beatrice Bishop Berle

Volume 20 THE SEPTAL NUCLEI
Edited by Jon F. DeFrance

Volume 21 COCAINE AND OTHER STIMULANTS
Edited by Everett H. Ellinwood, Jr. and M. Marlyne Kilbey

Volume 22 DISCRIMINATIVE STIMULUS PROPERTIES OF DRUGS
Edited by Harbans Lal

Volume 23 THE AGING BRAIN AND SENILE DEMENTIA
Edited by Kalidas Nandy and Ira Sherwin

Volume 24 CHOLINERGIC MECHANISMS AND PSYCHOPHARMACOLOGY
Edited by Donald J. Jenden

Volume 25 CHOLINERGIC MECHANISMS: Phylogenic Aspects, Central and Peripheral Synapses, and Clinical Significance
Edited by Giancarlo Pepeu and Herbert Ladinsky

Volume 26 CONDITIONING: Representation of Involved Neural Functions
Edited by Charles D. Woody

Volume 27 THE BASAL GANGLIA: Structure and Function
Edited by John S. McKenzie, Robert E. Kemm, and Lynette N. Wilcock

A Continuation Order Plan is available for this series. A continuation order will bring delivery of each new volume immediately upon publication. Volumes are billed only upon actual shipment. For further information please contact the publisher.

THE BASAL GANGLIA

Structure and Function

Edited by

John S. McKenzie

Robert E. Kemm

Lynette N. Wilcock

University of Melbourne
Parkville, Victoria, Australia

PLENUM PRESS • NEW YORK AND LONDON

Library of Congress Cataloging in Publication Data

Main entry under title:

The Basal ganglia.

(Advances in behavioral biology; v. 27)
"Based on proceedings of a symposium on the basal ganglia: structure and function, a satellite symposium of the 29th International Congress of the Union of Physiological Sciences, held August 28–September 2, 1983, in Lorne, Australia"—T.p. verso.
Includes bibliographies and index.
1. Basal ganglia—Congresses. I. McKenzie, John S. II. Kemm, Robert E. III. Wilcock, Lynette N. IV. International Union of Physiological Sciences. Congress (29th: 1983: Lorne, Vic.) V. Series.
QP383.3.B37 1984 599′.01′88 84-13343
ISBN 0-306-41808-8

Based on proceedings of a symposium on The Basal Ganglia: Structure and Function, a satellite symposium of the 29th International Congress of the Union of Physiological Sciences, held August 28–September 2, 1983, in Lorne, Australia

©1984 Plenum Press, New York
A Division of Plenum Publishing Corporation
233 Spring Street, New York, N.Y. 10013

All rights reserved

No part of this book may be reproduced, stored in a retrieval system, or transmitted in any form or by any means, electronic, mechanical, photocopying, microfilming, recording, or otherwise, without written permission from the Publisher

Printed in the United States of America

PREFACE

This volume arose out of the symposium: "The Basal Ganglia: Structure and Function," held at the beginning of September 1983 as a satellite of the 29th International Congress of Physiological Sciences. The symposium took place at Lorne, a village on the ocean 150km south-west of Melbourne in a former holiday guest-house situated beside the beach. The sounds of surf and winter rain on the iron roof provided a background to the proceedings. The symposium was a happy and productive event, among a small group of participants from twelve countries, undistracted by any competing activities in the out-of-season period. Over three days, there were formal papers with lively discussion, as well as posters displayed continuously and available for comment during coffee and lunch breaks. The more philosophical views on the basal ganglia were aired at informal evening discussions after dinner. At the symposium banquet on the final night, the participants voted to form the International Basal Ganglia Society (IBAGS); Malcolm Carpenter was elected Foundation President, with Richard Faull as Organizing Secretary.

The book comprises papers prepared by participants after returning home, so that they had opportunities for incorporating fruits of symposium discussions. Some anticipated contributors were finally unable to participate, and a few who presented data preferred not to submit papers for the book. Five participants - Lucy Brown, Ivan Divac, Irena Grofova, Gunilla Oberg, and Tony Phillips - were prevented by pressure of other commitments from preparing manuscripts in time for the publication. The omissions mean that some areas of investigation into structure, neural processes, and behavioural function are not represented. For example, we regret there is no systematic consideration of the ventral striatum - ventral pallidum component, nor of functional plasticity, learning, transplantation of brain tissue and recovery of function as they involve the basal ganglia. Ontogeny and phylogeny were well reviewed in the earlier publication, "The Neostriatum." Questions of etiology, histopathology and therapy of basal ganglia diseases were deliberately under - emphasized, despite their importance to the whole picture.

Otherwise the book broadly reviews the current state of knowledge about the basal ganglia, and we believe it presents a stimulating assembly of research findings, reviews and theory in the area. We could allow neither space nor time to include transcripts of discussions, but we trust they have influenced the authors in preparing their manuscripts; this certainly is true for the concluding article.

We wish to thank all who helped in conducting the symposium, especially Capri Stewart for her efficient organization of slide projection and tape recording of all proceedings. The organising work for the symposium and editorial preparation of camera - ready copy for the book were carried out in the course of our academic duties for the University of Melbourne, whose support we acknowledge. We are heavily indebted to the Physiology Department and its chairman, Sandford L. Skinner, for assistance and encouragement in many ways, large and small. We thank Carolyn, Heather and Pat for helping us to meet the manuscript deadline.

The symposium and editorial assistance were supported financially by the following donors, whose generous help we gratefully acknowledge:

The Australian Brain Foundation
The Ian Potter Foundation
The Van Cleef Foundation
The University Branch, National Australia Bank
Ansett Airlines of Australia
Merck, Sharp, Dohme (Australia)
S. Karger, A.G., Basel
Sandalford Wines.

John S. McKenzie
Robert E. Kemm
Lynette N. Wilcock

CONTENTS

ANATOMY OF THE SYSTEM - INTERCONNECTIONS, INPUTS AND OUTPUTS

NEURAL MECHANISMS - SYNAPTIC ACTIONS, TRANSMITTERS AND RECEPTORS, OUTPUT MECHANISMS

FUNCTIONS OF THE BASAL GANGLIA IN POSTURE, MOVEMENT AND BEHAVIOUR

INTERCONNECTIONS BETWEEN THE CORPUS STRIATUM AND BRAIN STEM NUCLEI

Malcolm B. Carpenter

Department of Anatomy
F. Edward Hébert School of Medicine
Uniformed Services University of the Health Sciences
Bethesda, Maryland 20814

INTRODUCTION

The basal ganglia are large subcortical nuclear masses classically considered to be telencephalic derivatives. Two major divisions of the basal ganglia have long been recognized and considered to be associated with distinctive functions. The corpus striatum, consisting of the putamen, caudate nucleus and globus pallidus, is regarded as primarily concerned with integrative aspects of somatic motor function. The phylogenetic older amygdaloid nuclear complex receives profuse olfactory inputs, has reciprocal connections with various parts of the hypothalamus and is regarded as an integral part of the limbic system. It appears to be concerned with visceral, endocrine and behavioral functions. Although the corpus striatum (neostriatum and paleostriatum) and the amygdala (archistriatum) have been considered to have separate and distinctive functions, recent evidence indicates a voluminous amygdalostriate projection distributed in a specific pattern (Kelley et al., 1982). These observations suggest that some functions of basal ganglion components may not be easily segregated and that the neostriatum may consist of limbic and non-limbic subdivisions (Figure 1).

The two components of the corpus striatum appear to have different embryological origins. The caudate nucleus and putamen are telencephalic derivatives which develop from different parts of the striatal ridge and follow separate migratory paths which account for their particular configuration (Hamilton and Mossman, 1972); continuity between these structures is preserved rostrally and ventrally. The globus pallidus, forming the smaller medial part of the lentiform nucleus, lies medial to the putamen throughout most of

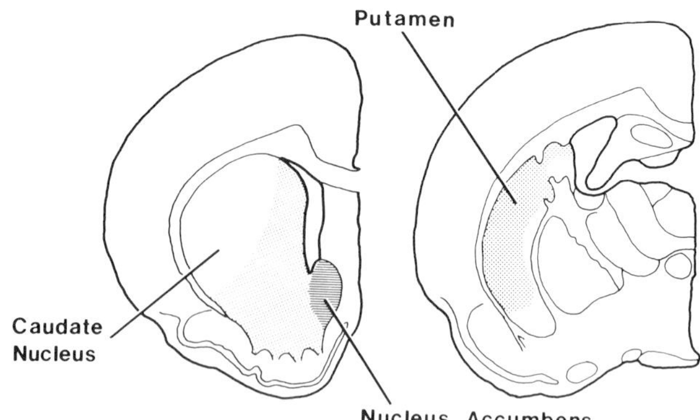

Fig. 1. Schematic diagram of amygdalostriate projections (shaded areas) in the rat which form a basis for dividing the striatum into "limbic" (shaded) and "non-limbic" parts (based on Kelley et al., 1982).

its extent. The two segments of the globus pallidus are separated from each other and from the putamen by medullary laminae. Both segments of the globus pallidus and the subthalamic nucleus are believed to be diencephalic derivatives arising from a dorsal lateral cell column of the hypothalamus referred to as the "subthalamic longitudinal zone" (Kuhlenbeck, 1948; Kuhlenbeck and Haymaker, 1949; Richter, 1965). The anlage of both pallidal segments and the subthalamic nucleus are arranged in column with the precursor of the lateral pallidal segment most rostral and that of the subthalamic nucleus most caudal. In the third fetal month the anlage of the lateral pallidal segment migrates rostrolaterally to contact the medial surface of the putamen. The entopeduncular nucleus follows the course of the lateral pallidal segment rostrolaterally and assumes an adjacent medial position, where it is called the medial pallidal segment. The most caudal part of the longitudinal hypothalamic cell column, the anlage of the subthalamic nucleus, is prevented from migrating laterally by the development of fibers of the hemispheric stalk. Although the subthalamic nucleus becomes separated from the two pallidal segments by the fibers of the internal capsule, anatomical connections between these structures of common origin is maintained by fibers which traverse the peduncular part of the internal capsule.

The two divisions of the corpus striatum appear to serve distinct and separate functions (Figure 2). The neostriatum, considered the receptive component, receives the massive inputs which originate from diverse sources. Striatal afferents originate from broad regions of the telencephalon, from the phylogenetic older diencephalic nuclei and from more than one mesencephalic structure.

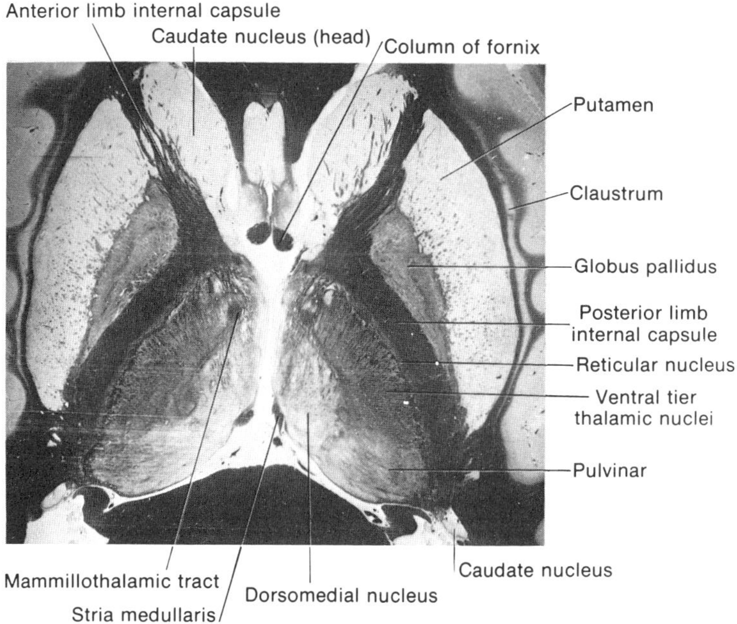

Fig. 2. Horizontal section of the human brain through the thalamus, internal capsule and corpus striatum. Pal Weigert stain (from Carpenter and Sutin (1983), Human Neuroanatomy, courtesy of Williams & Wilkins, Baltimore).

Most of the striatal afferent systems are associated with distinctive neurotransmitters. The neural activities of the neostriatum involve the largest number of distinctive cell types and the largest number of different neurotransmitters, which include acetylcholine, monamines, peptides and amino acids (Dahlstrom and Fuxe, 1964; Anden et al., 1966; Ungerstedt, 1971; Jacobwitz and Palkovits, 1974; DiFiglia et al, 1976, 1982; Pasik et al., 1979; Graybiel and Ragsdale, 1979; Pickel et al., 1980, Graybiel et al., 1981; Haber and Elde, 1981; Fibiger, 1982; Haber and Nauta, 1983). The striatum has a form of reciprocal connections with the distinctive cytological subdivisions of the substantia nigra in which fibers arise from and terminate on different cell populations.

Specific alterations of complex interrelationships between different neurotransmitters involved in striatal activities constitute one of the most critical aspects of the metabolic disturbances that underlie two forms of dyskinesia, namely, Parkinsonism and Huntington's disease (Hornykiewicz, 1966; Bird and Iversen, 1974).

The globus pallidus forming the smaller, most medial part of the lentiform nucleus consists of two cytologically similar segments that have common input systems with both common and distinctive neurotransmitters (Figs. 2 and 10). Each pallidal segment gives rise to an output system distributed differently in the brain stem. Curiously the medial pallidal segment, representing only 30% of the total pallidal volume (Thorner et al., 1975), gives rise to the major part of the output system extending beyond the corpus striatum. The projections of the medial pallidal segment are to ipsilateral thalamic nuclei which in turn have access to motor regions of the cerebral cortex (Nauta and Mehler, 1966; Kuo and Carpenter, 1973). The lateral pallidal segment, with a greater volume and cell density, projects primarily to portions of the subthalamic nucleus (Carpenter et al., 1968; Carpenter et al., 1981). Reciprocal connections interrelate the lateral pallidal segment and the subthalamic nucleus, but not on a point-to-point basis. The output systems of the corpus striatum arise from the medial pallidal segment and from the pars reticulata of the substantia nigra. There are striking similarities between the components of the output system arising separately from the medial pallidal segment and the pars reticulata of the substantia nigra in that: (1) cells and synaptic terminals bear strong morphological resemblances, (2) both receive major afferents from the striatum that have similar neurotransmitters, (3) both receive inputs from the subthalamic nucleus, (4) neither receives inputs from the cerebral cortex or thalamus and (5) both have major thalamic projections to distinctive ventral tier thalamic nuclei without overlap (Carpenter, 1981).

When Wilson (1912) introduced the term extrapyramidal system in his classic description of hepatolenticular degeneration, it was without definition. There seemed to be no question that the corpus striatum formed the centerpiece of this so-called system. Although this term has been defined, debated and interpreted in innumerable ways, it now appears generally accepted that the corpus striatum has meaningful connections with only a limited array of brain stem nuclei, chief among which are the substantia nigra, the subthalamic nucleus and the ventral tier thalamic nuclei. Whether the corpus striatum and these brain stem nuclei form a system can still be debated.

STRIATUM

The caudate nucleus and putamen appear as parts of a single anatomical and functional entity, partially separated in their spatial disposition for embryological reasons. The caudate nucleus has three parts: (1) a head which protrudes into the anterior horn of the lateral ventricle, (2) a body occupying a suprathalamic position and (3) a tail in the roof of the inferior horn of the lateral ventricle. The putamen, the largest part of the striatum in man and primates, is situated deep to the insular cortex where it lies between the external capsule and the lateral medullary lamina of the globus pallidus (Figs. 2 and 10). Lightly stained bundles of myelinated fibers traversing the putamen in a ventromedial course represent striopallidal fibers. Rostrally and ventromedially the putamen and the head of the caudate nucleus are continuous beneath the anterior limb of the internal capsule.

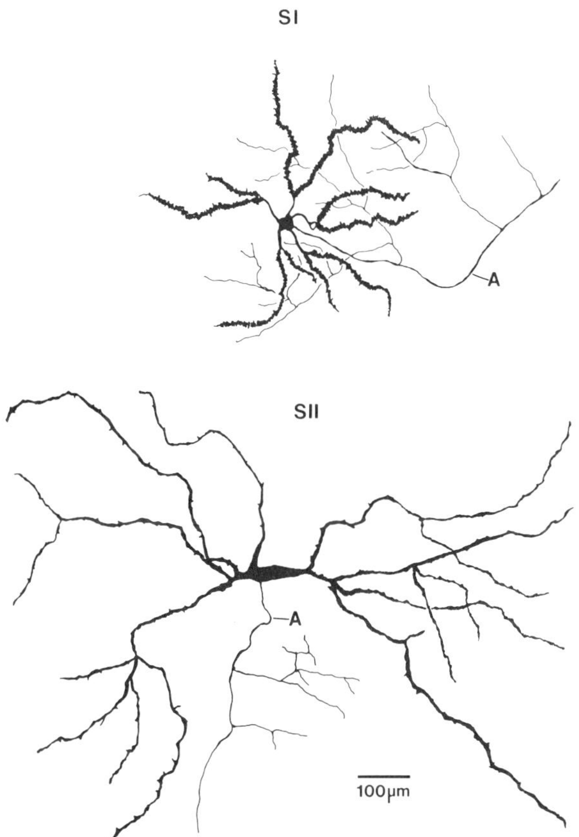

Fig. 3. Drawings of Spiny I (SI) and Spiny II (SII)) striatal efferent neurons. Spiny II neurons vary in size and shape and have dendrites that extend 600 μm from the somata. Spiny I neurons have dendrites that radiate into a spheroid domain of about 200 μm, but the first 20 μm of the dendritic stems are free of spines. GABA and enkephalin are considered to be the neurotransmitters of Spiny I neurons (Ribak et al., 1979; Pasik et al., 1979; DiFiglia et al., 1982). A, indicates axons.

Cytology

Cytologically the caudate nucleus and the putamen appear identical and are composed of enormous numbers of small cells with scant cytoplasm, small numbers of large cells with chromatic cytoplasmic granules and variable numbers of medium-sized cells which exhibit no lamination or special arrangements (Fox et al., 1971, 1971/1972; Kemp and Powell, 1971; Pasik et al., 1979). Attempts to subdivide the striatum into cytoarchitectonic fields have not been successful (Namba, 1957). Morphometric data indicate that small and medium-sized cells (11,000 cells/mm^3) outnumber large cells (65 cells/mm3) by ratios ranging from 130:1 to 258:1 (Schröder et al., 1975). There is reason to believe that the striatum may not be as uniform as it appears because in the developing striatum cells of different types migrate in clusters (Brand and Rakic, 1979), histochemical activity seems to have a patchy distribution (Graybiel and Ragsdale, 1980; Graybiel et al., 1981, 1981a) and efferent neurons are said to be grouped into complex geometrical configurations (Graybiel et al., 1979).

Golgi and electron microscopic studies indicate that striatal neurons fall into two categories: (1) those with spiny dendrites and (2) those with smooth dendrites (Adinolfi and Pappas, 1968; Fox et al., 1971, 1971/1972; Kemp and Powell, 1971; DiFiglia et al., 1976; Pasik et al., 1976, 1979). Spiny neurons, considered the most numerous striatal neuron, are round or oval, have relatively large nuclei, emit up to seven or eight primary dendrites covered with both sessile and pedunculated spines and have long axons which follow a tortuous course (Fig. 3). Two types of spiny striatal neurons are recognized (DiFiglia et al., 1976; Pasik et al., 1976, 1979). Type I Spiny neurons, present in enormous numbers, have smooth somata and proximal dendrites, but at about 20 μm from the soma the dendrites abruptly become laden with dendritic spines; dendrites of these neurones radiate outward into spherical space of about 200 μm (Kemp and Powell, 1971; Fox et al., 1971; Pasik et al., 1979). Axons of these cells arise from the somata or proximal dendrites and give rise to collateral networks which may be coextensive with the dendritic field; other collaterals form less extensive networks at greater distances from the cell body. Although the principal axon in Golgi preparations has not been followed beyond the borders of the striatum (Pasik et al., 1979), retrograde labeling technics suggest that the main axons reach both the globus pallidus and the substantia nigra (Grofova, 1975, 1979; Bak et al., 1978; Preston et al., 1980; Chang et al., 1981; Somogyi et al., 1981). Ultrastructurally Spiny type I neurons lack significant invaginations of the nucleus and have a rough endoplasmic reticulum but no formed Nissl bodies (Fox et al., 1971/1972; DiFiglia et al., 1980). Immunocytochemical localization of glutamic acid decarboxylase (GAD) in the soma of type I Spiny neurons and in synaptic endings in the globus pallidus and substantia

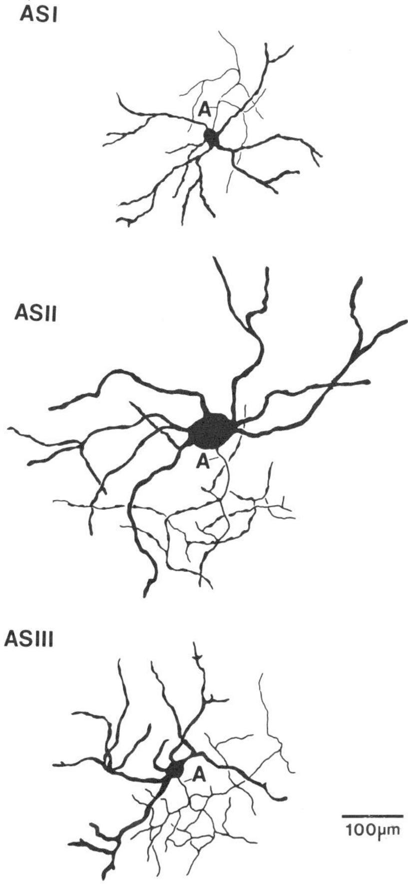

Fig. 4. Drawings of Aspiny striatal neurons whose processes are contained within the striatum. Aspiny I (ASI) neurons are the most frequently impregnated striatal cell in Golgi preparations; this local circuit neuron probably has GABA as its neurotransmitter (Ribak et al., 1979). Aspiny II (ASII) neurons correspond to a population of giant neurons considered to have acetylcholine as their neurotransmitter (Kimura et al., 1980;; Henderson, 1981). The neurotransmitter of Aspiny III (ASIII) has not been identified (based on Pasik et al., 1979). A, indicates axons.

nigra provide convincing evidence that large numbers of these striatal neurons contain γ-aminobutyric acid (GABA) which acts as their neurotransmitter (Jessell et al., 1978; Ribak et al., 1979, 1980, 1981; Ribak, 1981). Light and electron microscopic study of

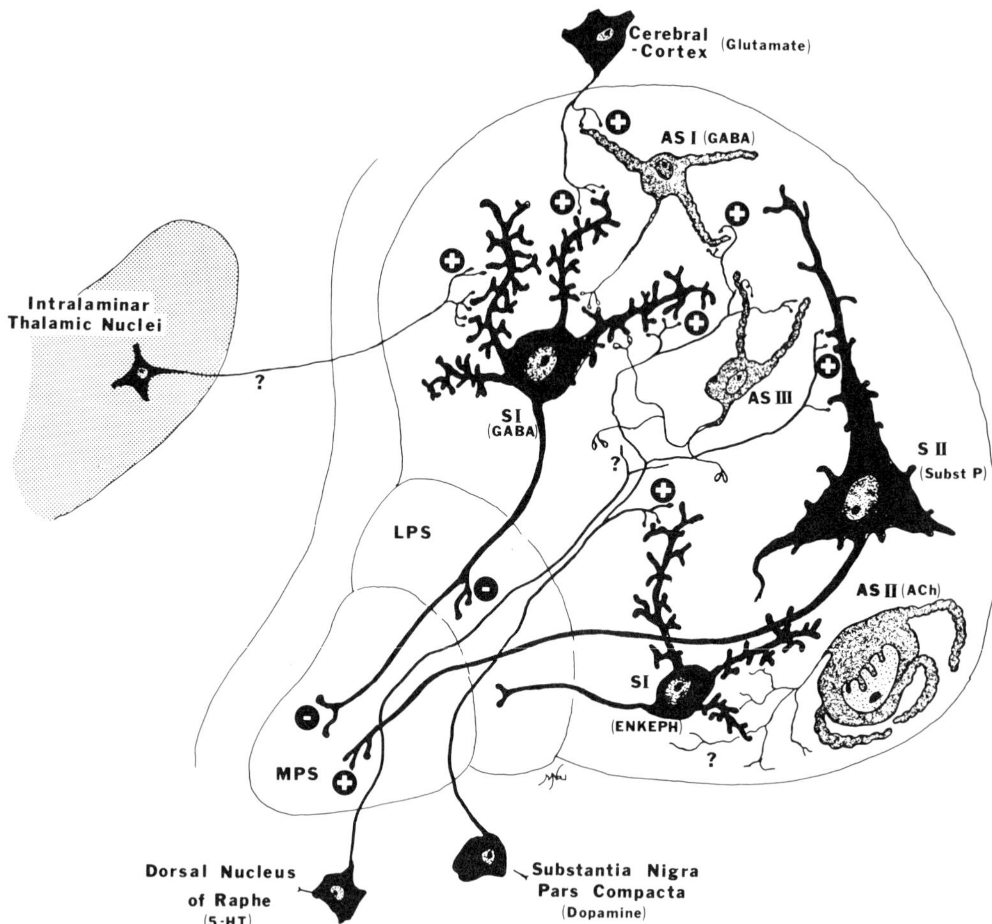

Fig. 5. Schematic drawing of striatal neurons with inputs and outputs, tentative indications of excitatory (+) or inhibitory (-) action and suspected neurotransmittors. Striatal inputs arise from: (1) cerebral cortex (glutamate), (2) intralaminar thalamic nuclei (glutamate?), (3) pars compacta of substantia nigra (SNC) (dopamine) and (4) dorsal nucleus of the raphe (5-HT, serotonin). Striatal outputs arise from Spiny neurons and project to both pallidal segments and pars reticulata of substantia nigra (SNR). Spiny I (SI) neurons have GABA and enkephalin as neurotransmitters. Spiny II (SII) probably have substance P as neurotransmitter. Aspiny I (ASI) and Aspiny II (ASII) local circuit neurons are considered to be GABAergic and cholinergic, respectively. See text for a more detailed description (based on Pasik et al., 1979 and Difiglia et al., 1982).

the localization of immunoreactive leu-enkephalin in the basal ganglia of the monkey, indicate that only striatal neurons contain this opioid pentapeptide (DiFiglia et al., 1982). Labeled medium-sized somata were identified in the caudate nucleus and ventromedial regions of the putamen. Leu-enkephalin labeled fibers were found in order of greatest density in the lateral pallidal segment, the pars reticulata of the substantia nigra and the striatum. At the electron microscopic level labeled striatal neurons had large, slightly indented nuclei, scant cytoplasm, smooth primary dendrites and distal spinous dendrites, characteristics which distinguish type I Spiny neurons. Thus Spiny type I striatal neurons may be associated with two different neurotransmitters (Fig. 5).

Spiny type II striatal neurons vary considerably in size and shape but commonly are large neurons with a spindle shape and spiny dendrites that extend 600 μm from the somata (Fig. 3). Spiny processes tend to be stubby and less dense than those on Spiny I neurons (Pasik et al., 1979; Groves, 1983). The main axons of Spiny II neurons are long and give off collaterals mainly near the somata. The ultrastructural appearance of Spiny II neurons is characteized by a centrally placed nucleus with numerous invaginations, a large volume of cytoplasm, a rough endoplasmic reticulum forming small Nissl bodies and an extensive Golgi system. While the neurotransmitter of Spiny II neuons has not been established, it has been suggested that it may be the source of excitatory striatal output to the medial pallidal segment and the pars reticulata of the substantia nigra (Jessell et al., 1978; Ljungdahl et al., 1978; Gauchey et al., 1979; DiFiglia et al., 1981). It seems likely that substance P is contained in Spiny II neurons and acts as their neurotransmitter (Gale et al., 1977; Hong et al., 1977; Kanazawa et al., 1977).

Three short-axoned Golgi type II striatal neurons are referred to as Aspiny neurons (Pasik et al., 1979). The Aspiny type I neuron is distinguished by its small size, varicose and recurring dendrites and a short, highly arborized axon (Fig. 4). Immunocytochemical localization of GAD suggests a localization in neurons with cytological features of Aspiny I cells (Ribak et al., 1979). In Golgi preparations Aspiny I neurons are the most frequently impregnated striatal cell (Pasik et al., 1979). Aspiny type II neurons have large cell bodies, eccentric nuclei and dendrites extending 250 um or more. These cells correspond to a subpopulation of giant striatal neurons seen in Nissl preparations. Immunohistochemical method demonstrate that Aspiny II neurons (Figs. 4 and 5) stain for choline acetyltransferase (Kimura et al., 1980) and acetylcholinesterase has been localized within these neurons in ultrastructural studies (Henderson, 1981). Aspiny type III striatal neurons have centrally placed nuclei, a thin rim of cytoplasm, rather straight, smooth dendrites approximately 150 um in length and a short axon with extensive arborizations. The neurotransmitter of this interneuron has not been identified.

Striatal Connections

The caudate nucleus and the putamen receive the principal afferent systems projecting to the corpus striatum. Afferent fibers arise from the cerebral-cortex, parts of the amygdala, the intralaminar thalamic nuclei, the substantia nigra and the dorsal nucleus of the raphe.

Corticostriate fibers. Virtually all regions of the neocortex project fibers to the striatum and all parts of the striatum receive fibers from the cortex (Carmen et al., 1963, 1965; Kemp and Powell, 1970). Cortical projections to the striatum are considered to be topographically organized in all dimensions. No part of the striatum is under the sole influence of one area of the neocortex. The projection to the striatum from the sensorimotor cortex is substantial while that from the visual cortex is small. Corticostriate projections from the association cortex of the frontal and parietotemporal lobes in the monkey are greater than in other mammals (Kemp and Powell, 1970).

Degeneration studies suggested that different regions of cortex projected to specific parts of the striatum (Kemp and Powell, 1970, 1971a). Although overlap of corticostriate projections was a common feature, this was a characteristic largely of adjacent cortical areas. Autoradiographic studies have revealed striking observations undetected by degeneration studies: (1) corticostriate axon terminals form mosaic-like patterns in the striatum (Fig. 6), (2) many cortical areas have widespread projections in several parts of the striatum and (3) widely separated cortical areas give rise to overlapping terminal fields in portions of the striatum (Kunzle, 1975, 1978; Goldman and Nauta, 1977; Jones et al., 1977; Yeterian and van Hoesen, 1978; van Hoesen et al., 1981). The terminal distribution of corticostriate fibers is extensive and characterized by a series of strips or clusters that separate from each other and then become confluent in other regions. This disjunctive pattern occurs after a single cortical injection in portions of both the putamen and caudate nucleus (Kunzle,, 1975; Goldman and Nauta, 1977; Jones et al., 1977; Yeterian and van Hoesen, 1978). The strips and clusters of striatal terminals arising from different cortical areas not only overlap each other, but also may overlap terminal projection zones of other striatal afferent systems. Autoradiographic data reveal that cortical areas reciprocally connected via cortico-cortical connections have their own unique corticostriate projection, and one shared by both cortical areas (Yeterian and van Hoesen, 1978; van Hoesen et al., 1981).

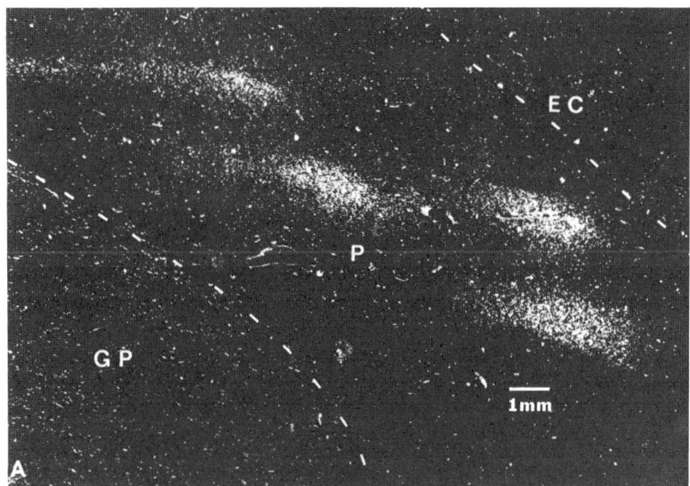

Fig. 6. Dark field autoradiograph of a portion of the putamen containing labeled terminals of corticostriate fibers. GP, globus pallidus; P, putamen; EC, external capsule (from Jones et al., 1977, courtesy of Dr. E. G. Jones and Alan Liss, New York).

Corticostriate fibers arising in the primary motor area (area 4) project bilaterally upon the putamen (Fig. 7) where patchlike terminations are greatest in lateral regions and extend nearly the length of putamen (Kunzle, 1975). The premotor area projects ipsilaterally to both the caudate nucleus and the putamen, while fibers from the prefrontal cortex project via an intranuclear trajectory to all parts of the caudate nucleus (Goldman and Nauta, 1977; Kunzle, 1978).

The laminar origins of the corticostriate system, the most massive projection system to the striatum, have been based upon the retrograde transport of horseradish peroxidase (HRP) using different chromogens. Studies in the monkey indicate that these fibers, both ipsilateral and contralateral, arise from cell populations distinct from other corticofugal systems (Jones and Wise, 1977; Jones et al., 1977). This massive projection is considered to arise from smaller pyramidal cells in the upper half of lamina V in all areas. The implication is that these fibers constitute a distinct projection, rather than collaterals of other descending systems. However, combined intracellular recording and injection of HRP suggest that axons of large pyramidal cells in lamina V which can be activated directly from the thalamus provide striatal collaterals (Jinnnai and Matsuda, 1979; Donoghue and Kitai, 1981). Other evidence suggests that the corticostriate fibers arise from both supragranular (laminae

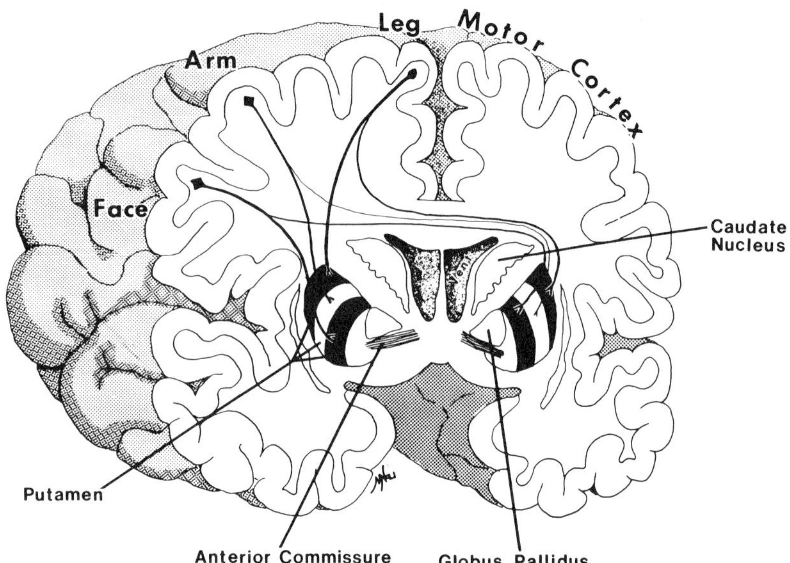

Fig. 7. Schematic diagram of bilateral somatotopically arranged corticostriate projections from the primary motor area (area 4) to the putamen in the monkey (based upon Kunzle, 1975).

II and III) and infragranular (laminae V and VI) layers of the cortex (Kitai et al., 1976; Oka, 1980; Royce, 1982). It appears likely that the corticofugal fibers arise only from the infragranular layers of the cortex. Striatal afferents from the cerebral cortex end principally upon the dendritic spines of Spiny I and Spiny II cells (Kemp,, 1968; Kemp and Powell, 1971b; Frotscher et al., 1981;; Somogyi et al., 1981). There is good evidence that corticostriate fibers are excitatory and have glutamate as their neurotransmitter (Buchwald et al., 1973; Spencer, 1976; Feger et al., 1979; Hattori et al., 1979; Dray, 1980; Kitai, 1981; Fonnum et al., 1981). Because the striatum receives inputs from virtually all areas of the cortex, there are grounds for believing that different regions may be concerned with different functions and that few regions are purely sensory or motor (Divac, 1972; Teuber, 1976).

Amygdalostriate fibers. Although degeneration and axoplasmic transport studies have indicated a modest projection from the amygdala to the ventrocaudal striatum (Nauta, 1961; Kettek and Price, 1978; Royce, 1978), recent data suggest a widespread amygdalostriate projection to the caudal (i.e. caudal to the anterior commissure) striatum (Kelley et al., 1982). Rostral to the anterior commissure this striatal projection is dense only in ventromedial regions. A modest symmetrical contralateral distribution reaches the striatum via the anterior commissure. The amygdalostriatal projection originates mainly from the basal lateral amygdaloid nucleus and appears to overlap striatal projections from the ventral tegmental area and the midbrain raphe nuclei. These observations have suggested a division of the striatum into "limbic" and "non-limbic" parts (Fig. 1). The smaller "non-limbic" striatum occupies an antero-dorsolateral sector and in the rat receives projections from the medial agranular sensorimotor cortex. This parcellation of the striatum into distinctive regions on the basis of amygdaloid projections suggests that the "limbic" striatum may be concerned with behavioural phenomenon.

Thalamostriate fibers. The intralaminar thalamic nuclei have for a long time been identified as an important source of striatal afferents (Vogt and Vogt, 1941; McLardy, 1948; Powell and Cowan, 1954, 1956).. Degeneration studies clearly indicated that the centromedian-parafascicular nuclear complex (CM-PF) projects to both the caudate nucleus and the putamen (Mehler, 1966; Powell and Cowan, 1967). Retrograde HRP studies indicated that injections of the putamen produced intense labeling of cells in the CM-PF and some labeling in cells in posterior parts of the central lateral nucleus (CL) (Jones and Leavitt, 1974); HRP injections in various cortical areas labeled corresponding thalamic relay nuclei intensely and lightly labeled portions of the intralaminar thalamic nuclei. These results were interpreted to mean that the intralaminar thalamic nuclei project profusely to the striatum and sparsely and diffusely, via collaterals, to broad cortical regions. HRP injections of the caudate nucleus revealed enzyme granules within all intralaminar thalamic nuclei, but most prominently in CM and CL (Royce, 1978); in addition, some labeled cells were found in the ventral anterior and mediodorsal thalamic nuclei. The projection from CL to the caudate nucleus involved only small cells in medial parts of the nucleus; large cells in lateral parts of CL were not labeled. Autoradiographic studies in monkey and cat showed that thalamostriate fibers from CM terminate, like corticostriate fibers, in a mosaic of disk-shaped aggregates or hollow rings in the caudate nucleus and putamen (Fig. 8) (Kalil, 1978; Royce, 1978). In the monkey CM had no projection to the caudate nucleus and labeling of the entire CL produced no transport to the striatum (Kalil, 1978). Physiological data indicate that cells in the central lateral and paracentral (PCN) nuclei contain two populations of neurons, one projecting to cortex and one projecting to the caudate nucleus; these different neural populations have different conduction velocities (Steriade and Glenn, 1982). Monosynaptic relays in CL and PCN from midbrain reticular efferents conduct impulses to many separate cortical areas and to the caudate nucleus. Thalamostriate fibers appear to terminate upon Spiny I neurons (Fox et al., 1971; Kemp and Powell, 1971b; Groves, 1983) and are thought to be excitatory (Buchwald et al., 1973; Dray, 1980; Kitai, 1981). The neurotransmitter utilized

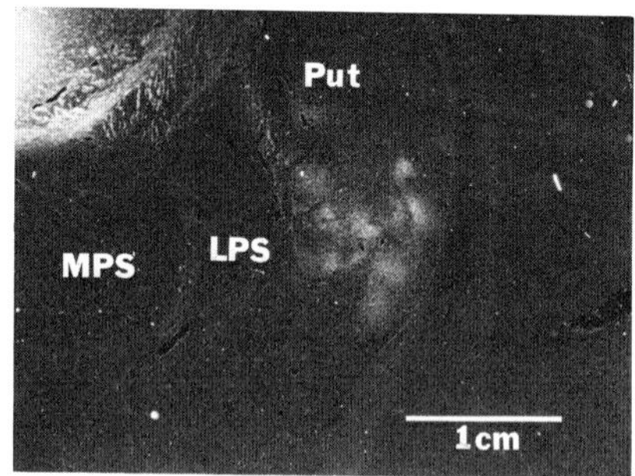

Fig. 8. Dark field autoradiograph of the mosaic pattern of termination of thalamostriate fibers in the putamen (PUT) of a monkey. MPS and LPS indicate medial and lateral pallidal segments.

by thalamostriate fibers is not known, but has been postulated to be glutamate (Fonnum et al., 1981).

Nigrostriatal fibers. Of the various striatal afferent systems, nigrostriatal fibers were the first to be defined (von Monakow, 1895). Attempts to trace nigrostriatal fibers in Marchi preparations were unsuccessful (Ranson et al., 1941; Fox and Schmitz, 1944). Fluorescent histochemical studies not only provided evidence that cells of the pars compacta project to the striatum but indicated that these cells convey dopamine to their terminals (Anden et al., 1964; Dahlstrom and Fuxe, 1964; Fuxe and Anden, 1966; Bedard et al., 1969; Moore et al., 1971; Ungerstedt, 1971). This technique indicated that dopamine, stored in varicosities in nerve terminals, is concentrated mainly in three forebrain areas: (1) the striatum, (2) the nucleus accumbens septi and (3) the olfactory bulb (Fuxe and Anden, 1966). Terminal varicosities in the striatum are fine, densely packed and exhibit a diffuse green fluorescence. Dopaminergic fibers form a matrix of fine varicose axons around both small and large striatal neurons (Siggins et al., 1976). In the developing striatum the glyoxylic acid histofluorescence method reveals a characteristic patchwork of dopamine "islands" in the striatum which is strikingly similar to the compartmentalization of acetylcholinesterase (Graybiel et al., 1981a). Although high-resolution autoradiography with [^{3}H] dopamine suggests displacement of the tracer during histological processing, double fixation with glutaraldehyde and osmium by vascular perfusion retains [^{3}H] dopamine *in situ* (Descarries et al., 1980, 1981). Using these methods dopamine axonal varicosities are of small caliber (0.5 μm), contain mostly clear synaptic vesicles with a few larger dense-core vesicles, and endings establish axo-dendritic synaptic junctions.

Studies of the organization of ascending striatal afferents based upon retrograde enzyme transport suggest cells in the rostral two-thirds of the pars compacta of the substantia nigra are related to the head of the caudate nucleus (Fig. 9), while those projecting to the putamen are more posterior (Szabo, 1980,, 1980a). An inverse dorsoventral relationship exists between the substantia nigra and the caudate nucleus, so that cells in the ventral pars compacta project to the dorsal caudate, and vice-versa. Lateral and posterior parts of the substantia nigra are related to dorsal and lateral parts of the putamen. Significant overlap exists in the projection fields of nigrostriatal fibers which may be due to differences in the size of these structures. Paranigral cell groups mainly supply the ventral striatum. Autoradiographic methods in the monkey indicate nigrostriate fibers arise from cells of the pars compacta primarily from central, medial and caudal regions of the substantia nigra (Carpenter et al., 1976). Rostral parts of the substantia nigra appear to project to the head of the caudate nucleus and rostral parts of the putamen. The distribution of silver grains in the striatum differs from that described for corticostriate and

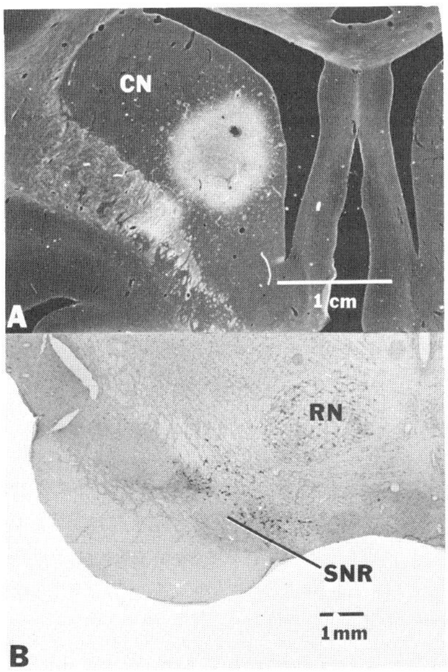

Fig. 9. (A), dark field photomicrograph of injection of WGA-HRP into the head of the caudate nucleus (CN) in a cat. (B), resulting retrograde and anterograde transport of the enzyme to the substantia nigra (bright field photomicrograph). RN, red nucleus; SNR, substantia nigra, pars reticulata.

thalamostriate systems in that they are evenly spread without forming clusters or patches.

Although the major nigrostriastal neurotransmitter is acknowledged to be dopamine, a consensus regarding its postsynaptic action has not emerged (see Groves, 1983). The inference that dopamine is excitatory is based upon: (1) intracellular recordings which show that electrical stimulation of the substantia nigra typically produces EPSPs or EPSP-IPSP sequences in striatal neurons and only rarely IPSPs (Kitai et al., 1976; Kitai, 1981; Wilson et al., 1982), and (2) anatomical data indicating asymmetrical synaptic contacts of dopaminergic terminals on dendritic spines of what are thought to be Spiny I neurons (Groves, 1983). The inference that dopamine is inhibitory is based upon extracellular recordings of striatal activity following application of dopamine or related agonists (Connor, 1970; Gonzalez-Vegas, 1974).

Studies using 6-OHDA or prior lesions in the medial forebrain bundle suggest that not all nigrostriatal fibers are dopaminergic (Fibiger et al., 1972. It has been postulated that some 20% of nigrostriatal fibers are non-dopaminergic. The cell of origin of this distinct striatal projection is uncertain. Non-dopaminergic nigrostriatal fibers are considered to form symmetrical synapses on Spiny striatal neurons (Hattori et al., 1973).

Striatal afferents from the raphe nuclei. Histofluorescent technics have demonstrated several ascending pathways originating from mesencephalic indolamine cell groups in the median raphe (Dahlstrom and Fuxe, 1964; Ternaux et al., 1977). Anatomical, physiological and biochemical investigations suggest that the dorsal (B7) and median (B8) raphe nuclei provide two distinct, but partially overlapping ascending serotininergic (5-HT) systems (Ungerstedt, 1971; Conrad et al., 1974; Miller et al., 1975; Bobillier et al., 1976; Dray et al., 1976; Ternaux et al., 1977; Royce, 1978). Pathways originating from the dorsal and median raphe nuclei, traced by a variety of technics, ascend in the medial forebrain bundle through the hypothalamic region,, but the specific projections of each nucleus differ (Parent et al., 1981). The serotoninergic projections arising from the dorsal nucleus of the raphe terminate mainly in ventrocaudal regions of the striatum (Bobillier et al., 1975; Ternaux et al., 1977; Dray et al., 1978; Dray, 1980). Fluorescent retrograde double labeling technics have demonstrated that the majority of neurons in the dorsal nucleus of the raphe project collaterals to both the striatum and the substantia nigra, although populations of cells within the nucleus have single targets (Kooy and Hattori, 1980). Stimulation of the dorsal nucleus of the raphe produces a strong, long-lasting inhibition of striatal neurons, whereas stimulation of the median raphe nucleus produces only a weak inhibition (Olpe and Koella, 1977). The type of striate neuron receiving projections from the raphe is unknown.

Identification of serotoninergic neurons in the dorsal nucleus of the raphe with [^{3}H] 5-HT in the rat revealed an estimate of 11,500 neurons; but twice as many in the nucleus were unreactive (nonserotoninergic) (Descarries et al., 1982).

Other Striatal Afferents

A number of other small striatal afferent systems include projections from the pontine and caudal mesencephalic reticular formation, the nucleus accumbens septi and the locus ceruleus (Dray, 1980). Axoplasmic transport studies suggest pallidal (i.e. lateral segment) and peripallidal neurons project to the striatum (Nauta, 1979; Staines et al., 1981; Jayaraman, 1983). The morphology and location of some of these neurons resemble those of the magnocellular nucleus basalis. The head of the caudate nucleus and the putamen are also said to receive ipsilateral projections from the subthalamic nucleus; labeled cells are diffusely distributed throughout the rostrocaudal extent of the nucleus (Beckstead, 1983).

GLOBUS PALLIDUS

The globus pallidus, forming the smaller and most medial part of the lentiform nucleus, lies medial to the putamen throughout its extent. A medial medullary lamina divides the pallidum into medial and lateral segments and a less distinct accessory medullary lamina divides the medial pallidal segment into inner and outer portions (Fig. 10). Some large cells belonging to the substantia innominata located along the ventral margin of the pallidum migrate dorsally into medial and lateral medullary laminae.

Cytology

In the human pallidum the volume and cell densities differ in the two segments (lateral segment: 70% of volume, cell density 437 cells/mm^3; medial segment: 30% of volume, cell density 327 cells/mm^3) (Thorner et al., 1975). Estimates of the number of pallidal neurons range from 465,000 to 540,000 for the lateral segment and from 143,000 to 171,000 for the medial segment.

Golgi studies reveal two types of pallidal neurons: (1) large ovoid or polygonal cells with long thick smooth dendrites and (2) small neurons with very few dendrites (Fox et al., 1974). Large pallidal neurons, forming the predominant cell type, have dendrites which extend nearly 900 μm and are entwined with arrays of longitudinally oriented afferent fibers bearing "boutons en passage". Nuclei of large pallidal neurons are irregular with deep infoldings. Golgi preparations do not reveal detectable morphological differences between large pallidal neurons in different segments (Fox et al., 1974). Large pallidal neurons intracellularly injected with HRP appear to be of two subtypes: (1) medially located neurons with

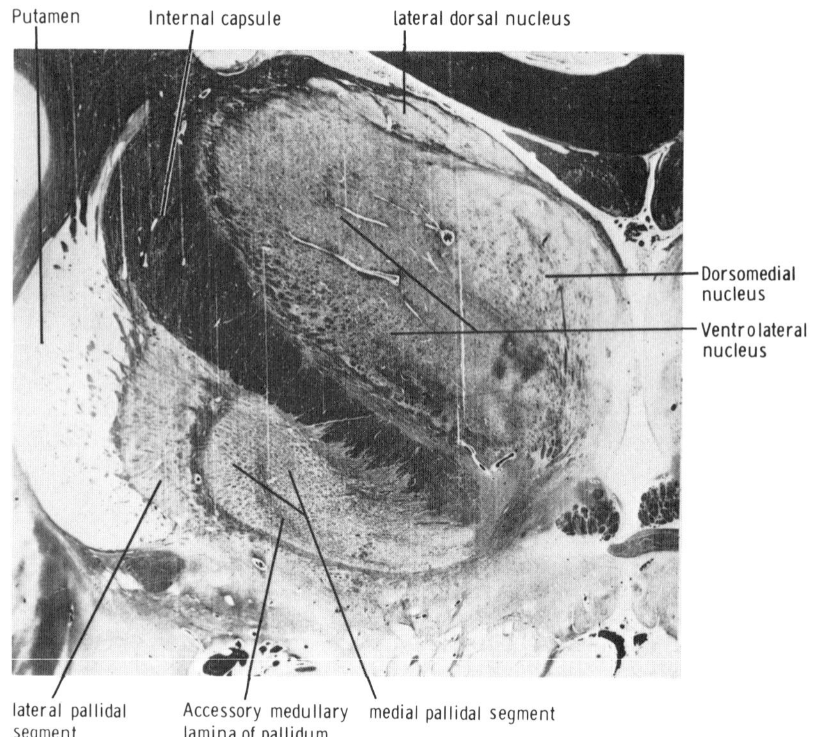

Fig.10. Photomicrograph of a transverse section of the human brain through the corpus striatum, internal capsule and thalamus. Segments of the globus pallidus are separated by the medial medullary lamina, and the accessory medullary lamina divides the medial pallidal segment into inner and outer parts. (Courtesy of George Thieme Verlag, Stuttgart)

mainly dorsoventral dendritic fields and (2) laterally situated neurons with disk-like dendritic fields extending both dorsoventrally and rostrocaudally (Park et al., 1982). Axons of only the second subtype of pallidal neurons possess collaterals.

Pallidal Afferent Connections

Major afferent fibers to the globus pallidus arise from the striatum and the subthalamic nucleus (Fig. 23). Unlike the striatum, the globus pallidus does not receive afferents from the cerebral cortex, the thalamus or the substantia nigra. In degeneration studies of corticostriate fibers, authors that comment upon the globus pallidus report no evidence of corticopallidal fibers (Webster, 1961; Petras, 1969). Autoradiographic studies in the monkey report no projections to the globus pallidus from cortical areas 4, 6, 8 and 9 or from the somatosensory cortex (Kunzle, 1975, 1977, 1978; Kunzle and Akert, 1977).

Striopallidal fibers. The most massive pallidal afferents arise from the striatum and project in an organized manner upon both segments of the globus pallidus. Pallidal afferents are topographically arranged in both dorsoventral and rostrocaudal sequences and radiate in parts of the pallidum like spokes of a wheel (Papez, 1942; Szabo, 1962, 1967, 1970; Cowan and Powell, 1966; Nauta and Mehler, 1966). The head of the caudate nucleus projects to dorsal and rostral parts of the pallidum; the putamen projects to more extensive ventral and caudal pallidal regions. The precommissural part of the putamen projects exclusively to the pallidum, while all other regions of the striatum project to both the globus pallidus and the pars reticulata of the substantia nigra. Bundles of myelinated striopallidal fibers, most numerous in medial regions of the putamen, are referred to as Wilson's pencils (Wilson, 1914). It has been suggested that striopallidal fibers emit collaterals to both pallidal segments, and as the collaterals become thin and nonmyelinated, they enter the "comb" bundle as strionigral projections (Fox and Rafols, 1975; Fox et al., 1975).

Ultrastructural studies demonstrate that pallidal dendrites are studded with axon terminals ensheathed with glia (Kemp, 1970). Large numbers of striopallidal fibers are considered to have γ-aminobutyric acid (GABA) as their neurotransmitter (McGeer et al., 1971; Hattori et al., 1973; Pycock et al., 1976). The globus pallidus is particularly rich in glutamic acid decarboxylase, the enzyme that synthesizes GABA (Okada et al., 1971; Balcom et al., 1975; Fahn, 1976; Fonnum et al., 1978; Ribak et al., 1979). GABA is conveyed to the globus pallidus via axons of Spiny I striatal neurons (Jessell et al., 1978; Pasik et a., 1979; Ribak, 1981).

Immunohistochemical technics reveal that enkephalin and substance P immunoreactive fibers in the globus pallidus are distributed in a specific pattern (Hughes et al., 1975; Elde et al., 1976; Cuello and Kanazawa, 1978; Ljungdahl et al., 1978; Wamsley et al., 1980). Studies in the monkey indicated that enkephalin immunoreactivity was dense in the lateral pallidal segment, where only sparse to moderate substance P staining was detected (Haber and Elde, 1981). Conversely substance P immunoreactivity was dense throughout the medial pallidal segment. Patterns of distribution of enkephalin and substance P immunoreactivity in the rat show similar quantitative features in homologous nuclei. There is a suggestion that regional differences in the distribution and concentration of these two peptides in the globus pallidus as defined by immunohistochemistry may be related to specific pallidal efferent systems (Kuo and Carpenter, 1973). In immunoreacted preparations for enkephalin and substance P the dendrites and cell bodies of unstained pallidal elements (called "woolly fibers") are enmeshed in a plexus of enkephalin or substance P-positive striopallidal fibers (Haber and Nauta, 1983). Using the pattern of immunoreactivity of enkephalin-like and substance P-like substances, these authors have attempted to redefine the borders of the globus pallidus. Enkephalin-like woolly fiber systems fill the entire lateral pallidal segment and extend rostrally and ventral into what is called the "ventral pallidum" (Heimer and Wilson, 1975). The enkephalin-positive region extends into the ventral striatum, where it includes the nucleus accumbens, dorsal regions of the amygdala and the bed nucleus of the stria terminalis. Substance P-like woolly fibers also fill the "ventral pallidum" and invade the olfactory tubercle, but avoid the lateral pallidal segment, the striatum and the amygdala. The term " ventral pallidum" has been defined as the region ventral and rostral to the anterior commissure and includes part of the substantia innominata (Heimer, 1978). In support of the view that this rostral region is an extension of the globus pallidus is the finding that efferents from the nucleus accumbens terminate in it (Nauta et al., 1978).. Immunoreactive striatal neurons containing enkephalin are considered to be Spiny I type neurons (Pickel et al., 1980; DiFiglia et al., 1982). Spiny type II neurons are thought to be the source of substance P-like immunoreactive pallidal fibers (Hong et al., 1977; Kanasawa et al., 1977; Gale et al, 1977, Pasik et al., 1979). Although a large part of the enkephalin-like and substance P-like immunoreactive striopallidal arises from limbic regions of the striatum, these afferents also arise from non-limbic regions (Kelley et al., 1982). Comparison of normal human brains with those from patients with Huntington's disease show substantial reductions in substance P (>80%)) and enkephalin (>50%) in the globus pallidus (both segments) and substantia nigra (Emson et al., 1980).

<u>Subthalamopallidal fibers</u>. Because it is virtually impossible to produce lesions in the subthalamic nucleus that do not involve adjacent fiber systems, degenerative methods contributed little to our understanding of the connections of this nucleus, although it was apparent early that its major projections were to the globus pallidus and substantia nigra (von Monakow, 1895; Balthasar, 1939; Glees and Wall, 1946; Whittier and Mettler, 1949; Knook, 1965). Autoradiographic studies have clearly shown subthalamopallidal fibers are topographically organized and project to both segments of the pallidum in arrays parallel to the medullary lamina (Nauta and Cole, 1978; Carpenter et al., 1981a). The largest number of terminals appear in the medial pallidal segment. Studies of the subthalamic nucleus based on the retrograde transport of HRP from various regions of individual pallidal segments indicate: (1) cells in medial parts of the caudal part of the nucleus project mainly to the medial pallidal segment, (2) cells in medial parts of the middle third of the nucleus project to the rostral part of the lateral pallidal segment, and (3) cells in the central parts of the rostral two-thirds of the nucleus project to the largest part of lateral segment adjacent to the putamen (Fig. 22A and B) (Carpenter et al., 1981).

Cells in the lateral third of the subthalamic nucleus project few fibers to any part of the globus pallidus. An inverse dorsoventral topographical relationship exists between cells in the subthalamic nucleus and the lateral pallidal segment which is similar to that described for the substantia nigra and the head of the caudate nucleus (Szabo, 1980, 1980a). Retrograde fluorescent labeling technics in the rat suggest that virtually all neurons of the subthalamic nucleus project to both the globus pallidus and the substantia nigra (Kooy and Hattori, 1980). In the monkey the projection from the subthalamic nucleus to the medial segment of the globus pallidus appears predominant (Fig. 23) (Carpenter et al., 1981a). Physiological studies in the rat also suggest that subthalamic neurons have branched axons reaching both the pallidum and the nigra, but different responses are recorded in each branch (Hammond et al., 1983). Stimulation of the subthalamic nucleus produces short latency inhibition of spontaneous activity in the pallidum and short latency excitation in the substantia nigra. These dual responses were considered to involve local interneurons.

Retrograde transport of [3H] GABA, from pallidum to subthalamic nucleus neurons, suggests that GABA may be the neurotransmitter of subthalamopallidal fibers (Nauta and Cuenod, 1982). Small localized injections of GABA antagonists (i.e. picrotoxin or bicuculline methiodide) into the subthalamic nucleus of alert monkeys and baboons via implanted cannulae produce contralateral violent choreoid dyskinesia, suggesting that GABA may be a neurotransmitter of subthalamopallidal fibers (Crossman et al.,, 1980; Crossman et al., 1984). This dyskinesia which appears after a short latency is identical with that resulting from small discrete electrolytic lesions in the subthalamic nucleus (Carpenter et a., 1950). Microiontophoretically applied GABA inhibits cells of the entopeduncular nucleus, and subthalamic nucleus-induced inhibition of the entopeduncular nucleus is reversed by iontophoretically applied bicuculline (Rouzaire-Dubois et al., 1983).

Pallidofugal fiber systems. Although cells in the medial and lateral pallidal segments appear identical and their inputs originate from the same sources, there are similarities (GABA) and differences (enkephalin and substance P) in the neurotransmitters conveyed by striopallidal fibers and each pallidal segment projects fibers to different brain stem nuclei. Fibers arising from cells of the medial pallidal segment project to thalamic nuclei, the lateral habenular nucleus, and via a descending tegmental bundle to a cell group in midbrain reticular formation (Fig. 14 and 15). Cells in the lateral pallidal segment project primarily to the subthalamic nucleus, although some terminate in parts of the substantia nigra (Figs. 22C,D).

Pallidothalamic projections. Fibers emerging from the medial pallidal segment via the ansa lenticularis and the lenticular

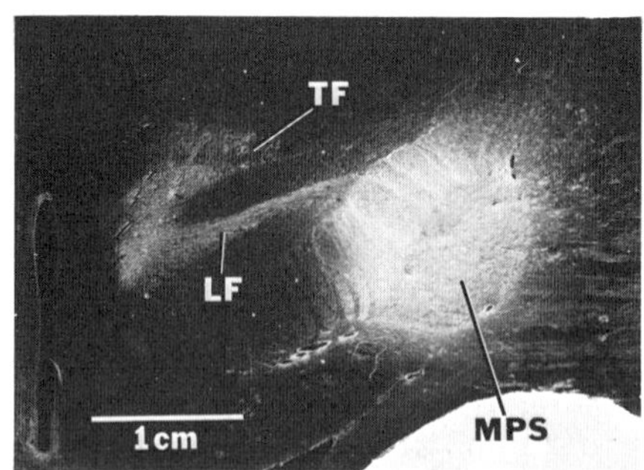

Fig.11. Autoradiograph of fibers in the lenticular fasciculus (LF) traversing the internal capsule, passing medial to Forel's field H and entering the thalamic fasciculus (TF) dorsal to the zona incerta. MPS indicates medial pallidal segment. (Courtesy of the American Physiological Society, Bethesda)

fasciculus follow distinctive courses but merge into Forel's field H; from this area they pass rostrally and laterally as a component of the thalamic fasciculus (Fig. 11). The full extent of this projection system was first appreciated in silver impregnation studies (Nauta and Mehler, 1966). Pallidothalamic fibers (Fig. 12) in the monkey project to the ventral anterior (VApc, pars principalis) and ventral lateral (VLo, pars oralis and VLm, pars medialis) thalamic nuclei and give off collaterals which terminate in the centromedian (CM) nucleus (Nauta and Mehler, 1966; Kuo and Carpenter, 1973; Kim et al., 1976; DeVito and Anderson, 1982). Pallidothalamic fibers terminating in the rostral ventral tier thalamic nuclei are topographically organized and suggest that neurons in caudal portions of the medial pallidal segment establish synaptic articulations mainly with neurons in VLo (Figs. 13 and 14) (Kuo and Carpenter, 1973). The outer portion of the medial pallidal segment gives rise to the ansa lenticularis which projects primarily to VApc.

There is increasing evidence that pallidothalamic terminations do not overlap cerebellar projections from the deep nuclei (Nauta and Mehler, 1966; Kuo and Carpenter, 1973; Kim et al., 1976; Percheron, 1977; Tracey et al., 1980; Asanuma et al., 1983, 1983a). Cerebellothalamic projections terminate in the so-called cell sparse zone which includes the ventral posterolateral nucleus (VPLo, pars oralis), the ventral lateral nucleus (VLc, pars caudalis and VLps, pars postrema) and area X of Olszewski (1952). The so-called cell sparse thalamic zone projects to area 4; VLo which receives pallidothalamic fibers projects to the premotor cortex (area 6) and the supplementary motor area (Asanuma et al., 1983a; Schell and Strick, 1984). Physiological findings indicate that stimulation of the globus pallidus in cats produces only monosynaptic IPSPs in the

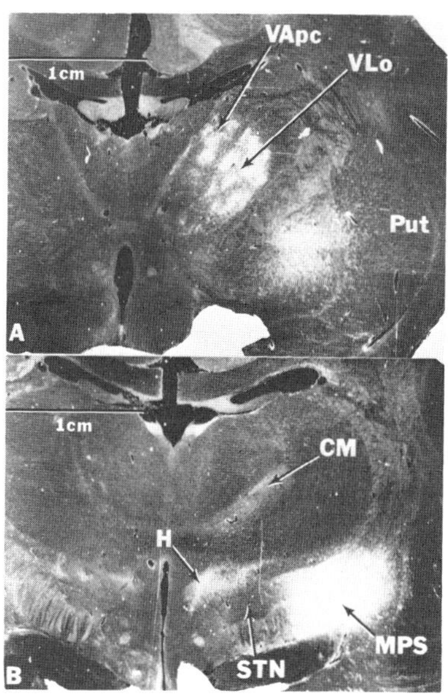

Fig.12. Dark field photomicrographs of autoradiographs of isotope transport from medial pallidal segment (MPS) to thalamic nuclei. (A), isotope transport from MPS to the ventral anterior nucleus (VApc) and ventral lateral nucleus (VLo) terminates in patchy formations. (B), shows central region of injection site in MPS, labeled fibers in Forel's field H (H), and relatively modest transport to the centromedian (CM) nucleus of the thalamus (from Kim et al., 1976, courtesy of Alan Liss, New York and the American Physiological Society, Bethesda).

VA-VL nuclear complex in cats anesthetized with sodium pentobarbital (Uno and Yoshida, 1975). No convergence of pallidal and cerebellar inputs was found on single thalamic neurons.

Pallidohabenular fibers. Fibers from the medial pallidal segment projecting to the lateral habenular nucleus have a complex course, but separate from the ansa lenticularis and the lenticular fasciculus near the apex of the pallidum and follow a dorsal, rostral course through and around the medial part of the internal capsule to enter the stria medullaris (Nauta and Mehler, 1966; Carpenter and Strominger, 1967; Kim et al., 1976). This projection is said to be significantly larger in the cat than the monkey (Larsen and McBride, 1979). Double labeling studies indicate that pallidohabenular fibers arise from a different cell population than pallidothalamic fibers; most of these cells are located in a peripallidal zone which encroaches upon the lateral hypothalamus (Parent and DeBellefeuille, 1982). In the habenular nuclear complex GAD is contained largely in

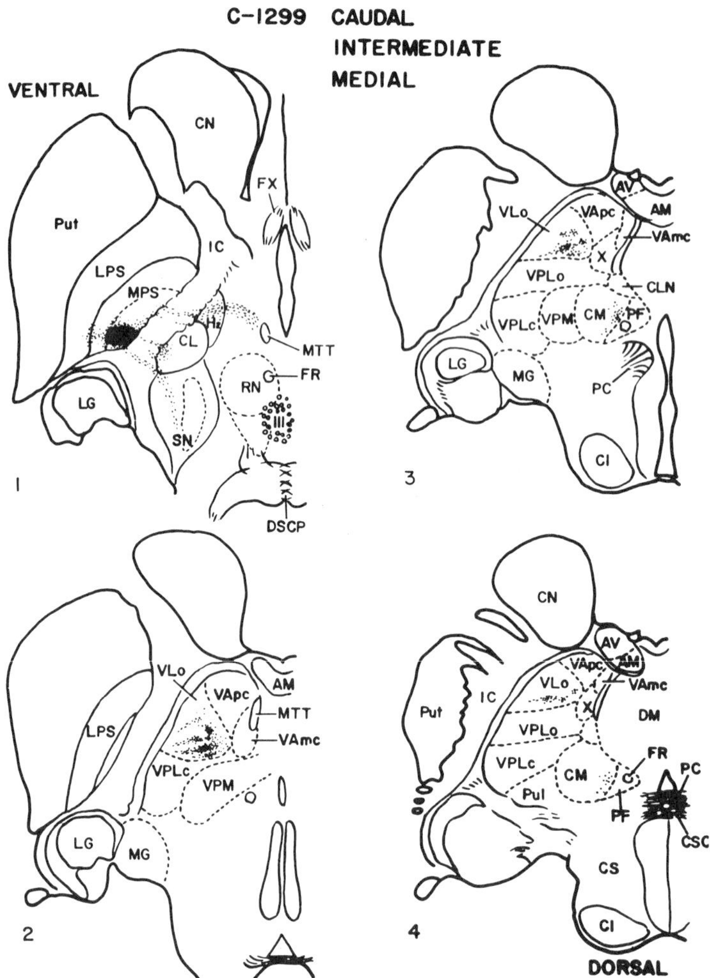

Fig.13. Projection drawings of horizontal sections through the corpus striatum and thalamus showing the terminal degeneration in the thalamus resulting from a small lesion in the caudal part of the medial pallidal segment medial to the accessory medullary lamina. Degeneration passed via the lenticular fasciculus to terminals in the ventral lateral (VLo) and centromedian (CM) thalamic nuclei. Abbreviations used: AM, anteromedial nucleus; AV, anteroventral nucleus; CN, caudate nucleus; CI, inferior colliculus; CL, subthalamic nucleus; CLN, central lateral nucleus; CS, superior colliculus; CSC, commissure of superior colliculus; DM, dorsomedial nucleus; FR, fasciculus retroflexus; FX, fornix; H2, lenticular fasciculus; IC, internal capsule; LG, lateral geniculate; LPS, lateral pallidal segment; MG, medial geniculate; MPS, medial pallidal segment; MTT, mammillothalamic tract;

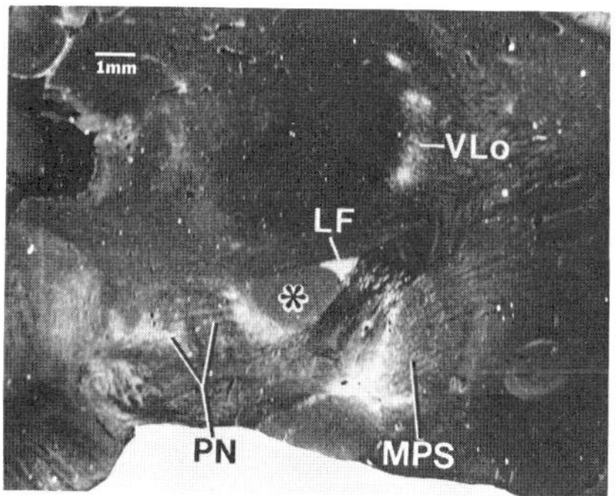

Fig.14. Sagittal dark field autoradiograph demonstrating isotope transport to ventral lateral nucleus of thalamus (VLo) from medial pallidal segment (MPS)) via lenticular fasciculus (LF) and via pallidonigral projections (PN), largely to parts of pars compacta (SNC). No transport of isotope was seen in subthalamic nucleus, indicated by asterisk (*) (from Kim et al., 1976, courtesy of Alan Liss, New York).

the lateral nucleus, suggesting that GABA may be the neurotransmitter of pallidohabenular fibers (Gottsfeld et al., 1977), but this is only one of several metabolic enzymes associated with the habenular nuclei.

Pallidotegmental fibers. The search for a descending pallidal projection in the brain stem has revealed one small pathway in which fibers could be followed only to caudal midbrain levels (Nauta and Mehler, 1966; Carpenter and Strominger, 1967; Kuo and Carpenter, 1973; Kim et al., 1976; Carpenter et al., 1981; DeVito and Anderson, 1982). These fibers descend from field H of Forel, ventral and lateral to the red nucleus and terminate in the compact portion of the pedunculopontine nucleus (PPN) (Figs. 15 and 16A). This nucleus in the monkey is partially embedded in fibers of the superior cerebellar peduncle. The projection from the entopeduncular nucleus to PPN in the cat and rat appears to be much smaller (Carter and

---

Fig. 13. (cont.) PC, posterior commissure; PF, parafascicular nucleus; PUT, putamen; RN red nucleus; SN, substantia nigra; VAmc, ventral anterior nucleus, pars magnocellularis; VAps, ventral anterior nucleus; VLo ventral lateral nucleus, pars oralis; VPLc, ventral posterolateral nucleus, pars caudalis; VPLo, ventral posterolateral nucleus, pars oralis; VPM, ventral posteromedial nucleus; X, area X; ZI, zona incerta. (From Kuo and Carpenter, 1973, courtesy of Alan Liss, New York).

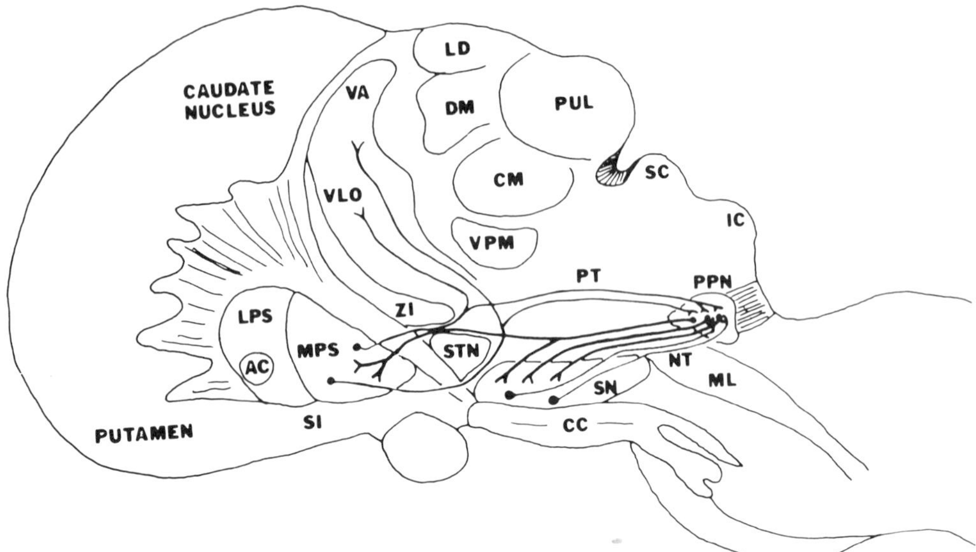

Fig.15. Schematic diagram of pallidotegmental (PT) and nigrotegmental (NT)) projections to pedunculopontine nucleus (PPN) in a sagittal-plane. Neurons in the medial pallidal segment (MPS) give rise to ascending branches that project to the ventral anterior (VA) and ventral lateral (VLo) thalamic nuclei and descending branches that end in PPN (Parent and DeBellefeuille, 1982). Cells in the pars reticulata of the substantia nigra (SN) also project to PPN. Ascending fibers from PPN project mainly to the pars compacta of SN, although small numbers of fibers project to the MPS and the subthalamic nucleus (STN) (Carpenter et al., 1981a). Abbreviations used: AC, anterior commissure; CC, crus cerebri; CM, centromedian nucleus; DM, dorsomedial nucleus; H, field H of Forel; LD, lateral dorsal nucleus; LPS, lateral pallidal segment; ML, medial lemniscus; PUL, pulvinar; SC, superior colliculus; SI, substantia innominata; VPM, ventral posteromedial nucleus; ZI, zona incerta.

Fibiger, 1978; Larsen and McBride, 1979; Nauta, 1979, Jackson and Crossman, 1981). Both physiological and double fluorescent labeling studies in the monkey support the thesis that cells in the central core of the medial pallidal segment project dichotomizing axons with the same signal to thalamic nuclei and to the pedunculopontine nucleus (Harnois and Filion, 1980; Parent and DeBellefeuille, 1982).

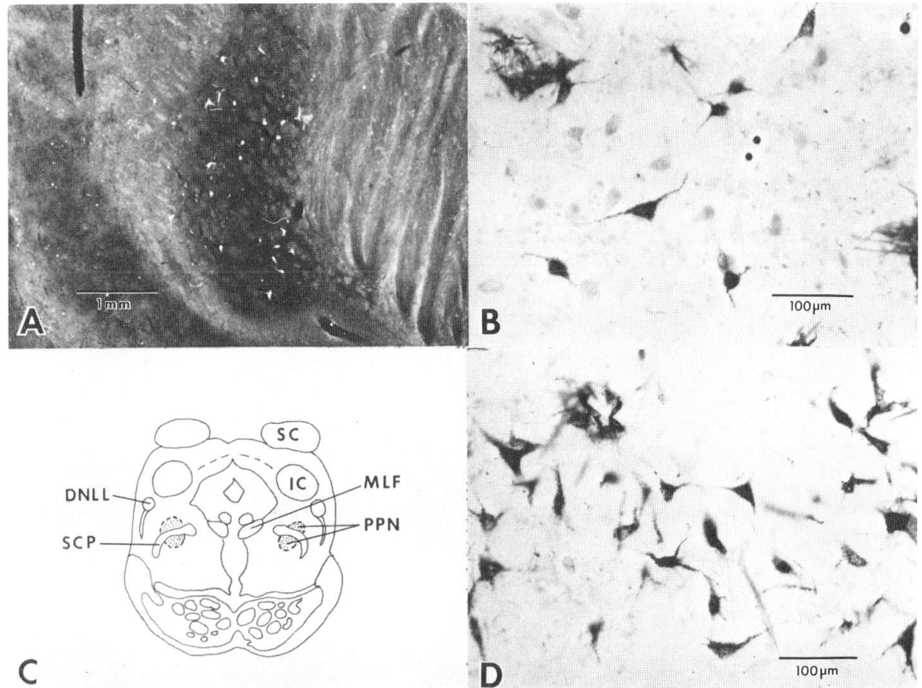

Fig.16. Some connections of the pedunculopontine nucleus (PPN) demonstrated by retrograde transport of HRP in the monkey. (C) indicates the position of the PPN in the caudal midbrain. An HRP injection of PPN retrogradely labeled cells in the MPS (A) and in the SNR (B). HRP injected into the substantia nigra retrogradely labeled cells in PPN (D). Abbreviations used: DNLL, dorsal nucleus of lateral lemniscus; IC, inferior colliculus; MLF, medial longitudinal fasciculus; MPS, medial pallidal segment; SC, superior colliculus; SNR, pars reticulata of substantia nigra [from Carpenter and Sutin (1983), Human Neuroanatomy, courtesy of Williams & Wilkins, Baltimore].

The pedunculopontine nucleus in addition receives inputs from a variety of specific sites, which include: (1) the motor cortex (Kuypers and Lawrence, 1967; Hartman-von Monakow et al., 1979), (2) the substantia nigra (Hedreen, 1971; Jayaraman et al., 1977; Beckstead et al., 1979; Carpenter et al., 1981a; Moon Edley and Graybiel, 1983), and (3) the subthalamic nucleus (Nauta and Cole, 1978; Jackson and Crossman, 1981; Moon Edley and Graybiel, 1983). There are obvious species differences which make comparisons unreliable (Jackson and Crossman, 1981; Moon Edley and Graybiel, 1983). Autoradiographic comparisons of pallidotegmental and nigrotegmental projections in the same monkey (C-1363; Fig. 18D) leave little doubt that the medial pallidal segment contributes a much larger number of fibers (Carpenter et al., 1981a). Data concerning the projections of PPN indicate it distributes fibers to the medial pallidal segment, the subthalamic nucleus, the substantia nigra and possibly to the intralaminar thalamic nuclei, the hypothalamus and the periaqueductal gray (Graybiel, 1977; DeVito et al., 1980; Nomura et al., 1980; Carpenter et al., 1981, 1981a; Moon Edley and Graybiel, 1983). In the monkey and cat the major projection of PPN is to the ipsilateral substantia nigra (Carpenter et al., 1981a; Moon Edley and Graybiel, 1983). Most available data suggest that descending projections from PPN are small and not directed towards any single region of the reticular formation (Moon Edley and Graybiel, 1983). The precise role of PPN in the activities of the globus pallidus, substantia nigra and the subthalamic nucleus remains undefined, but because this nucleus interrelates nuclei that either have no cortical input (e.g., globus pallidus and substantia nigra) or no reciprocal connections (e.g. substantia nigra and globus pallidus; substantia nigra and subthalamic nucleus), it may serve to integrate neuronal activities.

Pallidosubthalamic projections. The lateral pallidal segment projects mainly to the subthalamic nucleus (Ranson and Ranson, 1939; Nauta and Mehler, 1966; Carpenter and Strominger, 1967; Carpenter et al., 1968). Degeneration and autoradiographic data are in agreement that pallidosubthalamic projections are topographically organized (Carpenter et al., 1968; Carpenter et al., 1981, 1981a). The rostral division of the lateral pallidal segment (LPS) projects to the medial two-thirds of the rostral part of the subthalamic nucleus (STN) and to the central region of the middle third of the nucleus. Cells in the central division of the LPS (flanking the medial pallidal segment) project to the lateral third of the STN throughout most of its rostrocaudal extent (Fig. 22C and D).

Physiological studies in the monkey and cat indicate that stimulation of the lateral pallidal segment, or its equivalent, produces a depression of spontaneous activity in cells of the subthalamic nucleus (Frigyesi and Rabin, 1971; Tsubokawa and Sutin, 1972; Ohye et al., 1976). In the decorticated rat pallidal stimulation evoked hyperpolarizing potentials, usually followed by depolarizing potentials, considered to be monosynaptic in nature (Kita et al., 1983). GABA has been suggested as the probable neurotransmitter in the pallidosubthalamic projection in the cat and rat because the fall in GAD activity was proportional to the damage to, or blockade of, the LPS (Fonnum et al., 1978a; Rouzaire-Dubois et al., 1980). Kainic acid lesions of the LPS in the rat produced a loss of pallidal neurons and a drop in GAD and choline acetyltransferase in the pallidum and substantia nigra, but no change in these enzymes in the STN (Kooy et al., 1981). Presumptive evidence that neurons in the STN contain GABA as their neurotransmitter might account for this apparent discrepancy (Crossman et al., 1980, 1984; Nauta and Cuenod, 1982).

Pallidonigral fibers. Several kinds of evidence indicate that pallidal neurons project to substantia nigra. HRP injections of the pars reticulata (SNR) produce retrograde labeling of medium-sized spiny striate neurons and large pallidal neurons, mainly in the lateral segment (Grofova, 1975). According to this author, pallidonigral fibers do not reach the caudal pars compacta (SNC); the SNC is said to receive fibers from the entopenduncular nucleus in the cat. Although the results of Kanazawa et al. (1976) are similar, they found larger numbers of labeled pallidal neurons after HRP injections of the pars compacta. HRP injections in lateral parts of the SNR in the cat retrogradely labeled pallidal neurons mainly in ventral and caudal regions (McBride and Larsen, 1980). Medial parts of the substantia nigra appeared to receive projections largely from rostral parts of the LPS. Pallidonigral projections in the monkey were demonstrated autoradiographically by isotope injections confined to the medial pallidal segment (Kim et al., 1976); these fibers projected to caudal regions of the nigra which largely contain cells of the SNC (Fig. 14). Autoradiographic data in the rat revealed major accumulations of silver grains in the subthalamic nucleus and substantia nigra following isotope injections in the globus pallidus (Carter and Fibiger, 1978); HRP studies indicated no projections from the lateral pallidal segment to the thalamus. Electron microscopic analysis after intraventricular injections of 6-hydroxydopamine suggest that pallidonigral fibers terminate preferentially upon dopaminergic neurons, while strionigral fibers synapse upon the dendrites of neurons in the pars reticulata (Hattori et al., 1975).

Biochemical studies, revealing significant reductions in glutamic acid decarboxylase in the substantia nigra of the rat and cat following lesions in globus pallidus and entopeduncular nucleus, have been interpreted as supporting the existence of a pallidonigral pathway whose neurotransmitter is GABA (Hattori et al., 1973; Fonnum et al., 1974, 1978). The concentration of GAD was found to be highest in the medial part of the SNR and decreased in a mediolateral direction. Subcellular fractionation showed that 85% of the GAD was present in particles, probably synaptosomes, in unoperated animals (Fonnum et al., 1974). Electron microscopic comparisons of the tissue volume occupied by terminal boutons was 11.5% for the SNR and 5.9% for the SNC. These data support the existence of a pallidonigral pathway originating from both pallidal segments whose neurotransmitter probably is GABA. Immunohistochemical studies indicate that substance P-containing neurons within the globus pallidus project to the substantia nigra (Jessell et al., 1978). Substance P immunofluorescence, also observed in the entopeduncular nucleus neurons, is considered to have an excitatory action on nigral neurons.

SUBSTANTIA NIGRA

The substantia nigra (SN), the largest single mesencephalic nucleus, lies dorsal to the crus cerebri, extends the length of the mesencephalon and for descriptive purposes is divided into two parts: (1) the pars compacta, a cell rich region usually containing melanin pigment, and (2) the pars reticulata, a cell poor region. Some ultrastructural studies do not support the division of the SN into two distinct parts, even though cells vary in size, regional, distribution and orientation of dendrites (Bak, 1967; Hirosawa, 1968; Rinvik and Grofova, 1970).

Cytology

Cells in both the pars compacta and pars reticulata are triangular or fusiform and range from 15 to 50 μm in the cat (Rinvik and Grofova, 1970) and from 15 to 80 μm in the monkey (Schwyn and Fox, 1974). Pigment granules dispersed in the cytoplasm in large cells are seen in man, the squirrel monkey (Saimiri sciureus) and the spider monkey (Ateles geoffroyi), but do not occur in all large cells in the rhesus monkey (Macaca mulatta). In the cat all nigra neurons have a prominent rough endoplasmic reticulum and free ribosomes (Rinvik and Grofova, 1970). At the ultrastructural level neurones rich in cytoplasmic organelles appear to represent large neurons, while pale cells with a paucity of organelles appear to represent smaller neurons (Bak et al., 1975). Three types of neurons have been described in the substantia nigra in the rat: (1) large neurons distributed in the pars reticulata, (2) medium-sized neurons in the pars compacta, and (3) small, short axoned neurons found in both subdivisions (Gulley and Wood, 1971). Large neurons found throughout the SNR are especially prevalent in rostrolateral regions where they are embedded in a neuropil of fine unmyelinated fibers. Medium-sized neurons are closely grouped in the SNC, but separated by thin astrocytic sheaths. Small neurons (10 to 12 μm), considered interneurons, comprise about 10% of the SNC and 40% of the SNR.

Nigral neurons as seen in Golgi preparations give rise to long radiating, smooth dendrites with few branches (Rinvik and Grofova, 1970). Dendrites of cells in the SNC have primarily a dorsoventral orientation; dendrites of cell in the SNR have a rostrocaudal orientation and overlap those of the pars compacta. Dendritic surfaces are covered with boutons whose numbers increase with the distance from the soma. Although axons of nigral neurons are difficult to impregnate by the Golgi technic, axon-like structures give off thin collaterals at right angles to the parent axon. A prominent feature of the neuropil of the substantia nigra is an enormous number of thin poorly myelinated fibers coursing in a caudorostral direction.

Cells of the pars compacta contain high concentrations of dopamine as demonstrated by fluorescent histochemical studies (Anden et al., 1964; Dahlstrom and Fuxe, 1964; Hokfelt and Ungerstedt, 1969; Ungerstedt, 1971; Moore et al., 1971); these cells are recognized as the principal source of striatal dopamine. Lesions in the substantia nigra, or interruption of nigrostriatal projections, produce significant decreases in striatal dopamine (Poirier and Sourkes, 1965). The substantia nigra also contains the highest concentration in the nervous system of GAD (Fahn and Cote, 1968; Kim et al., 1971; Okada et al., 1971; Hattori et al., 1973; Fahn, 1976; Okada, 1976), which appears to be localized in nerve terminals (Fonnum et al., 1974; Ribak et al., 1976). Lesion studies indicate that GABA-containing terminals originate from ipsilateral striatal neurons and from the globus pallidus (McGeer et al., 1971, 1974; Hattori et al., 1973; Fonnum et al., 1974; Streit et al., 1979). Although the pallidonigral projection appears minor compared with the strionigral system, it has a greater distribution upon dopaminergic cells in the pars compacta (Kataoka et al., 1974; Hattori et al., 1975; Ribak et al., 1976; Fonnum et al., 1978). The highest concentration of GAD in the substantia nigra is in the medial part of the pars reticulata; subcellular fractionations indicate that 85% of the GAD in the substantia nigra is present in particles, probably synaptosomes (Fonnum et al., 1974). Immunocytochemical studies localize GAD to somata of Spiny type I and Aspiny type I striatal neurons; synaptic endings of Spiny type I neurons terminate in the globus pallidus and substantia nigra (Ribak et al., 1979, 1980, 1981; Ribak, 1981).

Enkephalin in the substantia nigra, demonstrated by immunoreactive methods in man, appears to be in the same range as reported for the striatum (Gramsch et al., 1979; Kubek and Wilber, 1980; Emson et al., 1980). Studies in the rat suggest lower enkephalin levels in the substantia nigra (Staines et al., 1980). It seems likely that immunoreactive terminals in the SNR originate from striatal neurons (DiFiglia et al., 1982). It is of particular interest that a reduction of about 50% of the enkephalin content of the substantia nigra occurs in Huntington's disease (Emson et al., 1980). Light and electron microscopic loclization of immunoreactive Leu-enkephalin in the monkey suggests that only Spiny type I striatal neurons contain this peptide and fibers of these neurons project to both the lateral pallidal segment and the pars reticulata of the substantia nigra (DiFiglia et al., 1982).

The substantia nigra also contains substance P, characterized as an undecapeptide with an excitatory action on neurons in many regions of the central nervous system (Konishi and Otsuka, 1974, 1974a; Hokfelt et al., 1977). The highest concentration of substance P in any brain region is found in the SN, where the substance is concentrated in nerve-ending particles (Powell et al., 1973; Duffy et al., 1975; Davies and Dray, 1976; Gauchey et al., 1979). Strionigral substance P fibers arise from different cells (Spiny II) than those having GABA (Spiny I) as their neurotransmitter, although they follow a parallel course. According to some studies substance P containing neurons tend to be found in rostral regions of the striatum (Davies and Dray, 1976; Gale et al., 1977; Jessell et al., 1978). Radioimmunoassay suggests that the highest concentration of substance P is found in the SNR where it is twice as great as in the SNC or pars lateralis (Gauchey et al., 1979). Electron microscopically substance P was localized in terminals of numerous myelinated and unmelinated fibers which contained granular vesicles and formed axodendritic synapses (DiFiglia et al., 1981). These morphological findings are consistent with the known excitatory effects of substance P upon nigral neurons. Because substance P has been identified in somata of pallidal neurons, part of the fibers with this neurotransmitter are considered to originate there (Hattori et al., 1975; Jessell et al., 1978).

The mesencephalic indolamine cell groups in the median raphe project to both the striatum and the substantia nigra (Kooy and Hattori, 1980). The substantia nigra contains serotonin and its synthesizing enzyme tryptophan 5-monooxygenase (Hajdu et al., 1973; Palkovitz et al., 1974; Brownstein et al., 1975), which are present within dense core vesicles in nerve terminals mainly within the SNR and make contact with dopaminergic neurons and their dendrites (Fuxe, 1965; Reubi and Emson, 1978). Most of the serotoninergic fibers arise from the dorsal nucleus of the raphe (Conrad et al., 1974); Bunney and Aghajanian, 1976; Dray et al., 1978; Kooy and Hattori, 1980; Carpenter et al., 1981a), but not all neurons of this nucleus are serotoninergic (Descarries et al., 1982). Raphe-nigral projections are monosynaptic, slowly conducting and evoke inhibition in the majority of nigral neurons (Dray, 1980).

Nigral Afferent Fibers

Although it had been presumed that the SN received abundant cortical projections (Levin, 1949), silver impregnation and electron microscopic studies have not substantiated this thesis (Rinvik, 1966; Rinvik and Walberg, 1969). Axoplasmic transport studies, however, suggest that a few projections from the prefrontal cortex terminate in the SNC (Bunney and Aghajanian, 1976; Kunzle, 1978). The substantia nigra receives afferent ribers from the striatum, the globus pallidus, the subthalamic nucleus, the dorsal nucleus of the raphe and the pedunculopontine nucleus. In addition, the nucleus accumbens projects topographically upon both subdivisions of the substantia nigra (Swanson and Cowan, 1975; Nauta et al., 1978).

Strionigral fibers. The extensive nature and topographical organization of strionigral fibers, evidenced by degeneration studies, indicated that fibers from: (1) the head of the caudate nucleus project to the rostral third of the SN (Fig. 9) and (2) the putamen terminate in the remaining caudal regions of the nigra (Voneida, 1960; Szabo, 1962, 1967, 1970). Dorsal regions of the putamen are related to lateral parts of the SN and there is similar correspondence between ventral regions of the putamen and medial parts of the nigra. Almost all strionigral fibers terminate in the SNR (Grofova and Rinvik, 1970; Kemp, 1970; Schwyn and Fox, 1974; Hattori et al., 1975). Retrograde transport studies confirm the topographical projection from the striatum to the SN (Bunney and Aghajanian, 1976). Strionigral projections arise from medium-sized Spiny striatal neurons (Grofova, 1975) that have GABA, enkephalin and substance P as their neurotransmitters (Ribak et al., 1976; Gale et al., 1977; Hong et al., 1977; Kanazawa et al., 1977; Jessell et al., 1978; Ribak et al., 1980; Emson et al., 1980; DiFiglia et al., 1981, 1982). Spiny type I striatal neurons appear related to two distinctive neurotransmitters, GABA and enkephalin (DiFiglia et al., 1982). Spiny type II striatal neurons appear to have substance P as their neurotransmitter (Pasik et al., 1979; Groves, 1983); fibers from these neurons have a course similar to those of the Spiny type I but may arise from different locations within the striatum (Figs. 2 and 5). Recent studies have demonstrated that electrical stimulation of the caudate nucleus leads to a marked increase in the release of [3H] GABA into the ipsilateral substantia nigra and some increase in the contralateral nigra (Kemel et al., 1983). Multi-unit recording revealed that neurons in the SNR were activated ipsilaterally and inhibited contralaterally. It has been suggested that nigrothalamic GABAergic neurons may be involved in transmission to the contralateral side.

Pallidonigral fibers. Projections from both the medial and lateral pallidal segments to the substantia nigra are considered to have GABA and substance P as their neurotransmitters (Hattori et al., 1973; Fonnum et al., 1974, 1978; Jessell et al., 1978; DiFiglia, 1981). This projection appears relatively minor compared with the pathway originating from the striatum. Fibers from the lateral pallidal segment project largely to the pars reticulata while the medial pallidal segment appears to end in the pars compacta (Grofova, 1975; Hattori et al., 1975; Bunney and Aghajanian, 1976; Kim et al., 1976; Carter and Fibiger, 1978). While nearly all strionigral fibers end in the SNR, pallidal afferents to the nigra have a greater distribution upon dopaminergic neurons in the SNC (Hattori et al., 1975; Carter and Fibiger, 1978).

Subthalamonigral fibers. Degeneration and axoplasmic transport studies indicate that the subthalamic nucleus projects fibers to the substantia nigra, most of which terminate in the pars reticulata (Whittier and Mettler, 1949; Knook, 1965; Kanazawa et al., 1976; Nauta and Cole, 1978; Deniau et al., 1978; Rinvik et al., 1979; Kooy and Hattori, 1980a; McBride and Larson, 1980; Ricardo, 1980; Carpenter ewt al., 1981a). Fluorescence retrograde double labeling indicates that virtually all neurons throughout the STN project to both the globus pallidus and the substantia nigra (Kooy and Hattori, 1980a). The latter finding has been confirmed physiologically in the rat (Deniau et al., 1978). Orthodromic and antidromic stimulations of the pathway from the subthalamic nucleus to the nigra support the thesis that these fibers end upon cells of the SNR and are excitatory (Hammond et al., 1978). As commented upon earlier under subthalamopallidal fibers, recent physiological observations suggest that stimulation of the same population of STN neurons exerts a short latency suppressive influence upon large numbers of entopeduncular neurons and a short latency excitation upon nigral neurons (Hammond et al., 1983). The suspected neurotransmitter in subthalamopallidal fibers is GABA (Nauta and Cuenod, 1982; Crossman et al., 1980, 1984). In the monkey, subthalamonigral fibers from rostral parts of the subthalamic nucleus descend for some distance along the dorsal border of the nigra before abruptly passing ventrally to end in the pars reticulata (Carpenter, 1981a).

Nigral afferents from the raphe and midbrain tegmentum. The SN receives a projection from the raphe nuclei of the midbrain, particularly the dorsal nucleus of the raphe. Lesions in the midbrain raphe nuclei in the rat greatly diminish yellow fluorescent terminals in the SNR (Kuhar et al., 1972). Electrical stimulation of the raphe nuclei depresses activity of neurons in the pars reticulata (Dray et al., 1976). While numerous studies indicate a direct projection from the dorsal nucleus of the raphe to the striatum (Nauta et al., 1974; Miller et al., 1975; Ternaux et al., 1977), several authors report that HRP injected into the substantia nigra is transported retrogradely to the dorsal raphe nucleus (Kanazawa et al., 1976; Bunney and Aghajanian, 1976; Fibiger and Miller, 1977; Carpenter et al., 1981a). Retrograde fluorescent double-labeling suggest that the raphe innervation of the nigra consists primarily of collateral branches of axons projecting to the striatum (Kooy and Hattori, 1980). Cells of the dorsal nucleus of the raphe projecting to the SN were situated in clusters in dorsal parts of the nucleus. The ascending projections from the dorsal nucleus of the raphe appear to modulate activity of both the substantia nigra and striatum. It must be recalled that not all neurons in the raphe nuclei are serotoninergic (Descarries et al., 1982).

The pedunculopontine nucleus, situated on both sides of the superior cerebellar peduncle at caudal midbrain levels, also projects fibers rostrad to the SN and to other structures (Fig. 15). Retrograde transport of HRP in the monkey suggests that the pedunculopontine nucleus projects more fibers to the substantia nigra than to the globus pallidus or the subthalamic nucleus (Carpenter et al., 1981, 1981a). According to Moon Edley and Graybiel (1983) deposits of labeled amino acids in the pedunculopontine nucleus, pars compacta, in the cat produce bilateral labeling in the pars compacta of the nigra and in the subthalamic nucleus with the densest label ipsilateral. As mentioned previously, this relatively small nucleus receives input from multiple sources. In the cat the pedunculopontine nucleus, not a cytoarchitecturally distinct cell group, is most readily identified by the terminal distribution of fibers originating from the substantia nigra (Moon Edley and Graybiel, 1983).

Nigral Efferent Fibers

Efferent fibers fall into two classes: those that are dopaminergic and those that are non-dopaminergic. The largest and most extensively studied projections are the nigrostriatal fibers distributed to all parts of the striatum (Figs. 17 and 18A). These dopaminergic fibers arise from the pars compacta. Cells of the pars compacta also give rise to fibers that descend and terminate in the dorsal nucleus of the raphe (Beckstead et al., 1979). Virtually all other nigrofugal fibers arise from cells of the pars reticulata. These nigral efferent fibers which constitute a large part of the output system of the corpus striatum can be grouped as: (1) nigrothalamic fibers, (2) nigrotectal fibers and (3) nigrotegmental projections.

Nigrothalamic fibers. One of the major projections of the pars reticulata is to thalamic nuclei. In silver degeneration studies terminals of nigrofugal fibers were described in the ventral anterior (pars magnocellularis), ventral lateral (pars medialis) and in the mediodorsal (pars paralaminaris) thalamic nuclei in the monkey (or their equivalents in other animals) (Carpenter and Strominger, 1967; Faull and Carman, 1968; Carpenter and Peter, 1972). The first indication that these fibers originated from the pars reticulata was the finding that lesions confined to the SNR produced only thalamic degeneration (Carpenter and Peter, 1972). This observation was confirmed by retrograde HRP transport (Rinvik, 1975). Autoradiographic studies confirmed the above described nigrothalamic projections and indicated that cells in the rostrolateral parts of the nigra (pars lateralis) preferentially projected to the thalamus (Figs. 17 and 18A, B) (Carpenter et al., 1976). According to Faull and Mehler (1978) the cells of origin of nigrothalamic, nigrotectal and nigrostriatal pathways in the rat are segregated into complementary longitudinal regions and each of these populations of neurons have somata exhibiting different cell sizes. Nigrothalamic neurons form longitudinal cell columns within lateral and central regions of the SNR and have large somata. Neurons giving rise to nigrotectal fibers form a ventral cell lamina within the pars reticulata composed of medium sized somata. Retrograde fluorescent double labeling studies in the rat indicated that a considerably population of neurons in the pars reticulata project divergent axon collaterals to both the thalamus and the superior colliculus (Bentivoglio et al., 1979). Branched collaterals of nigral neurons projecting to the thalamus and superior colliculus in the cat and rat constitute between 30 and 50% of the neurons in the SNR (Deniau et al., 1978; Anderson and Yoshida, 1980). Golgi studies are consistent

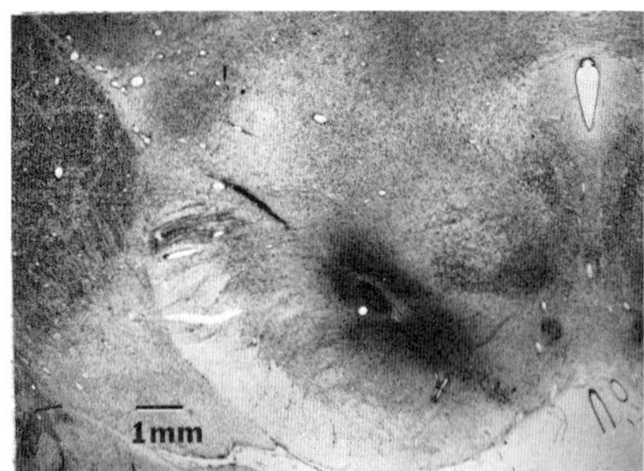

Fig.17. Dark field photomicrograph of an autoradiograph of a [^{3}H] amino acid injection largely into the pars compacta of the substantia nigra (SNC) (from Carpenter et al., 1976, courtesy of Alan Liss, New York).

with these observations indicating branched projections to thalamus and superior colliculus, as well as fibers which remain undivided (Juraska et al., 1977). Stimulation of the caudate nucleus produces inhibition of spontaneous activity in nigral neurons projecting to both the thalamic nuclei and the superior colliculus (Deniau et al., 1976; Anderson and Yoshida, 1980). Stimulation of the substantia nigra directly produces monosynaptic inhibition of thalamic neurons (Ueki et al., 1977). An electron microscopic autoradiographic comparison of the distribution of synaptic sites of pallidal and nigral efferent fibers on thalamocortical projection neurons has indicated that: (1) only boutons of pallidal terminal fibers form axosomatic synapses, (2) larger numbers of boutons of nigral fibers end on primary dendritic trunks and (3) both contact similar percentages of secondary dendrites (Kultas-Ilinsky et al., 1983). Pallidal and nigral projections to thalamic neurons are organized in a similar fashion so as to exert strong influences upon the firing patterns of thalamocortical projection neurons.

Nigrotectal projections. Although degenerated fibers from nigral lesions projecting to the deep layers of the superior colliculus were reported in early studies, this connection was regarded with suspicion (Carpenter and McMasters, 1964; Cole et al., 1964; Carpenter and Strominger, 1967). Acceptance of a nigrotectal projection, arising only from cells of the SNR was based upon axoplasmic transport studies (Graybiel and Sciascia, 1975; Hopkins and Niessen, 1976; Rinvik et al., 1976; Jayaraman et al., 1977; Graybiel, 1978). Autoradiographic data indicated that nigrotectal fibers terminated in bursts or clumps in the intermediate gray of the caudal two-thirds of the superior colliculus (Fig. 18C) (Jayaraman et al., 1977; Graybiel, 1978). There was a suggestion that nigrotectal fibers may be topographically arranged and that some nigral efferent fibers may be distributed bilaterally (Deniau et al., 1977; Jayaraman

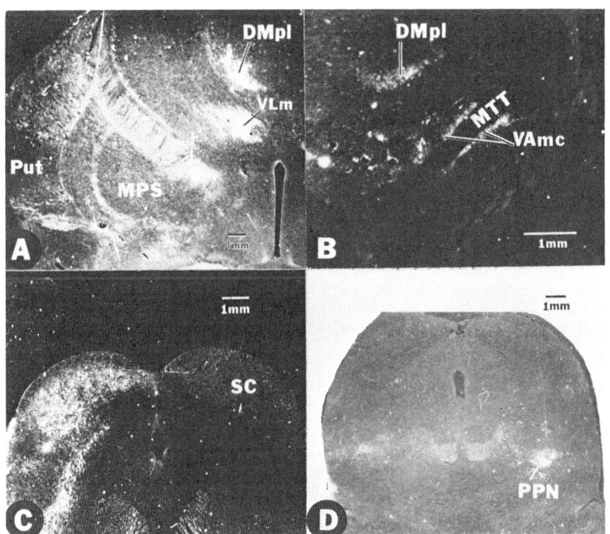

Fig. 18. Autoradiographs of isotope transport (A,B,C) from substantia nigra(SN). (A), transport via nigrostriatal fibers to dorsal regions of putamen and via nigrothalamic fibers to ventral lateral nucleus, pars medialis (VLm) and paralaminar part of dorsomedial nucleus (DMpl). (B), sagittal autoradiograph showing terminals of nigrothalamic fibers in ventral anterior nucleus (VAmc) surrounding the mammillothalamic tract (MTT) and in DMpl. (C), demonstration of nigrotectal fibers. (D), transport of isotope to pedunculopontine nucleus (PPN) from SN (left) and medial pallidal segment (right) (from Carpenter et al., 1976 and Jayaraman et al., 1977). All dark field photomicrographs. (Courtesy Alan Liss, New York, Elsevier, Amsterdam, & the American Physiological Society, Bethesda)

et al., 1977). Some nigrotectal fibers also projected to lateral regions of the central gray. As previously mentioned, a large number of cells in the SNR project divergent axonal collaterals that terminate in thalamic nuclei and the superior colliculus (Bentivoglio et al., 1979). According to Faull and Mehler (1978), nigrotectal neurons within the pars reticulata form a ventral lamina composed of medium sized cells. The fact that nigrotectal fibers arise from the pars reticulata does not exclude the possibility that they be dopaminergic, but it seems likely that another neurotransmitter may be involved (Rinvik et al., 1976). Although GABAergic projections to the superior colliculus have been suggested (Grofova et al., 1978), there is evidence suggesting that neurons in the superficial layers of the superiof colliculus are an intrinsic source of GABAergic terminals (Houser et al., 1983). Nigrotectal projections have been related to a variety of visual and auditory reponses concerned with orienting behavior, gaze shifting, gaze fixation and saccade responses (Hikosaka and Wurtz, 1983, 1983a, 1983b). Some evidence suggests that cells of the pars reticulata project to the brain stem reticular formation; part of this projection may be included in the nigrotegmental bundle (Rinvik et al., 1976; Hopkins and Niessen, 1976; Beckstead et al., 1979).

Nigrotegmental projections. Descending fibers from the SNC and the ventral tegmental area pass into medial midbrain regions and distribute fibers to the periaqueductal gray, the dorsal nucleus of the raphe and the parabrachial nuclei in the rat (Beckstead et al., 1979). Cells of the SNR give rise to a nigrotegmental bundle (Fig. 15). As previously mentioned, the pedunculopontine nucleus (PPN) receives descending fibers from multiple subcortical loci. In monkey (Fig. 16B) and cat a large number of afferents to PPN arise from the pars reticulata (Carpenter et al., 1981a; Moon Edley and Graybiel, 1983). Hedreen (1971) first identified the nigrotegmental pathway terminating in PPN and established its non-dopaminergic nature. The functional role of PPN remains elusive but most of its fibers project to nuclei that receive striatal efferents (Beckstead et al., 1979; De Vito et al., 1980; Carpenter et al., 1981; Moon Edley and Graybiel, 1983). Autoradiographic data in the cat suggest that ascending terminals of projections from PPN ending in the SNC and the subthalamic nucleus are distributed bilaterally, but with ipsilateral dominance (Moon Edley and Graybiel, 1983).

SUBTHALAMIC NUCLEUS

The subthalamic nucleus, or corpus Luysi, is a lens-shaped nucleus located on the dorsomedial surface of the peduncular part of the internal capsule (Fig. 19). Its caudal part lies dorsolateral to, and in contact with, the oral part of the substantia nigra. In man the subthalamic nucleus has an approximate volume of 160 mm^3, while the volume of this nucleus in the rhesus monkey is about 10 mm^3 (Whittier and Mettler, 1949; von Bonin and Shariff, 1951; Carpenter and Carpenter, 1951). This nucleus is considered to be derived from the most dorsocaudal part of the lateral hypothalamic cell column (Kuhlenbeck, 1948; Kuhlenbeck and Haymaker, 1949; Richter, 1965).

Lesions in the subthalamic nucleus, or its connections, in man produce contralateral hemiballism, a violent form of hemichorea involving proximal appendicular and axial musculature (Whittier, 1947; Carpenter, 1955). Discrete electrolytic lesions in the subthalamic nucleus of the monkey that destroy approximately 20% of the nucleus and preserve the integrity of surrounding pallidofugal pathways, produces large amplitude, patterned, violent dyskinesia contralaterally which resembles that seen in man; this abnormal involuntary activity has been called subthalamic dyskinesia (Whittier and Mettler, 1949a; Carpenter et al., 1950). Subthalamic dyskinesia

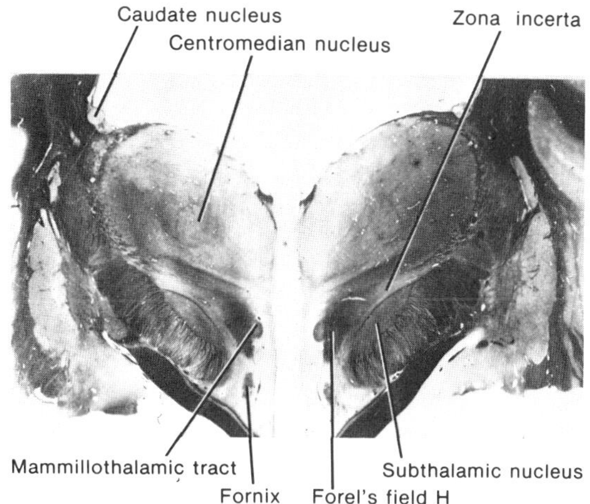

Fig.19. Transverse section of the human brain through central portions of the subthalamic nucleus. Pal Weigert stain. (Courtesy of Williams & Wilkins, Baltimore)

in the monkey is enduring, associated with distinct hypotonus and can be ameliorated or abolished contralaterally by subsequent stereotaxic lesions in the medial pallidal segment, the ventral lateral nucleus of the thalamus and the motor cortex (Carpenter et al., 1950; Carpenter and Mettler, 1951; Carpenter, 1961). Subthalamic dyskinesia produced in the rhesus monkey by discrete lesions of the corpus Luysi has been considered to be the physiological expression of the removal of inhibitory influences, generated in the STN, which normally act upon the medial segment of the globus pallidus. Lesions destroying up to 20% of the volume of the substantia nigra and producing extensive cell loss in the pars compacta had no effect upon subthalamic dyskinesia (Strominger and Carpenter, 1965). Subthalamic dyskinesia has been produced in the monkey by small injections of kainic acid into the STN; dyskinesia appeared on the third post-operative day and was identical to that resulting from electrolytic lesions (Hammond et al., 1979). The most illuminating development has been the production of subthalamic dyskinesia in the baboon and monkey by localized injections of GABA antagonists, picrotoxin and bicuculline methiodide, into awake animals via implanted cannulae (Crossman et al., 1980, 1984). The latency from injection of the GABA antagonist to appearance of unequivocal dyskinesia varied from 5 to 125 minutes and was shortest in well-targeted injections. The duration of the dyskinesia varied from two to four hours and was followed by an uneventful recovery without after effects. These data suggest that interruption, or pharmacological blockade, of subthalamopallidal fibers having GABA as their neurotransmitter may be the feature essential to the neural mechanism of subthalamic dyskinesia.

Cytology

Neurons of the subthalamic nucleus in the monkey are fairly large, round, polygonal or fusiform cells with large nuclei and nucleoli (Whittier and Mettler, 1949; Rafols and Fox, 1976). The cytoplasm is basophilic, but discrete Nissl bodies are not prominent. In some species lipofuscin granules are present in the cytoplasm. A quantitative and comparative analysis of the nucleus in primates, including man, indicated a wide range of cell sizes (from 10 x 10 µm to 30 x 6 µm in the Macaca, to 11 x 11 µm to 40 x 9 µm in man) (Yelnik and Percheron, 1979). All cells contribute to a rich neuropil. Several authors report regional differences in cell sizes and concentrations. Cells in medial parts of the nucleus were described as smaller and more numerous (Foix and Nicolesco, 1925; Kodama, 1928; Whittier and Mettler, 1949). Other authors consider that cell bodies only look more numerous in rostral and medial parts of the nucleus (Yelnik and Percheron, 1979). Rafols and Fox (1976) identify two types of principal subthalamic neurons and a small local interneuron with relative long dendrites. In the cat three Golgi type I neurons were described, all of which were quite large; no Golgi type II neurons wre identified (Iwahori, 1978). Statistical and comparative analyses strongly suggest there is only one variety of Golgi type I neuron in the subthalamic nucleus which is nearly identical in cat, monkey and man (Yelnik and Percheron, 1979). Fluorescent double labeling studies in the rat indicated that virtually all STN neurons (94%) are projection neurons (Kooy and Hattori, 1980). Variations in the cell shape and dendritic orientation were attributed to cell location within the nucleus. On average each cell gives rise to 7 dendritic stems which branch successively into an ellipsoidal domain, usually parallel with the rostrocaudal axis of the nucleus. Dendritic spines are not numerous, tend to be thin and pedunculated and most often are located distally.

Intracellularly labeled rat subthalamic nucleus neurons provide data which confirm Golgi studies and give additional details, especially regarding axons and their collaterals (Hammond and Yelnik, 1983; Kita et al., 1983). Dendritic morphology in the rat compared with the primate indicates: (1) a smaller number of dendritic stems in the rat (2-4) compared with the primate (4-9), (2) a larger number of dendritic branching points in the rat, and (3) similar ellipsoidal dendritic arborizations oriented in the rostrocaudal axis of the nucleus. The dendritic field of a single subthalamic neuron in the rat is calculated to cover the entire nucleus, while in the cat it would cover only half of the nucleus; the dendritic field would cover a fifth of the nucleus in the monkey and a ninth of the nucleus in man (Hammond and Yelnik, 1983). Thus, while subthalamic nucleus neurons in rat and primate are similar, their relationships to the nucleus as a whole have evolved to confer upon the primate the potential for a much more specific organization. Another important

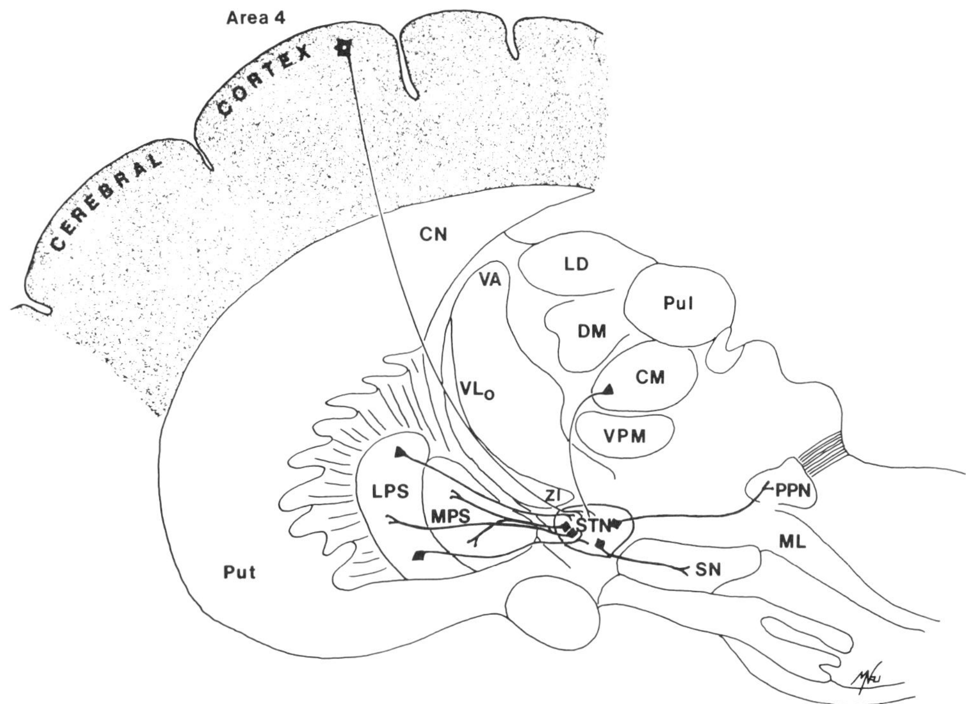

Fig.20. Schematic diagram of the connections of the subthalamic nucleus (STN) in a sagittal plane. Afferents to STN are derived from the motor cortex, the centromedian nucleus (CM) and the lateral pallidal segment (LPS). STN efferents project to both pallidal segments, the substantia nigra (SN) and the pedunduloppontine nucleus (PPN). In the rat single STN neurons project to both the pallidum and the SN (Kooy and Hattori, 1980a). See Fig. 15 for abbreviations.

feature disclosed by HRP intracellularly labeled STN neurons was the presence of intranuclear axon collaterals on one group of cells (Kita et al., 1983). On the basis of intranuclear axonal collaterals, STN neurons could be divided into two groups. It is believed that the type of STN neuron with intranuclear collaterals is fairly common and may be involved in a feed-forward circuitry. These data also demonstrated that axonal branches of STN neurons projecting to the globus pallidus are thicker than those descending to the substantia nigra, an observation consistent with physiological findings (Kita et al., 1983a). The evolution of STN neurons, as evidenced by significant differences in dendritic development (Hammond and Yelnik, 1983), suggests that a far more specific organization may exist in the primate subthalamic nucleus than in lower forms. One evidence

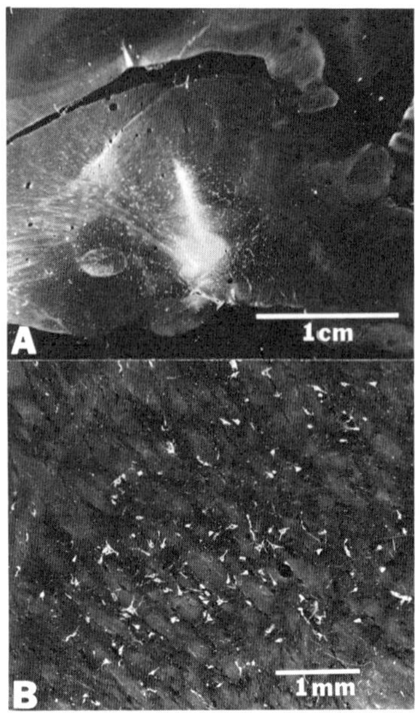

Fig. 21. Pallidal connections of the subthalamic nucleus (STN) in the monkey demonstrated by retrograde transport of HRP. (A), sagittal section of HRP injection into STN (dark field). (B), sagittal section of arrays of labeled neurons in the lateral pallidal segment (from Carpenter et al., 1981a). Dark field photomicrographs. (Courtesy of Elsevier Scientific Publishing Company, Amsterdam)

of this may be that subthalamic lesions in the cat are said to produce no dyskinesia (Adey et al., 1962).

Afferents to the Subthalamic Nucleus

The deep position and multiple surrounding pathways of the subthalamic nucleus severely limit the feasibility of determining the connections of this nucleus by degeneration technics (Glees and Wall, 1946; Whittier and Mettler, 1949; Carpenter and Strominger, 1967). Such studies reveal only the most massive connections and lack important detail. Axoplasmic transport methods have for the first time provided precise information concerning the multiple inputs to the STN (Fig. 20). Afferent projections to the subthalamic nucleus which have a firm anatomical basis arise from the motor, premotor and prefrontal cortex, the thalamus, the lateral pallidal segment and the pedunculopontine nucleus. The dominant input to the STN arises from the lateral pallidal segment (Figs. 21 and 22C,D).

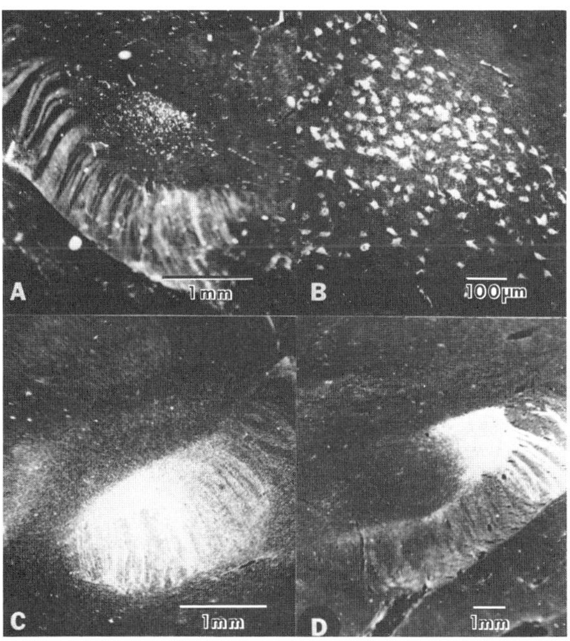

Fig. 22. Connections of the subthalamic nucleus (STN) in the monkey. (A) and (B), retrograde transport of HRP from ventral portions of the lateral pallidal segment (LPS) to cells in dorsal central parts of the STN. (C), pallidosubthalamic projections from the rostral part of LPS. (D), pallido-subthalamic fibers from the central part of the LPS, flanking the medial pallidal segment (from Carpenter et al., 1981, courtesy of Alan Liss, New York).

Corticosubthalamic fibers. Autoradiographic studies of corticofugal fibers from area 8 in the monkey were noted to produce low intensity labeling in the zona incerta and in the STN (Kunzle and Akert, 1977) which tended to support suspicions from degeneration studies (Mettler, 1947; Levin, 1949). More systematic analysis revealed an ipsilateral somatotopically organized projection from the precentral motor cortex to the STN, restricted to the lateral moiety (Fig. 20) (Hartmann-von Monakow et al., 1978). The somatotopic arrangement demonstrated a face region along the laterodorsal border of the nucleus, a leg region more centrally and an arm area between these representations. Area 6 projections, and those from areas 8 and 9, were more medial and ventral. Electron microscopic confirmation of corticosubthalamic fibers indicated terminations by asymmetrical synapses on small distal dendrites and spines (Romansky et al., 1979). Physiological identification of a corticosubthalamic pathway from the motor cortex in the rat followed (Kitai and Deniau, 1981). These data suggested that Golgi type I STN neurons receive excitatory monosynaptic inputs. Attempts to retrogradely label cortical neurons by HRP injections in the STN have been disappointing (Rinvik et al., 1979; McBride and Larsen, 1980; Carpenter et al., 1981a).

Other evidence suggests that the STN may have additional telecephalic inputs. Anterograde transport of wheat germ agglutinin conjugated to HRP indicated a striosubthalamic projection (Beckstead, 1983). The size of this pathway, which seems to be reciprocal, is undefined. This evidence is supported by physiological studies which reveal that striatal stimulation in awake monkeys evokes short latency spike discharges in STN neurons (Ohye et al., 1976). The significance of this apparently small reciprocal connection, also suggested by Nauta and Cole (1978), remains to be defined.

Thalamosubthalamic fibers. Both anterograde and retrograde transport studies in the cat and rat support a small projection from the centro-median parafascicular nuclear complex to the STN (Sugimoto et al., 1983). Cells labeled in CM-PF after HRP injections in the STN were scattered and not different from those which remain unlabeled. Fibers projecting from CM-PF to the STN were most concentrated in ventral and ventromedial portions of the rostral third of the nuleus; a few fibers were seen in the middle third of the nucleus, but none were found in the caudal third of the STN (Fig. 20). EM autoradiography confirmed these findings and showed that fiber terminations had asymmetrical boutons with round synaptic vesicles (Subimoto and Hattori, 1983). The authors suggest that cells of the CM-PF may project collateral fibers to both the striatum and the STN. Terminals of thalamosubthalamic projections do not overlap corticosubthalamic fibers (Hartmann-von Monakow et al., 1978). No other thalamic nuclei appear to provide inputs to the STN. Lesions in the intralaminar thalamic nuclei have no effect upon suthalamic dyskinesia in the monkey (Carpenter et al., 1965).

Pallidosubthalamic projections. The most massive input system to the STN arises from the lateral pallidal segment (Nauta and Mehler, 1966; Carpenter et al., 1968; Kim et al., 1976; Carpenter et al., 1981, 1981a). This topographically organized pathway is described in an earlier section concerning pallidal efferent fibers (Fig. 21). Data in the monkey show that even though the STN is a small nucleus it appears to have input and output regions (Carpenter et al., 1981). The lateral third of the STN receives the bulk of the inputs originating from the LPS and the motor cortex, the two largest afferent systems (Fig. 22D). The rostromedial part of the STN receives inputs from the rostral part of the LPS and from CM-PF (Fig. 22C). It is significant that cells in the medial and caudal parts of STN project predominantly to the medial pallidal segment and that few, if any, subthalamopallidal fibers arise from the lateral third of the STN throughout its length.

In a study of the ultrastructure of synaptic contacts in the STN, axodendritic synapses composed the largest group (80..3%), while axoaxonic (12.3%) and axosomatic (7.4%) synapses formed the remainder (Nakamura and Sutin, 1972). Following lesions in the lateral pallidal segment (LPS) in the cat, degenerated boutons were found to be most numerous in the lateral and intermediate parts of the nucleus. The most striking feature in cats with pallidal lesions was the absence of axoaxonic synaptic contacts. As mentioned previously stimulation of the LPS in rat, cat and monkey inhibits the spontaneous activity of STN neurons (Tsubokawa and Sutin, 1972; Ohye et al., 1976; Kita et al., 1983a). According to Kooy et al. (1981) neither GABA nor acetylcholine is the major neurotransmitter in the massive pallidosubthalamic pathway.

Tegmentosubthalamic fibers. The pedunculopontine nucleus which receives inputs from the cerebral cortex, the medial pallidal segment and the pars reticulata of the substantia nigra, is considered to project directly to the subthalamic nucleus. Both anterograde and retrograde transport studies in the cat have demonstrated this projection (Nomura et al., 1980; Moon Edley and Graybiel, 1983). Autoradiographic data also reveal projections from PPN to the zona incerta, the SNC and the intralaminar thalamic nuclei. Some of these ascending projections reach the controlateral side via Meynert's commissure. The subthalamic nucleus is considered to give rise to a small projection to PPN (Nauta and Cole, 1978; Jackson and Crossman, 1981; Moon Edley and Graybiel, 1983). Reciprocal connections between PPN and STN appear relatively small compared with those that interrelate PPN with the substantia nigra and with the medial pallidal segment.

Subthalamic Efferent Projections

The major efferents from the STN project to both segments of the globus pallidus and to the substantia nigra (Figs. 20, 22A,B and 23). Each of these pathways has been discussed separately in relation to the pallidal and nigral afferent systems.

Considerable evidence suggests that a single cell in the STN projects fibers to both pallidal segments and to the pars reticulata of the SN (Deniau et al., 1978; Kooy and Hattori, 1980; Kita et al., 1983). Branches of axon collaterals projecting to globus pallidus are thicker than those passing to the SN and have faster rates of conduction (Kita et al., 1983, 1983a). In the rat, subthalamo-pallidal collaterals of branches to the globus pallidus terminate within the entopeduncular nucleus, the equivalent of the primate medial pallidal segment. Stimulation of the STN in the rat produces short latency inhibition in the pallidum and short latency excitation

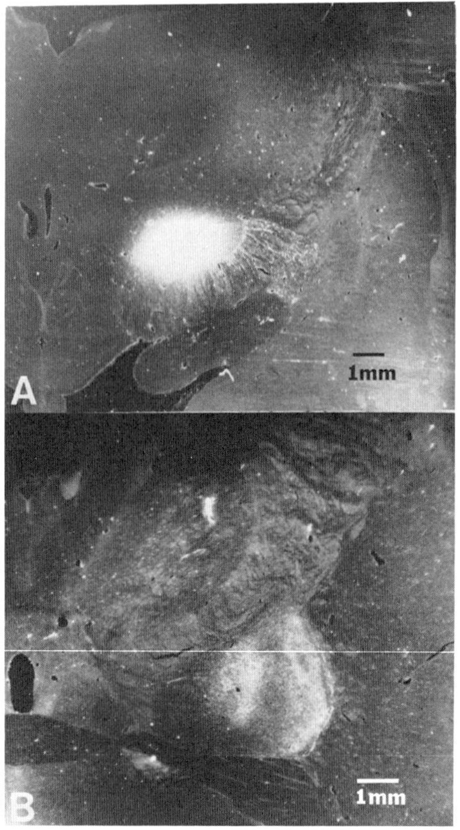

Fig. 23. (A), an isotope injection into the subthalamic nucleus showing labeled fibers crossing the internal capsule. (B), arrays of subthalamopallidal fibers parallel to the medullary laminae of the globus pallidus. Dark field autoradiographs (from Carpenter et al., 1981a).

in the substantia nigra (Hammond et al., 1978, 1983). Dual responses of a different nature are presumed to involve local interneurons.

Two recent reports suggest that the STN may project to the striatum and to the cerebral cortex. Retrograde studies using WGA-HRP injected into the head and body of the caudate nucleus reportedly label cells in STN (Beckstead, 1983). There is a rather crude mediolateral topographic organization in which STN neurons in medial parts of the nucleus are related to the head of the caudate nucleus and more laterally located cells project to the putamen. Nauta and Cole (1978) previously reported sparse projections from the STN to the putamen. In similar studies injections of HRP into the striatum of decorticated rats failed to retrogradely label cells in the STN and entopeduncular nucleus (Jackson and Crossman, 1981a). These same studies demonstrated that HRP injections confined to the dorsal cerebral cortex in the rat labeled STN neurons in the lateral half of the nucleus (Jackson and Crossman, 1981a).

SUMMARY

The corpus striatum, subthalamic nucleus, substantia nigra and pedunculopontine nucleus are interrelated with each other by an orderly linkage and with specific parts of the neuraxis that can modulate somatic motor activities. The striatum, representing the receptive component of the corpus striatum, receives inputs from broad regions of the cerebral cortex, the intralaminar thalamic nuclei, the substantia nigra and the midbrain raphe nuclei. Impulses from these sources are mediated by a variety of different neurotransmitters, most of which appear to be excitatory. Striatal output, originating from two types of Spiny neurons, projects to both segments of the globus pallidus and the pars reticulata of the substantia nigra. GABA is the principal neurotransmitter in striopallidal and strionigral fiber systems, but some fibers of both projections have enkephalin and substance P as their neurotransmitter.

The output systems of the corpus striatum arise from morphologically similar cells in the medial pallidal segment (MPS) and the pars reticulata of the substantia nigra. Thalamic projections of the MPS and SNR are distinctive without overlap; nuclear subdivisions of the thalamus receiving these outputs do not exert their major effects upon the primary motor cortex. Thalamic relay nuclei influenced by the output of the MPS appear to project to the premotor and supplementary motor areas. Most physiological data suggest that pallidothalamic and nigrothalamic projections exert primarily inhibitory effects upon thalamic neurons. It seems likely that striatal inhibition of neurons in the MPS and SNR projecting to the thalamus may result in a disinhibition of thalamic neurons which act upon the cortical neurons.

The substantia nigra receives inputs from both components of the corpus striatum (i.e., striatum and pallidum) and from all closely related subcortical nuclei (i.e., the subthalamic nucleus, the pedunculopontine nucleus and the dorsal nucleus of the raphe). The major output of the SN, exclusive of that to the thalamus and tectum, conveys dopamine to the striatum. The subthalamic nucleus in comparison to the SN receives major inputs from only two sources, the lateral pallidal segment and the motor cortex. A single type of STN neuron projects collateral branches to the globus pallidus and the SNR. The neurotransmitter of subthalamopallidal fibers appears to be GABA; subthalamonigral fibers appear to convey excitatory influences, but the neurotransmitters involved are unknown. Efferents of the STN appear organized in a manner to modulate the activities of output systems originating from the MPS and the SNR.

ACKNOWLEDGEMENTS

This work was supported by research grant 911C07005 from the Department of Defense, Uniformed Services University of the Health Sciences.

The experiments reported herein were conducted according to the principles set forth in the "Guide for the Care and Use of Laboratory Animals", NIH Pub. No. 80-23.

REFERENCES

Adey, W. R., Walter, D. O., and Lindsley, D. F., 1962, Subthalamic lesions: Effects on learned behavior and correlated hippocampal and subcortical slow-wave activity, Arch. Neurol., 6:194.

Adinolfi, A. M., and Pappas, G. D., 1968, The fine structure of the caudate nucleus of the cat, J. Comp. Neurol., 133:167.

Anden, N. -E., Carlsson, A., Dahlstrom, A., Fuxe, K., Hillary, N. -A., and Larsson, K., 1964, Demonstration and mapping out of nigrostriatal neurons, Life Sci., 3:523.

Anden, N. E., Fuxe, K., Hamberger, B., and Hökfelt, T., 1966, A quantitative study on the nigro-neostriatal dopamine neuron system in the rat, Acta Physiol. Scand., 67:306.

Anderson, M. E., and Yoshida, M., 1980, Axonal branching patterns and location of nigrothalamic and nigrocollicular neurons in the cat, J. Neurophysiol., 43:883.

Asanuma, C., Thach, W. T., and Jones, E. G., 1983, Cytoarchitectonic delineation of the ventral lateral thalamic region in the monkey, Brain Res. Rev., 5:219.

Asanuma, C., Thach, W. T., and Jones, E. G., 1983a, Distribution of cerebellar terminations and their relation to other afferent terminations in the ventral lateral thalamic region of the monkey, Brain Res. Rev., 5:237.

Bak, I. J., 1967, The ultrastructure of substantia nigra and caudate nucleus of the mouse and the cellular localization of catecholamines, Exp. Brain Res., 3:40.

Bak, I. J., Choi, W. B., Hassler, R., Usunoff, K. G., and Wagner, A., 1975, Fine structural synaptic organization of the corpus striatum and substantia nigra in rat and cat, in: "Dopaminergic Mechanisms," D. Calne, T. N. Chase and A. Barbeau, eds., Raven Press, New York.

Bak, I. J., Markham, C. H., Cook, M. L., and Stevens, J. G., 1978, Ultrastructural and immunoperoxidase study of striatonigral neurons by means of retrograde axonal transport of herpes simplex virus, Brain Res., 143:361.
Balcom, G. J., Lennox, R. H., and Meyerhoff, J. L., 1975, Regional γ-aminobutyric acid levels in rat brain determined by microwave fixation, J. Neurochem., 24:609.
Balthasar, K., 1939, Uber die Beteiligung des Globus Pallidus bei Athetose und Paraballismus, Deutsche. Ztschr. f. Nervenh., 148:243.
Beckstead, R. M., 1983, A reciprocal axonal connection between the subthalamic nucleus and the neostriatum in the cat, Brain Res., 275:137.
Beckstead, R. M., Domesick, V. B., and Nauta, W. J. H., 1979, Efferent connections of the substantia nigra and ventral tegmental area in the rat, Brain Res., 175:191.
Bédard, P., Larochelle, L., Parent, A., and Poirier, L. J., 1969, The nigrostriatal pathway: A correlative study based upon neuroanatomical and neurochemical criteria in the cat and monkey, Exp. Neurol., 25:365.
Bentivoglio, M., Kooy, D. van der, and Kuypers, H. G. J. M., 1979, The organization of the efferent projections of the substantia nigra in the rat. A retrograde fluorescent double labeling study, Brain Res., 174:1.
Bird, E. D., and Iversen, L. L., 1974, Huntington's chorea: Post-mortem measurement of glutamic acid decarboxylase, choline acetyltransferase and dopamine in basal ganglia, Brain, 97:457.
Bobillier, P., Segun, S., Petitjean, F., Salvert, D., Touret, M., and Jouvet, M., 1976, The raphe nuclei of the cat brain stem. A topographical atlas of their projections as revealed by autoradiography, Brain Res., 113:449.
Bonin, G. von, and Shariff, G. A., 1951, Extrapyramidal nuclei among mammals. A quantitative study, J. Comp. Neurol., 94:427.
Brand, S., and Rakic, P., 1979, Genesis of the primate neostriatum: [3H] thymidine autoradiographic analysis of the time of neuron origin in the rhesus monkey, Neurosci., 4:767.
Brownstein, M. J., Palkovitz, J. M., Saavedra, J. M., and Kizer, J. S., 1975, Tryptophan hydroxylase in the rat brain, Brain Res., 97:163.
Buchwald, N. A., Price, D. D., Vernon, L., and Hull, C. D., 1973, Caudate intracellular response to thalamic and cortical inputs, Exp. Neurol., 38:311.
Bunney, B. S., and Aghajanian, G. K., 1976, The precise localization of nigral afferents in the rat as determined by a retrograde tracing technique, Brain Res., 117:423.
Carman, J. B., Cowan, W. M., and Powell, T. P. S., 1963, The organization of corticostriate connexions in the rabbit, Brain, 86:525.

Carman, J. B., Cowan, W. M., Powell, T. P. S., and Webster, K. E., 1965, A bilateral corticostriate projection, J. Neurol. Neurosurg. Psychiat., 28:71.

Carpenter, M. B., 1955, Ballism associated with partial destruction of the subthalamic nucleus of Luys, Neurology, 5:479.

Carpenter, M. B., 1961, Brain stem and infratentorial neuraxis in experimental dyskinesia, A.M.A. Arch. Neurol., 5:504.

Carpenter, M. B., 1981, Anatomy of the corpus striatum and brain stem integrating systems, in: "Handbook of Physiology, The Nervous System, Motor Control," V. Brooks, ed., Amer. Physiol. Society, Bethesda, Vol. II.

Carpenter, M. B., Batton, R. R., Carleton, S. C., and Keller, J. T., 1981, Interconnections and organization of pallidal and subthalamic nucleus neurons in the monkey, J. Comp. Neurol., 197:579.

Carpenter, M. B., Carleton, S. C., Keller, J. T., and Conte, P., 1981a, Connections of the subthalamic nucleus in the monkey, Brain Res., 224:1.

Carpenter, M. B., and Carpenter, C. S., 1951, Analysis of somatotopic relations of the corpus Luysii in man and monkey. Relation between the site of dyskinesia and distribution of lesions within the subthalamic nucleus, J. Comp. Neurol., 95:349.

Carpenter, M. B., Fraser, R. A. R., and Shriver, J., 1968, The organization of the pallidosubthalamic fibers in the monkey, Brain Res., 11:522.

Carpenter, M. B., and McMasters, R. E., 1964, Lesions of the substantia nigra in the rhesus monkey. Efferent fiber degeneration and behavioral observations, Am. J. Anat., 114::293.

Carpenter, M. B., and Mettler, F. A., 1951, Analysis of subthalamic hyperkinesia in the monkey with special reference to ablations of agranular cortex, J. Comp. Neurol., 95:125.

Carpenter, M. B., Nakano, K., and Kim, R., 1976, Nigrothalamic projections in the monkey demonstrated by autoradiographic technics, J. Comp. Neurol., 165:401.

Carpenter, M. B., and Peter, P., 1972, Nigrostriatal and nigrothalamic fibers in the rhesus monkey, J. Comp. Neurol., 144:93.

Carpenter, M. B., and Strominger, N. L., 1967, Efferent fiber projections of the subthalamic nucleus in the rhesus monkey. A comparison of the efferent projections of the subthalamic nucleus, substantia nigra and globus pallidus, Amer. J. Anat., 121:41.

Carpenter, M. B., Strominger, N. L., and Weiss, A. H., 1965, Effects of lesions in the intralaminar thalamic nuclei upon subthalamic dyskinesia, Arch. Neurol., 13:113.

Carpenter, M. B., and Sutin, J., 1983, "Human Neuroanatomy," 8th Ed., Williams and Wilkins, Baltimore.

Carpenter, M. B., Whittier, J. R., and Mettler, F. A., 1950, Analysis of choreiod hyperkinesia in the rhesus monkey. Surgical and pharmacological analysis of hyperkinesia resulting from lesions of the subthalamic nucleus of Luys, J. Comp. Neurol., 92:293.

Carter, D. A., and Fibiger, H. C., 1978, The projections of the entopeduncular nucleus and globus pallidus in rat as demonstrated by autoradiography and horseradish peroxidase histochemistry, J. Comp. Neurol., 177:113.

Chang, H. T., Wilson, C. J., and Kitai, S. T., 1981, Single neostriatal efferent axons in the globus pallidus: a light and electron microscopic study, Science, 213:915.

Cole, M., Nauta, W. J. H., and Mehler, W. R., 1964, The ascending efferent projections of the substantia nigra, Tr. Am. Neurol. Assoc., 89:74.

Connor, J. D., 1970, Caudate nucleus neurones: correlation of the effects of substantia nigra stimulation with iontophoretic dopamine, J. Physiol. (Lond.), 208:691.

Conrad, L. C., Leonard, C. M., and Pfaff, D. W., 1974, Connections of the median and dorsal raphe nuclei in the rat. An autoradiographic and degeneration study, J. Comp. Neurol., 156:179.

Cowan, W. M., and Powell, T. P. S., 1966, Striopallidal projection in the monkey, J. Neurol. Neurosurg. Psychiat., 29:426.

Crossman, A. R., Sambrook, M. A., and Jackson, A., 1980, Experimental hemiballismus in the baboon produced by injection of a gamma-aminobutyric acid antagonist into the basal ganglia, Neurosci. Lett., 20:369.

Crossman, A. R., Sambrook, M. A., and Jackson, A., 1984, Experimental hemichorea/hemiballismus in the monkey: Studies on the intracerebral site of action in a drug-induced dyskinesia, Brain, (In press).

Cuello, A. C., and Kanazawa, I., 1978, The distribution of substance P immunoreactive fibers in the rat central nervous system, J. Comp. Neurol., 178:129.

Dahlstrom, A., and Fuxe, K., 1964, Evidence for the existence of monamine-containing neurons in the central nervous system. I. Demonstration of monamines in the cell bodies of brain stem neurons, Acta Physiol. Scand., 62 (Suppl. 232):1.

Davies, J., and Dray, A., 1976, Substance P in the substantia nigra, Brain Res., 107:623.

Deniau, J. H., Fèger, J., and LeGuyader, C., 1976, Striatal evoked inhibition of identified nigro-thalamic neurons, Brain Res., 104:152.

Deniau, J. M., Hammond, C., Chevalier, G., and Féger, J., 1978, Evidence for branched subthalamic nucleus projections to substantia nigra, ento-peduncular nucleus and globus pallidus, Neurosci. Lett., 9:117.

Deniau, J. M., Hammond, C., Rizk, A., and Féger, J., 1978, Electrophysiological properties of identified output neurons of the rat substantia nigra (pars compacta and pars reticulata): evidences for the existence of branched neurons, Exp. Brain Res., 32:49.

Deniau, J. M., Hammond-LeGuyader, C., Féger, J., and McKenzie, J. S., 1977, Bilateral projections of nigrocollicular neurons: An electrophysiological analysis in the rat, Neurosci. Lett., 5:45.

Descarries, L., Bosler, O., Berthelet, F., and DesRosiers, M. H., 1980, Dopaminergic nerve endings visualized by high-resolution autoradiography in adult neostriatum, Nature, 184:620.

Descarries, L., Bosler, O., Berthelet, F., and DesRosiers, M. H., 1981, Innervation dopaminergique du néostriatum: nouvelle possibilité d'investigation radioautographique, J. Physiol., Paris, 77:53.

Descarries, L., Watkins, K. C., Garcia, S., and Beaudet, A., 1982, The serotonin neurons in nucleus raphe dorsalis of adult rat: A light and electron microscope radioautographic study, J. Comp. Neurol., 27:239.

DeVito, J. L., and Anderson, M. E., 1982, An autoradiographic study of efferent connections of the globus pallidus in Macaca mulatta, Exp. Brain Res., 46:17.

DeVito, J. L., Anderson, M. E., and Walsh, K. E., 1980, A horseradish peroxidase study of afferent connections of the globus pallidus in Macaca mulatta, Exp. Brain Res., 38:65.

DiFiglia, M., Aronin, N., and Leeman, S. E., 1981, Immunoreactive substance P in the substantia nigra of the monkey: light and electron microscopic localization, Brain Res., 233:381.

DiFiglia, M., Aronin, N., and Martin, J. B., 1982, Light and electron microscopic localization of immunoreactive Leu-enkephalin in the monkey basal ganglia, J. Neurosci., 2:303.

DiFiglia, M., Pasik, P., and Pasik, T., 1976, A Golgi study of neuronal types in the neostriatum of monkeys, Brain Res., 114:245.

DiFiglia, M., Pasik, T., and Pasik, P., 1980, Ultrastructure of Golgi-impregnated and gold-toned spiny and aspiny neurons in the monkey neostriatum, J. Neurocytol., 9:471.

Divac, I., 1972, Neostriatum and functions of prefrontal cortex, Acta Neurobiol. Exp., 32:461.

Donoghue, J. P., and Kitai, S. T., 1981, A collateral pathway to the neostriatum from corticofugal neurons of the rat sensory-motor cortex: An intracellular study, J. Comp. Neurol., 201:1.

Dray, A., 1980, The physiology and pharmacology of mammalian basal ganglia, Progress in Neurol., 14:221.

Dray, A., Davies, J., Oakley, N. R., Tongroach, P., and Vellucci, S., 1978, The dorsal and medial raphe projections to the substantia nigra in the rat: electrophysiological, biochemical and behavioural observations, Brain Res., 151:431.

Dray, A., Gonye, T. J., Oakley, N. R., and Tanner, T., 1976, Evidence for the existence of a raphe projection to the substantia nigra in rat, Brain Res., 113:45.

Duffy, M. J., Mulhall, D., and Powell, D., 1975, Subcellular distributiion of substance P in bovine hypothalamus and substantia nigra, J. Neurochem., 25:305.

Elde, R., Hokfellt, T., Johansson, O., and Terenius, L., 1976, Immunohistochemical studies using antibodies to leucine-enkephalin: initial observations on the nervous system of the rat, Neurosci., 1:349.

Emson, P. C., Arregui, A., Clemont-Jones, V., Sandberg, B. E. B., and Rossor, M., 19800, Regional distribution of methionine-enkephalin and substance P-like immunoreactivity in normal human brain and in Huntington's disease, Brain Res., 199:147.

Fahn, S., 1976, Regional distribution studies of GABA and other putative neurotransmitters and their enzymes, in: "GABA in Nervous System Function," E. Roberts, ed., Raven Press, New York.

Fahn, S., and Côté, L. J., 1968, Regional distribution of γ-aminobutyric acid (GABA) in brain of the rhesus monkey J. Neurochem., 15:209.

Faull, R. L. M., and Carman, J. B., 1968, Ascending projections of the substantia nigra in the rat, J. Comp. Neurol., 132:73.

Faull, R. L. M., and Mehler, W. R., 1978, The cells of origin and nigrotectal, nigrothalamic and nigrostriatal projections in the rat, Neurosci., 3:989.

Féger, J., Deniau, J. M., de Champlain, J., and Feltz, P., 1979, A survey of electrophysiology and pharmacology of neostriatal input-output relations, in: "The Neostriatum," I. Divac and R. G. E. Oberg, eds., Pergamon Press, Oxford.

Fibiger, H. C., 1982, The organizatiion and some projections of cholinergic neurons of the mammalian forebrain, Brain Res. Rev., 4:327.

Fibiger, H. C., and Miller, J. J., 1977, An anatomical and electrophysiological investigation of the serotonergic projection from the dorsal raphe nucleus to the substantia nigra in the rat, Neurosci., 2:975.

Fibiger, H. C., Pudritz, R. E., McGeer, P. L., and McGeer, E. G., 1972, Axonal transport in nigro-striatal and nigrothalamic neurons: effects of medial forebrain bundle lesions and 6-hydroxydopamine, J. Neurochem., 19:1697.

Foix, C., and Nicolesco, J., 1925, Les Noyaux Gris Centraux et la Région Mésencéphalo-sous-optique, Masson et Cie, Paris.

Fonnum, F., Gottesfeld, Z., and Grofová, I., 1978, Distribution of glutamate decarboxylase, choline acetyltransferase and aromatic amino acid decarboxylase in the basal ganglia of normal and operated rats. Evidence for striatopallidal, striatoentopeduncular and striatonigral GABAergic fibers, Brain Res., 143:125.

Fonnum, F., Grofová, I., and Rinvik, E.., 1978a, Origin and distribution of glutamate decarboxylase in the nucleus subthalamicus of the cat, Brain Res., 153:370.

Fonnum, F. Grofová, I., Rinvik, E., Storm-Mathisen, J., and Walberg, F., 1974, Origin and distribution of glutamate decarboxylase in substantia nigra of the cat, Brain Res., 71:77.

Fonnum, F., Storm-Mathisen, J., and Divac, I., 1981, Biochemical evidence for glutamate as neurotransmitter in corticostriate and cortico-thalamic fibres in rat brain, Neurosci., 6:863.

Fox, C. A., Andrade, A. N., Hillman, D. E., and Schwyn, R. C., 1971, The spiny neurons in the primate striatum: A Golgi and electron microscopic study, J. f. Hirnforsch., 13:181.

Fox, C. A., Andrade, A. N., Lu Qui, I. J., and Rafols, J. A., 1974, The primate globus pallidus: A Golgi and electron microscopic study, J. Hirnforsch., 15:75.

Fox, C. A., Andrade, A. N., Schwyn, R. C., and Rafols, J. A., 1971/72, The aspiny neurons and the glia in the primate striatum. A Golgi and electron microscopic study, J. F.. Hirnforsch., 13:341.

Fox, C. A., and Rafols, J. A., 1975, The radial fibers in the globus pallidus, J. Comp. Neurol., 159:177.

Fox, C. A., Rafols, J. A., and Cowan, W. M., 1975, Computer measurements of axis cylinder diameters of radial fibers and "comb" bundle fibers, J. Comp. Neurol., 159:201.

Fox, C. A., and Schmitz, J. T., 1944, The substantia nigra and the entopeduncular nucleus in the cat, J. Comp. Neurol., 80:323.

Frigyesi, T., and Rabin, A., 1971, Basal ganglia, diencephalon synaptic relations in the cat. III. An intracellular study of ansa lenticularis, lenticular fasciculus and pallido-subthalamic projection activities, Brain Res., 35:67.

Frotscher, M., Rinne, U., Hassler, R., and Wagner, A., 1981, Termination of cortical afferents on identified neurons in the caudate nucleus of the cat, Exp. Brain Res., 41:329.

Fuxe, K., 1965, Evidence for the existence of monoamine neurones in the central nervous system. IV. Distribution of monoamine nerve terminals in the central nervous system, Acta Physiol. Scand., 64 (Suppl. 247):37.

Fuxe, K., and Andén, N. -E., 1966, Studies on the central monoamine neurons with special reference to the nigro-neostriatal dopamine neuron system, in: "Biochemistry and Pharmacology of the Basal Ganglia," E. Costa et al., eds., Raven Press, New York.

Gale, K., Hong, J. -S., and Guidotti, A., 1977, Presence of substance P and GABA in separate striatonigral neurons, Brain Res., 136:371.

Gauchey, C., Beaujouan, J. C., Besson, M. J., Kerdehue, B., Glowinski, J., and Michelot, R., 1979, Topographical distribution of substance P in the cat substantia nigra, Neurosci. Lett., 12:127.

Glees, P., and Wall, P. D., 1946, Fiber connections of the subthalamic region and the centromedian nucleus of the thalamus, Brain, 69:195.
Goldman, P. S., and Nauta, W. J. H., 1977, An intricately patterned prefrontocaudate projection in the rhesus monkey, J. Comp. Neurol., 171:369.
Gonzales-Vegas, J. A., 1974, Antagonism of dopamine-mediated inhibition in the nigrostriatal pathway: a mode of action of some catatonia-inducing drugs, Brain Res., 80:219.
Gottesfeld, Z., Massari, V. J., Muth, E. A., and Jacobwitz, D. M., 1977, Stria medullaris: a possible pathway containing GABAergic afferents to the lateral habenula, Brain Res., 130:184.
Gramsch, C., Höllt, V., Mehraein, P., Pasi, A., and Herz, A., 1979, Regional distributioin of methionine enkephalin-and beta-endorphin-like in human brain and pituitary, Brain Res., 177:261.
Graybiel, A. M., 1977, Direct and indirect preoculomotor pathways of the brain stem: An autoradiographic study of the pontine reticular formation in the cat, J. Comp. Neurol., 175:37.
Graybiel, A. M., 1978, Organization of the nigrotectal connection: An experimental tracer study in the cat, Brain Res., 143:339.
Graybiel, A. M., Pickel, V. M., Joh, T. J., Reis, D. J., and Ragsdale, C. W., 1981a, Direct demonstration of a correspondence between the dopamine islands and acetylcholinesterase patches in the developing striatum, Proc. Natl. Acad. Sci., 78:5871.
Graybiel, A. M., and Ragsdale, C. W., 1979, Histochemically distinct compartments in the striatum of human, monkey and cat demonstrated by acetylcholinesterase staining, Proc. Natl. Acad. Sci., 75: 5723.
Graybiel, A. M., and Ragsdale, C. W., 1980, Clumping of acetylcholinesterase activity in the developing striatum of the human fetus and young infant, Proc. Natl. Acad. Sci., 77:1214.
Graybiel, A. M., Ragsdale, C. W., and Moon Edley, S., 1979, Compartments in the striatum of the cat observed by retrograde cell labeling, Exp. Brain Res., 34:189.
Graybiel, A. M., Ragsdale, C. W., Jr., Yonkeaka, E. S., and Elde, R. P., 1981, An immunohistochemical study of enkephalins and other neuropeptides in the striatum of the cat with evidence that opiate peptides are arranged to form mosaic patterns in register with striosomal compartments visible by acetycholinesterase staining, Neurosci., 6:377.
Graybiel, A. M., and Sciascia, T. R., 1975, Origin and distribution of nigrotectal fibers in the cat, Neurosci. Abstr., 1:271.
Grofová, I., 1975, The identification of striatal and pallidal neurons projecting to substantia nigra. An experimental study by means of retrograde axonal transport of horseradish peroxidase, Brain Res., 91:286.

Grofová, I., 1979, Extrinsic connections of the neostriatum, in: "The Neostriatum," I. Divac, and R. G. E. Oberg, eds., Pergamon Press, Oxford.

Grofová, I., Ottersen, O. P., and Rinvik, E., 1978, Mesencephalic and diencephalic afferents to the superior colliculus and periaqueductal gray substance demonstrated by retrograde axonal transport of horseradish peroxidase in the cat, Brain Res., 146:205.

Groves, P. M., 1983, A theory of the functional organization of the neostriatum and the neostriatal control of voluntary movement, Brain Res. Rev., 5:109.

Gulley, R. L., and Wood, R. L., 1971, The fine structure of the neurons in the rat substantia nigra, Tissue Cell, 3:675.

Haber, S., and Elde, R., 1981, Correlatiion between met-enkephalin and substance P immunoreactivity in the primate globus pallidus, Neurosci., 6:1291.

Haber, S. N., and Nauta, W. J. H., 1983, Ramifications of the globus pallidus in the rat as indicated by patterns of immunohistochemistry, Neorosci., 9:243.

Hajdu, F., Hassler, R., and Bak, I. J., 1973, Electron microscopic study of the substantia nigra and the strionigral projection in the rat, Z. Zellforsch, Mikrosk, Anat., 146:207.

Hamilton, W. J., and Mossman, H. W., 1972, "Human Embryology," Williams and Wilkins, Baltimore.

Hammond, E., Deniau, J. M. Rizk, A., and Féger, J., 1978, Electrophysiological demonstration of an excitatory subthalamo-nigral pathway in the rat, Brain Res., 235.

Hammond, C., Féger, J., Bioulac, B., and Souteyrand, J. P., 1979, Experimental hemiballism in the monkey produced by unilateral kainic acid lesion in corpus Luysii, Brain Res., 171:577.

Hammond, C., Shibazaki, T., and Rouzaire-Dubois, B., 1983, Branched output neurons of the rat subthalamic nucleus: electrophysiological study of the synaptic effects on identified cells in the two main target nuclei, the entopeduncular nucleus and the substantia nigra, Neurosci., 9:511.

Hammond, C., and Yelnik, J., 1983, Intracellular labelling of rat subthalamic neurones with horseradish peroxidase: computer analysis of dendrites and characterization of axon arborization, Neurosci., 8:781.

Harnois, C., and Filion, M., 1980, Pallidal neurons branching to the thalamus and to the midbrain in the monkey, Brain Res., 186:222.

Hartmann-von Monakow, K., Akert, K., and Kunzle, H., 1978, Projections of the precentral motor cortex and other cortical areas of the frontal lobe to the subthalamic nucleus in the monkey, Exp. Brain Res., 33:395.

Hartmann-von Monakow, K., Akert, K., and Kunzle, H., 1979, Projections of precentral and premotor cortex to the red nucleus and other midbrain areas in Macaca fascucularis, Exp. Brain Res., 34:91.

Hattori, T., Fibiger, H. C., and McGeer, P. L., 1975, Demonstration of a pallido-nigral projection innervating dopaminergic neurons, J. Comp. Neurol., 162:487.

Hattori, T., McGeer, P. L., Fibiger, H., C., and McGeer, E. G., 1973, On the source of GABA-containing terminals in the substantia nigra. Electron microscopic, autoradiographic and biochemical studies, Brain Res. 54:103.

Hattori, T., McGeer, E. G., and McGeer, P. L., 1979, Fine structural analysis of the cortico-striatal pathway, J. Comp. Neurol., 185:347.

Hedreen, J. C., 1971, Separate demonstration of dopaminergic and non-dopaminergic projections of substantia nigra in the rat, Anat. Rec., 169:338.

Heimer, L., 1978, The olfactory cortex and the ventral striatum, in: Limbic Mechanisms, K. E. Livingston and O. Hornykiewicz, eds., Plenum Press, New York.

Heimer, L., and Wilson, R. D., 1975, The subcortical projections of the allocortex: similarities in the neural associations of the hippocampus, the piriform cortex, and the neocortex, in: Golgi Centennial Symposium, M. Santini, ed., Raven Press, New York.

Henderson, Z., 1981, Ultrastructure and acetylcholinesterase content of neurones forming connections between the striatum and substantia nigra of rat, J. Comp. Neurol., 197:185.

Hikosaka, O., and Wurtz, R. H., 1983, Visual and oculomotor functions of monkey substantia nigra pars reticulata. I. Relation of visual and auditory responses to saccades, J. Neurophysiol., 49:1230.

Hikosaka, O., and Wurtz, R. H., 1983a, Visual and oculomotor functions of monkey substantia nigra pars reticulata. II. Visual responses related to fixation of gaze, J. Neurophysiol., 49:1254.

Hikosaka, O., and Wurtz, R. H., 1983, Visual and oculomotor functions of monkey substantia nigra pars reticulata. III. Memory-contingent visual and saccade responses, J. Neurophysiol., 49:1268.

Hirosawa, K., 1968, Electron microscopic studies on pigment granules in substantia nigra and locus coeruleus of the Japanese monkey (Macaca fuscata yaku), Z. Zellforsch. Mikrosk. Anat., 88:187.

Hoesen, G. W. van, Yeterian, E. H., and Lavizzo-Mourey, R., 1981, Widespread corticostriate projections from temporal cortex of rhesus monkey, J. Comp. Neurol., 199:205.

Hökfelt, T., Johansson, O., Kellerth, J. -O., Ljungdahl, A., Nilsson, G., Nygords, A., and Pernow, B., 1977, Immunohistochemistical distribution of substance P, in: "Substance P," U. S. von Euler and B. Pernow, eds., Raven Press, New York.

Hökfelt, T., and Ungerstedt, U., 1969, Electron and fluorescence microscopical studies on the nucleus caudatus putamen of the rat after unilateral lesions of ascending nigro-neostriatal dopamine neurons, Acta Physiol. Scand., 76:415.

Hong, J. S., Yang, H. -Y., Racagni, G., and Costa, E., 1977, Projections of substance P containing neurons from the neostriatum to the substantia nigra, Brain Res., 122:541.

Hopkins, D. A., and Niessen, L. W., 1976, Substantia nigra projections to the reticular formation, superior colliculus and central gray in the rat, cat and monkey, Neurosci. Lett., 2:253.

Hornykiewicz, O., 1966, Metabolism of brain doipamine in human parkinsonism: neurochemical and clinical aspects, in: "Biochemistry and Pharmacology of the Basal Ganglia," E.. Costa et al., eds., Raven Press, Hewlett, N. Y.

Houser, C. R., Lee, M., and Vaughn, J. E., 1983, Immunocytochemical localization of glutamic acid decarboxylase in normal and deafferented superior colliculus: Evidence for reorganization of γ-aminobutyric acid synapses, J. Neurosci., 3:2030.

Hughes, J., Smith, T. W., Kosterlitz, H. W., Fothergill, L. A., Morgan, B. A., and Morris, H. R., 1975, Identification of two related pentapeptides from brain with potent opiate agonist activity, Nature, Lond., 258:577.

Iwahori, N., 1978, A Golgi study on the subthalamic nucleus of the cat, J. Comp. Neurol., 182:382.

Jackson, A., and Crossman, A. R., 1981, Subthalamic projection to nucleus tegmenti pedunculopontinus in the rat, Neurosci. Lett., 22:17.

Jackson, A., and Crossman, A. R., 1981a, Subthalamic nucleus efferent projections to the cerebral cortex, Neurosci., 6:2367.

Jacobowitz, D. M., and Palkovits, M., 1974, Topographic atlas of catecholamine and acetylcholinesterase-containing neurons in the rat brain. I. Forebrain (telencephalon, diencephalon), J. Comp. Neurol., 157:13.

Jayaraman, A., 1983, Topographic organization and morphology of peripallidal and pallidal cells projecting to the striatum in cats, Brain Res., 275:279.

Jayaraman, A., Batton, R. R., and Carpenter, M. B., 1977, Nigrotectal projections in the monkey: An autoradiographic study, Brain Res., 135:147.

Jessell, T. M., Emson, P. C., Paxinos, G., and Cuello, A. C., 1978, Topographical projections of substance P and GABA pathways in the striato- and pallido-nigral system: A biochemical and immunohistochemical study, Brain Res., 152:487.

Jinnai, K., and Matsuda, Y., 1979, Neurons of the motor cortex projecting commonly on the caudate nucleus and the lower brain stem in the cat, Neurosci. Lett., 13:121.

Jones, E. G., Coulter, J. D., Burton, H., and Porter, R., 1977, Cells of origin and terminal distribution of corticostriatal fibers arising in the sensory-motor cortex of monkeys, J. Comp. Neurol., 173:53.

Jones, E. G., and Leavitt, R. Y., 1974, Retrograde axonal transport and the demonstration of non-specific projections to the cerebral cortex and striatum from thalamic intralaminar nuclei in the rat, cat and monkey, J. Comp. Neurol., 154:349.

Jones, E. G., and Wise, S. P., 1977, Size, laminar and columnar distribution of efferent cells in the sensory-motor cortex of monkeys, J. Comp. Neurol., 175:391.

Juraska, J. A., Wilson, C. J., and Groves, P. M., 1977, The substantia nigra of the rat: A Golgi study, J. Comp. Neurol., 172:585.

Kalil, K., 1978, Patch-like termination of thalamic fibers in the putamen of the rhesus monkey: an autoradiographic study, Brain Res., 140:333.

Kanazawa, I., Emson, P. C., and Cuello, A. C., 1977, Evidence for the existence of substance P-containing fibres in striato-nigral and pallido-nigral pathways in rat brain, Brain Res., 119:447.

Kanazawa, I. Marshall, G. R., and Kelly, J. S., 1976, Afferents to the rat substantia nigra studied with horseradish peroxidase, with special reference to fibres from the subthalamic nucleus, Brain Res., 115:485.

Kataoka, K., Bak, I. J., Hassler, R., Kim, J. S., and Wagner, A., 1974, L-Glutamate decarboxylase and choline acetyltransferase activity in the substantia nigra and striatum after surgical interruption of strio-nigral fibers of the baboon, Exp. Brain Res., 19:217.

Kelley, A. E., Domesick, V. B., and Nauta, W. J., 1982, The amygdalostriatal projection in the rat - an anatomical study by anterograde and retrograde tracing methods, Neurosci., 7:615.

Kemel, M. L., Gauchy, C., Romo, R., Glowinski, J., and Besson, M. J., 1983, In vivo release of [3H] GABA in cat caudate nucleus and substantia nigra. I. Bilateral changes induced by a unilateral nigral application of muscimol, Brain Res., 272:331.

Kemp, J. M., 1968, An electron microscopic study of the terminations of afferent fibers in the caudate nucleus, Brain Res., 11:464.

Kemp, J. M., 1970, The termination of strio-pallidal and strio-nigral fibers, Brain Res., 17:125.

Kemp, J. M., and Powell, T. P. S., 1970, The corticostriate projection in the monkey, Brain, 93:525.

Kemp, J. M., and Powell, T. P. S., 1971, The structure of the caudate nucleus of the cat: light and electron microscopy, Phil. Trans. Roy. Soc. Lond. (Biol.), 262:383.

Kemp, J. M., and Powell, T. P. S., 1971a, The connexions of the striatum and globus pallidus: synthesis and speculation, Phil. Trans. Roy. Soc. Lond. (Biol.), 262:442.

Kemp, J. M., and Powell, T. P. S., 1971b, The site of termination of afferent fibres in the caudate nucleus, Phil. Trans. Roy. Soc. Lond. (Biol.), 262:413.

Kim, J. S., Bak, I. J., Hassler, R., and Okada, Y., 1971, Role of γ-aminobutyric acid (GABA) in the extrapyramidal motor system. 2. Some evidence of a type of GABA-rich strionigral neuron, Exp. Brain Res., 14:95.

Kim, R., Nakano, K., Jayaraman, A., and Carpenter, M. B., 1976, Projections of the globus pallidus and adjacent structures: An autoradiographic study in the monkey, J. Comp. Neurol., 169:263.

Kimura, H., McGeer, P. L., Peng, F., and McGeer, E. G., 1980, Choline acetyltransferase-containing neurons in rodent brain demonstrated by immunohistochemistry, Science, 208:1057.

Kita, H., Chang, H. T., and Kitai, S. T., 1983, The morphology of intracellularly labeled rat subthalamic neurons: A light microscopic analysis, J. Comp. Neurol., 215:245.

Kita, H., Chang, H. T., and Kitai, S. T., 1983a, Pallidal inputs to subthalamus: Intracellular analysis, Brain Res., 264:255.

Kitai, S. T., 1981, Anatomy and physiology of the neostriatum, in: "GABA and the Basal Ganglia, G. DiChiara and G. L. Gessa, eds., Raven Press, New York.

Kitai, S. T., and Deniau, J. M., 1981, Cortical inputs to the subthalamus: intracellular analysis, Brain Res., 214:411.

Kitai, S. T., Kocsis, J. D., and Wood, J., 1976, Origin and characteristics of the cortico-caudate afferents: An anatomical and electrophysiological study, Brain Res., 118:137.

Knook, H. L., 1965, "The Fibre-Connections of the Forebrain," Van Gorcum & Co., N. V., Assen, The Netherlands.

Kodama, S., 1928, Beiträge zur normalen anatomie des corpus luysi beim Menschen. Arbeiten a. d. Anat. Institute, Sendai, 13:221.

Konishi, S., and Otsuka, M., 1974, The effects of substance P and other peptides on spinal neurons of the frog, Brain Res., 65:397.

Konishi, S., and Otsuka, M., 1974a, Excitatory action of hypothalamic substance P on spinal motoneurones of newborn rats, Nature (Lond.), 252:734.

Kooy, van der, D., and Hattori, T., 1980, Dorsal raphe cells with collateral projections to the caudate-putamen and substantia nigra: A fluorescent retrograde double labeling study in the rat, Brain Res., 186:1.

Kooy, van der, D., and Hattori, T., 1980a, Single subthalamic nucleus neurons project to both the globus pallidus and substantiaa nigra in rat, J. Comp. Neurol., 192:751.

Kooy, van der, D., and Hattori, T., Shannak, K., and Hornykiewicz, O., 1981, The pallido-subthalamic projection in rat: Anatomical and biochemical studies, Brain Res., 204:253.

Krettek, J. E., and Price, J. L., 1978, Amygdaloid projections to subcortical structures within the forebrain and brain stem in the rat and cat, J. Comp. Neurol., 178:225.

Kubek, M. J., and Wilber, J. F., 1980, Regional distribution of leucine-enkephalin in hypothalamus and extrahypothalamic loci of the human nervous system, Neurosci. Lett., 18:155.

Kuhar, M. J., Aghajanian, G. K., and Roth, R. H., 1972, Tryptophan hydroxylase activity and synaptosomal uptake of serotonin in discrete brain regions after midbrain raphe lesions: correlations with serotonin levels and histochemical fluorescence, Brain Res., 44:165.

Kuhlenbeck, H., 1948, The derivatives of the thalamus ventralis in the human brain and their relation to the so-called subthalamus, Mil. Surgeon, 102:433.

Kuhlenbeck, H., and Haymaker, W., 1949, The derivatives of the hypothalamus in the human brain; their relation to the extrapyramidal and autonomic systems, Mil. Surgeon, 105:26.

Kultas-Ilinsky, K., Ilinsky, I., Warton, S., and Smith, K. R., 1983, Fine structure of nigral and pallidal afferents in the thalamus: An EM autoradiographic study in the cat, J. Comp. Neurol., 216:390.

Kunzle, H., 1975, Bilateral projections from precentral motor cortex to the putamen and other parts of the basal ganglia, Brain Res., 88:195.

Kunzle, H., 1978, An autoradiographic analysis of the efferent connections from premotor and adjacent prefrontal regions (areas 6 and 9) in Macaca fascicularis, Brain, Behavior Evolution, 15:185.

Kunzle, H., and Akert, K., 1977, Efferent connections of cortical area 8 (frontal eye field) in Macaca fascicularis. A reinvestigation using the autoradiographic technique, J. Comp. Neurol., 173:147.

Kuo, J. S., and Carpenter, M. B., 1973, Organization of pallidothalamic projections in the rhesus monkey, J. Comp. Neurol., 151:201.

Kuypers, H. G. J. M., and Lawrence, D. G., 1967, Cortical projections to the red nucleus and the brain stem in the rhesus monkey, Brain Res., 4:151.

Larsen, K. D., and McBride, R. L., 1979, The organization of feline entopenduncular nucleus projections: Anatomical studies, J. Comp. Neurol., 184:293.

Levin, P. M., 1949, Efferent fibers, in: "The Precentral Motor Cortex," P. C. Bucy, ed., University of Ill. Press, Urbana, Ill.

Ljungdahl, A., Hökfelt, T., and Nilsson, G., 1978, Distribution of substance P-like immunoreactivity in the central nervous system of the rat. I. Cell bodies and nerve terminals, Neurosci., 3:861.

McBride, R. L., and Larsen, K. D., 1980, Projections of the feline globus pallidus, Brain Res., 189:3.

McGeer, P. L., Fibiger, H. C., Maler, L., Hattori, T., and McGeer, E., 1974, Evidence for descending pallidonigral GABA-containing neurons, Adv. Neurol., 5:153.

McGeer, P. L., McGeer, E. G., Wada, J. A., and Jung, E., 1971, Effects of globus pallidus lesions and Parkinson's disease on brain glutamic acid decarboxylase, Brain Res., 32:425.

McLardy, T., 1948, Projection of the centromedian nucleus of the human thalamus, Brain, 71:290.
Mehler, W. R., 1966, Further notes on the center median nucleus of Luys, in: "The Thalamus," D. P. Purpura and M. D. Yahr, eds., Columbia University Press, New York.
Mettler, F. A., 1947, Extracortical connections of the primnate frontal cerebral cortex. II. Corticofugal connections, J. Comp. Neurol., 86:119.
Miller, J. J. Richardson, T. L., Fibiger, H. C., and McLennan, H., 1975, Anatomical and electrophysiological identification of a projection from the mesencephalic raphe to the caudate putamen in the rat, Brain Res., 97:133.
Monakow, von, C., 1895, Experimentelle and pathologischenatomische Untersuchungen über die Haubenregion, Schhügel und die Regio subthalamica, Arch. Psychiat. Nervenk., 27:1.
Moon Edley, S., and Graybiel, A. M., 1983, The afferent and efferent connections of the feline nucleus tegmenti pedunculopontinus, pars compacta, J. Comp. Neurol., 217:187.
Moore, R. Y., Bhatnagar, R. K., and Heller, A., 1971, Anatomical and chemical studies of a nigro-neostriatal projection in the cat, Brain Res., 30:119.
Nakamura, S., and Sutin, J., 1972, The pattern of termination of pallidal axons upon cells of the subthalamic nucleus, Exper. Neurol., 35:254.
Namba, von, M., 1957, Cytoarchitectonische Untersuchungen am Striatum, J. Hirnforsch., 3:24.
Nauta, H. J. W., 1979, Projections of the pallidal complex: An autoradiographic study in the cat, Neurosci., 4:1853.
Nauta, H. J. W., and Cole, M., 1978, Efferent projections of the subthalamic nucleus: An autoradiographic study in monkey and cat, J. Comp. Neurol., 180:1.
Nauta, H. J. W., and Cuenod, M., 1982, Perikaryal cell labeling in the subthalamic nucleus following the injection of $[^3H]$-γ-aminobutyric acid into the pallidal complex: An autoradiographic study in cat, Neurosci., 7:2725.
Nauta, H. J. W., Pritz, M. B., and Lasek, R. J., 1974, Afferents to the rat caudoputamen studied with horseradish peroxidase. An evaluation of a retrograde neuroanatomical research method, Brain Res., 67:219.
Nauta, W. J. H., 1961, Fibre degeneration following lesions of the amygdaloid complex in the monkey, J. Anat., Lond., 95:515.
Nauta, W. J. H., and Mehler, W. R., 1966, Projections of the lentiform nucleus in the monkey, Brain Res., 1:3.
Nauta, W. J. H., Smith, G. P., Faull, R. L. M., and Domesick, V. B., 1978,, Efferent connections and nigral afferents to the nucleus accumbens septi in the rat, Neurosci., 3:385.
Nomura, S., Mizuno, N., and Sugimato, T., 1980, Direct projections from the pedunculopontine tegmental nucleus to the subthalamic nucleus in the cat, Brain Res., 196:223.

Ohye, C., LeGuyader, C., and Féger, J., 1976, Responses of subthalamic and pallidal neurons to striatal stimulation: An extracellular study on awake monkeys, Brain Res., 111:241.

Oka, H., 1980, Organization of the cortico-caudate projections. A horse-radish peroxidase study in the cat, Exp. Brain Res., 40:203.

Okada, Y., 1976, Role of GABA in the substantia nigra, in: "GABA in Nervous System Function," E. Roberts, T. N. Chase and D. B. Tower, eds., Raven Press, New York.

Okada, Y., Nitsch-Hassler, C., Kim, J. S., Bak, I. J., and Hassler, R., 1971, The role of γ-aminobutyric acid (GABA) in the extrapyramidal motor system. I. Regional distribution of GABA in rabbit, rat and guinea pig brain, Exp. Brain Res., 13:514.

Olpe, H. R., and Koella, W. P., 1977, The response of striatal cells upon stimulation of the dorsal and median raphe nuclei, Brain Res., 122:357.

Olszewski, J., 1952, "The thalamus of the Macaca mulatta. An Atlas for Use with the Stereotaxic Instrument," S. Karger, Basel and New York.

Palkovitz, M., Brownstein, M., and Saavedra, J. M., 1974, Serotonin content of the brain stem nuclei in the rat, Brain Res., 80:237.

Papez, J. W., 1942, A summary of fiber connections of the basal ganglia with each other and with other portions of the brain, Res. Publ. Assoc. Nerv. Ment. Dis., 21:21.

Parent, A., and DeBellefeuille, L., 1982, Organization of efferent projections from the internal segment of the globus pallidus in primate as revealed by fluorescence retrograde labeling method, Brain Res., 245:201.

Parent, A., Descarries, L., and Beaudet, A., 1981, Organization of ascending serotonin systems in the adult rat brain. A radioautographic study after intraventricular administration of [3H] 5-hydroxytryptamine, Neurosci., 6:115.

Park, M. R., Falls, W. M., and Kitai, S. T., 1982, An intracellular HRP study of the rat globus pallidus. I. Responses and light microscopic analysis, J. Comp. Neurol., 211:284.

Pasik, P., Pasik, T., and DiFiglia, M., 1976, Quantitative aspects of neuronal organization in the neostriatum of the Macaque monkey, in: The Basal Ganglia, M. C. Yahr, ed., Raven Press, New York.

Pasik, P., Pasik, T., and DiFiglia, M., 1979, The internal organization of the neostriatum in mammals, in: "The Neostriatum," I. Divac, ed., Pergamon Press, Oxford.

Percheron, G., 1977, The thalamic territory of cerebellar afferents and the lateral region of the thalamus of the macaque in stereotaxic ventricular coordinates, J. Hirnforsch., 18:375.

Petras, J. M., 1969, Some efferent connections of the motor somato-sensory cortex of simian primates and felid, canid and procyonid carnivores, Ann. N. Y. Acad. Sci., 167:469.

Pickel, V. M., Sumai, K. K., Beady, S. C., Miller, R. J., and Reis, D. J., 1980, Immunocytochemical localization of enkephalin in the neostriatum of rat brain: A light and electron microscopic study, J. Comp. Neurol., 189:721.

Poirier, L. J., and Sourkes, T. L., 1965, Influences of the substantia nigra on catecholamine content of the striatum, Brain, 88:181.

Powell, D. S., Leeman, S. E., Tregear, G. W., Niall, H. D., and Potts, J. T., 1973, Radioimmunoassay for substance P, Nature, Lond. New Biol., 241:252.

Powell, T. P. S., and Cowan, W. M., 1954, The connexions of the midline and intralaminar nuclei of the thalamus of the rat, J. Anat., Lond., 88:307.

Powell, T. P. S., and Cowan, W. M., 1956, A study of thalamo-striate relations in the monkey, Brain, 79:364.

Powell, T. P. S., and Cowan, W. M., 1967, The interpretation of the degenerative changes in the intralaminar nuclei of the thalamus, J. Neurol., Neurosurg. and Psychiat., 30:140.

Preston, R. J., Bishop, G. A., and Kitai, S. T., 1980, Medium spiny neuron projection from the rat striatum: An intracellular horseradish peroxidase study, Brain Res., 183:253.

Pycock, C., Horton, R. W., and Marsden, C. D., 1976, The behavioural effects of manipulating GABA function in the globus pallidus, Brain Res., 116:353.

Rafols, J. A., and Fox, C. A., 1976, The neurons in the primate subthalamic nucleus: A Golgi and electron microscopic study, J. Comp. Neurol., 168:75.

Ranson, S. W., and Ranson, M., 1939, Pallidofugal fibers in the monkey, Arch. Neurol. Psychiat., 42:11059.

Ranson, S. W., Ranson, S. W. Jr., and Ranson, M., 1941, Fiber connections of the corpus striatum as seen in Marchi preparations, Arch. Neurol. Psychiat., 46:230.

Reubi, J. C., and Emson, P. C., 1978, Release and distribution of endogenous 5-HT in rat substantia nigra, Brain Res., 139:164.

Ribak, C. E., 1981, The GABAergic neurons of the extrapyramidal system as revealed by immunocytochemistry, in: "GABA and the Basal Ganglia," G. DiChiara and G. L. Gessa, eds., Raven Press, New York.

Ribak, C. E., Vaughn, J. E., and Barber, R. P., 1981, Immunocytochemical localization of GABAergic neurones at the electron microscopical level, Histochem. J., 13:555.

Ribak, C. E., Vaughn, J. E., and Roberts, E., 1979, The GABA neurons and their axon terminals in rat corpus striatum as demonstrated by GAD immunocytochemistry, J. Comp. Neurol., 187:261.

Ribak, C. E., Vaughn, J. E., and Roberts, E., 1980, GABAergic nerve terminals decrease in the substantia nigra following hemitransections of the striatonigral and pallidonigral pathways, Brain Res., 192:413.

Ribak, C. E., Vaughn, J. E., Saito, K., Barber, R., and Roberts, E., 1976, Immunocytochemical localization of glutamate decarboxylase in rat substantia nigra, Brain Res., 116:287.

Ricardo, J. A., 1980, Efferent connections of the subthalamic region in the rat. I. The subthalamic nucleus of Luys, Brain Res., 202:257.

Richter, E., 1965, Die Entwichlung des Globus Pallidus und des Corpus Subthalamicum, Springer-Verlag, Berlin.

Rinvik, E., 1966, The corticonigral projection in the cat. An experimental study with silver impregnation methods, J. Comp. Neurol., 126:241.

Rinvik, E., 1975, Demonstration of nigro-thalamic connections in the cat by retrograde axoxnal transport of horseradish peroxidase, Brain Res., 90:313.

Rinvik, E., and Grofová, I., 1970, Observations on the fine structure of the substantia nigra in the cat, Exp. Brain Res., 11:229.

Rinvik, E., Grofová, I., Hammond, C., Féger, J., and Deniau, J. M., 1979, A study of the afferent connections of the subthalamic nucleus in the monkey and the cat using the HRP technique, Advances in Neurol., (Raven Press), 24:53.

Rinvik, E., Grofová, I., and Ottersen, O. P., 1976, Demonstration of nigrotectal and nigroreticular projections in the cat by axonal transport of protein, Brain Res., 112:388.

Rinvik, E., and Walberg, F., 1969, Is there a cortico-nigral tract? A comment based on experimental electron microscopic observations in the cat, Brain Res., 11:742.

Romansky, K. V., Usunoff, K. G., Ivanov, D. P., and Galabov, G. P., 1979, Corticosubthalamic projection in the cat: An electron microscopic study, Brain Res., 163:319.

Rouzaire-Dubois, B., Hammond, C., Hamon, B., and Féger, J., 1980, Pharmacological blockade of the globus pallidus-induced inhibitory response of subthalamic cells in the rat, Brain Res., 200:321.

Rouzaire-Dubois, B., Scarnati, E., Hammond, C., Crossman, A. R., and Shibazaki, T., 1983, Microiontophoretic studies on the nature of the neurotransmitter in the subthalamo-entopeduncular pathway of the rat, Brain Res., 271:11.

Royce, G. J., 1978, Cells of origin of subcortical afferents to the caudate nucleus. A horseradish peroxidase study in the cat, Brain Res., 152:465.

Royce, G. J., 1978a, Autoradiographic evidence for a discontinuous projection to the caudate nucleus from the centromedian nucleus in the cat, Brain Res., 146:145.

Royce, G. J., 1982, Laminar origin of cortical neurons which project upon the caudate nucleus: A horseradish peroxidase investigation in the cat, J. Comp. Neurol., 205:8.

Schnell, G. R., and Strick, P. L., 1984, The origin of thalamic inputs to the arcuate premotor and supplementary motor areas, J. Neurosci., (in Press).

Schröder, K. F., Hopf, A., Lange, H., and Thorner, G., 1975, Morphometrisch-statische Strukturanalysen des Striatum, Pallidum und Nucleus Subthalamicus beim Menschen, J. Hirnforsch., 16:333.
Schwyn, R. C., and Fox, C. A., 1974, The primate substantia nigra: a Golgi and electron microscopic study, J. Hirnforsch., 15:95.
Siggins, G. r., Hoffer, B. J., Bloom, F. E., and Ungerstedt, U., 1976, cytochemical and electrophysiological studies of dopamine in the caudate nucleus, Res. Publ. Assoc. Res. Nerv. Ment. Dis., 55:227.
Somogyi, P., Bolam, J. P., and Smith, A. D., 1981, Monosynaptic cortical input and local axon collaterals of identified striatonigral neurons. A light and electron microscopic study using the Golgi-peroxidase transport-degeneration procedure, J. Comp. Neurol., 195:567.
Spencer, H. JK., 1976, Antagonism of cortical excitation of striatal neurons by glutamic acid diethyl ester: evidence for glutamic acid as an excitatory transmitter in the rat striatum, Brain Res., 102:91.
Staines, W. A., Atmadja, S., and Fibiger, H. C., 1981, Demonstration of a pallidsotriatal pathway by retrograde transport of HRP-labeled lectin, Brain Res., 206:446.
Staines, W. A., Nagy, J. I., Vincent, S. R., and Fibiger, H. C., 1980, Neurotransmitters contained in the efferents of the striatum, Brain Res., 1194:391.
Steriade, M., and Glenn, L. L., 1982, Neocortical and caudate projections of intralaminar thalamic neurons and their synaptic excitation from midbrain reticular core, J. Neurophysiol., 48:352.
Streit, P., Knecht, E., and Cuenod, M., 1979, Transmitter-specific retrograde labelling in the striato-nigral and raphe-nigral pathways, Science, 205:306.
Strominger, N. L., and Carpenter, M. B., 1965, Effects of lesions in the substantia nigra upon subthalamic dyskinesia in the monkey, Neurol., 15:587.
Sugimoto, T., Hattori, T., 1983, Confirmation of thalamosubthalamic projections by electron microscopic autoradiography, Brain Res., 267:335.
Sugimoto, T., Hattori, T., Mizuno, N., Itoh, K., and Sato, M., 1983, Direct projections from the centre median-parafascicular complex to the subthalamic nucleus in the cat and rat, J. Comp. Neurol., 214:209.
Swanson, L. W., and Cowan, W. M., 1975, A note on the connections and development of the nucleus accumbens, Brain Res., 92:324.
Szabo, J., 1962, Topical distribution of striatal efferents in the monkey, Exper. Neurol., 5:21.
Szabo, J., 1967, The efferent projections of the putamen in the monkey, Exp. Neurol., 19:463.
Szabo, J., 1970, Projections from the body of the caudate nucleus in the rhesus monkey, Exper. Neurol., 27:1.
Szabo, J., 1980, Organization of the ascending striatal afferents in monkeys, J. Comp. Neurol., 189:307.

Szabo, J., 1980a, Distribution of striatal afferents from the mesencephalon in the cat, Brain Res., 188:3.

Ternaux, J. P., Hery, F., Bourgoin, S., Adrien, J., Glowinski, J., and Hamon, M., 1977, The topographical distribution of serotoninergic terminals in the neostriatum of the rat and the caudate nucleus of the cat, Brain Res., 121:311.

Teuber, H. L., 1976, Complex functions of basal ganglia, in: "The Basal Ganglia," M. D. Yahr, ed., Raven Press, New York.

Thorner, G., Lange, H., and Hopf, A., 1975, Morphometrisch - statistische Strukturanalysen des Striatum, Pallidum und Nucleus Subthalamicus beim Menschen, J. Hirnforsch., 16:401.

Tracey, D. J., Asanuma, C., Jones, E. G., and Porter, R., 1980, Thalamic relay to motor cortex: Afferent pathways from brain stem, cerebellum and spinal cord in monkeys, J. Neurophjysiol., 44:532.

Tsubokawa, T., and Sutin, H., 1972, Pallidal and tegmental inhibition of oscillatory slow waves and unit activity in the subthalamic nucleus, Brain Res., 41:101.

Ueki, A. M., Uno, M., Anderson, M., and Yoshida, M., 1977, Monosynaptic inhibition of thalamic neurons produced by stimulation of the substantia nigra, Experientia, 33:1480.

Ungerstedt, U., 1971, Stereotaxic mapping of the monamine pathways in the rat brain, Acta Physiol. Scand. (Suppl.), 367:1.

Uno, M., and Yoshida, M., 1975, Monosynaptic inhibition of thalamic neurons produced by stimulation of the pallidal nucleus in cats, Brain Res., 99:377.

Vogt, C., and Vogt, O., 1941, Thalamusstudien I-III, J. Psychol. Neurol., 50:32.

Voneida, T. J., 1960, An experimental study of the course and destination of fibers arising in the head of caudate nucleus in the cat and monkey, J. Comp. Neurol., 115:75.

Wamsley, J. K., Young, W. S., and Kuhar, M. J., 1980, Immunohistochemical localization of enkephalin in rat forebrain, Brain Res., 190:153.

Webster, K. E., 1961, Cortico-striate interrelations in the albino rat, J. Anat., 95:532.

Whittier, J. R., 1947, Ballism and the subthalamic nucleus (nucleus hypothalamicus corpus Luysii), Arch. Neurol. Psychiat., 58:672.

Whittier, J. R., and Mettler, F. A., 1949, Studies on the subthalamus of the rhesus monkey. I. Anatomy and fiber connections of the subthalamic nucleus of Luys, J. Comp. Neurol., 90:281.

Whittier, J. R., and Mettler, F. A., 1949a, Studies on the subthalamus of the rhesus monkey. II. Hyperkinesia and other physiological effects of subthalamic lesions with special reference to the subthalamic nucleus of Luys, J. Comp. Neurol., 90:319.

Wilson, C. J., Chang, H. T., and Kitai, S. T., 1982, Origins of post-synaptic potentials evoked in identified rat neostriatal neurons by stimulation in substantia nigra, Exp. Brain Res., 45:157.

Wilson, S. A. K., 1912, Progressive lenticular degeneration: a familial nervous disease associated with cirrhosis of the liver, Brain, 34:295.

Wilson, S. A. K., 1914, An experimental research into the anatomy and physiology of the corpus striatum, Brain, 36:427.

Yelnik, J., and Percheron, G., 1979, Subthalamic neurons in primates: A quantitative and comparative analysis, Neurosci., 4:1717.

Yeterian, E. H., and van Hoesen, G. W., 1978, Cortico-striate projections in the rhesus monkey. The organization of certain cortico-caudate connections, Brain Res., 139:43.

THALAMOSTRIATE PROJECTIONS - AN OVERVIEW

A. Jayaraman

Department of Neurology
Louisiana State University
School of Medicine
1542 Tulane Avenue
New Orleans, LA 70112, U.S.A.

The thalamus provides a major source of afferents to the striatum with a density second only to the corticostriatal projections (Graybiel and Ragsdale, 1979). Among the different thalamic nuclei, the nonspecific intralaminar nuclei provide the majority of the thalamic afferents to the striatum (Jones and Leavitt, 1974). The intralaminar nuclear complex has diffuse but significant reciprocal connections with the cerebral cortex (see Macchi and Bentivoglio, 1982 for review). Stimulation of the intralaminar nuclei elicits recruiting and augmenting responses in the cerebral cortex, and they have been suggested to play a role in arousal and attentional mechanisms (Jasper, 1961). Hassler (1982) proposed that conscious perception requires the simultaneous activation of the same cortical sensory or integrative field by the non-specific intralaminar nuclei and by the specific thalamic projections. While such theories may suggest a role of the cortical projections of the intralaminar nuclei, it is ironic that the major target of the intralaminar nuclei is the striatum and not the cerebral cortex and that as yet we do not have a testable hypothesis for the role of the thalamostriate projections. Hassler (1978) suggested that the function of the putamen is to focus the attention, excitability and emotional participation on a single event by simultaneously suppressing all other attention attracting events. It is possible that the attention focusing role of the striatum may be facilitated by the projections from the intralaminar thalamic nuclei to the striatum, similar to their proposed role of cerebral cortical activation. The anatomical and physiological studies have emphasized the projection pattern of the posterior intralaminar nuclei, especially of centre median (CM) and the parafascicular (PF) nuclei,

(See Graybiel and Ragsdale, 1979; Dray, 1980; Carpenter, 1981; Mehler, 1982 for reviews), but the sources and pattern of projection from other subdivisions of the intralaminar nuclei have not been studied in detail. A review of the literature suggests that the striatum also receives projections from thalamic nuclei other than the intralaminar nuclear complex. The object of this report is to briefly review the literature of the thalamostriate projections and to present some preliminary observations from a study to identify the different sources of thalamic nuclei projecting to the caudate nucleus in cats, with wheat germ agglutinin-conjugated horseradish peroxidase (WGA-HRP) as the neuroanatomical tracer.

SOURCE OF THALAMOSTRIATE AFFERENTS

Intralaminar nuclei

Centre median nucleus (CM). Among the intralaminar nuclei, the projection from CM-PF complex has been studied extensively in monkeys and man. Despite initial reservations (Walker, 1938; Mettler, 1943, 1945) later studies (Mettler, 1947; Simma, 1951; Nauta and Whitlock, 1954) showed definite evidence for a projection from CM to the striatum. Studies of neuropathological material from human brain (Vogt and Vogt, 194; Freeman and Watts, 1947; McLardy, 1948) also have suggested that CM projects to the neostriatum. Neuroanatomical tracer techniques have confirmed beyond doubt, the projection from CM to the striatum (Jones and Leavitt, 1974; Royce, 1978; Kalil, 1978; Sato et al., 1979; Royce, 1983).

Vogt and Vogt (1941) showed that the ventrolateral small cell region of human CM projected selectively to the putamen. This was subsequently confirmed in human brains (Powell, 1952; Pribram and Bagshaw, 1953) and in monkeys with retrograde cell degeneration technique by Powell and Cowan (1956). The literature concerning the afferent and efferent projections of CM was later reviewed by Mehler (1966), and he provided additional evidence for a selective projection of CM to the putamen in monkeys with the Nauta method. Recent studies with autoradiographic techniques in monkeys (Kalil, 1978) have confirmed that CM projects specifically to lateral regions of the putamen. The degeneration and autoradiographic studies (Mehler, 1966; Kalil, 1978; Royce, 1978) show that the fibers from CM course rostrally through the lateral aspect of ventral anterior (VA) nucleus and traverse the internal capsule to terminate within the putamen. Some of these fibers also pass through the laminae separating the pallidal segments and the lamina separating the putamen from the pallidum to terminate within the putamen.

In contrast to the monkey, the CM of cats projects to both the putamen and the caudate nucleus (Johnson, 1961; Mehler, 1966; Royce, 1978). After injections of triated proline in CM of cats, Royce

(1973) observed prominent anterograde labeling in lateral areas of the putamen and the head and tail of caudate nucleus. In rats the CM-PF projects profusely to the striatum (Jones and Leavitt, 1974; van der Kooy, 1979) but details of the topography of this projection are not available.

Powell and Cowan (1956) showed that the projection of CM within the putamen is anteroposteriorly and mediolaterally organized. Royce (1978) showed that after injections of triated proline in dorsal CM, anterograde labeling was seen in the dorsal caudate nucleus, while ventral injections led to terminal labeling of ventral areas of the caudate nucleus, thereby arguing for a dorsoventral organization within the striatum of cats.

Parafascicular nucleus (PF). LeGros Clark and Russell (1939) noted that the entire PF was preserved after lesions of putamen in human brain. In monkeys PF projects primarily to the caudate nucleus, but a small projection to the dorsal and medial regions of the putamen has also been shown (Powell and Cowan, 1956). The medial PF has been suggested to project to the nucleus accumbens (NA) by Simma (1951) and Knook (1966). The HRP studies of Groenewegen et al., (1980) showed that the dorsal and rostral-most region of medial PF projects to NA. The neurons projecting to NA are located medial and lateral to, and surrounding the fibers of the fasciculus retroflexus (Groenewegen et al., 1980).

Central lateral nucleus (CL). The projection from CL to striatum was suggested by degeneration studies. Studies using HRP techniques (Royce, 1978; Sato et al., 1979) and neurophysiological methods (Kunze et al., 1979; Jinnai and Matsuda, 1981; Steriade and Glenn, 1982) have confirmed this projection in rats, cats and monkeys. On the basis of cytoarchitectonics, CL in cats has been separated into anterior and posterior divisions, and each of these divisions has been shown to have a small-cell medial region and a lateral region consisting of large cells (Niimi and Kuwahara, 1973). The study of Royce (1978) suggests that the caudate projection arises solely from the small-cell region of the nucleus. The large-cell regions of CLa and CLp, which receive projections from the spinothalamic system (Mehler, 1966); Jones and Burton, 1974; Berkley, 1980) may not project to the striatum.

Paracentral nucleus (PC). PC has been shown to project to the striatum in rats (Jones and Leavitt, 1974; van der Kooy, 1979) and cats (Steriade and Glenn, 1982).

Rostral intralaminar nuclei. The projections from the rostral intralaminar nuclei to the striatum are poorly understood. The borders of the different subdivisions, unlike the posterior intralaminar nuclei, are not clearly defined and the cytoarchitectonics of this area has not been studied in detail in different animals. The degeneration studies either involved several subdivisions of the rostral intralaminar nuclei along with other thalamic nuclei or invariably interrupted the fiber systems in which the neurons of the intralaminar system reside. Since the introduction of neuroanatomical tracer techniques a clearer understanding of the projection from the rostral intralaminar nuclei to the striatum is emerging. HRP studies of thalamocaudate projections have shown retrograde labeling of neurons in the lamina separating the anterior thalamic nuclei and VA (Sato et al., 1979; Macchi and Bentivoglio, 1982). More caudally labeled neurons within the rostral intralaminar region surround the lateral, dorsal and medial borders of medial dorsal (MD) nucleus along with a few cells within MD (Sato et al., 1979). Some of these subdivisions, at least in cats, coincide with Niimi and Kuwahara's (1973) central anterior (CA), central dorsal (CD), CL, and the dorsal division of the parafascicular nucleus.

Midline thalamic nuclei

The existence of a projection from the nucleus rhomboideus (Rh) to the striatum was shown with degeneration techniques in rats (Hiddema and Fortuyun, 1960) and with HRP methods in cats (Sato et al., 1979; Royce, 1983) and in rats (Veening et al., 1980). The nucleus centralis medialis (CeM) has been shown to project to the caudate nucleus (Powell and Cowan, 1956; Sato et al., 1979; Macchi and Bentivoglio, 1982), but this projection appears to be meager. Autoradiographic (Herkenham, 1978) and HRP (Groenewegen et al., 1980) studies suggest that the nucleus reuniens (Re) projects sparsely to the caudate nucleus.

The paraventricular (PV) and parataenial (PT) nuclei

A projection from PT to NA was suggested by the studies of Simma (1951). After lesions of NA and the medial caudate nucleus, retrograde cell degeneration was shown in PV and PT in rabbits (Cowan and Powell, 1955) and monkeys (Powell and Cowan, 1956). Injections of tritiated proline bilaterally within PV and PT in rats by Swanson and Cowan (1975) resulted in anterograde transport of the marker to NA. HRP studies of Groenewegen et al., (1980) in cats confirmed the projections from PV and PT to NA. Injections of HRP in the head of the caudate nucleus failed to show any retrogradely labeled cell in these two nuclei (Royce, 1978; Sato et al., 1979), suggesting that the projections of these two nuclei may be restricted to NA. It is not known whether PV and PT project to the putamen.

Ventral anterior nucleus (VA)

McLennan and York (1966) and Buchwald et al., (1973) showed that an excitatory potential could be elicited in the caudate nucleus after stimulation of VA in cats. Golgi studies by Scheibel and

Scheibel (1967) showed axons of VA neurons passing through the caudate nucleus, but a termination of these axons within the caudate nucleus could not be visualized. Studies of efferent projections of VA in monkeys by Carmel (1971) failed to show any projections to the striatum. However, modern tracing methods have confirmed these electrophysiological findings. HRP studies by Royce (1978) and Macchi and Bentivoglio (1982) show a diffuse projection from VA to the caudate nucleus. Neurophysiological studies by Jinnai and Matsuda (1981), and neuroanatomical studies using fluorescent tracers by Royce (1983) show cells bordering the ventral anterior-ventrolateral (VA-VL) region project to both the caudate nucleus and the cerebral cortex by axon collaterals. These studies suggest that VA projects to the caudate nucleus, but the mediolateral and rostrocaudaal topography of this projection to the caudate nucleus is not known. It is not known whether the putamen also receives any projections from VA. Anterograde tracing studies using either autoradiographic techniques or WGA-HRP should provide answers to some of these questions.

Lateral posterior complex (LP)

During a study of the ascending connections of the posterior group of thalamic nuclei in cats with Nauta-Gygax and Fink-Heimer methods, Heath and Jones (1971) observed degenerated axons in the striatum after large lesions involving the suprageniculate nucleus and the adjacent posterior thalamic nuclei. This observation led them to conclude that "the main efferent connections of these parts of the posterior group appear to be with the striatum". A similar study by Graybiel (1973) showed that after lesions of LP and posterior nuclear group, degenerated fibers could be seen within the striatum. She concluded that a central posterior nuclear group-recipient zone may be flanked by a LP-recipient zone within the putamen. She also observed degenerated axons in the caudate nucleus after lesions of LP, but not of posterior nuclear group, thereby suggesting a differential pattern of projection of LP and the posterior thalamic nuclei to the caudate and putamen in cats. Neurophysiological studies of Buchwald et al., 1973) show that after stimulation of LP, an EPSP-IPSP sequence can be recorded within the striatum. Retrograde labeling of cells within LP was also observed by Royce (1978) after an injection of HRP in the caudate nucleus. Studies of the connections of the LP complex in cats have shown that this nucleus can be divided into a medial division (LPm), a visual cortical recipient lateral division (LPl) and a tectorecipient intermediate division (LPl) (Updyke, 1977). Symonds et al., (1981) noted that injections of tritiated proline involving all three subdivisions of LP lead to anterograde labeling of caudal and ventrolateral regions of the head and the body of the caudate nucleus. When injections were done in pulvinar the autoradiographic labeling was not seen in the caudate nucleus. Their study did not demonstrate any projection from LP to the putamen. A projection from LP to the striatum has not been reported in rats or monkeys.

Ventrolateral nucleus (VL)

Several neurophysiological studies suggest that VL projects to the striatum. As with other thalamic nuclei, stimulation of VL results in an EPSP response in the striatum (Purpura and Malliani, 1967; Buchwald et al., 1973; Hull et al., 1973). Neurons of VL could also be antidromically activated by stimulation of the putamen (Rasminsky et al., 1973). It is not clear whether this EPSP response within the striatum is mediated through the intralaminar projection to the cortex, or due to activation of collateral fibers of VL passing through the putamen to the motor cortex. The possibility that cells of VL may project to the striatum was suggested by recent neurophysiological studies of Steriade and Glenn (1982). Physiological studies of Jinnai and Matsuda (1981) have shown that an area adjacent to VA-VL border projects to the motor cortex and to the striatum by axon collaterals. With double labeling techniques, using fluorescent tracers, Royce (1983) showed that VL projects to the caudate nucleus and the cerebral cortex by axon collaterals.

Medial dorsal nucleus (MD)

Among the different thalamic nuclei, the projection from MD to the striatum has been the most controversial. Sachs (1909) first described a projection from MD to the caudate nucleus using Marchi techniques. Since that observation, a projection from MD to the striatum has been reported in different animals with studies using Marchi techniques (Showers, 1958), Fink-Heimer technique (Leonard, 1971; Tobias and Ebner, 1974), HRPP (Royce, 1978; Sato et al., 1979) and autoradiographic techniques (Tobias, 1975). Physiological studies by Buchwald et al. (1973), and Kunze et al. (1979), have also suggested the existence of MD projection to the striatum. In contrast, degenerations studies by Nauta and Whitlock (1954), Khalifeh et al. (1965), and a more recent study using HRP technique by Groenewegen et al. (1980), have failed to observe a projection from MD to the caudate nucleus. The possibility that MD projects to the striatum cannot be excluded.

Nucleus subparafascicularis

Groenewegen et al., (1980) have noted retrograde labeling of the nucleus subparafascicularis in one of their cases with an injection of HRP in NA.

Anterior thalamic nuclei

Buchwald et al. (1973), reported an EPSP response in the caudate nucleus after stimulation of AM. Sato et al. (1979), observed a few retrogradely labeled neurons within the AM after injections of HRP in

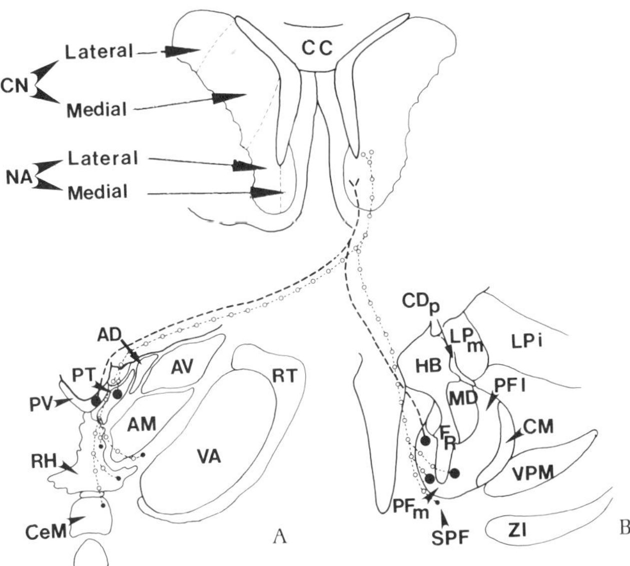

Fig. 1. The pattern of thalamic projections to the medial and lateral nucleus accumbens.

the rostral caudate nucleus. These observations have not been confirmed by other workers using similar techniques.

PRELIMINARY OBSERVATIONS WITH WGA-HRP TECHNIQUE

In an effort to resolve some of the controversies concerning the source of thalamic projections to the striatum, a study using WGA-HRP was initiated. In 27 adult cats stereotaxic injections of 0.05 to 0.1 microliters of 2% WGA-HRP were made into the nucleus accumbens (NA) and various regions of the caudate nucleus. After a survival of 24 to 36 hours, the animals were sacrificed, and the brains were subsequently processed according to the benzidine dihydrochloride (BDHC) modification of HRP histochemistry of Mesulam (1976). Sections were then counterstained with neutral red and studied under dark and brightfield microscopy. The classification of the thalamic nuclei as defined by Niimi and Kuwahara (1973) was followed.

After small injections in medial NA, WGA-HRP was transported to PV (Fig. 1A) and to the rostral aspects of the medial parafascicular (PFm) nucleus (Fig. 1B). Labeled cells in PFm were seen close to the ventromedial border of MD (Fig. 1B) surrounding the fibers of the fasciculus retroflexus.

Small injections in lateral NA led to prominent labeling of PT, lateral aspects of Rh, CeM, few cells of CA (Fig. 1A) and the nucleus subparafascicularis (Fig. 1B). The most prominent transport was to medial PF, surrounding the fibers of the fasciculus retroflexus (Fig. 1B).

After injections in medial caudate nucleus the different subdivisions of the midline thalamic nuclei and the rostral intralaminar nuclei were profusely labeled (Fig. 2A&B). The projection from Rh was very prominent and bilateral with a predominance of ipsilateral projections (Fig. 2B). The projection from CeM was minor. A mediolaterally organized projection from VA was also seen in all these cases. Prominent projections were seen from CA, and the anterior divisions of CD (Fig. 2B). Significant retrograde labeling of LPm (Fig. 2B) and the nucleus limitans (Fig. 2D) was also seen. The most prominent labeling was seen in the rostral CM (Fig. 2C) and the lateral division of PF (Fig. 2D).

Injections restricted to the extreme lateral caudate nucleus resulted in retrograde labeling of the magnocellular ventral subdivision of anteroventral nucleus (AV), a few cells of VA and the ventrolateral nucleus (VL) near the VA-VL border (Fig. 2B). Small cells of posterior division of CL, cells of posterior division of CD (Fig. 2C) and few cells of PC nucleus and PFl were also labeled. The most prominent transport of the marker was to the parvocellular caudal CM (Fig. 2D).

The results from the study show that in addition to the much-emphasized projection from CM-PF complex, the caudate nucleus also receives significant projections from the CA, CD and CL subdivisions of the intralaminar nuclei, and that these subdivisions project to specific areas of the striatum. Among the midline thalamic nuclei, RH and CeM but not the nucleus reuniens, project to the striatum. Besides these sources the caudate nucleus receives a mediolaterally organized projection from VA, and sparse projections from VL, LPm and the nucleus subparafascicularis. The study also demonstrates that the nucleus limitans is yet another source of thalamic afferent to the caudate nucleus and confirms the projection from PV and PT to NA. The study also confirms the observation that the thalamostriate projections are rostrocaudally organized.

PATTERN OF TERMINATION

Thalamostriate neurons and their axon collaterals

There is ample evidence to show that thalamic nuclei that project to the striatum also project to the cerebral cortex by axon collaterals of the same neuron. This possibility was suggested by

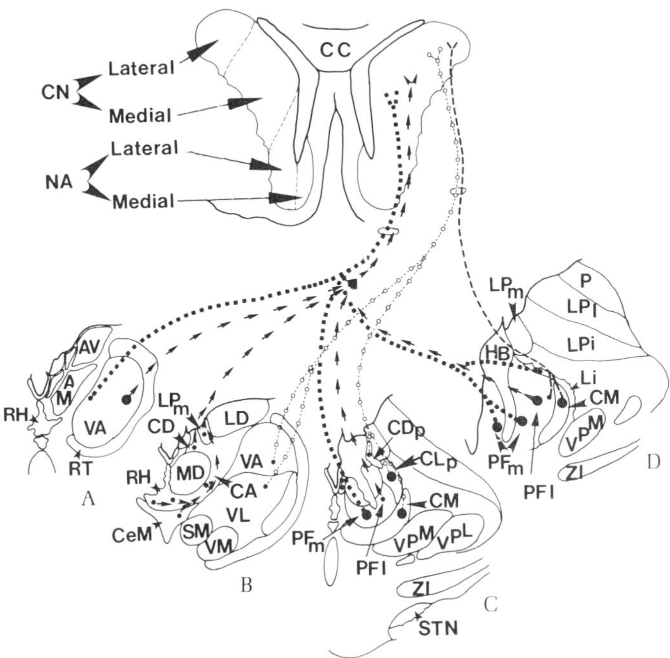

Fig. 2. Thalamic projection pattern to the medial and lateral areas of the head of the caudate nucleus. Different symbols have been used to denote the mediolateral organization of the thalamic projections within these areas. The sizes of the closed circles indicate the density of the projection from that thalamic nucleus to the caudate nucleus.

the anatomical studies of Murray (1966) and Jones and Leavitt (1974) and by the physiological studies of Rasminsky et al. (1973). Injections of HRP in the motor cortex of rats, cats and monkey (Jones and Leavitt, 1974) resulted in retrograde transport of the tracer to the posterior intralaminar nuclei and injections of HRP in the striatum led to labeling of neurons of the same posterior intralaminar nuclei. These authors suggested that the intralaminar nuclei may project to the striatum and the cerebral cortex by axon collaterals. This issue was directly addressed by Jinnai and Matsuda (1979) who, with neurophysiological techniques, showed that cells near the VA-VL border and CL and CM-PF neurons project to the striatum and the cerebral cortex by axon collaterals. Anatomical studies using double labeling techniques by Cesaro et al. (1981) and Royce (1983) have further confirmed this.

The number of cells projecting to both the striatum and the cerebral cortex varies from one thalamic nucleus to the other

(Jinnai and Matsuda, 1979; Royce, 1983). Royce (1983) showed that the number of VA and rostral VL cells which project to both the cerebral cortex and the caudate nucleus are more numerous than those which project to either the cortex or the caudate alone. Caudal VL cells project only to the cerebral cortex. Rostral intralaminar nuclei also contained significant number of cells which project to the caudate and cortex by axon collaterals. In the caudal intralaminar nuclei, only cells of PF had double labels suggesting that PF cells projected to the cortex and caudate by axon collaterals. Most of the cells of CM projected only to the caudate, and very few only to the cortex (Royce, 1983). This observation concerning CM cells is in contrast with that of Jinnai and Matsuda (1979) who found that CM projects to the striatum and the cerebral cortex by axon collaterals.

Thalamostriate Projections terminate in a Patchy Fashion

The striatal projection of CM-PF complex has been shown to terminate in a patchy manner in rats (Herkenham and Pert, 1981), cats (Royce, 1978) and monkeys (Kalil, 1978). The similarities of these puffs of terminal labeling to the pattern of corticostriate projections (Goldman and Nauta, 1977) are striking. The sizes and shapes of these patches vary within different regions of the striatum (Royce, 1978; Herkenham and Pert, 1981). It is not known whether other thalamic nuclei also terminate in a patchy fashion.

Herkenham and Pert (1981) showed that the patchy distribution of PF terminations within the caudate nucleus of rats interdigitated with the opioid receptor patches. Immunohistochemical studies in rats (Herkenham and Pert, 1981) and in cats (Graybiel et al., 1981) have shown that the opioid receptors are organized, similar to the corticostriatal and the thalamostriatal projections, in a patchy manner. When adjacent sections are stained for acetylcholinsterase (AchE), the patches of opioid receptors correspond to areas of the striatum that stain poorly for AchE. The projections from PF avoid the opioid rich and AchE poor clusters and terminate in areas of AchE rich patches (Herkenham and Pert, 1981).

Neurotransmitters

The neurotransmitter of the thalamostriate projection is not known. Acetylcholine was suggested to be the candidate for the projections from VA to the caudate nucleus (McLennan and York, 1966) and for CM-PF projections to the striatum (Simke and Saelens, 1978; Kim, 1978). But several studies (McGeer et al., 1971; Fonnum and Waalas, 1979; Fibiger, 1982) have failed to show a cholinergic projection from the thalamus to the striatum. Immunohistochemical studies using antibodies to choline acetyltransferase in cats (Kimura et al., 1981) and AchE histochemical technique in cats pretreated with DFP (Parent and O'Reilly Fromentin, 1982) have failed to show any cells staining positive for either choline acetyltransferase or acetylcholinesterase within VA or the intralaminar nuclei.

The Mode of Termination and Neurophysiology

Hassler (1978) suggested that the caudate nucleus receives more of the thalamic afferents than does the nucleus accumbens or the putamen. The thalamostriate projection has been shown to terminate, along with cortical and nigral afferents, on the dendritic spines of the spiny striatal neurons (Kemp, 1968; Fox et al., 1971; Wilson et al., 1983). Electronmicroscopic studies (Hassler, 1978) have shown that the thalamostriate projections terminate either as long, divided, curved, strongly asymmetric axo-spinous synapses (type IV of Hassler) or as synapses with densely packed small rounded vesicles and an asymmetric junction (type VI of Hassler). Stimulation of the thalamic nuclei causes either EPSP or an EPSP-IPSP sequence (see Dray, 1980 and Kitai, 1981 for reviews).

CONCLUSIONS

It is obvious from this review of the literature and the preliminary results from a study using WGA-HRP reported here, that the striatum receives significant projections from several thalamic nuclei which are not conventionally included in the intralaminar system. At least some of these nuclei (ventral anterior, nucleus limitans, nucleus reuniens, rhomboideus, and the paraventricular nuclei) have been shown to cause "recruiting" responses in the cerebral cortex and are included in the diffuse non-specific thalamocortical projection system (Jasper, 1961). These results raise the question whether the parataenial nucleus, on the basis of its profuse connectivity with the nucleus accumbens, should also be included within the non specific diffuse thalamic nuclear group. Further physiological studies of the paraventricular, parataenial and the midline nuclei are needed to answer this question.

In their study of efferent projections of the intralaminar nuclei in rats, cats and monkeys, Jones and Leavitt (1974) observed that the intralaminar nuclei were labeled more intensely and in greater number after HRP injections in the striatum than those in the cerebral cortex. They speculated that: 1. the intralaminar nuclei send direct axons to the striatum and minor collaterals to the cortex; or 2. that the intralaminar nuclei have two different neuronal populations, one of which projected densely to the striatum, and another diffusely to the cerebral cortex; or 3. that both the first and the second possibilities exist. The results of the WGA-HRP study confirm their observation that the density and the number of retrogradely labeled cells within the intralaminar nuclei, especially CM-PF, are significantly higher than in other thalamic nuclei, e.g.

VA, VL and LP. The findings of Royce (1983) that CM projected primarily to the striatum and that VA, VL and the rostral intralaminar nuclei project to both the striatum and the cerebral cortex favor the hypotheses of Jones and Leavitt (1974).

Studies with modern neuroanatomical tracer techniques have supported the view of Lorente de No (1938) that the pattern of termination of the cortical projections of different thalamic nuclei may vary. Using autoradiographic techniques in rats, Herkenham (1980) suggested that the thalamic nuclei can be classified into four groups. The ventroposterior, medial geniculate and lateral geniculate, and the medial dorsal nuclei constituted a group of thalamic nuclei which projected primarily to layers III and IV of the cerebral cortex. The second group consisted of the posterior intralaminar nuclei (CL, PC, CM and PF) which projected selectively to layers V and VI. (See Jones and Leavitt, 1974; Steriade and Glenn, 1982 for evidence for a layer I projection of CL). Groups 3 and 4 consisted of ventral anterior, lateral posterior, lateral dorsal, nucleus reuniens and the magnocellular medial geniculate nucleus: a group of thalamic nuclei located adjacent to the intralaminar nuclei, which projects profusely to layer I and in some cases to other layers of the cerebral cortex. The results from the study using WGA-HRP reported here, when combined with the results of Royce (1983), suggest that those thalamic nuclei that terminate in layers III and IV do not project to the striatum, whereas the posterior intralaminar group, which terminate primarily in layers V and VI, project mostly to the striatum, and those thalamic nuclei which project to layer I project to both the striatum and the cerebral cortex. The functional significance of this aspect of thalamostriate organization remains to be explored.

Behavioral and neurophysiological studies performed after selective lesioning of different thalamic nuclei with neurotoxins should provide additional clues towards the understanding of the role of the thalamostriate projection.

ACKNOWLEDGEMENTS

I am greatly indebted to Dr Bruce V. Updyke for his suggestions and criticisms. Supported by NS 16609.

ABBREVIATIONS USED IN FIGURES

AD: anterior dorsal nucleus
AM: anterior medial nucleus
AV: anterior ventral nucleus
CA: central anterior nucleus
CC: corpus callosum

CD: central dorsal nucleus
CDp: central dorsal nucleus, posterior division
CeM: centralis medialis nucleus
CLp: central lateral nucleus, posterior division
CM: centre median nucleus
CN: caudate nucleus
FR: fasiculus retroflexus
HB: habenular nucleus
LD: lateral dorsal nucleus
Li: nucleus limitans
LPi: lateral posterior nucleus, interjacent division
LPl: lateral posterior nucleus, lateral division
LPm: lateral posterior nucleus, medial division
MD: medial dorsal nucleus
NA: nucleus accumbens
P: pulvinar
PFl: parafascicular nucleus, lateral division
PFm: parafascicular nucleus, medial division
PT: parataenial nucleus
PV: paraventricular nucleus
RH: nucleus rhomboideus
RT: reticular nucleus of the thalamus
SM: submedial nucleus
SPF: subparafascicular nucleus
STN: subthalamic nucleus
VA: ventral anterior nucleus
VL: ventrolateral nucleus
VM: ventromedial nucleus
VPL: ventroposterior nucleus, lateral division
VPM: ventroposterior nucleus, medial division
ZI: zona incerta

REFERENCES

Berkley, K. J., 1980, Spatial relationships between the terminations of somatic sensory and motor pathways in the rostral brainstem of cats and monkeys. I. Ascending somatic sensory inputs to lateral diencephalon, J. Comp. Neurol., 193:283.

Buchwald, N. A., Price, D. D., Vernon, L., and Hull, C. D., 1973, Caudate intracellular response to thalamic and cortical inputs, Exp. Neurol., 38:311.

Carmel, P. W., 1970, Efferent projections of the ventral anterior nucleus of the thalamus in the monkey, Am. J. Anat., 128:159.

Carpenter, M. B., 1981, Anatomy of the corpus striatum and brain stem integrating systems, in: "Handbook of Neurophysiology," V. B. Brooks, ed., Williams and Wilkins, Baltimore.

Cesaro, P., Nguyen-Legros, J., Berger, B., Alvarez, C., and Albe-Fessard, D., 1979, Double labeling of branched neurons in the central nervous system of the rat by retrograde transport of horseradish peroxidase and iron-dextran complex, Neurosci. Lett., 15:1.

Clark, W. E. Le Gros, and Russell, W. R., 1939, Observations of the efferent connections of the centre median nucleus, J. Anat., Lond., 73:255.

Cowan, W. M., and Powell, T. P. S., 1955, The projection of the midline and intralaminar nuclei of the thalamus of the rabbit, J. Neurol. Neurosurg. Psychiat., 18:266.

Dray, A., 1980, The physiology and pharmacology of mammalian basal ganglia, Prog. in Neurobiology, 14:221.

Fibiger, H. C., 1982, The organization and some projections of cholinergic neurons of the mammalian forebrain, Brain Res. Rev., 4:327.

Fonnum, F., and Waalas, I., 1979, Localization of neurotransmitter candidates in neostriatum, in: "The Neostriatum," I. Divac and R. G. E. Oberg, eds., Pergamon Press, Oxford.

Fox, C. A.., Andrade, A. N., Hillman, D. E., and Schwyn, R. C., 1971, The spiny neurons in the primate striatum: A Golgi and electron microscopic study, J. Hirnforsch, 13:181.

Freeman, W., and Watts, J. W., 1947, Retrograde degeneration of the thalamus following prefrontal lobotomy, J. Comp. Neurol., 86:65.

Goldman, P. S., and Nauta, W. J. H., 1977, An intricately patterned prefrontocaudate projection in the rhesus monkey, J. Comp. Neurol., 171:369.

Graybiel, A. M., 1973, The thalamo-cortico projection of the so-called posterior nuclear group: a study with anterograde degeneration methods in the cat, Brain Res., 49:229.

Graybiel, A. M., and Ragsdale, C. W. Jr., 1979, Fiber connections of the basal ganglia, in: "Progress in Brain Research," M. Cuenod, G.. W. Kreutzberg, and F. E. Bloom, eds., Elsevier, Amsterdam.

Graybiel, A. M., Ragsdale, C. W. Jr., Yoneoka, E. S., and Elde, R. P., 1981, An immunocytochemical study of enkephalins and other neuropeptides in the striatum of the cat with evidence that the opiate peptides are arranged to form mosaic patterns in register with the striosomal compartments visible by acetylcholinesterase staining, Neuroscience, 3:377.

Groenewegen, H. J., Becker, N.E.H.M., and Lohman, A.H.M., 1980, Subcortical afferents of the nucleus accumbens septi in the cat, studied with retrograde axonal transport of horseradish peroxidase and bisbenzimid, Neuroscience, 5:1903.

Hassler, R., 1978, Striatal control of locomotion, intentional actions, and of integrating and perceptive activity, J. Neurol. Sci., 36:187.

Hassler, R., 1982, The possible pathophysiology of schizophrenic disturbances of perception and their relief by sterotactic coagulation of intralaminar mediothalamic nuclei, in: "Psychobiology of Schizophrenia," M. Namba and H. Kaiya, eds., Pergamon Press, Oxford.

Heath, C. J., and Jones, E. G., 1971, An experimental study of ascending connections from the posterior group of thalamic nuclei in the cat, J. Comp. Neurol., 141:397.

Herkenham, M., 1978, The connections of the nucleus reuniens thalami: evidence for a direct thalamo-hippocampal pathway in the rat, J. Comp. Neurol., 177:589.

Herkenhamm, M., and Pert, C. B., 1981, Mosaic distribution of opiate receptors, parafascicular projections and acetylcholinesterase in rat striatum, Nature, 291:415.

Herkenham, M., 1980, Laminar organization of thalamic projections to the rat neocortex, Science, 207:532.

Hiddema, F., and Fortuyun, J. D., 1960, The projections of the intermediary nuclear system of the thalamus and of the parataenial and the paraventricular nucleus in the rat, Psychiat. Neurol. Neurochir., Amst., 63:8.

Hull, C. D., Bernardi, G., Price, D. D., and Buchwald, N. A., 1973, Intracellular responses of caudate neurons to temporally and spatially combined stimuli, Exp. Neurol., 38:324.

Jasper, H., 1961, Thalamic reticular system, in: "Electrical Stimulation of the Brain," D. E. Sheer, ed., University of Texas Press, Austin.

Jinnai, K., and Matsuda, Y., 1981, Thalamocaudate projection neurons with a branching axon to the motor cortex, Neurosci. Lett., 26:95.

Johnson, T. N., 1961, Fiber connections of the dorsal thalamus and the corpus striatum in the cat, Exp. Neurol, 3:556.

Jones, E. G., and Leavitt, R. Y., 1974, Retrograde axonal transport and the demonstration of non-specific projections to the cerebral cortex and striatum from the thalamic intralaminar nuclei in the rat, cat and monkey, J. Comp. Neurol., 154:349.

Jones, E. G., and Burton, H., 1974, Cytoarchitecture and somatic sensory connectivity of thalamic nuclei other than the ventrobasal complex in the cat, J. Comp. Neurol., 154:395.

Kalil, K., 1978, Patch-like termination of thalamic fibers in the putamen of the monkey: an autoradiographic study, Brain Res., 140:33.

Kemp, J. M., 1968, An electron microscopic study of the terminations of afferent fibers in the caudate nucleus, Brain Res., 11:464.

Khalifeh, R. R., Kaelber, W. M., and Ingram, W. R., 1965, Some efferent connections of the nucleus medialis dorsalis: an experimental study in the cat, Amer. J. Anat., 116:341.

Kim, J. S., 1978, Transmitters for the afferent and efferent systems of the neostriatum and their possible interactions, in: "Dopamine, Advances in Biochemical Psychopharamacology," P. J. Roberts, G. N. Woodruff, and L. L. Iversen, eds., Raven Press, New York.

Kimura, H., McGeer, P. L., Peng, J. H., and McGeer, E. G., 1981, The central cholinergic system studied by choline acetyltransferase immunohistochemistry in the cat, J. Comp. Neurol., 200:151.
Kitai, S. T., 1981, Electrophysiology of the corpus striatum and brain stem integrating system, in: "Handbook of Neurophysiology," V. B., Brooks, ed., Williams and Wilkins, Baltimore.
Knook, H. L., 1966, The fibre-connections of the forebrain, F. A. Davis, Philadelphia.
Kunze, W., McKenzie, J. S., and Bendrups, A. P., 1979, An electrophysiological study of thalamo-caudate neurons in the cat, Exp. Brain Res., 36:233.
Leonard, C. M., 1969, The prefrontal cortex of the rat. I. Cortical projections of the mediodorsal nucleus. II. Efferent connections, Brain Res., 12:321.
Lorento de No, R., 1938, The cerebral cortex: architecture, intracortical connections and motor projections, in: "Physiology of the Nervous System," J. F. Fulton, ed., Oxford University Press, New York.
Macchi, G., and Bentivoglio, M., 1982, The organization of the efferent projections of the thalamic intralaminar nuclei: past, present and future of the anatomical approach, Ital. J. Neurol. Sci., 2:83.
McGeer, P. L., McGeer, E. G., Fibiger, H. C., and Wickson, V., 1971, Neostriatal cholineacetylase and cholinesterase following selective brain lesions, Brain Res., 35:308.
McLardy, T., 1948, Projections of the centromedian nucleus of the human thalamus, Brain, 71:290.
McLennan, H., and York, D. H., 1966, Cholinergic mechanisms in the caudate nucleus, J. Physiol., Lond., 187:163.
Mehler, W. R., 1966, Further notes on the center median nucleus of Luys, in: "The Thalamus," D. P. Purpura and M. D., Yahr, eds., Columbia University Press, New York.
Mehler, W. R., 1982, The basal ganglia - Circa 1982. A Review and commentary, Appl. Neurophysiol., 44:261.
Mesulam, M. M., 1976, The blue reaction product in the horseradish peroxidase neurohistochemistry: incubation parameters and visibility, J. Histochem. Cytochem., 24:1273.
Mettler, F. A., 1943, Extensive unilateral cerebral removals in the primate. Physiologic effects and resultant degeneration, J. Comp. Neurol., 79:185.
Mettler, F. A., 1945, Fiber connections of the corpus striatum of the monkey and baboon, J. Comp. Neurol., 82:169.
Mettler, F. A., 1947, Extracortical connections of the primate frontal cerebral cortex. I. Thalamo-cortical connections, J. Comp. Neurol., 86:95.
Murray, M., 1966, Degeneration of some intralaminar thalamic nuclei after cortical removals in the cat, J. Comp. Neurol., 127:344.

Nauta, W. J. H., and Whitlock, D. G., 1954, An anatomical analysis of the nonspecific thalamic projection system, in: "Brain Mechanisms and Consciousness," J. F. Delafresnaye, C. C. Thomas, ed., Springfield, Ill.

Niimi, K., and Kuwahara, E., 1973, The dorsal thalamus of the cat and comparison with monkey and man, J. Hirnforsch., 14:303.

Parent, A., and O'Reilly-Fromentin, J., 1982, Distribution and morphological characteristics of acetylcholinesterase containing neurons in the basal forebrain of the cat, Brain Res. Bull., 8:183.

Powell, T. P. S., 1952, Residual neurones in the human thalamus after hemidecortication, Brain, 75:571.

Powell, T. P. S., and Cowan, W. M., 1956, A study of the thalamo-striate relations in the monkey, Brain, 79:364.

Pribram, K. H., and Bagshaw, M., 1935, Further analysis of the temporal lobe syndrome utilizing fronto-temporal ablantions, J. Comp. Neurol., 99:347.

Purpura, D. P., and Malliani, A., 1967, Intracellular studies of the corpus striatum. I. Synaptic potentials and discharge characteristics of caudate neurons activated by thalamic stimulation, Brain Res., 6:325.

Rasminsky, M., Mauro, A. J., and Albe-Fessard, D., 1973, Projections of medial thalamic nuclei to putamen and cerebral frontal cortex in the cat, Brain Res., 61:69.

Royce, G. J., 1978, Cells of origin of subcortical afferents to the caudate nucleus: a horseradish peroxidase study in the cat, Brain Res., 153:465.

Royce, G. J., 1978, Autoradiographic evidence for a discontinuous projection to the caudate nucleus from the centromedian nucleus in the cat, Brain Res., 146:145.

Royce, G. J., 1983, Single thalamic neurons which project to both the rostral cortex and caudate nucleus studied with the fluorescent double labeling method, Exp. Neurol., 79:773.

Sachs, E., 1909, On the structure and function relations of the optic thalamus, Brain, 32:7.

Sato, M., Itoh, K., and Mizuno, N., 1979, Distribution of the thalamo-caudate neurons in the cat as demonstrated by horseradish peroxidase, Exp. Brain Res., 196:228.

Scheibel, M. E., and Scheibel, A. B., 1967, Structural organization of nonspecific thalamic nuclei and their projection toward cortex, Brain Res., 6:60.

Showers, M. J. C., 1958, Correlation of medial thalamic nuclear activity with cortical and subcortical neuronal arcs, J. Comp. Neurol., 109:261.

Simma, K., 1951, Zur Projektion des Centrum medianum und Nucleus parafascicularis thalami beim Menschen, Meschr. Psychiat. Neurol., 122:32.

Simke, J. P., and Saelens, J. K., 1977, Evidence for a cholinergic fiber tract connecting the thalamus with the head of the striatum of the rat, Brain Res., 106:189.

Steriade, M., and Glenn, L. L., 1982, Neocortical and caudate projections of intralaminar thalamic neurons and their synaptic excitation from midbrain reticular core, J. Neurophysiol., 48:352.

Swanson, L. W., and Cowan, W. M., 1975, A note on the connections and development of the nucleus accumbens, Brain Res., 92:324.

Symonds, L. L., Rosenquist, A. C., Edwards, S. B., and Palmer, L. A., 1981, Projections of the pulvinar-lateral posterior complex to the visual cortical areas in the cat, Neuroscience, 6:1995.

Tobias, T. J., 1975, Afferents to prefrontal cortex from the thalamic mediodorsal nuclei in the rhesus monkey, Brain Res., 83:191.

Tobias, T. J., and Ebner, F. F., 1975, Thalamocortical projections from the medialdorsal nucleus in the Virginia opossum, Brain Res., 52:79.

Updyke, B. V., 1977, The functional organization and cytoarchitecture of the feline lateral posterior complex, with observations on adjoining cell groups, J. Comp. Neurol., 219:143.

Van der Kooy, D., 1979, The organization of the thalamic, nigral and raphe cells projecting to the medial vs lateral caudate-putamen in rat. A fluorescent retrograde double labeling study, Brain Res., 169:381.

Veening, J. G., Cornelissen, F. M., Lieven, P.A.J.M., 1980, The topical organization of the afferents to the caudatoputamen of the rat. A horseradish peroxidase study, Neuroscience, 5:1253.

Vogt, C., and Vogt, O., 1941, Thalamusstudien I-III, J. Psychol. Neurol., 50:32.

Walker, A. E., 1938, The primate thalamus, Univ. of Chicago Press, Chicago.

Wilson, C. J., Chang, H. T., and Kitai, S. T., 1983, Origins of post synaptic potentials evoked in spiny projection neurons by thalamic stimulation in the rat, Exp. Brain Res., 51:217.

THE PRIMATE STRIATO-PALLIDO-NIGRAL SYSTEM: AN INTEGRATIVE SYSTEM FOR CORTICAL INFORMATION

G. Percheron, J. Yelnik, and C. Francois

INSERM U3, Hôpital de la Salpêtrière
47 Bd de l'Hôpital
75013, Paris, France

"It is essential in discussing the physiology of the corpus striatum to think anatomically". Wilson (1912).

The notion of basal ganglia has not been developed without difficulty. The term "basal" was mainly used to differentiate the "base of the brain" from the cortex. It is only recently that entities such as the thalamus, the striatum and the pallidum have been identified within the "optostriate bodies". In his historical record (I-368), Déjerine (1895) quotes Willis as the first to separate the corpus striatum from the thalamus, Vicq d'Azyr (1786) as the first to link the caudate nucleus and the putamen as the two parts of the striatum, and Burdach (1819) as the first to isolate the globus pallidus from the putamen inside the nucleus lentiformis. This latter distinction took time to be accepted. Luys (1865) described a "yellow nucleus" of the "striate body" but did not consider it any more in 1882. Forel (1877), and Monakow (1895) among many others, numbered lenticulate "segments" from 1 for the medial pallidal segment to 3 for the putamen. Kolliker (1896) erroneously located pallidal neurons in the striatum. Brissaud (1893) used the term globus pallidus for the external segment only (Mettler, 1968) and Ramon y Cajal (1911) failed to recognize it in his "noyau central à cellules géantes" (Fig. 1). This is really surprising when one looks at the striking contrast between the striatum and the globus pallidus. The difference is evident in almost all respects: the macroscopic aspect, the cytoarchitecture (Foix and Nicolesco, 1925; Feremutsch, 1961), the dendroarchitecture and ultrastructure (Fox et al., 1966), or the cytohistochemistry (Marchand et al., 1979). Conversely, these criteria show a strong similarity of the pallidum to the pars reticulata of the substantia nigra (Mirto, 1966; Fox et al., 1974). The name globus pallidus,

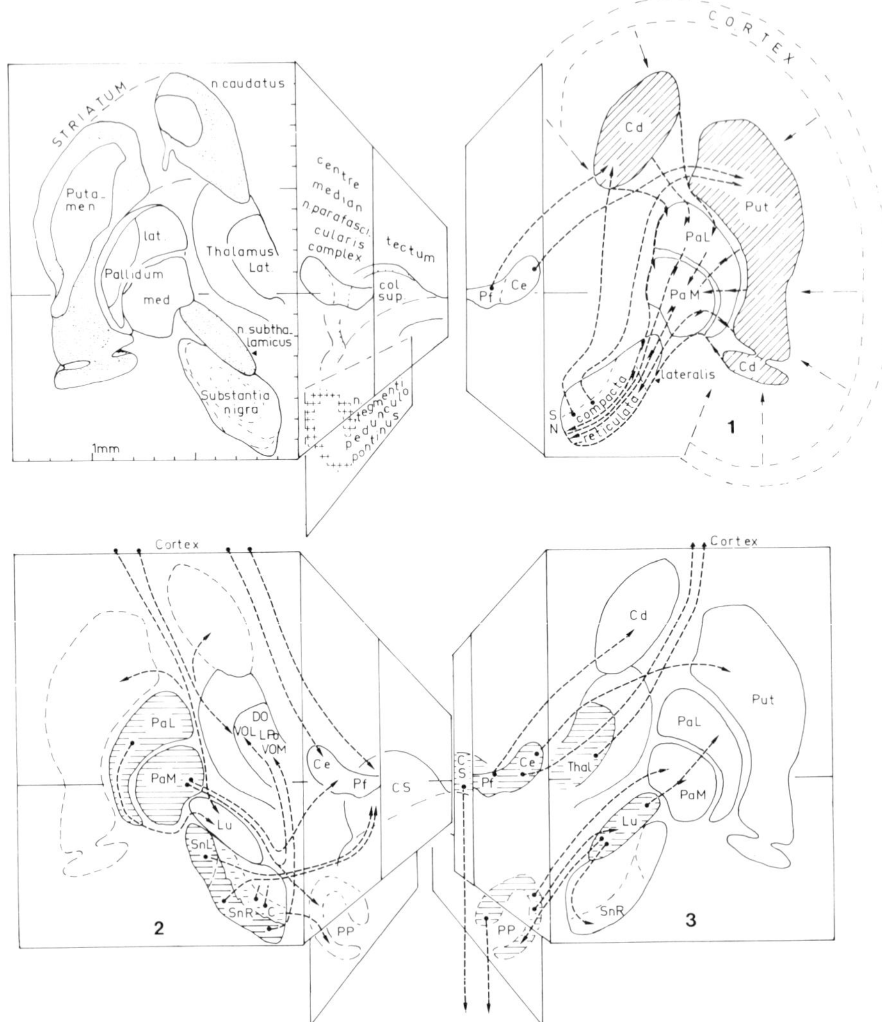

Fig. 1. The basal ganglia of macaque in transverse sections studied in relation to ventricular landmarks. The transverse reconstructions are observed from front to back. Their admitted connections (see text) are analyzed in terms of successive relays: 1, the striatum (nucleus caudatus and putamen), 2, the pallido-nigral complex (nucleus lateralis and medialis of the globus pallidus and the substantia nigra), 3, the targets of the pallido-nigral complex (nucleus subthalamicus, nucleus parafascicularis- "centre médian" complex, rostral part of the lateral mass of the thalamus, nucleus tegmenti pedunculo-pontinus and superior colliculus). Abbreviations: see list.

retained by the Nomina Anatomica (1955), is often replaced by the shorter "pallidum" (Foix and Nicolesco, 1925; Feremiutsch, 1961). The globus pallidus is separated from the striatum by the "lamina medullaris externa". It is subdivided into two nuclei by the "lamina medullaris interna": the nucleus lateralis (PaL), pars lateralis, pars externa, pallidum laterale, lateral segment (Papez, 1941), or crus II (Mettler, 1948); and the nucleus medialis (PaM), pars medialis, pars interna, pallidum mediale, inner or medial segment, or crus I. Very often, in man but also in the macaque, a "lamina medullaris accessoria" (Feremutsch, 1961), "lames médullaires supplémentaires" (Dejerine, 1895), or "lamina medullaris incompleta" (Mettler, 1948) incompletely subdivides the nucleus medialis. Another subdivision, a "pallidum ventralis" (Heimer et al., 1982) is actually only an appearance created by the commissura anterior. It is, in primates at least, only a part of the lateral nucleus.

The term basal ganglia, thus, grouped together cerebral regions whose list has greatly changed in history. The lack of a definition could be one reason why there is yet no commonly accepted list. A present-day definition could be: "the set constituted by the striato-pallido-nigral system and the cerebral regions which receive direct afferent input from it". The accepted connections are shown in Figure 1, where they are interpreted in terms of successive relays (see also Carpenter, 1976). Inputs from the raphe are not represented. The pallido-habenular connection, which deeply changed in evolution and which appears in primates to be mainly a hypothalamo-habenular connection (Parent et al., 1981), is also not considered. In addition to the striatum, the pallidum and the substantia nigra, this definition would include the nucleus subthalamicus, the nucleus parafascicularis- "centre médian" complex of the thalamus, and the nucleus pedunculo-pontinus, which all give feed-back loops to other basal ganglia. The definition would, however, add two other cerebral regions: the superior colliculus and the rostral part of the lateral mass of the thalamus. There is a consensus to include the nigrotectal and nigrothalamic connections within the basal ganglia, but certainly not the superior colliculus itself. Thus, another implicit criterion could be used to establish that a cerebral region belongs to the basal ganglia, namely that the cerebral region must send a feedback loop towards other basal ganglia. Consideration of this criterion therefore excludes the rostral part of the lateral mass of the thalamus from the basal ganglia, despite the fact that through its cortical relay it can send feedback information towards the striatum. A third criterion could be the topographic requirement which is included in the term basal ganglia i.e. to be located in the base of the brain. This would eliminate the nucleus pedunculo-pontinus, formerly a part of the reticular formation, the superior colliculus and also the substantia nigra which are both mesencephalic. This criterion is unsatisfactory. The type of analysis performed in terms of successive relays, which is usual in sensory and in pyramidal systems, is also inadequate but an analysis in terms of loops offers no better result. The above proposed anatomical definition could be accepted if one adds that the basal ganglia constitute a set whose core, the striato-pallido-nigral system, is easily identifiable but whose limits remain unclear. Such a definition gives to the analysis of this core a character of priority from which the basal ganglia can be better understood. Its logic led us to concentrate on an interpretation in terms of cortical information, rather than in terms of loops.

Our study was limited to primates (and essentially to macaques) and to a purely spatial analysis of the striato-pallido-nigral system. Such an analysis combines two sets of methods: topographic methods and computer three-dimensional analyses of arborizations. Our topographical studies always started from ventricular coordinates based on two commissural points, CA and CP and the mid-sagittal plane (Percheron, 1975), (see ventriculogram, Fig. 2). Each series of transverse sections was cut perpendicular to the CA-CP plane. Thus there is a constant angle of section which is of cardinal importance. All sections are analyzed using a calibrated XY plotter and transformed into scaled maps precisely located in relation to the ventricular system of coordinates. These maps may thus be used for reconstructing cerebral regions in the two other dimensions. The superimposition of such sagittal and horizontal reconstructions (Fig. 2) of the striatum of 6 macaques shows that ventricular landmarks offer stable enough positions, shapes and dimensions to allow reliable comparisons of data from different individuals. Topographic reconstructions of the contours of the striatum show the classical anatomical opposition between two striatal masses of about the same volume (the head of the caudate nucleus and the putamen). The study of the striatal distribution of corticostriate axons leads to another, anatomofunctional, subdivision of the striatum. The earlier results obtained after anterograde degeneration (Kemp and Powell, 1970; Petras, 1971) differed from those based on autoradiographic studies, probably due to the relative scarcity and the thin diameter of corticostriatal axons. Figure 3 shows the data from available autoradiographic studies (Kunzle, 1975, 1977 and 1978; Kunzle and Akert, 1977; Jones et al, 1977; Goldman and Nauta, 1977; Yeterian and Van Hoesen, 1978; Van Hoesen et al., 1981) on a topographic atlas of the macaque's striatum. The striatal territories of projections of the motor and somatosensory cortex coincide. Together, they constitute a striatal sensorimotor territory somatotopically organized into obliquely piled up somatotopic (leg, L, arm, A and face, F, Fig. 3) strips. The whole sensorimotor territory is purely putaminal. Frontal and horizontal (Fig. 7) reconstructions show its important rostrocaudal extent. There are, however, no sensorimotor afferent inputs anteromedially or posteriorly. The caudate nucleus and the remaining part of the putamen receive their cortical inputs from

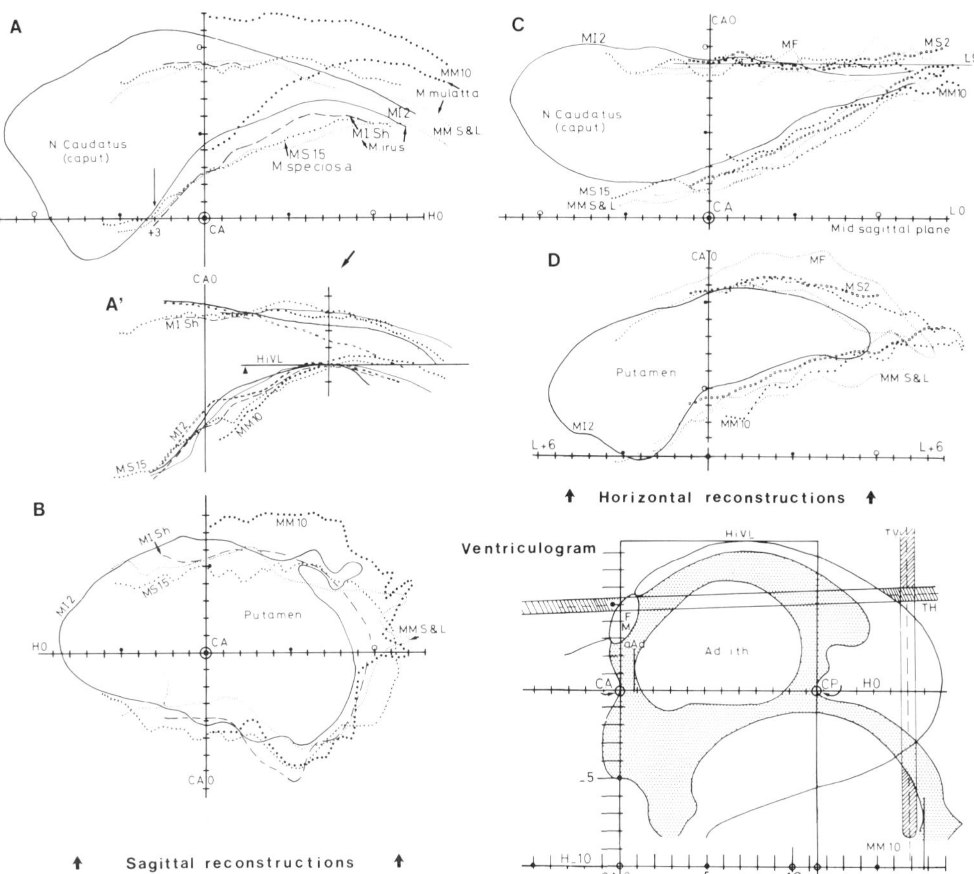

Fig. 2. Topographic analyses of macaque striatum according to ventricular coordinates. The two commissural points, CA and CP, are shown on a ventriculogram (lower right). A and B are superimpositions of the sagittal reconstructions and C and D are superimpositions of the horizontal reconstructions of the head of the caudate nucleus (A and C) and of the putamen (B and D) in 6 macaques belonging to four species (MF - Macaca fuscata, MI - Macaca fascicularis, MM - Macaca mulatta, MS - Macaca speciosa). Abbreviations: see list.

other cortical regions (Fig. 3). Prefrontal, temporal, parietal and anterior cingulate regions do not have their own territory. Their striatal islands in fact intermingle. Together, they thus constitute a common striatal "associative" territory. The opposition between a sensorimotor and an associative territory is shown by comparing their reciprocal boundaries. There is almost no

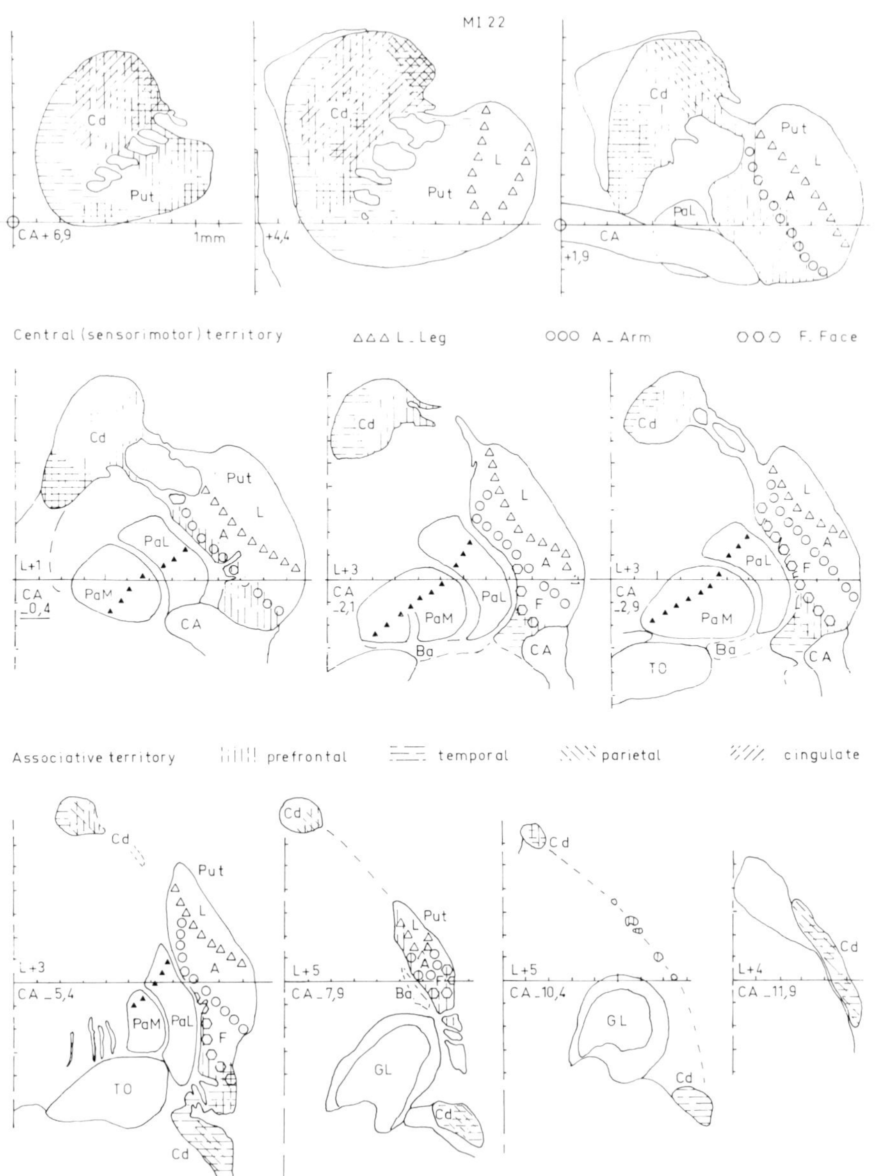

Fig. 3. Transfer from data of literature showing the striatal distribution of corticostriate afferents on to a topographic atlas of the macaque striatum established according to ventricular coordinates in a Macaca fascicularis. The opposition between a sensorimotor and an associative striatal territory may be observed by comparing their respective boundaries. Abbreviations: see list.

overlap. Rather than the classical anatomical opposition between caudate nucleus and putamen, one is thus led to prefer a functional opposition between a sensorimotor and an associative territory of about the same striatal volume.

From the striatum, all of the striatal axons constitute the striato-pallido-nigral bundle which may be followed in the pallidum and in the pars lateralis and reticulata of the substantia nigra (Fox et al., 1975; Fox and Rafols, 1976; Francois et al., 1981). Its funnel shape, number of axons and extreme density make it an extraordinary bundle. Groups of striatal axons gather to constitute radial fascicles perpendicular to the two pallidal laminae medullares and parallel to the inferolateral border of the substantia nigra. Our recent study (Percheron et al., 1984) confirms all previous data (Voneida, 1960; Szabo, 1962, 1967, 1970 and 1972; Nauta and Mehler, 1966; Cowan and Powell, 1966) on the internal topographic organization of the bundle. Caudate and putaminal axons do not mix. The first are oral and superior in the pallidum and oral and medial in the substantia nigra. The latter are more caudal and ventral in the pallidum and caudo-lateral in the substantia nigra. The two components fill an equal pallidal and nigral volume. The pallidonigral complex is thus as much "associative" as it is "sensorimotor".

The study of the three-dimensional geometry of the neurons which receive the bundle necessitates preliminary technical comments. The analysis of the three-dimensional morphology of dendritic arborizations is limited by three technical drawbacks. Histological sectioning leads to incomplete pictures. Pallidonigral dendrites are very long. Even very thick sections show only a restricted part of any neuron. Microscopic observation gives projected pictures where angles and lengths are distorted. The plane of section is not necessarily the one that best shows the distinctive geometry of the neuron. The first two drawbacks were overcome by using a video-computer system, introducing the position of each dendritic point in the depth of each section as the third dimension and reorganizing the data obtained in the successive sections into continuous dendritic sequences (Yelnik et al., 1981). The solution of the third drawback was more difficult. We resorted to principal component analysis to bring out the plane (deduced from the distinctive geometry of the neuron and thus independent of the plane of section) that gives the most significant projection of a neuron (Yelnik et al., 1983b). In their initial planes, pallidal dendritic arborizations are much more extended than previously shown (Fox et al., 1974) without serial reconstructons of sections. They have various shapes. Their analysis in a new system of coordinates based on their principal planes shows (Fig. 4) that all pallidal dendritic arborizations, in man and in macaque, in the medial as well as in the lateral pallidum, are alike. They statistically belong to a single population and may be described as large flat

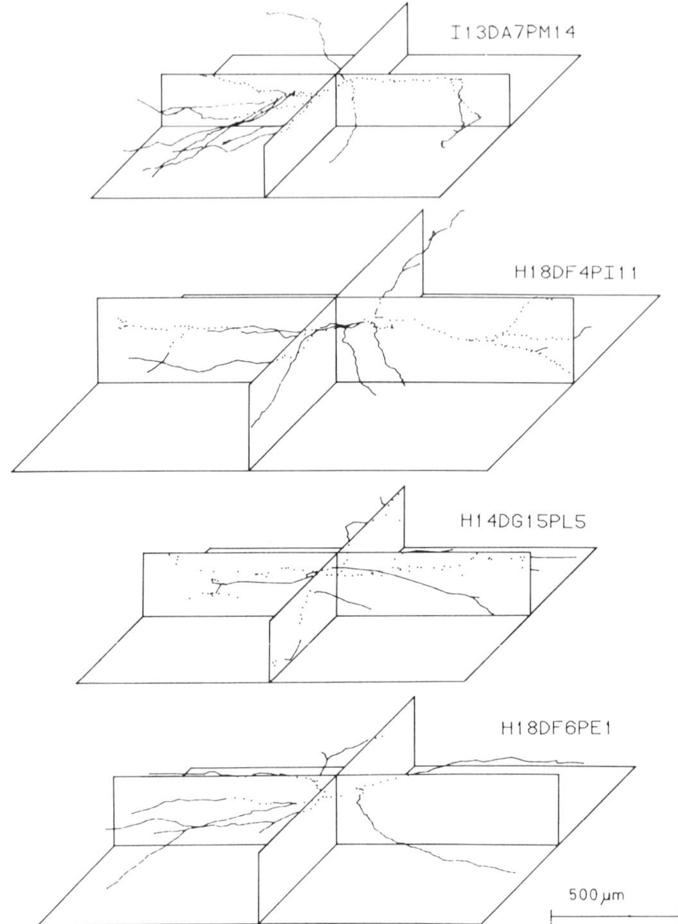

Fig. 4. Four pallidal neurons (successively from the medial pallidum in a macaque and a man and from the lateral pallidum in two men) computer analyzed according to their principal planes. They all appear to be flat. They belong statistically to one population which may be modeled as large flat discs.

discs whose mean dimensions are 1500 x 1000 x 250 μm (Fig. 6). This shows that the three dimensions are not equivalent and that there is one preferential plane. The spatial analysis must orient the principal planes in relation to the three-dimensional geometry of the nucleus in which the discs are contained. The orientation is based on the determination of two angles in relation to the initial plane: the polar angle and the azimuth (Yelnik et a., 1983b). The transfer of these data to topographic maps, where each neuron is precisely localized by means of an XY plotter, shows the relation between the three-dimensional geometry of neurons and that of the nucleus.

Our recent study showed (Yelnik et al., 1984) that pallidal dendritic discs are parallel to each other, parallel to the lateral border of the pallidum and thus perpendicular to striatal axons. Neurons of the pars lateralis and pars reticulata of the substantia nigra have the same topological (Percheron, 1982) and metrical characteristics as pallidal neurons and, in this respect, may be said to belong to the same neuronal population (Yelnik et al, 1983a). Contrary to pallidal neurons, however, principal component analysis indicates that their overall shapes and dimensions remain variable. Thus they cannot be statistically modeled.

The interpretation of the relations between the geometry of the striato-pallido-nigral axons and that of the pallido-nigral dendritic arborizations used theoretical premises that must be explained. Let us consider two cerebral regions (Fig. 5a), R1 and R2. R1 receives its inputs from the bundle B1. R2 receives its inputs from the axons of R1 constituting the bundle B2 and sends its output by the bundle B3. We assume that B1 conveys four different items of information, A, B, C and D. These items may be somatotopic (for instance A = foot, B = arm, C = leg and D = tongue representations), or may be items of modality (for instance A = somesthetic, B = visual, C = auditory and D = gustatory). In order that the four items remain discriminated after their passage through R2, it is necessary that axons conveying the different items contact separated neurons. This condition implies particular relations between the geometry and dimensions of region R2 and that of the axonal arborizations of B2 and that of the dendritic arborizations in R2. This may be demonstrated by using two examples. In Figure 5b, the geometric conditions have maintained the segregation of the four items in R1 but the axonal arborization of each R1 neuron may contact any dendritic arborization of R2 neurons. Each of these R2 neurons can be contacted by axons conveying all four items. Its axon conveys a combination of the four items and is not specific to any one of them. In Figure 5c, geometric conditions have maintained the separation between the four items in R1 and B2 and also in the axonal arborizations in R2. Here, R2 dendritic arborizations fill the entire volume of R2. In this case, too, each R2 neuron can be contacted by axons conveying the four items and its output also loses its specificity. Neither dendritic nor axonal arborizations fill solid spaces. They intermingle. It may be asserted that there is no interference only in the case of non-overlapping spaces. Any analysis in terms of a number of possible items should thus be based only on numbers of non-overlapping arborizations. Such a type of analysis requires that one is able to statistically determine the mean volumes of the space containing axonal (Vol aa) or dendritic (Vol da) arborizations. The latter quantifies the "grain" of a

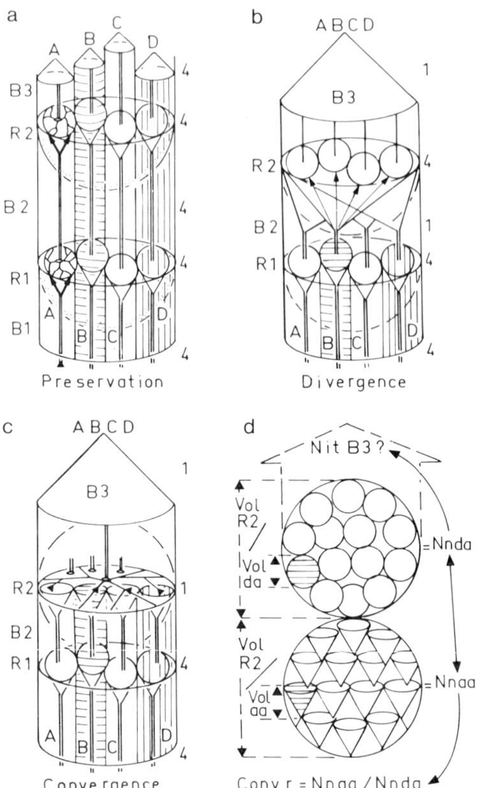

Fig. 5. Theoretical interpretations of (a) preservation of items, (b) divergence, and (c) convergence. See text.

cerebral region. The number of possible items (N it) may be computed by dividing the volume of R2 (Vol R2) by Vol aa and by Vol da and by selecting the smaller of the two results. The amount of convergence or divergence can be determined by dividing one figure by the other (Fig. 5d). Such a system of analysis rests on an hypothesis of random contacts between axons and dendrites (or "random connectivity", Szentagothai and Arbib, 1974). One has thus to verify that this is the case in the studied cerebral region. Random connectivity is what appeared to be most probable in our Golgi study of the primate pallidum (Francois et al., 1984). According to the type of analysis initiated by Scholl (1956), one must take into account not only the comparison between axonal and dendritic spaces, but also an estimation of their chance of contact, by introducing the axonal and dendritic densities of these spaces. The low density of dendritic membranes inside each pallido-nigral dendritic arborization is compensated by a very high density of crossing striatal axons (Fig. 6, upper left). The small striatal dendritic spheres (Fig. 6) give rise to axons that cross the pallidum perpendicularly to its two laminae. A striatal axon is able to cross successively hundreds of pallidal discs. This is a very particular (linear and sequential) pattern of divergence. Pallidal discs are parallel to the lateral border of the pallidum and thus perpendicular to striatal axons to which they present their maximal extent. The general principles used for the quantitative determination of neuronal convergence only considered non-directed spaces where neuronal elements equally radiate in all three dimensions. In the pallidum, the direction of striatal axons appears to "polarize" the spatial organization. Only planes perpendicular to this direction are discriminant. The surface of pallidal discs of which the face is crossed by striatal axons is very large. Each pallidal disc is then able to be crossed by axons from a wide striatal region. Due to the direct intrastriatal route followed by striatal axons, the striatal region which emits axons which are able to cross the most laterally located pallidal discs (neurons 1, Fig. 6), is almost cylindrical with a diameter equal to that of a disc. Such a volume contains thousands of striatal neurons. The volume of the striatal region, and thus the number of striatal neurons, increases mediad. Neurons 2 (Fig. 6) for instance, may be crossed by axons coming from wide truncated conical regions. The striking change in geometry between the striatum and the pallidum thus appears as a device which makes the striatal neurons converge intensively on pallidal neurons. Due to the spatial "polarization" of the pallidal structure, the amount of this convergence and/or the number of possible items, can only be analyzed by comparing the main surface of pallidal discs to those of the pallidum and of the striato-pallido-nigral bundle (which in fact coincide) in the curved planes perpendicular to the direction of striatal axons. The funnel shape of the system also obliged us to consider successive planes, lateral to medial. The lateral border of the pallidum corresponds to the area of penetration of all striatal axons. In macaques, only 5 to 6 non-overlapping discs cover this surface in the transverse dimension and 10 to 11 in the rostrocaudal dimension. The values concerning more medially located planes are indicated in Figure 6. If one brings together the numbers of 5 to 6 million striatal axons given by Verhaart (1980) to the number of non-overlapping most medially located pallidal dendritic arborizations, one finds that these could be crossed by 200,000 to 500,000 striatal axons. This is certainly not a common feature in the brain. In the substantia nigra, the shape of dendritic arborizations is irregular but the overall dimensions are the same as those of pallidal arborizations, while the bundle is much thinner here. A single arborization suffices to cover the thickness of the bundle in the transverse dimension but 4 to 5 arborizations are needed in the rostrocaudal dimension. Nigral neurons, as well as pallidal neurons, may receive very convergent information from the striatum.

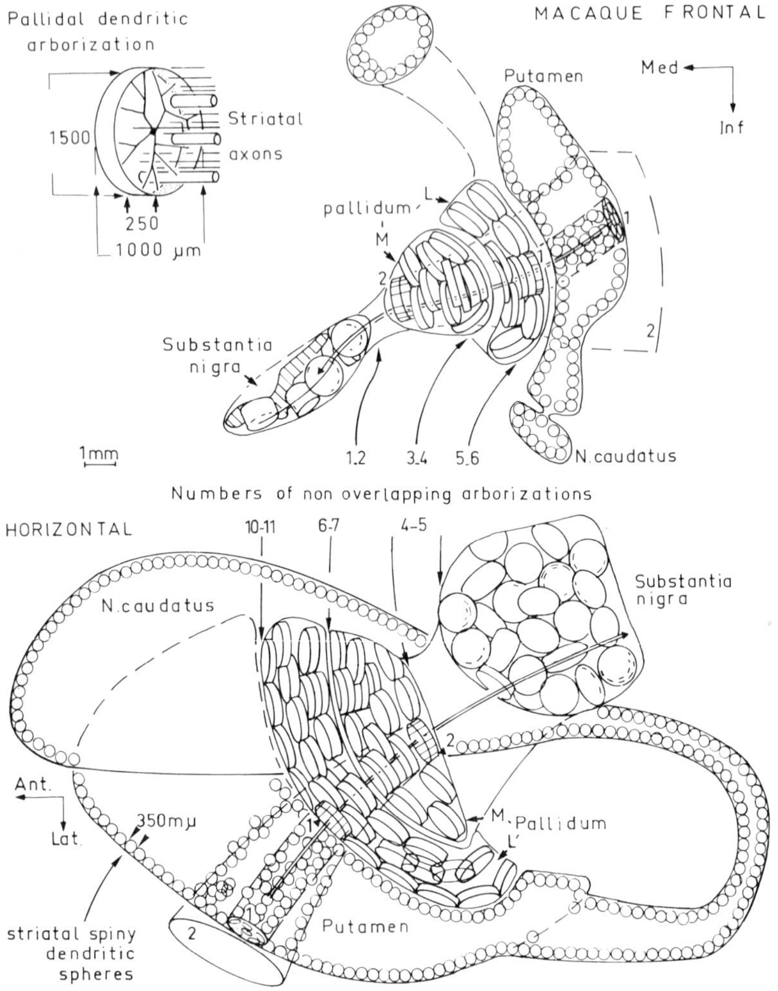

Fig. 6. Semi-schematic spatial description of the striato-pallido-nigral system. See text.

Such a pattern has two apparent effects in relation to the two components of the striato-pallido-nigral bundle. The first (Fig. 7) concerns the somatotopic representation. The minor somatotopic organization which remained in the striatum is further strongly mixed up. The direction of the radial fascicles is indeed almost perpendicular to the three somatotopic strips. Fascicle 1, for instance (Fig. 7), may then mix axons from leg and arm strips, fascicle 3 from arm and face strips and fascicle 2 from all three. There is a contrast between the large rostrocaudal extent of the

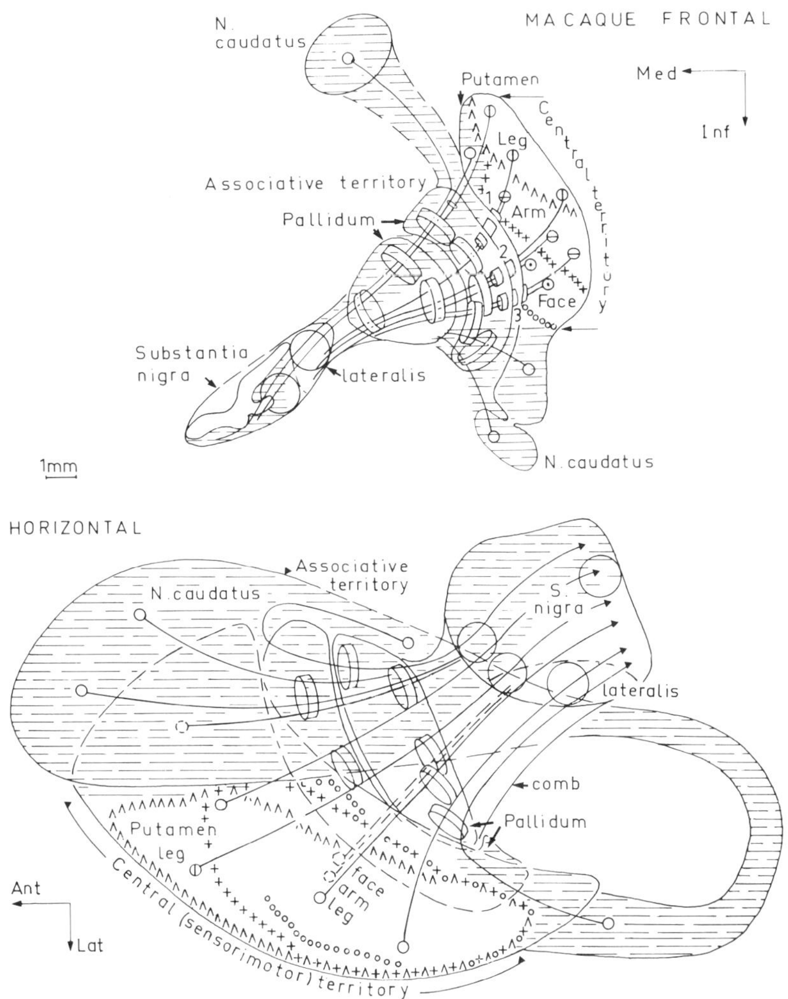

Fig. 7. Comparison between the spatial model and the distribution of the two components, sensorimotor and associative, of the striato-pallido-nigral system. See text.

sensorimotor territory and the lack of somatotopic differentiation in this dimension. There can be no more than 2 to 3 somatotopic items inside the pallidum (which is compatible with physiological data) and only one inside the substantia nigra. The second effect linked to the orientation and to the large dimensions of dendritic arborizations of pallido-nigral neurons is that many of these neurons may receive convergent information from the two components, sensorimotor and associative, of the bundle. It should be recalled that the latter is already a mixture of frontal, temporal, parietal and cingular information. However, the convergence of sensorimotor and associative information on the same pallido-nigral neuron should not be seen as a general feature explaining a fundamental function of pallido-nigral neurons. Indeed many of them may receive only sensorimotor or associative information. The striato-pallido-nigral bundle could thus, in addition to a role of convergence, play a role of selective topographic distribution. This seems particularly apparent at the level of the pars lateralis of the substantia nigra which is the main origin of nigrotectal neurons (Francois et al., 1983) and which seems to receive exclusively axons from the sensorimotor striatal territory.

1 - The contrast in geometry between the elements of the striatum and the pallidum which is the sign of a major transformation of information processing, the extraordinary density of the striato-pallido-nigral bundle, the evidence for an intensive convergence and for particular redistributions of cortical information seem to verify the hypothesis that the striato-pallido-nigral system could be the core of basal ganglia. The anatomical definition that we gave in the introduction could lead to an anatomofunctional definition of the basal ganglia as the set of cerebral regions which elaborate or directly receive a particular, cortico-striato-pallido-nigral, type of information. This offers the advantage of distinguishing the set of basal ganglia from all sensory systems and also from other "motor" systems (the cortico-spinal and cortico-ponto-cerebellar systems).

2 - The deliberate concentration on interpretation in terms of cortical information allowed us to show that the organization of the core is not a replica of that of the cortex. A strong reorganization opposes two components, sensorimotor versus associative, of equal volume. The core, and therefore probably the whole set of basal ganglia, cannot be considered as merely a motor system.

3 - The sensorimotor component undergoes a major transformation. The somatotopic representation that comes from the cortex seems to be systematically destroyed. A more global level of function could thus be achieved.

4 - The associative component raises a major question: what type of information does the associative cortex send to the basal ganglia?

"In the corpus striatum there would thus appear to be an integration of the various centres which are differentiated in the cortex". Ferrier, 1873 (quoted by Wilson, 1912).

LIST OF ABBREVIATIONS

A - Striatal arm strip
aAd - Most anterior point of the Adhaesio interthalamica
Ad.ith. - Adhaesio interthalamica
Ba - Nucleus basalis of Meynert
CA - Commissura anterior
CAO - Transverse plane of the CA ventricular point
Cd - Nucleus caudatus
Ce - "Centre médian". Nucleus centralis thalami
CS - Colliculus superior
DO - Nucleus dorsalis oralis thalami
F - Face strip
FM - Foramen of Monroe
GL - Nucleus geniculatus lateralis thalami
HiVL - Horizontal plane of the superior ventricular point
HO - Intercommissural plane
L - Leg strip
LO - Mid-sagittal plane
LPo - Nucleus lateralis polaris thalami
Lu - Nucleus subthalamicus. Corpus Luysii
PaL - Nucleus lateralis of the globus pallidus
PaM - Nucleus medialis of the globus pallidus
Pf - Nucleus parafascicularis thalami
PP - Nucleus tegmenti pedunculo-pontinus
Put - Putamen
SN - Substantia nigra: SnC - pars compacta, SnL - pars lateralis, SnR - pars reticulata
Thal - Thalamus
TO - Tractus opticus
VOL - Nucleus ventralis oralis lateralis thalami
VOM - Nucleus ventralis oralis medialis thalami

REFERENCES

Brissaud, E., 1893, "Anatomie du Cerveau de l'Homme," Masson et Cie, Paris.

Burdach, K. F., 1819, "Von Baue und Leben des Gehirns," Dyk, Leipzig.

Carpenter, M. B., 1976, Anatomical organization of the corpus striatum and related nuclei, in: "The Basal Ganglia," M. D. Yahr, ed., Raven Press, New York.

Cowan, W. M., and Powell, T. P. S., 1966, Strio-pallidal projection in the monkey, J. Neurol. Neurosurg. Psychiat., 29:426.

Déjerine, J., 1895, "Anatomie des Centres Nerveux," Rueff, Paris, réédition Masson, Paris, 1980.

Feremutsch, K., 1961, Basalganglien, in: "Primatologia, Vol. II," H. Hofer, A. H. Schultz, and D. Starck, eds., Karger, Basel.

Foix. C., and Nicolesco, J., 1925, "Anatomie Cérébrale. Les Noyaux Gris Centraux et la Région Mésencéphalo-Sous-Optique," Masson, Paris.

Forel, A., 1877, Untersuchungen über die Haubenregion und ihre oberen Verknüpfungen im Gehirne des Menschen und einiger Säugethiere, mit Beittragen zu den Methoden der Gehirnuntersuchung, Arch. Psychiatr. Nervenkr., 7:393.

Fox, C.A., Hillman, D. E., Siegesmund, K. A., and Sether, L. A., 1966, The primate globus pallidus and its feline and avian homologues: a Golgi and electron microscopic study, in: "Evolution of the Forebrain," R. Hassler, and H. Stephan, eds., G. Thieme, Stuttgart.

Fox, C. A., Andrade, A. N., LuQui, I. J., and Rafols, J. A., 1974, The primate globus pallidus: a Golgi and electron microscopic study, J. Hirnforsch., 15:75.

Fox, C. A., Rafols, J. A., and Cowan, W. M., 1975, Computer measurements of axis cylinder diameters of radial fibers and "comb" bundle fibers, J. Comp. Neurol., 159:201.

Fox, C. A., and Rafols, J. A., 1976, The striatal efferents in the globus pallidus and in the substantia nigra, in: "The Basal Ganglia," M. D. Yahr, ed., Raven Press, New York.

Francois, C., Nguyen-Legros, J., and Percheron, G., 1981, Topographical and cytological localization of iron in rat and monkey brains, Brain Res., 215:317.

Francois, C., Percheron, G., and Yelnik, J., 1983, Compared localizations of nigrotectal and nigrothalamic neurons studied in ventricular coordinates in macaques, in: "The Basal Ganglia - Structure and Function," J. S. McKenzie, R. E. Kemm and L. N. Wilcock, eds., A satellite symp. XXIXth internat. Cong. Internat. Union Physiol. Sci., Lorne.

Francois, C., Percheron, G., Yelnik, J., and Heyner, S., 1984, A Golgi analysis of the primate globus pallidus. I. Particular processes of large neurons. Other neuronal types. Afferent axons. To be published in J. Comp. Neurol.

Goldman, P. S., and Nauta, W. J. H., 1977, An intricately patterned prefrontocaudate projection in the Rhesus monkey, J. Comp. Neurol., 171:369.

Heimer, L., Switzer., R. D., and Van Hoesen, G. W., 1982, Ventral striatum and ventral pallidum. Components of the motor system?, Trends Neurosci., 5:83.

Jones, E. G., Coulter, J. D., Burton, H., and Porter, R., 1977, Cells of origin and terminal distribution of corticostriatal fibers arising in the sensory-motor cortex of monkeys, J. Comp. Neurol., 173:53.

Kemp, J. N., and Powell, T. P. S., 1970, The cortico-striate projection in the monkey, Brain, 93:525.

Kölliker, A., 1986, "Handbuch der Gewebelehre des Menschen und der Thiere," Engleman, Leipzig.

Kunzle, H., 1975, Bilateral projections from precentral motor cortex to the putamen and other parts of the basal ganglia. An autoradiographic study in Macaca fascicularis, Brain Res., 88:195.

Kunzle, H., 1977, Projections from the primary somatosensory cortex to basal ganglia and thalamus in the monkey, Exp. Brain Res., 30:481.

Kunzle, H., 1978, An autoradiographic analysis of the efferent connections from premotor and adjacent prefontal regions (areas 6 and 9) in Macaca fascicularis, Brain, Behav. Evol., 15:185.

Kunzle, H., and Akert, K., 1977, Efferent connections of cortical, area 8 (frontal eye field) in Macaca fascicularis. A reinvestigation using the autoradiographic technique, J. Comp. Neurol., 173:147.

Luys, J. B., 1865, "Recherches sur le système nerveux cérébrospinal," Baillère, Paris.

Luys, J. B., 1882, "Le Cerveau et ses Fonctions," Baillère, Paris.

Marchand, O. R., Poirier, L. J., and Parent, A., 1979, Cytohistochemical study of the primate basal ganglia and substantia nigra, in: "The Extrapyramidal System and its Disorders," L. J. Poirier, T. L. Sourkes, and P. J. Bédard, eds., Adv. Neurol., 24:13.

Mettler, F. A., 1948, "Neuroanatomy," Mosby, Saint Louis.

Mettler, F. A., 1968, Anatomy of the basal ganglia, in: "Diseases of the Basal Ganglia. Handbook of Clinical Neurology," P. J. Vinken, and G. W. Bruyn, eds., North Holland, Amsterdam.

Mirto, D., 1896, Contributo alla fina anatomia della substantia nigra di Sommering e del peduncolo cerebrale dell'uomo, Riv. Speriment. Fren. Med. Legale, 22:197.

Monakow, C. von, 1895, Experimentelle und pathologisch-anatomische Untersuchungen Uber die Haubenregion, des Sehhügel und die Regio subthalamica nebst beiträgen zur Kenntnis früh erworbener Gross- und Kleinhirndefecte, Arch. Psychiatr. Nervenkr., 27:1 and 27:386.

Nauta, W. J. H., and Mehler, W. R., 1966, Projections of the lentiform nucleus in the monkey, Brain Res., 1:2.

Nomina Anatomica, 1955, Williams and Wilkins, Baltimore.

Papez, J. W., 1941, A summary of fiber connections of the basal ganglia with each other and with other portions of the brain, Res. Publ. Ass. nerv. ment. Dis., 21:21.

Parent, A., Gravel, S., anc Boucher, R., 1981, The origin of forebrain afferents to the habenula in rat, cat and monkey, Brain Res. Bull., 6:23.

Percheron, G., 1975, Ventricular landmarks for thalamic stereotaxy in Macaca, J. med. Primatol., 4:217.

Percheron, G., 1982, Principles and methods of the graph-theoretical analysis of natural binary arborescences, J. Theoret. Biol., 99:509.

Percheron, G., Yelnik, J., and Francois, C., 1984, A Golgi analysis of the primate globus pallidus. III. Spatial organization of the striato-pallidal complex. To be published in J. Comp. Neurol.
Petras, J. M., 1971, Connections of the parietal lobe, J. Psychiatr. Res., 8:189.
Ramon y Cajal, S., 1911, "Histologie du Système Nerveux de l'Homme et des Vertébrés," Maloine, Paris.
Sholl, D. A., 1956, "The Organization of the Cerebral Cortex," Hafner, New York.
Szabo, J., 1962, Topical Distribution of the Striatal Efferents in the Monkey, Exp. Neurol., 5:21.
Szabo, J., 1967, The efferent projections of the putamen in the monkey, Exp. Neurol., 19:463.
Szabo, J., 1970, Projections from the body of the caudate nucleus in the rhesus monkey, Exp. Neurol., 27:1.
Szabo, J. 1972, The course and distribution of efferents from the tail of the caudate nucleus in the monkey, Exp. Neurol., 37:562.
Szentagothai, J., and Arbib, M. A., 1974, Conceptual models of neural organization, Neurosci. Res. Prog. Bull., 12:307.
Van Hosen, G. W., Yeterian, E. H., and Lavizzo-Mourey, R., 1981, Widespread corticostriate projections from temporal cortex of the rhesus monkey, J. Comp. Neurol., 199:205.
Verhaart, W. J. C., 1950, Fiber analysis of the basal ganglia, J. Comp. Neurol., 93:425.
Vicq d'Azyr, R., 1786, "Traité d'Anatomie et de Physiologie avec des Planches Colorées," Didot, Paris.
Voneida, T. J., 1960, An experimental study of the course and destination of fibers arising in the head of the caudate nucleus in the cat and monkey, J. Comp. Neurol., 115:75.
Wilson, K., 1912, An experimental research into the anatomy and physiology of the corpus striatum, Brain, 36:427.
Yelnik, J., Percheron, G., Perbos, J., and Francois, C., 1981, A computer-aided method for the quantative analysis of dendritic arborizations reconstructed from serial sections, J. Neurosci. Methods, 4:347.
Yelnik, J., Francois, C., and Percheron, G., 1983a, Comparative analysis of pallidal and nigral (pars reticulata) dendritic arborizations on the basis of quantitative 3-dimensional parameters, in: "The Basal Ganglia - Structure and Function," J. S. McKenzie, R. E. Kemm and L. N. Wilcock, eds., A satellite symp. XXIXth internat. Cong. Internat. Union Physiol. Sci., Lorne.
Yelnik, J. Percheron, G., Francois, C., and Burnod, Y., 1983b, Principal component analysis: a suitable method for the 3-dimensional study of the shape, dimensions and orientation of dendritic arborizations, J. Neurosci. Methods, in press.

Yelnik, J., Percheron, G., and Francois, C., 1984, A Golgi analysis of the primate globus pallidus. II. Quantitative morphology and spatial disposition of dendritic arborizations. To be published in J. Comp. Neurol.
Yeterian, E. H., and Van Hosen, G. W., 1978, Cortico-striate projections in the rhesus monkey: the orginization of certain cortico-caudate connections, Brain Res., 139:43.

THE TERMINATION OF STRIATONIGRAL FIBRES ON NIGROTECTAL NEURONS IN THE RAT - A PRELIMINARY REPORT

R. L. M. Faull and M. N. Williams

Department of Anatomy, School of Medicine
University of Auckland
Auckland, New Zealand

INTRODUCTION

Light microscopic axoplasmic transport studies in the rat brain have demonstrated that the dorsal striatum projects on to the ventral region of the substantia nigra pars reticulata (Nauta and Domesick, 1979; Faull, Nauta and Domesick, 1980) which gives origin to the nigrotectal projection (Faull and Mehler, 1978; Beckstead, Edwards and Frankfurter, 1981). The synaptic details of this striato-nigro-tectal pathway have been investigated in the rat with the electron microscope by combining the techniques of (i) anterograde degeneration of synaptic boutons following the placement of kainic acid lesions in the striatum and (ii) retrograde labelling of nigrotectal neurons following injections of horseradish peroxidase in the superior colliculus. This paper presents our preliminary findings in this study and a more detailed account will be presented elsewhere.

MATERIALS AND METHODS

Stereotaxically guided iontophoretic or pressure (15-60 nl) injections of 1% kainic acid (a neurocytotoxin known to spare fibres of passage, McGeer et al., 1978) were made into the dorsal region of the right striatum in 8 anaesthetized adult albino rats. Two to three days after the kainic acid injections, the animals were re-anaesthetized and stereotaxically guided pressure injections (50-150 nl) of horseradish peroxidase (HRP; either Sigma type VI, 25% or Sigma HRP-lectin product No. L-2384, 5%) were made into the right superior colliculus. The animals were allowed to recover and 24 hours later were anaesthetized and perfused through the heart with

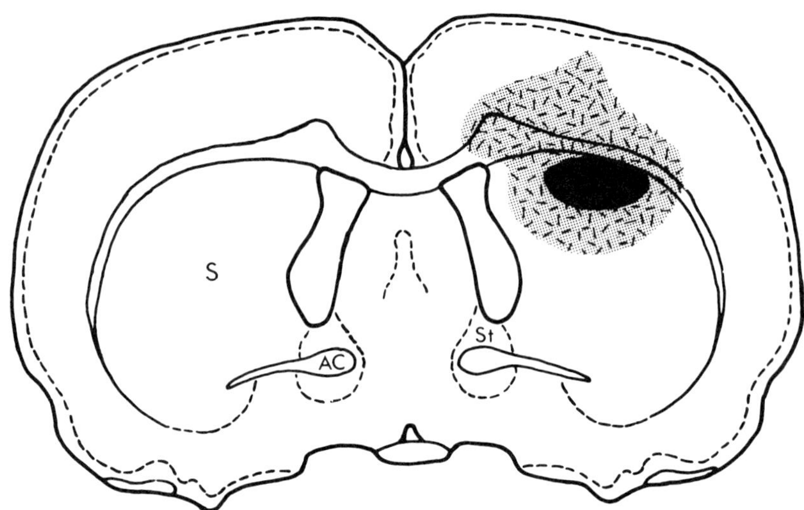

Fig. 1. The kainic acid lesions in experiments SNC2 (stippled) and SNC3 (dark-tone). The kainic acid lesions were placed in the right dorsal striatum 3 days (SNC2) and 2 days (SNC3) before placement of the horseradish peroxidase injections in the superior colliculus (see Fig. 2).

phosphate buffered (pH 7.4) 2.5% glutaraldehyde and 0.5% paraformaldehyde. Serial frontal 80-120 μm tissue slices through the substantia nigra (SN) were cut and processed for the visualization of HRP using o-tolidine as the chromogen according to the method of Somogyi, Hodgson and Smith (1979) and processed for electron-microscopy using standard procedures. HRP labelled perikarya and dendrites were identified by light microscopic examination of "thick" (1 μm) sections of the SN pars reticulata (SNr) and adjacent thin sections were examined in the elctron microscope after staining with uranyl acetate and lead citrate. In each experimental case, the site and extent of the kainic acid lesion in the striatum, and the HRP injection in the superior colliculus were determined by light microscopic examination of 50 μm frozen sections through the striatum (Nissl stained) and 40 μm frozen sections through the superior colliculus (after processing with either tetramethylbenzidine, (Mesulam, 1982) or diaminobenzidine (Graham and Karnovsky, 1966).

In addition, the SN was examined by electron microscopy in several control animals where kainic acid injections had been placed in the region of the cortex which was sometimes accidentally involved in the cases of striatal kainic acid lesions.

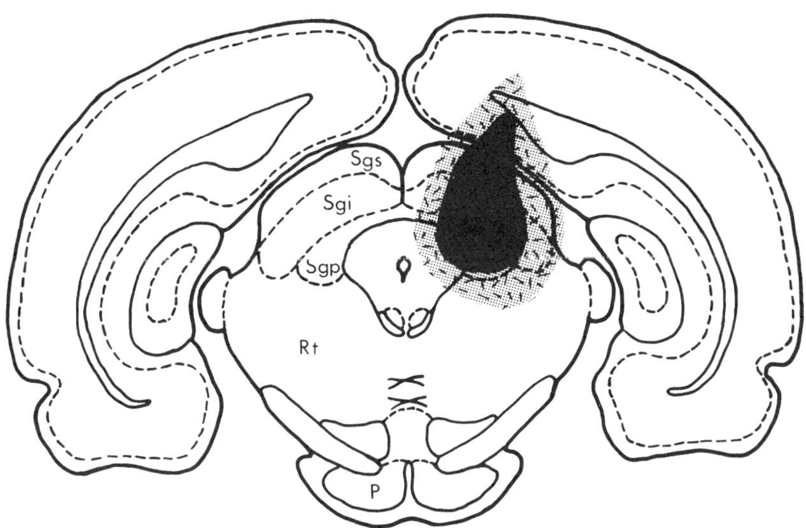

Fig. 2. The horseradish peroxidase injections which were placed in the right superior colliculus 24 hours before death in experiments SNC2 (dark-tone) and SNC3 (stippled).

RESULTS

The kainic acid lesions involved the dorsal region of the right striatum lying immediately ventral to the corpus callosum (Fig. 1). In several cases the lesion was restricted to the striatum (e.g. SNC3, Fig. 1), but in some cases the adjacent sensorimotor cortex was additionally involved (e.g. SNC2, Fig. 1). The HRP injections in the midbrain consistently involved regions of the caudal two-thirds of the right superior colliculus (Fig. 2); in some cases the injections were virtually confined to the superior colliculus (e.g. SNC2, Fig. 2) but in other cases there was also spread of HRP to the central gray and to the adjacent reticular formation (e.g. SNC3, Fig. 2).

Degenerating axons and terminal boutons in the substantia nigra showed the typical dark reaction (db, Figs. 3a, 4) and were therefore easily distinguished from normal axons and terminations. SNr neurons projecting to the superior colliculus showed variable labelling with HRP reaction product when viewed with light microscopy (LM; Fig. 3b); moderately and heavily labelled perikarya and dendrites of nigrotectal neurons which were identified at the LM level were also easily identified by electron microscopy because of the electron density of the HRP reaction product. Degenerating striatonigral fibres were seen making synaptic contact on the cell body (Fig. 3a), and on the proximal and distal (Fig. 4) dendritic segments of HRP-labelled and unlabelled neurons in the ventral region of the SNR. Multiple degenerating synaptic boutons were sometimes

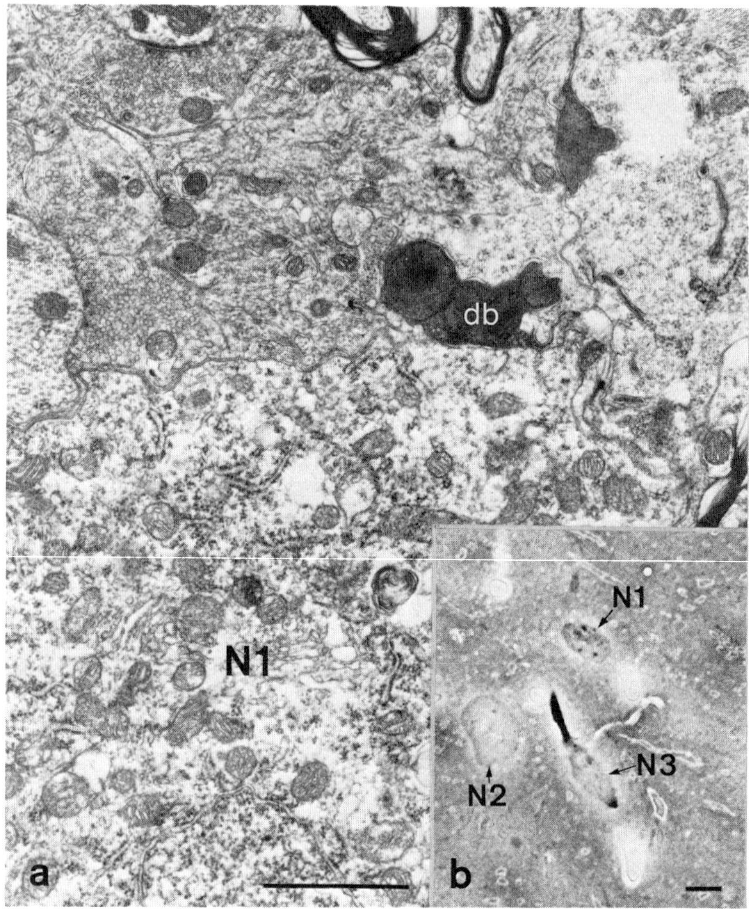

Fig. 3. (a) An electron micrograph showing a degenerating bouton (db) making synaptic contact with the soma of an HRP labelled neuron (N1; same neuron as N1 in b) in experiment SNC3 (see Figs. 1 and 2 for experimental details; magnification bar = 1 μm).
(b) A light micrograph of a 1 μm "thick" section lightly counterstained with toluidine blue and viewed with phase contrast illumination showing 3 HRP labelled neurons (N1, N2 and N3) in experiment SNC3; N2 is lightly labelled, N1 is moderately labelled and N3 is heavily labelled especially in the main-stem dendrite (Magnification bar = 10 μm).

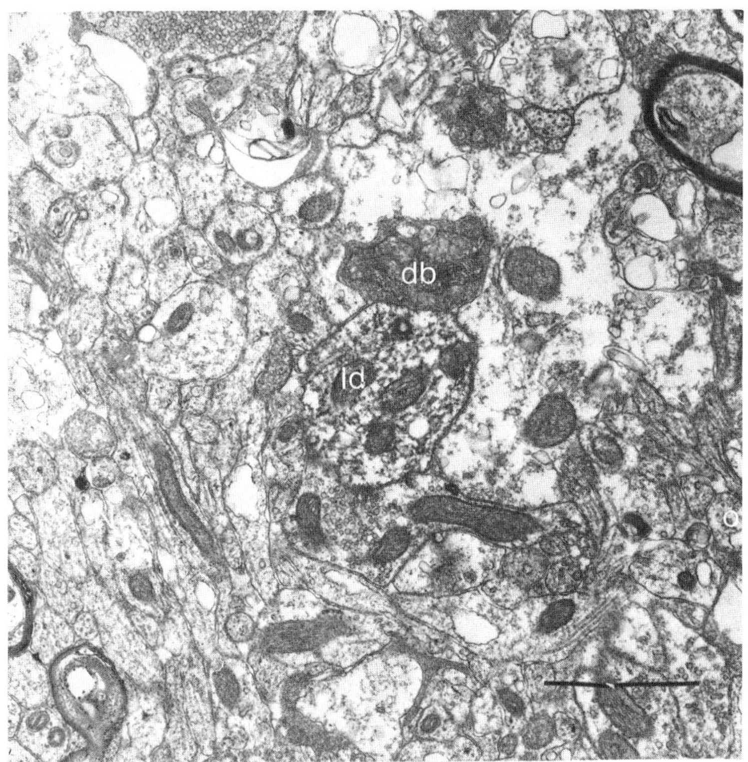

Fig. 4. An electron micrograph showing a degenerating bouton (db) making synaptic contact with an HRP labelled small peripheral dendrite (ld) in experiment SNC2 (see Figs 1 and 2 for experimental details; magnification bar = 1 μm).

seen contacting the soma or dendrites of individual labelled neurons. In this material it was often difficult to determine the synaptic type involved because the dark reaction of the degenerating terminals and the intensity of staining of the HRP reaction product often obscured the fine details of the contact region.

In the control experiments with kainic acid lesions restricted to the sensorimotor cortex, no degenerating terminals were seen in the substantia nigra.

DISCUSSION

The findings in this study provide anatomical evidence for a monosynaptic input from the dorsal striatum to nigrotectal projection neurons in SNr and thus demonstrate the existence of a bineuronal pathway from the striatum through the substantia nigra to the superior colliculus.

These findings confirm and extend light microscopic studies on the topography of the striatonigral projection (Nauta and Domesick, 1979; Faull et al., 1980) viz: dorsal striatum projects to and establishes synaptic contact with neurons in the ventral region of SNr. The finding that striatal efferents form synaptic contacts with the soma and dendritic processes of nigral neurons is in accord with the results of previous electron microscopic studies in the rat (Hajdu, Hassler and Bak, 1973; Somogyi, Hodgson and Smith, 1979) and cat (Grofova and Rinvik, 1970; Kemp, 1970).

Concerning the possible functional significance of the striato-nigro-tectal pathway, it is interesting to note that the nigrotectal pathway terminates in the deeper "motor" layers of the superior colliculus (Graybiel, 1975, 1978; Jayaraman, Batton, and Carpenter, 1977; Beckstead, Domesick, and Nauta, 1979) suggesting that the striato-nigral-tectal link may enable the basal ganglia to contribute to visuomotor functions of the superior colliculus. Furthermore, since the visual cortex projects onto the dorsal region of the striatum (Faull, Nauta, and Domesick, 1980) and, in addition, corticostriate axons have been shown to establish direct monosynaptic contacts with striatonigral neurons (Somogyi, Bolam and Smith, 1981), then these findings when taken together with the present observations suggest the possible existence of an indirect visual cortico-tectal connection through the basal ganglia mediated via a trineuronal cortico-striato-nigro-tectal pathway.

ACKNOWLEDGEMENTS

This study was supported by grants from the New Zealand Neurological Foundation and the Medical Research Council of New Zealand.

ABBREVIATIONS

AC Anterior commissure
P Pontine nuclei
Rf Reticular formation
S Striatum (caudoputamen)
Sgi Stratum griseum intermedium colliculi superioris
Sgp Stratum griseum profundum colliculi superioris
Sgs Stratum griseum superficiale colliculi superioris
St Bed nucleus of stria terminalis

REFERENCES

Beckstead, R. M., Domesick, V. B., and Nauta,, W. J. H., 1979, Efferent connections of the substantia nigra and ventral tegmental area in the rat, Brain Res., 175:191.

Beckstead, R. M., Edwards, S. B., and Frankfurter, A.., 1981, A comparison of the intranigral distribution of nigrotectal neurons labeled with horseradish peroxidase in the monkey, cat and rat, J. Neuroscience, 1:121.

Faull, R. L. M., and Mehler, W. R., 1978, The cells of origin of nigrotectal, nigrothalamic and nigrostriatal projections in the rat, Neuroscience, 3:989.

Faull, R. L. M., Nauta, W. J. H., and Domesick, V. B., 1980, The visual cortico-striato-nigral relay in the rat, J. Anat., 130:199.

Graham, R. C., and Karnovsky, M. J., 1966, The early stages of absorption of injected horseradish peroxidase in the proximal tubules of mouse kidney: ultrastructural cytochemistry by a new technique, J. Histochem. Cytochem., 14:291.

Graybiel, A. M., 1975, Anatomical organizatiion of retinotectal afferents in the cat: an autoradiographic study, Brain Res., 96:1.

Graybiel, A. M., 1978, Organization of the nigrotectal connection: an experimental tracer study in the cat, Brain Res., 143:339.

Grofova, I., and Rinvik, E., 1970, An experimental electron microscopic study on the striatonigral projection in the cat, Exp. Brain Res., 11:249.

Hajdu, F., Hassler, R., and Bak., I. J., 1973, Electron microscopic study of the substantia nigra and the strio-nigral projection in the rat, Z. Zellforsch., 146:207.

Jayaraman, A., Batton, R. R., and Carpenter, M. B., 1977, Nigrotectal projections in the monkey: an autoradiographic study, Brain Res., 135:147.

Kemp,. J. M., 1970, The termination of strio-pallidal and strio-nigral fibres, Brain Res., 17:125.

McGeer, E. G., Olney, J. W., and McGeer, P. L., eds., 1978, "Kainic Acid as a Tool in Neurobiology," Raven Press, New York.

Mesulam, M. M., ed., 1982, "Tracing Neural Connections with Horseradish Peroxidase," John Wiley and Sons, Chichester.

Nauta, W. J. H., and Domesick, V. B., 1979, The anatomy of the extrapyramidal system, in: "Dopaminergic Ergot Derivatives and Motor Function," K. Fuxe and D. B. Calne, eds., Pergamon Press, Oxford.

Somogyi, P., Bolam, J. P., and Smith, A. D., 1981, Monosynaptic cortical input and local axon collaterals of identified striatonigral neurons. A light and electron microscopic study using the Golgi-peroxidase transport degeneration procedure, J. comp. Neurol., 195:567.

Somogyi, P., Hodgson, A. J., and Smith, A. D.., 1979, An approach to tracing neuron networks in the cerebral cortex and basal ganglia. Combination of Golgi staining, retrograde transport of horseradish peroxidase and anterograde degeneration of synaptic boutons in the same material, Neuroscience, 4:1805.

SEROTONINERGIC INNERVATION OF THE MONKEY BASAL GANGLIA: AN IMMUNOCYTOCHEMICAL, LIGHT AND ELECTRON MICROSCOPY STUDY

Pedro Pasik[1,2], Tauba Pasik[1], Gay R. Holstein[1] and Jorge Pecci Saavedra[3]

[1]Department of Neurology, Mount Sinai School of Medicine CUNY, New York, NY

[2]Department of Anatomy, Mount Sinai School of Medicine CUNY, New York, NY

[3]Instituto de Biologia Celular, Facultad de Medicina Universidad de Buenos Aires, Argentina

INTRODUCTION

Normal functioning of the basal ganglia system depends upon the proper balance among several substances controlling information transfer within the constituent neuronal circuits. Most of the components of this system are considerably richer than other regions of the central nervous system in these putative neurotransmitters and/or neuromodulators. This quality makes the basal ganglia an optimal group of structures in which to study the position of the neuropil elements with respect to each other, and the alterations induced by surgical, toxicological or pharmacological manipulations. The advent of immunocytochemical techniques which allow observations at the ultrastructural level (Moriarty and Halmi, 1972; Sternberger, 1979; Pickel, 1981) has superseded, to some extent, the autoradiographic method based on the uptake of radioactive exogenous substances. Thus, major strides have been made in the localization of dopamine, gamma amino butyric acid and acetylcholine through the use of antibodies to their corresponding synthesizing enzymes (tyrosine hudroxylase, glutamic acid decarboxylase and choline acetyltransferase), and more directly with antibodies to certain peptides such as the enkephalins and substance P (Ribak et a., 1979; Pickel et al., 1980, 1981; Kimura et al., 1980; Cuello et al., 1981, 1982; DiFiglia et al., 1982a; Somogyi et al., 1982; Bolam et al., 1983). Less attention has been directed to serotonin, despite its

Table I

Mean Serotonin Content	(ng/mg protein)*
Midbrain raphe nuclei+	22.8
Substantia nigra	19.9
Globus pallidus	19.0
Zona incerta	16.0
Facial nucleus	13.6
Oculomotor nucleus	12.3
Central grey matter+	12.1
Superior colliculus	11.7
Red nucleus	11.3
Subthalamic nucleus	11.2
Caudate-putamen	11.0

Notes
* Data obtained from Palkovits et al., 1974. Nuclei are ranked according to the amount of serotonin.

+ Mean of several subdivisions

The content of other 34 brain structures sampled ranged from 10.6 to 2.0 ng/mg protein.

high concentration (Table I) and presumably prominent role in aforementioned chemical balance necessary for the physiologic performance of the basal ganglia (Hassler and Bak, 1969).

It is well known that most serotonin immunoreactive neurons are located in the raphe nuclei (Steinbusch, 1981; Pecci Saavedra et al., 1983). The axons of these neurons may travel long distances and arborize extensively in their terminal fields, thus providing innervation to broad expanses of the CNS (Azmitia and Segal, 1978). The basal ganglia receive these afferents from the midbrain raphe, namely n. raphe dorsalis and n. centralis superior (Bobillier et al. 1976). The terminal fields have been visualized mostly in the rat b-formaldehyde-induced fluorescence (Fuxe, 1965), uptake of radioactive serotonin (Parizek et al., 1971; Arluison and de la Manche, 1980), and immunofluorescence (Steinbusch, 1981). The goal of the present series of studies was to define the mode of termination of these fibers in the neostriatum, pallidum and substantia nigra of monkeys by means of electron microscopic immunocytochemistry. This article contains a review of previously published data on the neostriatum (Pasik and Pasik, 1982) and pallidum (Pasik et al., 1983), as well as new information on the substantia nigra.

MATERIALS AND METHODS

Adolescent cynomolgus monkeys (Macaca fascicularis) were prepared for immunocytochemistry according to a protocol published in detail elsewhere (Pasik et al., 1982). In short, the animals were perfused under deep barbiturate narcosis with a buffered mixture of 4% pure formaldehyde and 0.25% purified glutaraldehyde, after pretreatment with pargyline and L-tryptophan. Brain slices, 5 mm thick, were left in the fixative for a total of 2.5 hours from the start of the perfusion. Blocks containing the structures under study were cut with a vibrating microtome at 40 um, and the sections were treated for 15 minutes in a 0.2% solution of Triton X-100. This was followed by successive incubations in 3% normal sheep serum, 30 minutes; rabbit antiserotonin serum, 1:800 to 1:1000, 12-18 hours; sheep antirabbit serum, 1:30, 30 minutes; rabbit peroxidase-anti-peroxidase complex, 1:50, 30 minutes; 0.05% diaminobenzidine plus 0.01% hydrogen peroxide, 5-8 minutes. All steps were carried out at $4^{o}C$, with intermediate washes in tris-saline. Fragments of these sections were osmicated, stained with aqueous uranyl acetate, dehydrated and flat embedded in Epon-Araldite between plastic coverslips. The portions containing immunoreactive elements, visualized by light microscopy, were dissected, attached to blank blocks of resin, machine trimmed further, and cut at 80-100 nm. These sections were mounted on formvar coated slot grids and examined with the electron microscope before and after staining with uranyl acetate and lead acetate.

The primary antibody was obtained from Immuno Nuclear Corp., Stillwater, Minnesota. It was derived from rabbits immunized with a serotonin-bovine serum albumin conjugate. This antigen was used to preabsorb the serum for treating control sections which resulted in no staining (Pecci Saavedra et al., 1983). It is probable that the antibody had a certain degree of crossreactivity with dopamine and norepinephrine. In pilot experiments with dilutions below 1:500, there was also a light staining of neurons in the substantia nigra pars compacta and in the locus coeruleus. No such reaction was present with the higher dilutions used in this study.

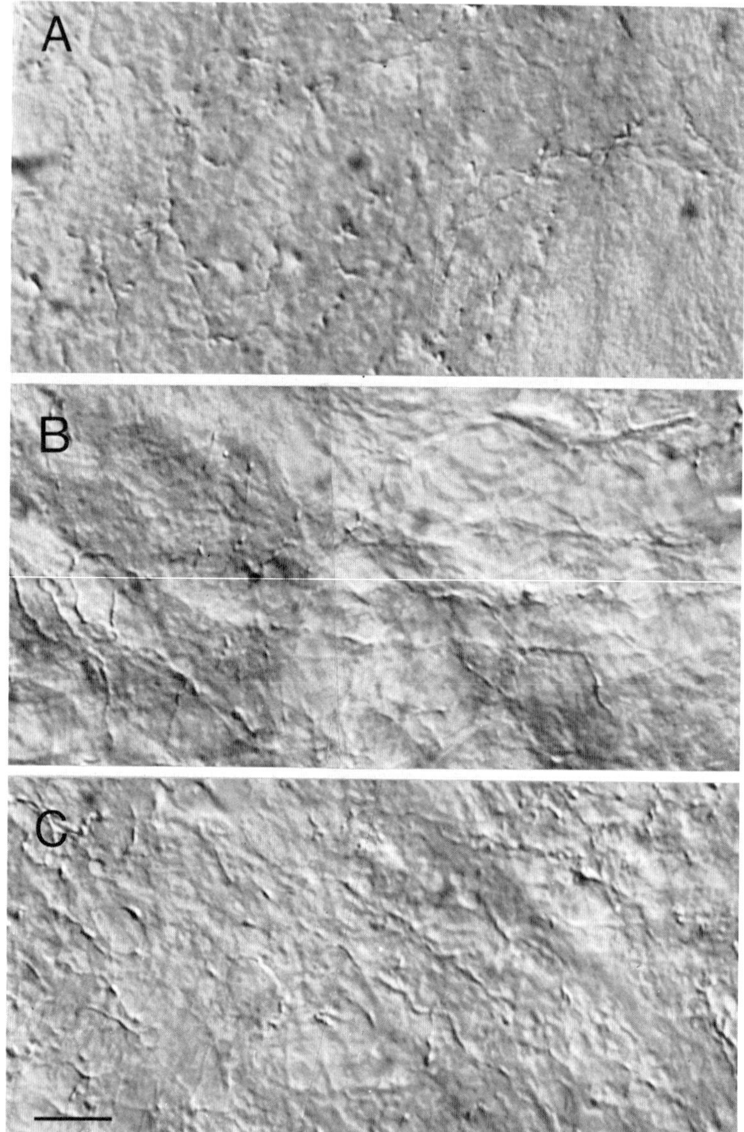

Fig. 1. Immunoreactive fibers in head of caudate nucleus (A), medial globus pallidus (B), and substantia nigra pars reticulata (C). Antiserotonin serum, 1:1000 dilution, PAP technique, Nomarski optics. Scale bar: length, 20 μm; width, 0.4 μm.

RESULTS

Light Microscopy

Immunoreactive elements were present in the caudate nucleus, putamen, both segments of the globus pallidus, and the substantia nigra. Their general appearance was similar in all of these structures (Fig. 1), and consisted of fine fibers which gave rise to a profusion of thinner branches. Fusiform varicosities, 0.7-1.5 μm in size, spaced irregularly 1-5 μm apart, were visible along all these axonal processes. Some of these axons could only be seen with the aid of differential interference contrast enhancement, suggesting diameters thinner than 0.2 μm. Occasionally these processes were observed to outline neuronal somata. There was some regional variation such that fibers tended to be thinnest in the neostriatum and thickest in the substantia nigra. The density of the axonal plexi was also different, although the possibility that this heterogeneity might be due in part to variations in antibody penetration could not be ruled out. The fibers seen in the caudate nucleus were more numerous in ventromedial regions of the head, and decreased progressively toward more dorsal and lateral regions. The distribution in the putamen was more uniform and of medium density. They appeared more compactly arranged in the medial than in the lateral segment of the globus pallidus, and predominated in the pars reticulata as compared with the pars compacta of the substantia nigra. No immunostained somata were found in any of these structures with dilutions of 1:800 - 1:1000 (see Material and Methods section).

Electron Microscopy

Neostriatum. The material was derived from areas of maximal density of labeled fibers (see above). The rich striatal neuropil is characterized by an array of small dendritic and axonal profiles, and the predominance of axospinous synapses (Pasik et al., 1976). In the midst of these structures, there were few immunostained elements which conformed to the light microscopic observations, i.e. thin profiles with strictures down to 0.1 μm in caliber and dilations reaching up to 1 μm (Fig. 2A). These profiles were filled with electron dense material which remained within the confines of the axolemma with no spillage into the surrounding structures. Most of the label appeared in the shape of round bodies, 15-35 nm in diameter, although smaller particles attached to the outer surfaces of microtubules and mitochondria were also visible. Few of the varicosities seen in single sections formed synapses, and did so exclusively with small dendritic spines. Examination of serial sections demonstrated that such spines received only the one labeled

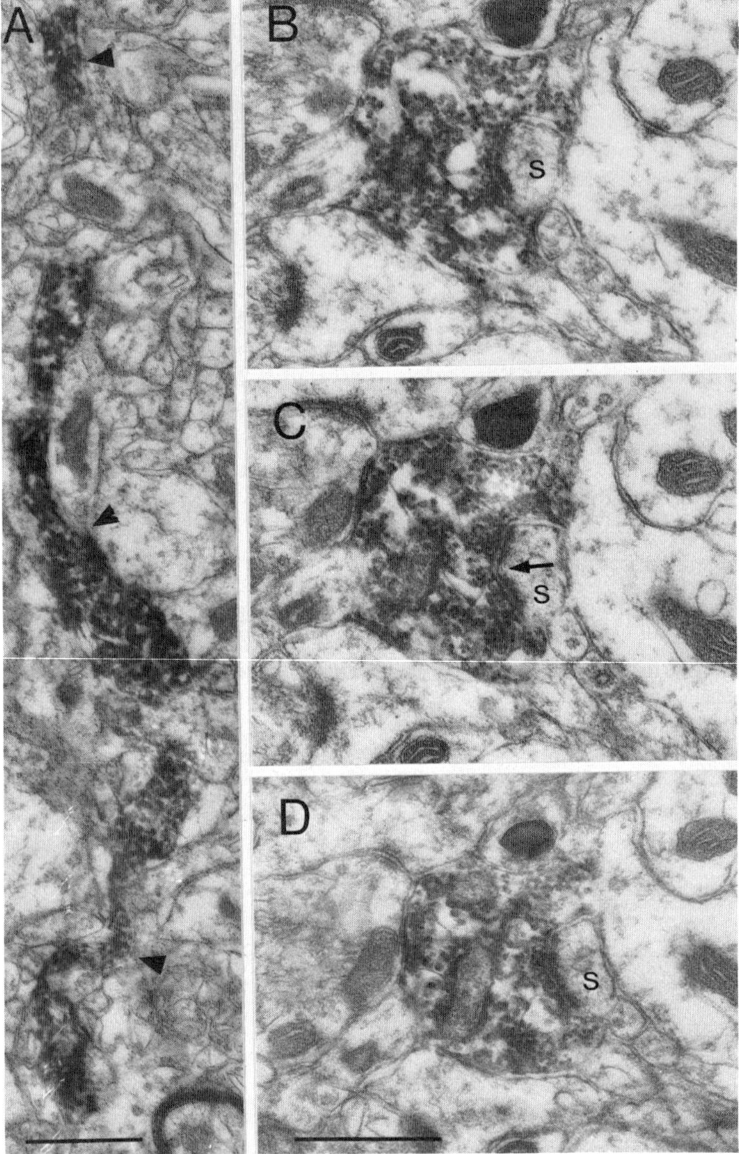

Fig. 2. Electron micrograph of striatal neuropil. A: three serotonin-labeled profiles (arrowheads) which, in serial sections, proved to belong to a single fine fiber. Scale bar: 0.5 µm. B, C and D: serial sections of an asymmetric axospinous synapse formed by a labeled axonal profile and a small spine (s). The synaptic plaque shows a pinhole at arrow in B. Scale bar in D: 0.5 µm (also for B and C). (From Pasik and Pasik, 1982, courtesy Akademiai Kiado, Budapest)

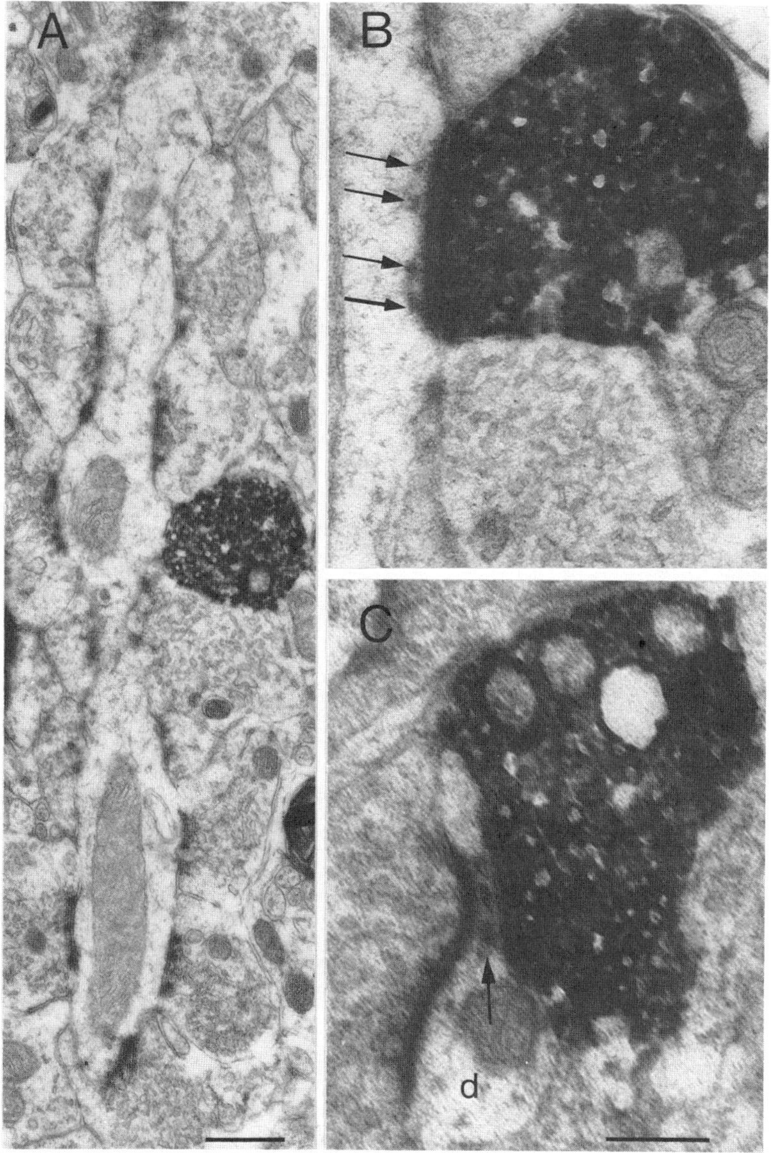

Fig. 3. Electron micrograph of pallidal neuropil (lateral segment). A: longitudinally cut dendrite totally covered with synapsing boutons, one labeled for serotonin. Scale: 0.5 μm. B: another section from a series of the same profile, showing an extensive asymmetric synapse and several subjunctional dense bodies (arrows). C: labeled profile represents one element of a twin synapse with dendritic crest emerging from small dendrite (d); crest contains a dense core formed by sequence of dense bodies (arrow). Scale bar in C: 0.25 μm (also for B). (From Pasik et al., 1983, courtesy Raven Press, N. Y.)

terminal, and there was never a spinule projecting into the presynaptic element, a feature commonly seen in the striatal neuropil (Pasik et al., 1976). The membrane specializations were strongly asymmetric. Although no quantitative data were available, the frequency of finding morphologically defined synapses was low, suggesting that many of the varicosities were not in synaptic contact with other elements.

Globus pallidus. The essential feature of the pallidal neuropil, both of the medial and lateral segments, is the presence of long and thick dendrites covered almost entirely by axonal boutons forming synapses (Fox et al., 1974; DiFiglia et al., 1982b). Immunostained fibers, similar in shape and content to those just described in the striatum, followed a complicated course through the neuropil, only unraveled by the examination of serial sections. In contrast to the striatum, however, many of the labeled axons terminated in boutons, up to 1.5 μm in diameter, which formed markedly asymmetric synapses with dendritic shafts and crests (Fig. 3). Again, the proportion of these serotonin immunoreactive elements was low. In Figure 3A, for example, it is of the order of 1:20 compared with the unlabeled boutons. Serial sections revealed the common occurrence of subjunctional dense bodies associated with the postsynaptic membrane of dendritic trunks (Fig. 3A and B), and a sequence of dense granules in the core of dendritic crests (Fig. 3C). The axonal profile forming a synapse on the oposite aspect of the crest was always unlabeled. Immunostained profiles wre never observed to interdigitate with other elements, or to participate in the formation of triadic synapses. These two features are rather common in the pallidal neuropil (DiFiglia et al., 1982b).

Substantia nigra. The samples used for electron microscopy derived from the pars reticulata where the densest plexus was apparent at the light microscopic level. The ultrastructure of the substantia nigra is remarkably similar to that of the pallidum (Schwyn and Fox, 1974). Accordingly, our material also showed large dendrites, oriented in parallel to fine unmyelinated axon collaterals, forming synapses along the dendritic shafts, and orthogonally to the course of the myelinated fibers from which the collaterals arose. In cross-section, the dendritic surfaces appeared to be ensheathed by fine caliber axons and their terminal boutons. Glial processes usually surrounded the longitudinal axodendritic complexes.

Serial sections showed fine fibers with strictures and dilations, containing dark particulate reaction products, about 50 nm in diameter (Fig. 4A). Small mitochondria and round vesicles, 50-100 nm in diameter, were scattered throughout the immunoreactive elements. Synaptic contacts between labeled terminals and large dendritic elements wre also observed (Fig. 4B). Often the surfaces of the dendrites were completely covered with synaptic membrane specializations, only one of which involved a labeled axon. Most of

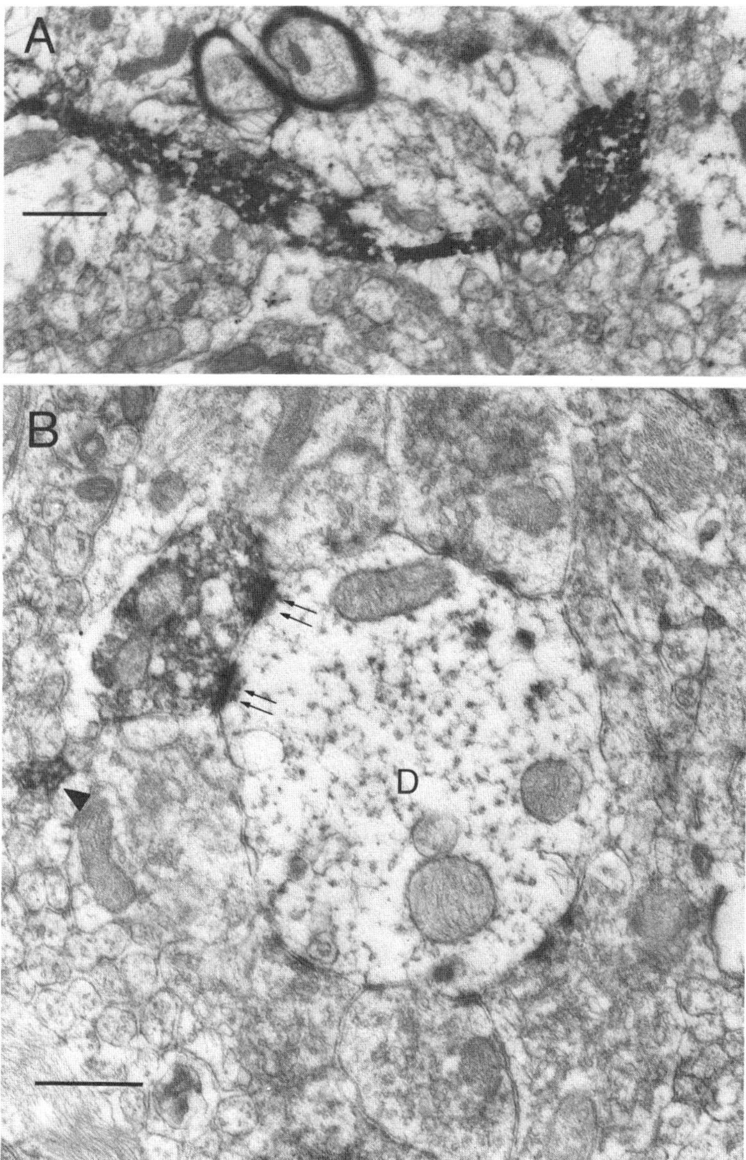

Fig. 4. Electron micrograph of nigral neuropil (pars reticulata). A: a fine caliber immunoreactive fiber with strictures and varicosities. B: a transversely sectioned dendrite (FD), totally covered with boutons, one of which is labeled and forms asymmetric synapses with subjunctional dense bodies (arrows). A fine immunoreactive fiber (arrowhead) is also seen within a bundle of labeled axons of similar caliber. Scale bar: 0.5 μm.

the synaptic articulations appeared asymmetrical, and occasionally subjunctional dense bodies were visible under the postsynaptic membrane (Fig. 4B). Immunostained profiles were also seen forming the same type of synapses with dark dendrites.

DISCUSSION

The preceding findings provide positive identification of serotoninergic afferentation to three major components of the basal ganglia system, namely the neostriatum, pallidum and substantia nigra. Moreover, they offer clues for the location of serotonin containing terminals within the neuronal circuits of these structures. The results lend a firm basis to earlier descriptions derived from formaldehyde-induced fluorescence (Fuxe, 1965), specific uptake of tritiated serotonin (Parizek et al., 1971; Arluison and de la Manche, 1980), and immunofluorescence (Steinbusch, 1981).

The morphology of serotoninergic fibers corresponds to the type 3 extrinsic axons described in Golgi preparations of monkey neostriatum (DiFiglia et al., 1978; Pasik and Pasik, 1982), and to similar axons present in the pallidum (Pasik et al., 1983). No equivalent descriptions are available for the substantia nigra. It is unclear whether single raphe neurons provide innervation to the three structures of this study, through either branching axons near their origin, or collaterals of a passing axon. The latter possibility is suggested by our finding that the caliber of the fibers diminishes from the substantia nigra where they may attain up to 1 μm, to maximum diameters of only 0.3-0.5 μm in the pallidum and striatum. There is also evidence from fluorescent double labeling studies that at least some of the fibers innervating the substantia nigra are collaterals of the raphe-striatal axons (van der Kooy and Hattori, 1980). However, the principal sources for these pathways may be different, the n. raphe dorsalis providing input primarily to pallidum and striatum, and n. centralis superior sending projections to the substantia nigra (Dray, 1981). These two possibilities are not incompatible and the issue must remain open.

The striatal cells receiving the serotoninergic input apparently belong to the spiny class which has been shown to consist of long-axoned efferent cells. Whether they are of the type I and/or II variety (Pasik et al., 1979) is uncertain. The pallidal and nigral neurons, which are postsynaptic to serotonin immunostained profiles, are also of the long-axoned efferent type and receive many additional inputs particularly from the striatum, as well as from the subthalmic nucleus (Fox et al., 1974; DiFiglia et al., 1982b; Pasik and Pasik, 1983). There is some controversy regarding the relative magnitude of the serotoninergic input to the two main subdivisions of the substantia nigra. Some investigators report it to be greater in the pars compacta (Bobillier et al., 1975; Fibiger and Miller, 1977;

Asmitia and Segal, 1978), and others in the pars reticulata (Fuxe, 1965; Dray et al., 1976; Reubi and Emson, 1978). Our findings support the latter contention; it is the pars reticulata that exhibits the denser plexus of immunoreactive fibers. There are also preliminary data regarding thre interaction of serotonin and dopamin containing elements in the substantia nigra and ventral tegmental areas (Pickel et al., 1975; Hervee et al., 1983). Apparently varicosities which show uptake of triated 5-HT form synapses on tyrosine hydroxylase immunoreactive dendrites.

The functional significance of the serotoninergic input to the basal ganglia is not clear. The morphological evidence offered by the present study favors an excitatory action, at least regarding the initial changes in the neuronal membrane. Although the classification of synapses according to the symmetry of the membrane specializations (Gray, 1959) and their functional correlations may have a large margin of uncertainty, there is little doubt that strongly asymmetric synapses are indicative of excitatory processes. We found that all well-defined synapses made by serotonin immunoreactive terminals are of this latter type, in accord with the majority of contacts labeled by uptake are of this latter type, in accord with the majority of contacts labeled by uptake of tritiated 5-HT (Arluison and de la Manche, 1980). Intracellular recordings from striatal (Vandermaelen et al., 1979) and nigral (Karabelas and Purpura, 1979) neurons show that stimulation of dorsal raphe nuclei elicits membrane depolarization and excitatory postsynaptic potentials as initial phenomena. It has also been demonstrated that this action is mediated by serotonin, at least in the striatum (Park et al., 1982). Yet, multiple studies with extracellular recordings indicate that dorsal raphe stimulation or serotonin microintophoretic application cause a net inhibitory effect on striatal as well as on nigral neurons (for review see Dray, 1981). No data is available concerning the pallidum but probably a likely set of findings could be expected. It is noteworthy that similar controversies exist regarding the action of dopamine, particularly in the neostriatum. It has been reported, however, that the membrane depolarization caused by dopamine ejection onto striatal cells is accompanied by a marked decrease in the firing rate (Bernardi et al., 1978; Herrling and Hull, 1980), suggesting a more complex action for dopamine than that of just an excitatory or inhibitory substance. Whether these findings are also applicable to serotonin awaits further experimentation. As discussed earlier (Pasik and Pasik, 1982), there is still the possibility of two separate effects of 5-HT in the basal ganglia, as suggested by the presence of synaptic and non-synaptic varicosities particularly in the neostriatum, as well as by the existence of more than one type of serotonin receptors (Peroutka et al., 1981).

ACKNOWLEDGEMENTS

The authors are grateful to M. Feliciano, M. Ilvento and V. Rodriguez for their skilful assistance. The research was supported by NINCDS Grants #NS-11631 and F32NS-06954, and the American Parkinson's Disease Association.

REFERENCES

Arluison, M., and de la Manche, I. S., 1980, High resolution radioautographic study of the rat corpus striatum after intraventricular administration of ^{3}H 5-hydroxytryptamine, Neuroscience, 5:229.

Azmitia, E., and Segal, M., 1978, An autoradiographic analysis of the differential ascending projections of the dorsal and median raphe nuclei in the rat, J. Comp. Neurol., 179:641.

Bernardi, G., Marciani, M. G., Morocutti, C., Pavone, F., and Stanzione, P., 1978, The action of dopamine on rat caudate neurones intracellularly recorded, Neurosci. Lett., 8:235.

Bobillier, P., Petitjean, F., Salvert, D., Ligier, M., and Seguin, S., 1975, Differential projections of the nucleus raphe dorsalis and nucleus raphe centralis as revealed by autoradiography, Brain Res., 85:205.

Bobillier, P., Seguin, S., Petitjean, F., Salvert, D., Touret, M. and Jouvet, M., 1976, The raphe nuclei of the cat brain stem: a topographical atlas of their efferent projections as revealed by autoradiography, Brain Res., 113:449.

Bolam, J. P., Somogyi, P., Takagi, H., Fodor, I., and Smith, A.D., 1983, Localization of substance P-like immunoreactivity in neurons and nerve terminals in the neostriatum of the rat: a correlated light and electron microscopic study, J. Neorocytol., 12:325.

Cuello, A. C., Del Fiacco, M., Paxinos, G., Somogyi, P., and Priestley, J. V., 1981, Neuropeptides in striato-nigral pathways, J. Neural Transmission, 51:83.

DiFiglia, M., Aronin, N. and Martin, J. B., 1982a, Light and electron microscopic localization of immunoreactive leu-enkephalin in monkey basal ganglia, J. Neurosci., 2:303.

DiFiglia, M., Pasik, P., and Pasik, T., 1982b, A Golgi and untrastructural study of the monkey globus pallidus, J. Comp. Neurol., 212:53.

DiFiglia, M., Pasik, T., and Pasik, P., 1978, A Golgi study of afferent fibers in the neostriatum of monkeys, Brain Res., 152:341.

Dray, A., 1981, Serotonin in the basal ganglia: functions and interactions with other neuronal pathways, J. Physiol. (Paris), 77:393.

Fibiger, H. C., and Miller, J. J., 1977, An anatomical and electrophysiological investigation of the serotonergic projections from the dorsal raphe nucleus to the substantia nigra in the rat, Neuroscience, 2:975.
Fox, C. A., Andrade, A. N., Lu Qui, I. J., and Rafols, J. A., 1974, The primate globus pallidus: A Golgi and electron microscopic study, J. Hirnfirsch, 15:75.
Fuxe, K., 1965, Evidence for the existence of monoamines in the central nervous system. IV. Distribution of monoamine nerve terminals in the central nervous system, Acta Physiol. Scand. Suppl., 247:37.
Gray, E. G., 1959, Axo-somatic and axo-dendritic synapses of the cerebral cortex: an electron microscope study, J. Anat. (London), 93:420.
Hassler, R., and Bak, I. J., 1969, Unbalanced ratios of striatal dopamine and serotonin after experimental interruption of strionigral connections in rat, in: "Third Symposium on Parkinson's Disease,"" F. J. Gillingham, and I. M. L. Donaldson, eds., E. and S. Livingstone Ltd, Edinburgh.
Herrling, P. L., and Hull, C. D., 1980, Iontophoretically applied dopamine depolarizes and hyperpolarizes the membrane of cat caudate neurons, Brain Res., 192:441.
Herve, D., Joh, T. H., Pickel, V. M., and Beaudet, A., 1983, Serotonin innervation of dopamine neurons in rat ventral tegmental area, Soc. Neurosci. Abstr., 9:713.
Karabelas, A. B., and Purpura, D. P., 1979, Functional properties of dorsal raphe-substantia nigra projections in the cat, Soc. Neurosci. Abstr., 5:73.
Kimura, H., McGeer, P. L., Pengf, F., and McGeer, E. G., 1980, Choline acetyltransferase-containing neurons in rodent brain demonstrated by immunohistochemistry, Science, 208:1057.
Moriarty, G. C., and Halmi, N. S., 1972, Electron microscopic study of the adrenocorticotropin producing cell with the use of unlabeled antibody and the soluble peroxidase-antiperosidase complex, J. Histochem. Cytochem, 20:590.
Palkovits, M., Brownstein, M., and Saavedra, J. M., 1974, Serotonin content of the brain stem nuclei in the rat, Brain Res., 80:237.
Parizek, J., Hassler, R., and Bak, I. J., 1971, Light and electron microscopic autoradiography of substantia nigra of rat aftrer intraventricular administration of tritium labelled norepinephrine, dopamine, serotonin and the precursors, Z. Zellforsch., 115:137.
Park, M. R., Gonzales-Vegas, J. A., and Kitai, S. T., 1982, Serotonergic excitation from dorsal raphe stimulation recorded intracellularly from rat caudate-putamen, Brain Res., 243:49.
Pasik, P., Pasik, T., and DiFiglia, M., 1976, Quantitative aspects of neuronal organization in the neostriatum of the macaque monkey, in: "The Basal Ganglia," Res. Publ. Assoc. Res. Nerv. Ment. Dis., M. D. Yahr, ed., Raven Press, New York.

Pasik, P., Pasik, T., and DiFiglia, M., 1979, The Internal organization of the neostriatum in mammals, in: "The Neostriatum," E. Divac, and R. G. E. Orberg, eds., Pergamon Press, Oxford, New York.
Pasik, P., Pasik, T. and Pecci Saavedra, J., 1982. Immunocytochemical localization of serotonin at the ultrastructural level, J. Histochem. Cytochem., 30:760.
Pasik, P., Pasik, T. and Pecci Saavedra, J.,Holstein, G. R., and Yahr, M. D., 1983, Serotonin in pallidal neuronal circuits: An immunocytochemical study in monkeys, in: "Parkinson-specific Motor and Mental Disorders, Role of the Pallidum: Pathophysiological, Biochemical and Therapeutic Aspects," R. G. Hassler, and J. F. Christ, eds., Raven Press, New York.
Pasik, T., and Pasik, P., 1982, Serotoninergic afferents in the monkey neostriatum, Acta Biol. Acad. Sci. Hung., 33:277.
Pasik, T., and Pasik, P., 1983, The internal organization of the pallidum in mammals, J. Neural Transmission Suppl., 19:13.
Pecci Saavedra, J., Pasik, T., and Pasik, P., 1983, Immunocytochemistry of serotoninergic neurons in the central nervous system of monkeys, in: "Neural Transmission, Learning and Memory," IBRO Monograph Series, Vol. 10, R. Caputto, and C. Ajmone-Marsan, eds., Raven Press, New York.
Peroutka, S. J., Lebovitz, R. M., and Snyder, S. H., 1981, Two distinct central serotonin receptors with different physiological functions, Science, 212:827.
Pickel, V. M., 1981, Immunocytochemical methods, in: "Neuroanatomical Tract-Tracing Methods," L. Heimer, and M. J. Robards, eds., Plenum Press, New York.
Pickel, V. M., Beckley, S. C., Joh, T. H., and Reis, D. J., 1981, Ultrastructural immunocytochemical localization of tyrosine hydroxylase in the neostriatum, Brain Res., 225:373.
Pickel, V. M., John, T. H., and Reis, D. J., 1975, Immunocytochemical demonstration of a serotonergic innervation of catecholamine neurons in locus coeruleus and substantia nigra, Soc. Neurosci. Abstr., 1:320.
Pickel, V. M., Sumal, K. K., Beckley, S. C., Miller, R. J., and Reis, D. J., 1980, Immunocytochemical localization of enkephalin in the neostriatum of rat brain: A light and electron microscope study, J. Comp. Neurol., 189:721.
Reubi, J.-C., and Emson, P. C., 1978, Release and distribution of endogenous 5-HT in rat substantia nigra, Brain Res., 139:164.
Ribak, C. E., Vaughn, J. E., and Roberts, E., 1979, The GABA neurons and their axon terminals in rat corpus striatum as demonstrated by GAD immunocytochemistry, J. Comp. Neurol., 187:261.
Schwyn, R. C., and Fox, C. A., 1974, The primate substantia nigra: A Golgi and electron microscopic study, J. Hirnforsch., 15:95.

Somogyi, P., Priestley, J. V., Cuello, A. C., Smith, A. D., and Takagi, H., 1982, Synaptic connections of enkephalin-immunoreactive nerve terminals in the neostriatum: A correlated light and electron microscopic study, J. Neurocytol., 11:779.

Steinbusch, H. W. M., 1981, Distribution of serotonin-immunoreactivity in the central nervous system of the rat-cell bodies and terminals, Neuroscience, 6:557.

Sternberger, L. A., 1979, "Immunocytochemistry," John Wiley and Sons, New York.

Van der Kooy,, D., and Hattori, T., 1980, Dorsal raphe cells with collateral projections to the caudate-putamen and substantia nigra: a fluorescent retrograde double labeling study in the rat, Brain Res., 186:1.

Vandermaelen, C. P., Bonduki, A. C., and Kitai, S. T., 1979, Excitation of caudate-putamen neurons following stimulation of the dorsal raphe nucleus in the cat, Brain Res., 175:356.

FLUORESCENT DOUBLE LABELING STUDIES OF THALAMOSTRIATAL AND CORTICOSTRIATAL NEURONS

G. James Royce and Sarah Bromley

Department of Anatomy
University of Wisconsin
Madison, Wisconsin 53706
U. S. A.

For many years, one of most intriguing questions in neuroanatomy concerned the hodological relationships between the striatum (caudate nucleus and putamen), the thalamus, and the cerebral cortex. While degeneration methods had readily demonstrated that most thalamic nuclei project to the cortex, other thalamic nuclei, particularly those with connections to the striatum, appeared to lack cortical connections.

For example, Powell (1952) noted that after hemidecortication in a human case, the neurons in most thalamic nuclei had undergone complete retrograde cell degeneration, whereas those in the intralaminar nuclei had survived. Similar results were reported in experimental studies in monkeys (Powell and Cowan, 1956, 1967). In support of these findings, studies which had used the anterograde degeneration method found little or no evidence for cortical projections of the intralaminar nuclei, but instead reported that these nuclei project massively upon the striatum (Nauta and Whitlock, 1954; Powell and Cowan, 1956; Mehler, 1966).

With the advent of the horseradish peroxidase method it became clear that the intralaminar nuclei _do_ project to the cortex, as well as to the striatum (Ralston and Sharp, 1973; Jones and Leavitt, 1974; and others, see Royce, 1983b for review). The question therefore arose: Why were these nuclei refractory to earlier attempts to reveal their cortical connections? A perceptive clue to answering this question was contained in the earlier work of Rose and Woolsey (1943, 1949), who found that the intralaminar nuclei in cats and rabbits degenerated, but only when cortical lesions were combined with ablations of subcortical structures that included the striatum. Rose and Woolsey therefore introduced the concept of "sustaining collaterals" which proposed that the neuronal cell body may survive, following a lesion to its primary axon, if collaterals from its axon project to a different, undamaged structure. The reverse may also be true, i.e. a lesion to a collateral may not affect the cell body if the primary axon is intact. On the other hand, both branches of an axon may have developed simultaneously and be of similar diameter, and therefore neither process may properly be designated as "collateral" or "primary". However, the principal of sustaining collaterals probably applies to any set of two axonal branches, regardless of their developmental history.

In order to obtain direct evidence for branching axons of thalamofugal neurons, we used the method of retrograde fluorescent microscopy, as originally pioneered by Kuypers et al. (1977). In this study, Evans Blue was injected into the caudate nucleus and Fast Blue was injected into the prefrontal cortex (Royce, 1983b). Following these injections, and appropriate tissue processing (see Royce, 1983b,c for procedural details) doubly labeled neurons were found within most members of the intralaminar nuclei, including the central medial, paracentral, central lateral and parafascicular nuclei. However, doubly labeled neurons were curiously absent from the centromedian nucleus in this study, despite the presence of large numbers of singly labeled thalamostriatal neurons and a few labeled thalamocortical neurons. Also, doubly labeled neurons were found in several other thalamic nuclei which lie outside the intralaminar complex, including the ventral anterior, ventral lateral and mediodorsal nuclei. Labeled thalamostriatal neurons lying outside the intralaminar complex had been reported in a previous horseradish peroxidase study (Royce, 1978), and such inputs have also been shown in physiological studies (Buchwald et al., 1973; Kunze et al., 1979; Steriade and Glenn, 1982).

Our study of the collateralizing projections of thalamic neurons to the striatum and cortex involved only a relatively small cortical region (the prefrontal cortex). However, it seems likely that doubly labeled neurons will be found in the thalamus (perhaps within the centromedian nucleus) following dual injections into both the caudate nucleus and other cortical areas, and experiments are currently in progress to explore this possibility.

We can only speculate upon the function of such branched axons projecting to both the cortex and striatum. In accordance with Rose and Woolsey's (1943, 1949) concept of sustaining collaterals, it may be important for the survival of the organism for these thalamic neurons to survive traumatic processes, such as decortication, thus emphasizing their importance. Alternatively, there may be a functional utility for the striatum to sample thalamic information destined for the cortex. As will be commented upon subsequently, a similar type of sampling function may be operating in the corticostriatal system.

In addition to finding branched axons of thalamostriatal neurons, we have also been investigating collateralizing axons of neurons which reside in the cerebral cortex. Our studies of doubly projecting corticostriatal neurons were inspired by our findings that cortical neurons which project to the caudate nucleus arise from both the supragranular and the infragranular layers, in contrast to all other known cortical projections to subcortical structures which appear to arise only from the infragranular layers (Royce, 1982). This finding led us to speculate that the dispersion of the cells of origin of the corticostriatal system throughout many cortico layers may exist because many of the fibers in this system are collaterals of axons which project elsewhere. Rather than being tightly segregated into individual layers or sublayers, as are most cortical projections to subcortical structures, the corticostriatal system appears to be widely dispersed throughout the cortical layers which may allow the striatum to sample, by collaterals, the entire corticofugal output. What also makes this concept plausible is the demonstration in numerous studies that the entire topographical extent of the cerebral cortex projects to the striatum (Webster, 1965; Yeterian and Van Hoesen, 1978; Royce, 1982).

Evidence to support the existence of collateralizing corticostriatal axons has accumulated over the years. In Golgi stained material, Ramón y Cajal (1909) found axons of passage which sent collaterals into the striatum, which he speculated were of cortical origin. More recently, physiological studies (Endo et al., 1973; Oka and Jannai, 1978; Jinnai and Matsuda, 1979; Donoghue and Kitai, 1981) and anatomical studies based on anterograde degeneration methods (Glees, 1944; Webster, 1965; Martin and Hamel, 1967) and intracellular horseradish peroxidase staining (Donoghue and Kitai, 1981) have all presented evidence in support of this concept.

We became interested in the possibility that corticostriatal axons may collateralize to the thalamus after noting the remarkable similarity between the topographical distributions of corticostriatal neurons which project to the head of the caudate nucleus (Royce, 1982) and of corticothalamic neurons which project to the centromedian-parafascicular complex (Royce, 1983a). Therefore, we injected Evans Blue into the caudate nucleus and Fast Blue into the centromedian and parafascicular thalamic nuclei. Following such injections, doubly labeled cortical neurons were found within the inferior bank of the cruciate sulcus, in the anterior limbic area, in the cingulate and anterior sylvian gyri and within the hidden cortex of the presylvian sulcus (Royce, 1983c).

These doubly labeled cortical neurons were all medium-sized pyramidal cells, and were found exclusively within layers V and VI. They were relatively few in number, especially when compared to the more numerous singly labeled cortico-thalamic neurons and the abundantly present singly labeled cortico-striatal neurons. These singly labeled corticostriatal neurons were present in layers II, III, V and VI, thus confirming our previous results obtained with the horseradish peroxidase method (Royce, 1982).

The bilateral nature of the corticostriatal projection has been well established in numerous studies in which anterograde tracing methods were used (Carman et al., 1965; Webster, 1965; Kunzle, 1975; Ragsdale and Graybiel, 1981; Tanaka et al., 1981). Our retrograde HRP study of this system established that the neurons which project contralaterally to the caudate nucleus, like their ipsilateral counterparts, arise from both the supragranular and infragranular cortical layers (Royce, 1982). These data raised the possibility that some of these cortical neurons may project both ipsilaterally, and contralaterally, by axonal collaterals. The findings in this regard to be presented subsequently have not been published previously.

In order to study the intrahemispheric connections of the corticostriatal neurons, different fluorescent tracers were injected into the right and left caudate nuclei of the same animal (in cats). In five cats, Evans Blue was injected into the left caudate nucleus, while Fast Blue was injected into the right caudate nucleus. In two additional animals, Evans Blue was injected into the left caudate nucleus, and the right caudate nucleus was injected with SITS (4-acetamido, 4'-isothiocyanostilbene-2,2'disulfonic acid). The rationale for using SITS in these experiments was that this tracer has been reported not to be taken up by fibers of passage (Schmued and Swanson, 1982).

Equal quantities of the different tracer substances were injected into both hemispheres (between 0.1-0.3 μl of 10% Evans Blue, 10% SITS and/or 3% Fast Blue). Following different survival periods (4-14 days) the animals were perfused either with 10% formalin or with the perfusate solution recommended by DeOlmos and Heimer (1980). After the tissue was cut (at 40 μ) and mounted on subbed slides, it was examined on a Nikon Fluorophot microscope, and the location of labeled cells was recorded with an X-Y plotter.

Our most successful experiment was Cat 201, in which the tracer combination was Fast Blue and Evans Blue, and in which the survival period was four days. Figure 1 illustrates the injection sites of Evans Blue (on the left, at levels 281, 301 and 321) and of Fast Blue

⟶

Fig. 1. A series of line drawings of frontal sections in Cat 201 showing the injection sites of Evans Blue on the left side, and Fast Blue on the right side at levels 281, 301 and 321, and also the location of doubly-labeled neurons at several levels, indicated by triangles.

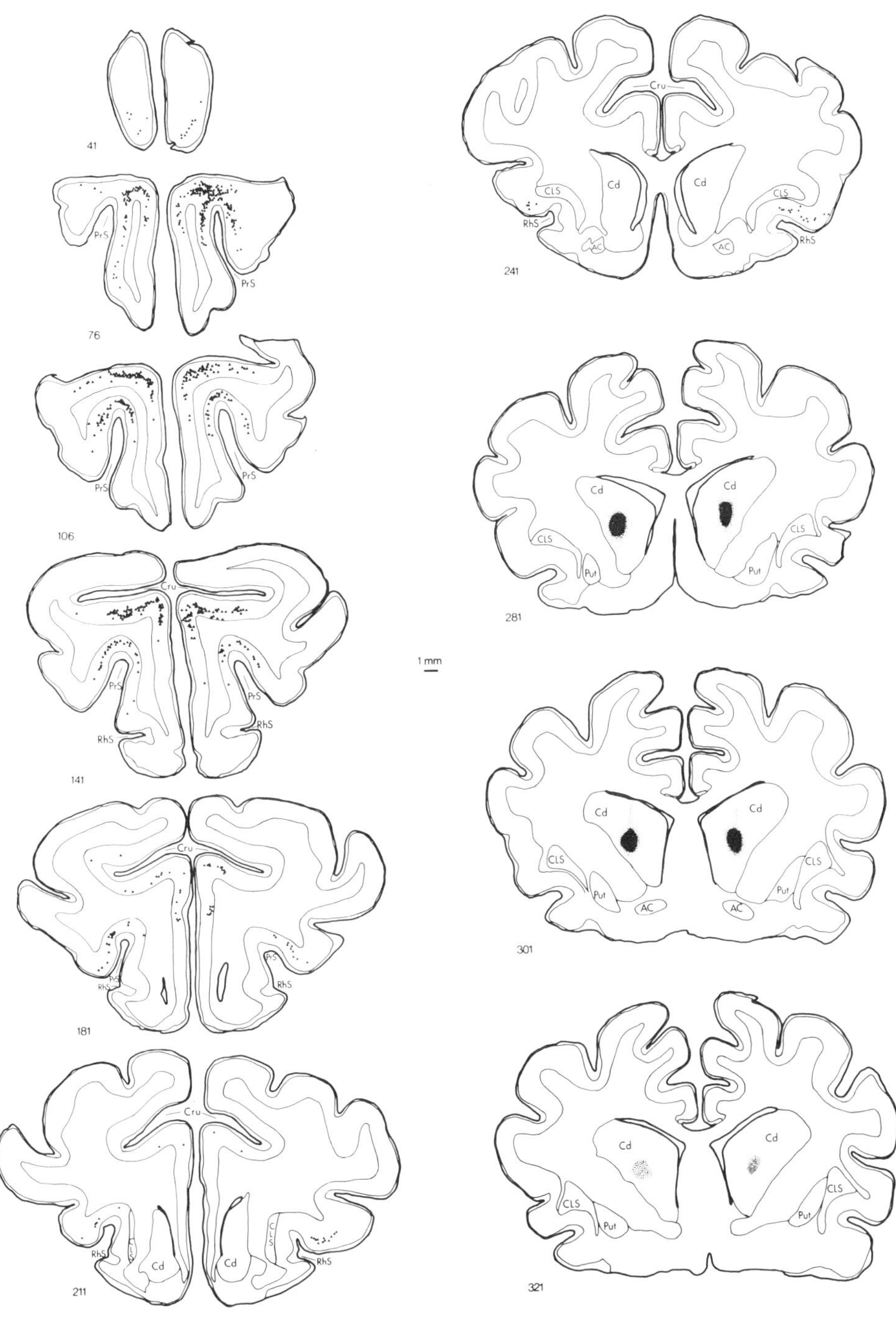
41
76
PrS
106
141
Cru
RhS
181
211
Cd
CLS
241
AC
281
Put
301
321
1 mm

(on the right, at the same levels), and the location of doubly labeled neurons within the cortex (indicated by triangles). As shown in Figure 1, doubly labeled cells were found in both hemispheres, and their topographical distributions were similar on each side.

In Figure 1, and in the reconstruction drawings of the doubly labeled neurons in the right hemisphere (Fig. 3) and left hemisphere (Fig. 4), it is apparent that these doubly labeled neurons were confined to a rather restricted cortical sector which, in the lateral cortex, included the depths of the presylvian sulci and the rostral gyrus proreus; and in the medial cortex included the precruciate region, parts of the anterior limbic area, and the gyrus proreus. These doubly labeled neurons were medium in size and were found in both the supragranular (layers II and III) and infragranular (layers V and VI) cortical layers, although the majority of such cells were located in layers II and V. Two examples of such doubly labeled neurons are shown in the photomicrographs in Figure 5, A and B.

In certain regions, particularly in the motor cortex, these doubly labeled neurons were quite abundant. For example, in a single section through areas, 4, 6 and the presylvian regions (Fig. 2C) by actual count there were 192 doubly labeled neurons (in both hemispheres). In comparison, a total of only 58 doubly labeled cortical neurons were found in the entire cortex of a representative experiment in a different fluorescent study, mentioned previously (Royce, 1983c).

However, the number of doubly labeled neurons that projected bilaterally to both hemispheres was relatively small when compared to the total number of corticostriatal neurons which were present. For example, in the same single section where counts were made of doubly labeled neurons, there was a total of 1270 cells labeled with Fast Blue (Fig. 2A) and 7121 cells labeled with Evans Blue (Fig 2B). Therefore, the doubly labeled neurons only represented 15.1% of the total number of Fast Blue labeled cells, and 2.7% of the total number of Evans Blue labeled cells (in that particular section).

In the experiments in which SITS was used as a tracer, doubly labeled neurons were also found. Such cells were present in the same cortical lamina and in the same topographical regions as in the other experiments. Occasionally, the cells labeled with SITS were quite conspicuous, as shown in Figure 5C and therefore the identification of doubly labeled neurons could be made with certainty (compare Figs. 5C and D). However, the fluorescence luminosity of most SITS labeled cells was quite low, and subjective judgement calls were often necessary to determine which cells were doubly labeled and which were not. Therefore, our results with the use of SITS as a tracer have been rather disappointing.

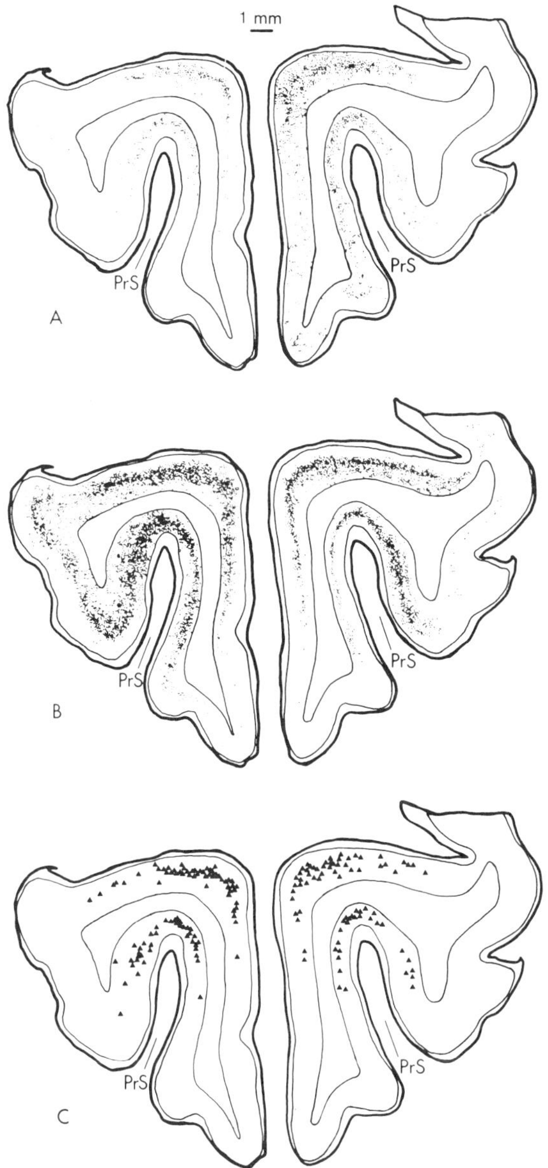

Fig. 2 This figure illustrates level 106 of Cat 201 where all labeled Fast Blue cells are shown in A, all labeled Evans Blue cells are shown in B, and all doubly-labeled cells are shown in C. Doubly-labeled cells at level 106 are also shown in Fig. 1.

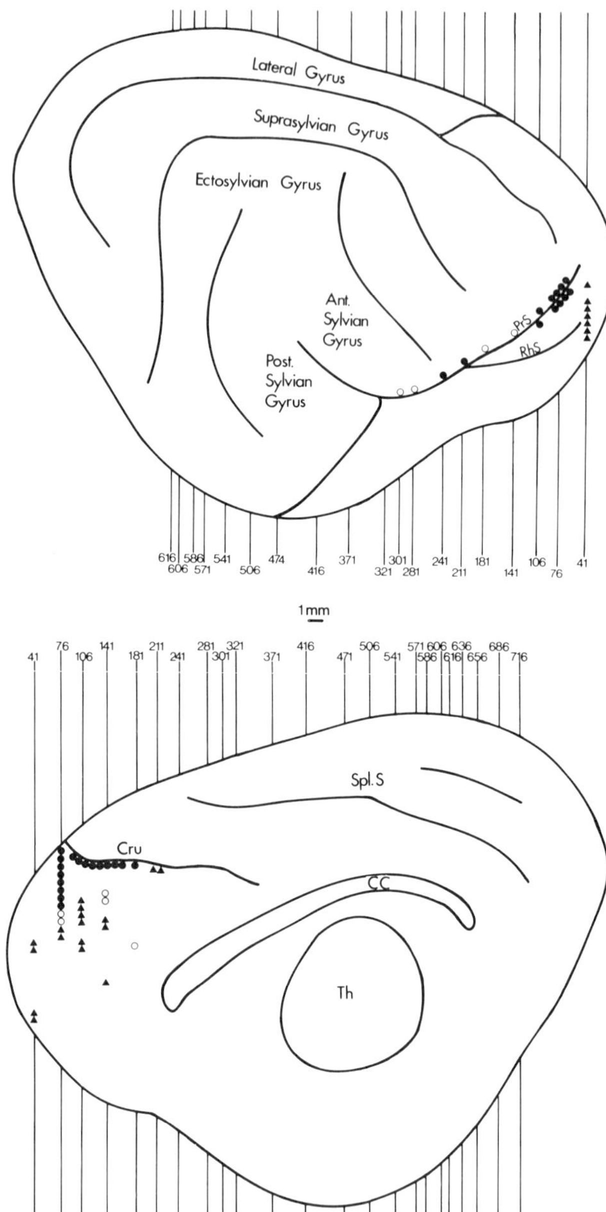

Fig. 3 Reconstructions of medial and lateral aspects of <u>right</u> hemisphere in Cat 201. Figs. 3 and 4 show location of doubly-labeled cortical neurons (only). Triangle: one labeled cell; open circle: 2-9 labeled cells; solid circle: 10 labeled cells. Triangles and circles contiguous with sulci indicate that labeled cells lie within sulci.

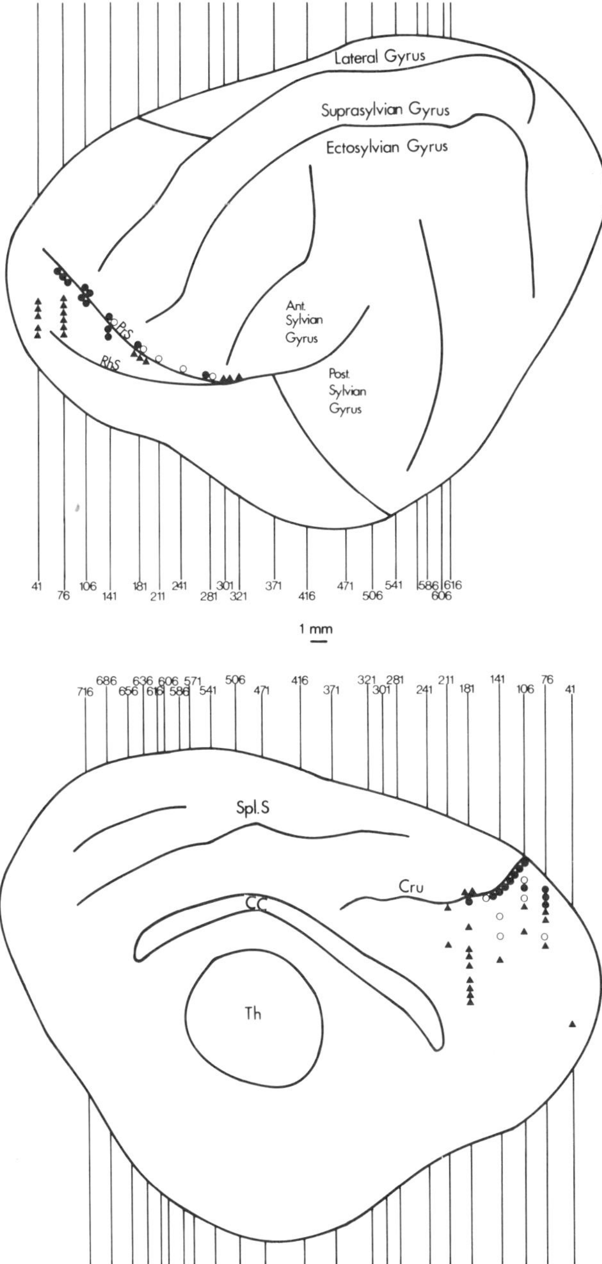

Fig. 4 Reconstruction drawings of double-labeled neurons in the medial and lateral aspects of the <u>left</u> hemisphere of Cat 201.

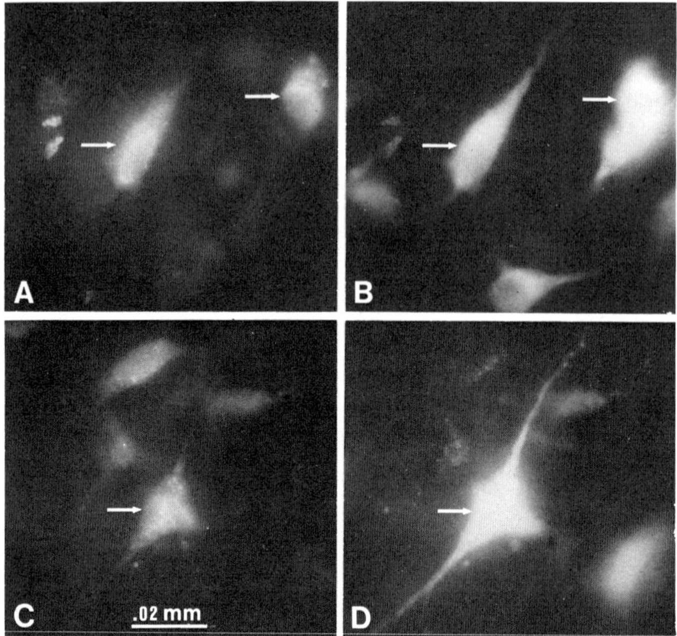

Fig. 5 A and B show corresponding fields in area 6 of Cat 201 where doubly-labeled neurons are present (at arrows). Fast Blue and Evans Blue are fluorescing in A and B, respectively. C and D are corresponding fields in a similar experiment, but in which SITS was used as a tracer (C) in combination with Evans Blue (D). Arrows point to the doubly-labeled neuron. The scale in C applies to the entire figure.

In the first article entirely devoted to the bilateral nature of the corticostriatal system, Carman et al. (1965) suggested that this arrangement may function to coordinate motor activity, since the cortex also projects bilaterally to other structures which are also known to have motor functions, such as the spinal cord, reticular formation and pontine nuclei. In support of this concept, they cited the work of Poggio et al (1957), who obtained bilateral responses in the putamen after the production of epileptic foci in the ipsilateral cortex of monkeys.

More recently, there have been numerous biochemical, pharmacological, and electrophysiological studies which have indicated that the activities of the two striata are interhemispherically coordinated (Conner et al., 1967; Chandu-Lall et al., 1970; Hull et al., 1974; Levine et al., 1974; Levine et al., 1977; Nieoullon et al., 1977; Leviel et al., 1979; Hahn et al., 1979; Anderson et al., 1980; Besson et al., 1982; Chesselet et al., 1983). Many of these studies seem to implicate either the nigrostriatal or striatonigral systems in these interhemispheric interactions. Such interactions could conceivably be mediated by the crossed nigrostriatal pathway, which has been shown to exist in both the cat (Royce, 1978a) and rat (Fass and Butcher, 1981), and which consists at least partially of collateralizing axons of nigrostriatal neurons which project bilaterally (Pritzel et al., 1983). Also, our recent studies (with the horseradish peroxidase method), have shown that all subdivisions of the substantia nigra project both contralaterally as well as ipsilaterally, including both the pars compacta and the pars reticularis (Royce and Laine, in Press).

However, the bilateral nature of the nigrostriatal pathway seems inadequate to explain many aspects of striatal interhemispheric coordination. The ventrolateral-ventral anterior (VA-VL) and centromedian-parafascicular (CM-Pf) thalamic complexes have been implicated in the bilateral striatal interactions by some investigators (Levine et al., 1977). Also, the globus pallidus is apparently involved in intrastriatal coordination (Hahn et al., 1979). While our recent findings have shown that both the VA-VL and CM-Pf nuclei project ipsilaterally to the striatum (Royce, 1978a,b, 1983b), we have no evidence to suggest that either complex projects to the contralateral striatum. On the other hand, recent studies with the horseradish peroxidase method in our laboratory have shown that the CM-Pf complex receives a prominent bilateral input from most of its ascending afferent sources, including those arising from the cerebellum and all portions of the reticular formation (Royce, submitted for publication).

Some studies have suggested that interhemispheric coupling occurs by direct connections between the striatal masses. For example, Mensah and Deadwyler (1974) used the degeneration method to show direct connections between the two caudate nuclei in the cat. Also, electrophysiological evidence has been advanced to support this concept (Medina and Pazo, 1981; Paxo and Medina, 1982). However, we have never found labeled neurons in the contralateral caudate nucleus after ipsilateral horseradish peroxodase injections; nor contralateral autoradiographic label following tritiated tracer injections into the ipsilateral caudate nucleus (Royce, 1978a; Royce and Laine, in Press). Therefore, we are doubtful of the existence of direct interstriatal connections.

In the physiological studies mentioned above (Medina and Pazo, 1981; Pazo and Medina, 1982) field potential data was used to demonstrate caudate-caudate connections. We suggest that such findings could have been mediated through the corticostriatal neurons which our findings have shown project bilaterally, i.e. by antidromic conduction through one collateral and orthodromic conduction through the other collateral to the opposite hemisphere.

Despite the fact that recent findings have suggested several additional routes for interstriatal interactions, it is our contention that the primary route for such interactions is through the bilateral corticostriatal connections. This seems especially probable in view of the considerable body of data (much of which appears in other articles in this volume), which suggests that the eventual target of most activities of the basal ganglia and associated structures is the cerebral cortex. Our findings that a population of corticostriatal neurons exists whose axons branch to both the ipsilateral and contralateral caudate nuclei add further emphasis to the importance of this system.

Finally, a brief consideration should be given to the topographical location of the doubly labeled corticostriatal neurons in our study. Such neurons might reasonably be expected to be found in regions where there is the greatest need for fine control of motor activity, if indeed they do function in this capacity. Their location appears to be consistent with this concept, since they are located in the anterior pericruciate cortex (areas 4γ and $6\alpha\beta$ of Hassler and Muhs-Clement, 1964), the lateral part of which constitutes a portion of Motor I to the extremities of the cat (Woolsey, 1958). Also, these doubly labeled neurons are located in regions which may correspond to the frontal eyefields of other species. For example, conjugate eye movements have been elicited in area $6\alpha\beta$, in the anterior limbic area, and within the presylvian sulcus (Schlag and Schlag-Rey, 1970; Guitton and Mandl, 1978; Pascuzzo and Skeen, 1982), all of which areas contain doubly labeled neurons in our experiments.

ACKNOWLEDGEMENTS

We wish to acknowledge Donna Washa for typing the manuscript. This work was supported by a grant from the National Institutes of Health, NS13453.

ABBREVIATIONS IN THE FIGURES

AC	anterior commissure
CC	corpus callosum
Cd	caudate nucleus
CLS	claustrum
Cru	cruciate sulcus
PrS	presylvian sulcus
Put	putamen
RhS	rhinal sulcus
Spl.S	splenial sulcus
Th	thalamus

REFERENCES

Andersson, K., Schwarcz, R., and Fuxe, K., 1980, Compensatory bilateral changes in dopamine turnover after striatal kainate lesion, Nature, 283:94.

Besson, M. J., Kemel, M. L., Gauchy, C., and Glowinski, J., 1982, Bilateral asymmetrical changes in the nigral release of (^{3}H) GABA induced by unilateral application of acetylcholine in the cat caudate nucleys, Brain Res., 241:241.

Buchwald, N. A., Price, D. D., Vernon, L., and Hull, C. D., 1973, Caudate intracellular response to thalamic and cortical inputs, Exp. Neurol., 38:311.

Carman, J. B., Cowan, W. M., Powell, J. P. S., and Webster, K. E., 1965, A bilateral cortico-striate projection, J. Neurol. Neurosurg. Psychiat., 28:71.

Chandu-Lall, J. A., Haase, G. R., Zivanovic, and Szekeley, E. G., 1970, Dopamine interdependence between the caudate nuclei, Exp. Neuron., 29:101.

Chesselet, M. F., Chéramy, A., Reisine, T. D., Lubetzki, C., Desban, M., and Glowinski, J., 1983, Local and distal effects induced by unilateral striatal application of opiates in the absence or in the presence of naloxone on the release of dopamine in both caudate nuclei and substantiae nigrae of the cat, Brain Res., 258:229.

Conner, J. D., Rossi, G., and Baker, W., 1967, Antagonism of intracaudate carbachol tremor by local injections of catecholamines, J. Pharm. Exp. Therap., 155:545.

DeOlmos, J., and Heimer, L., 1980, Double and triple labeling of neurons with fluorescent substances; the study of collateral pathways in the ascending raphe system, Neuroscience Letts., 19:7.

Donoghue, J. P., and Kitai, S. T., 1981, A collateral pathway to the neostriatum from corticofugal neurons of the rat sensory-motor cortex: an intracellular HRP study, J. Comp. Neurol., 201:1.

Endo, K., Araki, T., and Yagi, N., 1973, The distribution and pattern of axon branching of pyramidal tract cells, Brain Res., 57:484.

Fass, B. and Butcher, L. L., (1981), Evidence for a crossed nigrostriatal pathway in rats, Neurosci. Lett., 22:109.

Glees, P., 1944, The anatomical basis of cortico-striate connections, J. Anat., 78:47.

Guitton, D., and Mandl, G., 1978, Frontal "oculomotor" area in alert cat. I. Eye movements and neck activity evoked by stimulation, Brain Res., 149:295.

Hahn, Z., Lénárd, L., and Karádi, 1979, Contralateral decrease of neostriatal dopamine concentration after unilateral pallidal lesion in the rat, Neurosci. Lett., 13:91.

Hassler, R., and Muhs-Clement, K., 1964, Architektonischer Aufbau des sensormotorischen und parietalen Cortex der Katze, J. Hirnforsch., 6:377.

Hull, C. D., Levine, M. S., Buchwald, N. A., Heller, A., and Browning, R. A., 1974, The spontaneous firing pattern of forebrain neurons. I. The effects of dopamine depleting lesions on caudate unit firing patterns, Brain Res., 73:241.

Jinnai, K., and Matsuda, Y., 1979, Neurons of the motor cortex projecting commonly on the caudate nucleus and the lower brain stem in the cat, Neurosci. Lett., 13: 121.

Jones, E. G., and Leavitt, R. Y., 1974, Retrograde axonal transport and the demonstration of non-specific projections to the cerebral cortex and striatum from thalamic intralaminar nuclei in the rat, cat and monkey, J. Comp. Neurol., 154:349.

Kunze, W., McKenzie, J. S., and Bendrups, A. P., 1979, An electrophysiological study of thalamo-caudate neurones in the cat, Exp. brain Res., 36:233.

Künzle, H., 1975, Bilateral projections from precentral motor cortex to the putamen and other parts of the basal ganglia. An autoradiographic study in Macaca fascicularis, Brain Res., 88:195.

Kuypers, H. G. J. M., Catsman-Benevoets, C. E., and Padt, R. E., 1977, Retrograde axonal transport of fluorescent substances in the rat's forebrain, Neurosci. Lett., 6:127.

Leviel, V., Chéramy, A., Nieoullon, A., and Glowinski, J., 1979, Symmetric bilateral changes in dopamine release from the caudate nuclei of the cat induced by unilateral nigral application of glycine and GABA-related compounds, Brain Res., 175:259.

Levine, M. S., Hull, C. D., Buchwald, N. A., Garcia-Rill, E., Heller, A., and Erinoff, L., 1977, The spontaneous firing patterns of forebrain neurons. III. Prevention of induced asymmetries in caudate neuronal firing rates by unilateral thalamic lesions, Brain Res., 131:215.

Levine, M. S., Hull, C. D., Buchwald, N. A., and Villablanca, J. R., 1974, The spontaneous firing patterns of forebrain neurons. II. Effects of unilateral caudate ablations, Brain Res., 78:411.

Martin, G. F., and Hamel, E. G., 1967, The striatum of the opossum, Didelphis virginiana. Description and experimental studies, J. Comp. Neurol., 131: 491.

Medina, J. H., and Pazo, J. H., 1981, Electrophysiological evidence for the existence of caudate-caudate connections, Int. J. Neurosci., 15:99.

Mehler, W. R., 1966, Further notes on the center median nucleus of Luys, in: "The Thalamus", D. P. Purpura and M. D. Yahr, eds., Columbia Univ. Press, New York.

Mensah, P., and Deadwyler, S., 1974, The caudate nucleus of the rat: cell types and the demonstration of a commissural system, J. Anat., 117:281.

Nauta, W. J. H., and Whitlock, D. G., 1954, An anatomical analysis of the non-specific thalamic projection system, in: "Brain Mechanisms and Consciousness," J. F. Delafresnaye, ed., Blackwell, Oxford.

Nieoullon, A., Chéramy, and Glowinski, J., 1977, Interdependence of the nigrostriatal dopaminergic systems on the two sides of the brain in the cat, Science, 198:416.

Oka, H., and Jannai, H., 1978, Common projection of the motor cortex to the caudate nucleus and the cerebellum, Exp. Brain Res., 31:31.

Pascuzzo, G. J., and Skeen, L. C., 1982, Brainstem projections to the frontal eye field in cat, Brain Res., 241:341.

Pazo, J. H. and Medina, J. H., 1982, Somatotopic organization of caudate-caudate relationships in the cat, Int. J. Neurosci., 16:143.

Poggio, G. F., Ribstein, M., Udvarhelyi, G. B., and Walker, A. E., 1957, Subcortical responses to focal epileptic discharges in the monkey motor cortex, Electroenceph. Clin. Neurophysiol., 9:164.

Powell, T. P. S., 1952, Residual neurons in the human thalamus following hemidecortication, Brain, 75:571.

Powell, T. P. S., and Cowan, W. M., 1956, A study of thalamo-striate relations in the monkey, Brain, 79: 364.

Powell, T. P. S., and Cowan, W. M., 1967, The interpretation of the degenerative changes in the intralaminar nuclei of the thalamus, J. Neurol. Neurosurg. Psychiat., 30:140.

Pritzel, M., Sarter, M., Morgan, S. and Houston, J. P., (1983), Interhemispheric nigrostriatal projections in the rat: bifurcating nigral projections and loci of crossing in the diencephalon, Brain Res. Bull. 10:385.

Ragsdale, C. W., and Graybiel, A. M., 1981, The fronto-striatal projection in the cat and monkey and its relationship to inhomogeneities established by acetylcholinesterase histochemistry, Brain Res., 208:259.

Ralston, H. J., and Sharp, P. V., 1973, The identification of thalamocortical relay cells in the adult cat by means of retrograde axonal transport of horseradish peroxidase, Brain Res., 62:273.

Ramon y Cajal, S., 1909, Histologie du Système Nerveux de L'Homme et Des Vertebres. Vol. II. As published in 1955, Instituto Ramón y Cajal.

Rose, J. E., and Woolsey, C. N., 1943, A study of thalamo-cortical relations in the rabbit, Bull. Johns Hopkins Hosp., 73:65.

Rose, J. E., and Woolsey, C. N., 1949, Organization of the mammalian thalamus and its relationships to the cerebral cortex, E.E.G. and Clin. Neurophysiol., 1:391.

Royce, G. J., 1978a, Cells of origin of subcortical afferents to the caudate nucleus: a horseradish peroxidase study in the cat, Brain Res., 153:465.

Royce, G. J., 1978b, Autoradiographic evidence for a discontinuous projection to the caudate nucleus from the centromedian nucleus in the cat, Brain Res., 146:145.

Royce, G. J., 1982, Laminar origin of cortical neurons which project upon the caudate nucleus: a horseradish peroxidase investigation in the cat, J. Comp. Neurol., 205:8.

Royce, G. J., 1983a, Cells of origin of corticothalamic projections upon the centromedian and parafascicular nuclei in the cat, Brain Res., 258:11.

Royce, G. J., 1983b, Single thalamic neurons which project to both the rostral cortex and caudate nucleus studied with the fluorescent double labeling method, Exp. Neurol., 79:773.

Royce, G. J., 1983c, Cortical neurons with collateral projections to both the caudate nucleus and the centromedian-parafascicular thalamic complex: a fluorescent retrograde double labeling study in the cat, Exp. Brain Res., 50:157.

Royce, G. J., Subcortical projections to the centromedian and parafascicular thalamic nuclei, as studied in the cat with the horseradish peroxidase method, Submitted to J. Comp. Neurol.

Royce, G. J., and Laine, E. J., Efferent connections of the caudate nucleus, including cortical projections of the striatum and other basal ganglia: an autoradiographic and horseradish peroxidase investigation in the cat, In Press, J. Comp. Neurol.

Schlag, J., and Schlag-Rey, M., 1970, Induction of oculomotor responses by electrical stimulation of the prefrontal cortex in the cat, Brain Res., 22:1.

Schmeud, L. C., and Swanson, C. W., 1982, SITS: a covalently bound fluorescent retrograde tracer that does not appear to be taken up by fibers-of-passage, Brain Res., 249:137.

Steriade, M., and Glenn, L. L., 1982, Neocortical and caudate projections of intralaminar thalamic neurons and their synaptic excitation from midbrain reticular core, J. Neurophysiol., 48:352.

Tanaka, D., Gorska, T., and Dutkiewicz, 1981, Corticostriate projections from the primary motor cortex, Brain Res., 209:287.

Woolsey, C. N., 1958, Organization of somatic sensory and motor areas of the cerebral cortex, in: "Biological and Biochemical Bases of Behaviour", H. E. Harlow and C. N. Wollsey, eds., Univ. of Wisconsin press, Wisconsin.

Webster, K. E., 1965, The cortico-striatal projection in the cat, J. Anat. (London), 99:329.

Yeterian, E. H., and Van Hoesen, G. W., 1978, Cortico-striate projections in the rhesus monkey: the organization of certain cortico-caudate connections, Brain Res., 139:43.

THE OUTPUT ORGANIZATION OF THE PALLIDUM AND SUBSTANTIA NIGRA IN PRIMATE AS REVEALED BY A RETROGRADE DOUBLE-LABELING METHOD

A. Parent, Y. Smith and L. Bellefeuille

Laboratoire de Neurobiologie et Département d'Anatomie
Faculté de Médecine
Université Laval
Québec, Qué., Canada G1K 7P4

INTRODUCTION

The present paper summarizes the findings of our most recent series of neuroanatomical investigations whose aim was to define the exact cellular origin and to evaluate the degree of collateralization of the efferent projections of the two major output structures of the basal ganglia in primates: the pallidum and the substantia nigra (Parent and De Bellefeuille, 1982, 1983; Parent et al., 1983a, b). For such a purpose we have chosen to use the fluorescence retrograde double-labeling technique recently introduced by Kuypers and Van der Kooy (see Van der Kooy et al., 1978). This procedure, which is based on the retrograde transport of two or more substances that fluoresce maximally at different wave-lengths, is particularly suitable for the study of neuronal systems having multiple axonal processes terminating in different brain areas. All our studies were performed in the squirrel monkey (Saimiri scieureus) and we have used three different combinations of fluorescent tracers: (1) Evans blue (EB) and DAPI-primuline (DP), (2) Fast blue (FB) and Nuclear yellow (NY), and (3) True blue (TB) and Nuclear yellow (NY). The stereotaxic coordinates for the injection of the fluorescent tracers were chosen according to the atlas of Emmers and Akert (1963). For technical details regarding the number of animals, the injection sites, the survival times and the fluorescent tracer combinations used, the reader is referred to Table 1.

Table I. Number of animals, injection sites and survival times for the double-labeling experiments with Evans blue (EB), DAPI-primuline (DP), Fast blue (FB), True blue (TB) and Nuclear yellow (NY).

Number of Animals	Injection sites (tracer, survival times)		
7	VA/VL (EB, 4-6 days)	+	HB (DP, 4-6 days)*
1	VA/VL (DP, 7 days)	+	HB (EB, 7 days)
9	VA/VL (EB, 6 days)	+	TPP (DP, 6 days)
1	VA/VL (NY, 18 hrs)	+	TPP (FB, 2 days)
1	VA/VL (NY, 18 hrs)	+	TPP (TB, 6 days)
4	VA/VL (EB, 6 days)	+	CM/Pf (DP, 6 days)
8	CM/Pf (EB, 6 days)	+	TPP (DP, 6 days)
9	SN (DP, 6 days)**		
3	VA/VL (EB, 6 days)	+	SC (DP, 6 days)
1	VA/VL (FB, 2 days)	+	SC (NY, 18 hrs)
1	VA/VL (TB, 6 days)	+	SC (NY, 18 hrs)
1	SC (NY), 18 hrs)	+	TPP (TB, 6 days)
2	PUT (EB, 6 days)	+	CD (DP, 6 days)
1	PUT (DP, 6 days)	+	CD (EB, 6 days)
1	PUT (FB, 3 days)	+	CD (NY, 12 hrs)
1	CD (FB, 2 days)	+	PUT (NY, 12 hrs)
1	CD (TB, 3 days)	+	PUT (NY, 1 day)

* In all double-labeling experiments, the 2 injections were made on the same side of the brain

**Unilateral injections

RESULTS

The Pallidum

Since the pioneering study of Nauta and Mehler (1966) it is well established that the internal segment of the globus pallidus (GPi) in primates (or its equivalent, the entopenduncular nucleus, in non-primates) projects to: (1) the ventral anterior (VA) and ventral lateral (VL) thalamic nuclei, (2) the lateral nucleus of the habenula (HB), (3) the Centre médian/Parafascicular complex (CM/Pf), and (4) the pedunculopontine nucleus in the midbrain tegmentum (TPP: nucleus tegmenti pedunculopontinus, subnucleus compactus) (see review by Carpenter, 1981, and his contribution in the present volume). Furthermore, besides its well-known reciprocal connection with the subthalamic nucleus (which will not be dealt with here), the globus pallidus has also been shown to project directly to the substantia nigra in the rat (Hattori et al., 1975; Bunney and Aghajanian, 1976; Kanazawa et al., 1976; Carter and Fibiger, 1978) and in the cat (Grofova, 1975; McBride and Larsen, 1980). Therefore, in order to study these various efferent projections of the pallidum in primate, fluorescent tracer injections in the different pallidal target structures were made according to the following combinations: (1) thalamus-habenula, (2) thalamus-midbrain, (3) thalamus-intralaminar nuclei, (4) intralaminar nuclei - midbrain, and (5) substantia nigra alone.

(1) Thalamus (VA/VL) and habenula (HB) injections. In this first group of squirrel monkeys the injection of one fluorescent tracer (EB or DP) was made in the VA/VL thalamic nuclei whereas the complementary tracer was delivered in the habenula on the same side of the brain. After these injections a multitude of neurons retrogradely-labeled with the tracer injected in the thalamus are found in the central two-thirds of the ipsilateral internal pallidum (Fig. 1A). In contrast, the pallidal neurons containing the tracer injected in the habenula are few in number and most of them lie along the peripheral border of the internal pallidum (Fig. 1A). Along the ventromedial aspect of the pallidum, these pallidohabenular neurons merge with a large population of similarly labeled cells present in the lateral hypothalamus, which thus stands out as one of the major forebrain sources of afferents to the habenula in the monkey. Very few double-labeled neurons can be found at pallidal levels after thalamus-habenula injections. These findings indicate that the pallidothalamic and pallidohabenular projections in primates arise largely from two different neuronal populations each having a preferential distribution in the internal pallidum.

(2) Thalamus (VA/VL) and midbrain (TPP) injections. Strikingly different results are obtained in monkeys in which one fluorescent tracer (EB or NY) is injected in VA/VL nuclei and the complementary tracer (DP, FB or TB) is delivered in the area of the pedunculopontine nucleus (TPP) in the midbrain tegmentum. In such cases a multitude of double-labeled cells can be seen in the internal pallidum (Fig. 1B). It is estimated that the number of double-labeled cells observed after thalamus-midbrain injection amounts approximately to 70% of all the labeled pallidal neurons. A significant number of neurons projecting to thalamus and to midbrain are also present in the contralateral internal pallidum. These results reveal that the pallidothalamic and pallidotegmental pathways in primates arise largely from the same neurons in the core of the internal pallidum, and that these projections are partly crossed.

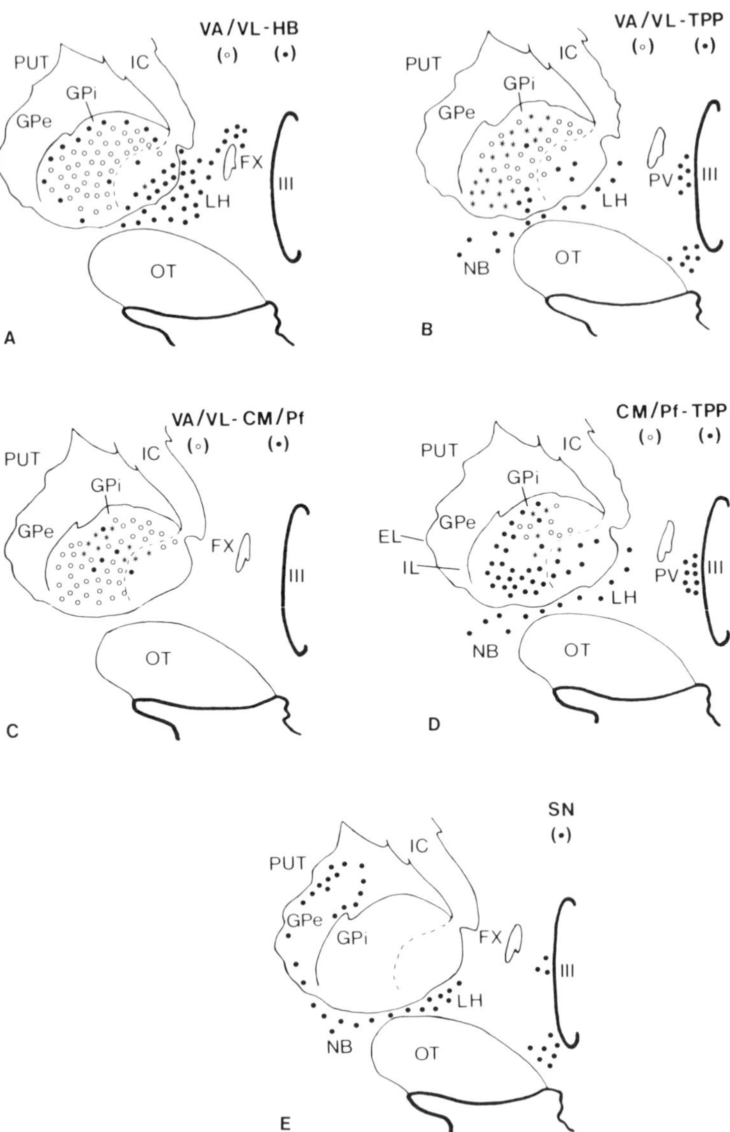

Fig. 1. Semi-schematic drawings of a transverse section through the globus pallidus and adjacent areas in the squirrel monkey to illustrate the distribution of retrogradely labeled neurons in a typical case of each group in the first experimental series (A-E). The double-labeled cells are represented by asterisks.

(3) Thalamus (VA/VL) and intralaminar nuclei (CM/Pf) injections. In a third group of squirrel monkeys one fluorescent tracer (EB) was injected again in VA/VL nuclei and the complementary tracer (DP) delivered in CM/Pf complex. After VA/VL-CM/Pf injections numerous neurons containing the tracer injected in VA/VL are found within the central portion of the internal pallidum compared to a much smaller number of cells labeled with the tracer injected in CM/Pf complex. We estimate that the number of pallidointralaminar neurons amounts to approximately 30-35% that of the pallidothalamic neurons. The pallidointralaminar neurons are mostly concentrated in two somewhat continuous cell clusters: one located dorsolaterally in the internal pallidum, the other lying ventromedially along the accessory medullary lamina (Fig. 1C). As much as 80% of the neurons forming these clusters are double-labeled after VA/VL-CM/Pf injection. These results suggest that the pallidointralaminar and pallidothalamic pathways arise largely from the same neurons clustered in the dorsolateral and central portion of the internal pallidum. However, the pallidointralaminar pathway appears to be much less prominent than the pallidothalamic pathway.

(4) Intralaminar nuclei (CM/Pf) and midbrain (TPP) injections. The distribution of pallidal cells projecting to the CM/Pf complex and to the pedunculopontine nucleus disclosed in this experimental group is overall similar to that reported above for groups 2 and 3, respectively. The pallidointralaminar neurons form two small clusters located centrally in the internal pallidum, whereas the pallidotegmental neurons abound in the ventromedial half of this structure (Fig 1D). However, at variance with the results obtained after thalamus-intralaminar nuclei injection, only approximately 15-20% of the neurons forming the two clusters are double-labeled after intralaminar nuclei-midbrain injections. These findings reveal that the pallidointralaminar projection is not composed, to a significant extent, of axonal branches of the more prominent pallidotegmental pathway.

(5) Substantia nigra (SN) injections. In the monkeys of this experimental group the DP injections were rather massive involving both the pars reticulata (SNr) and the pars compacta (SNc) of the substantia nigra, but sparing the subthalamic nucleus. After such injections a moderate number of labeled cells occur in the external segment of the globus pallidus (GPe) whereas no retrograde cell labeling is observed in the internal pallidum itself (Fig. 1E). The internal pallidum is nevertheless closely surrounded ventromedially by numerous retrogradely-labeled cells belonging to the magnocellular component (nucleus basalis of Meynert) of the substantia innominata (Fig. 1E). These results suggest that a distinct but rather discrete pallidonigral projection exists in the monkey and that it arises from the external pallidum and not from the internal pallidum.

The Substantia Nigra

Besides its well-established projection to the striatum arising from the dopaminergic neurons of the pars compacta (see Anden et al., 1964; Poirier and Sourkes, 1964; Faull and Mehler, 1978; Beckstead et al., 1979; Bentivoglio et al., 1979; Szabo, 1980a, b), the substantia nigra is also known to send a direct projection via the pars reticulata neurons to: (1) the thalamus (VA/VL nuclei and the paralamellar part of the mediodorsal nucleus), (2) the tectum (middle grey layer of the superior colliculus), and (3) the midbrain tegmentum (pedunculopontine nucleus) see Rinvik et al., 1976; Anderson and Yoshida, 1977; Deniau et al., 1978; Bentivoglio et al., 1979; Beckstead et al., 1981, 1982; Carpenter, 1981). Therefore, in order to study the exact cellular origin and the degree of collateralization of these nigral projections, fluorescent tracer injections were made in the various nigral recipient structures according to the following combinations: (1) thalamus - midbrain tegmentum, (2) thalamus - superior colliculus, (3) superior colliculus - midbrain tegmentum, and (4) caudate nucleus - putamen (see Table 1).

(1) Thalamus (VA/VL) and midbrain (TPP) injections. After thalamus-midbrain injections, cells containing the tracer (EB or NY) injected in VA/VL nuclei (nigrothalamic neurons) and others labeled with the tracer (DP, FB or TB) delivered in the pedunculopontine nucleus (nigrotegmental neurons) are found in large numbers along the entire rostrocaudal and mediolateral extents of the substantia nigra pars reticulata (SNr), ipsilateral to the injection sites. Overall, however, the nigrothalamic neurons tend to predominate in the rostolateral portion of SNr whereas the nigrotegmental neurons abound particularly in the caudomedial aspect of SNr. In the middle third of SNr the nigrothalamic neurons form 3 to 4 characteristic cell clusters of different sizes along the dorsal border of the cerebral peduncle (Fig. 2A). It is estimated that the nigrothalamic and nigrotegmental neurons occur in about equal numbers in SNr, and that as many as 60% of all these neurons are double-labeled after thalamo-tegmental injections. These data reveal that a large proportion of SNr neurons in primates send axonal branches to both VA/VL nuclei and midbrain tegmentum.

(2) Thalamus (VA/VL) and superior colliculus (SC) injections. After thalamo-collicular injections numerous nigrothalamic cells occur in SNr and their distribution is similar to that described above for the animals of the first experimental group. A moderate number of nigrocollicular neurons is also found particularly in the middle third of SNr. These neurons tend to predominate in the ventrolateral portion of SNr where a significant number of double-labeled cells occur (Fig. 2B). It is estimated that of all the labeled cells present within SNr, 60% contain the tracer injected in the thalamus versus 40% labeled with the tracer delivered in the

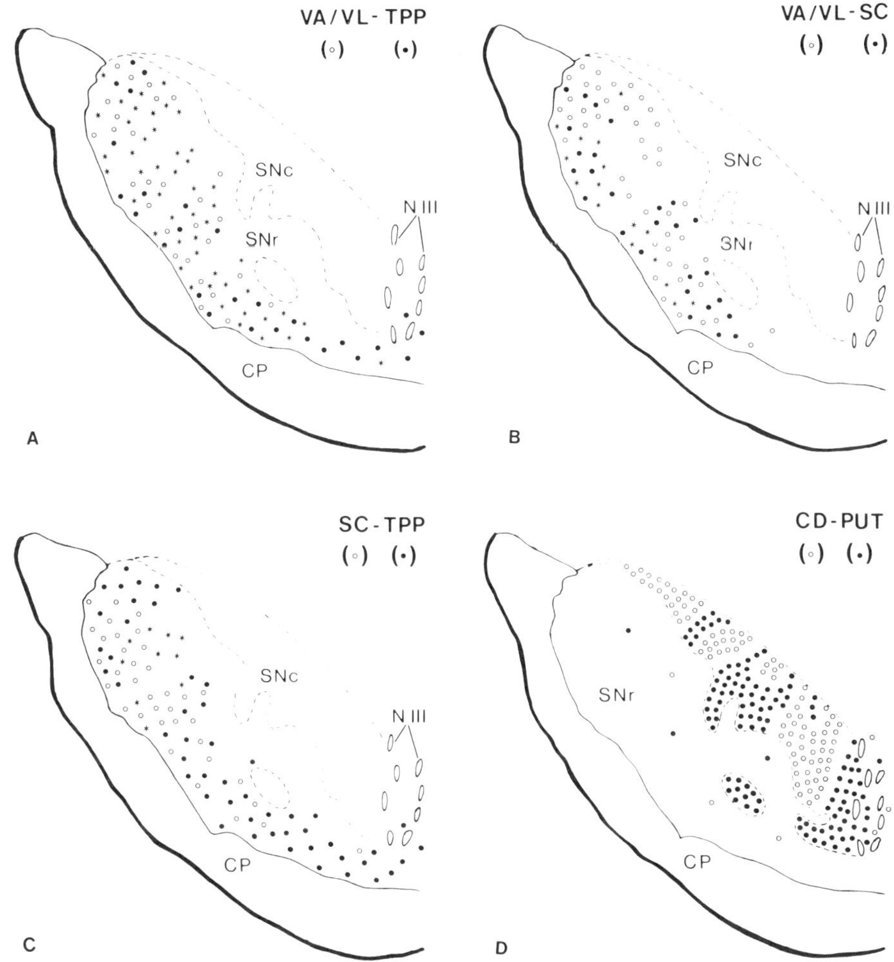

Fig. 2. Semi-schematic drawings of a transverse section through the substantia nigra in the squirrel monkey to illustrate the distribution of retrogradely labeled neurons in a typical case of each group in the second experimental series. The double-labeled cells are represented by asterisks.

superior colliculus. Approximately 20-25% of all positive SNr neurons contain both fluorescent tracers, and this proportion climbs to 30-40% in sections taken through the middle third of SNr. Also noteworthy is the presence of nigrocollicular neurons in contralateral SNr where they number approximately 15% of the nigrocollicular neurons found ipsilaterally. Therefore these various findings reveal that a significant proportion of SNr neurons in the monkey send axonal branches to both thalamus and superior colliculus and that some SNr cells also project to the contralateral superior colliculus.

(3) Superior colliculus (SC) and midbrain (TPP) injections. The topographical distribution as well as the relative proportions of the nigrotegmental and nigrocollicular neurons found after superior colliculus-midbrain injection are similar to what has been seen in the monkeys of the two previous experimental groups. About 60% of all labeled cells contain the tracer injected in midbrain tegmentum in comparison to 40% of SNr neurons labeled with the tracer delivered in the superior colliculus. However, at variance with the results obtained after thalamus-midbrain and thalamus-superior colliculus injections, only about 10% of SNr cells are double-labeled after superior colliculus midbrain tegmentum injection (Fig. 2C). Hence, only a small proportion of SNr neurons appears to project to both superior colliculus and midbrain tegmentum in the primate.

(4) Caudate nucleus (CD) and putamen (PUT) injections. After the injection of one fluorescent tracer in the caudate nucleus and the delivery of the complementary tracer in the putamen on the same side of the brain, numerous cells retrogradely labeled with the tracer injected in CD (nigrocaudate cells) and a multitude of others containing the tracer delivered in PUT (nigroputamen cells) occur ipsilaterally at all rostocaudal levels of the substantia nigra pars compacta (SNc). The nigrocaudate cells tend to be more numerous rostrally and dorsally in SNc, whereas the nigroputamen neurons are slightly more abundant caudally and ventrally. However, by far the most characteristic feature of the pattern of distribution of striatal afferent neurons in SNc is that both nigrocaudate and nigroputamen cells form clusters of various sizes which are closely intermingled but never really overlap (Fig. 2D). These clusters are distributed according to a very complex pattern that varies along the rostrocaudal extent of SNc. Although clearly separated from one another in the transverse plane, each of the nigrocaudate and nigroputamen cell clusters appears to run in continuity from section to section when examined along the rostrocaudal axis. Thus, these clusters may in fact be part of a complex three-dimensional array of tubular neuronal aggregates. Surprisingly, very few double-labeled cells can be visualized after caudate-putamen injections. These rare double-labeled neurons occur most often at the junction between the SNc proper and the ventral tegmental area, which contains a few scattered nigrocaudate and nigroputamen cells. Also worth noting is the presence of a small number of isolated nigrocaudate and nigroputamen neurons in SNr ipsilaterally (Fig. 2D), and in SNc contralaterally.

CONCLUSION AND SUMMARY

This paper has summarized the results of our recent double-labeling investigation of the cellular origin and degree of

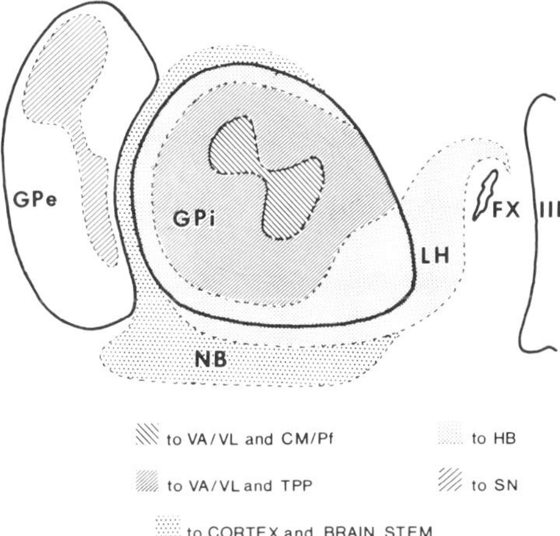

Fig. 3. Diagram summarizing the output organization of the primate globus pallidus as revealed after study of its main efferent projections by means of fluorescence retrograde double-labeling method. The various stippled and hatched areas represent the zones of origin of the major pallidal efferents.

collateralization of the efferent projections of the pallidum and substantia nigra in the squirrel monkey (Saimiri sciureus). Such a study has allowed us to delineate some hitherto unknown organizational features of these two major output structures of the basal ganglia in the primate. The models of the output organization of the pallidum and substantia nigra that we propose here take into account most of the aforementioned findings.

The pallidum (Fig. 3). In regards to its thalamic, habenular and midbrain efferents, the internal segment of the globus pallidus in the primate is organized according to a complex pattern which differs markedly from that disclosed in the rat entopenduncular nucleus (see Van der Kooy and Carter, 1981). This pattern can best be described as consisting of 3 concentric zones: (1) a large central 'motor' zone where most neurons send axonal branches to both thalamus and midbrain, (2) a small peripheral 'limbic' zone which encroaches largely upon the lateral hypothalamus and whose neurons project only to the habenula, and (3) a peripallidal 'reticular' zone composed of large cholinergic cells belonging to the nucleus basalis and projecting diffusely to the neocortex. The latter zone was not described here (for details see Parent and De Bellefeuille, 1982) but is of fundamental importance. This prominent cholinergic cortical input is known to undergo severe degeneration in parallel with cognitive function disorders encountered in cases of senile dementia of the Alzheimer type or in demented parkinsonism (see Coyle et al., 1983). In regard to the pallidal efferents to the Centre median-Parafascicular complex and to the substantia nigra, our findings suggest that the pallidointralaminar projection in the monkey is less prominent than the pallidothalamic (to VA/VL) and pallidotegmental pathways, and that it is largely composed of collaterals of pallidothalamic (but not pallidotegmental) axons whose parent cell bodies form two well-defined clusters in the large central 'motor' zone of the internal pallidum. Our data also reveal that a distinct pallidonigral projection exists in primates. This rather modest pallidal output appears to arise from neurons located in the external pallidum, and not from the internal pallidum.

The substantia nigra (Fig. 4). The neurons of the substantia nigra pars reticulata (SNr) display a high degree of collateralization. It is estimated that the largest number of SNr branching neurons (about 60%) are those projecting to thalamus and midbrain; that a moderate number (30-40%) send collaterals to thalamus and superior colliculus; whereas only about 10% of SNr neurons project to both superior colliculus and midbrain tegmentum. By comparison, the neurons of the pars compacta (SNc) display only a modest degree of axonal branching. For instance, very few SNc neurons appear to project to both caudate nucleus and putamen. Instead, numerous clusters of nigrocaudate and nigroputamen cells occur at all

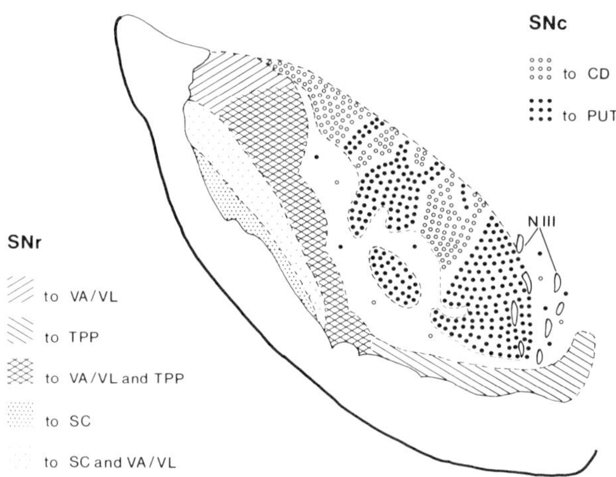

Fig. 4. Diagram summarizing the output organization of the primate substantia nigra as revealed after study of its main efferent projections by means of fluorescence retrograde double-labeling method. The various stippled and hatched areas represent the zones of origin of the major nigral efferents arising from both the pars compacta (SNc) and the pars reticulata (SNr).

rostrocaudal levels of SNc. These clusters are closely interlocked and distributed according to a complex mosaic-like pattern. Although clearly separated from one another in transverse plane, each of the nigrocaudate and nigroputamen cell clusters appear to run in continuity from section to section when examined along the rostro-caudal axis. This suggests that these clusters may in fact be part of a complex three-dimensional array of tubular neuronal aggregates. Thus, our findings reveal that in regard to their efferent projections the two major subdivisions of the substantia nigra in the primate are organized quite differently. Whereas SNc cells occur in the form of nigrocaudate and nigroputamen neuronal subunits that display a complex mosaic-like arrangement, SNr neurons clearly appear to be more multipotential, and not constrained to such a rigid topographical organization.

The high degree of axonal branching is a neuronal characteristic that the SNr elements obviously share with those of the internal pallidum. In fact, the output organization of the SNr and the internal pallidum is such that single neurons in both of these key structures of the basal ganglia have direct access to the same thalamic and brain stem nuclei. The functional significance of such a remarkable arrangement, which allows a large amount of redundancy to occur in the processing of 'motor' information within the extrapyramidal system, remains to be investigated.

ACKNOWLEDGEMENT

This research was made possible through a grant (MT-5781) from the Medical Research Council of Canada.

ABBREVIATIONS

CD : Caudate nucleus
CM/Pf : Centre médian - parafascicular complex
CP : Cerebral peduncle
DP : DAPI-Primuline
EB : Evans blue
EL : External medullary lamina
FB : Fast blue
FX : Fornix
GPe : Globus pallidus, external segment
GPi : Globus pallidus, internal segment
HB : Habenula
IC : Internal capsule
IL : Internal medullary lamina
LH : Lateral hypothalamus

NB : Nucleus basalis
NY : Nuclear yellow
NIII : Oculomotor nerve root fibers
OT : Optic tract
PUT : Putamen
SC : Superior colliculus
SN : Substantia nigra
SNr : Pars reticulata of SN
SNc : Pars compacta of SN
TB : True blue
TPP : Pedunculopontine nucleus
VA/VL : Ventral anterior - Ventral lateral thalamic nuclei
III : Third ventricle

REFERENCES

Anden, N. E., Carlsson, A., Dahlstrom, A., Fuxe, K., Hillarp, N. A., and Larsson, K., 1964, Demonstration and mapping out of nigroneostriatal dopamine neurons, Life Sci., 3:523.

Anderson, M. E., and Yoshida, M., 1977, Electrophysiological evidence for branching nigral projections to the thalamus and the superior colliculus, Brain Res., 137: 361.

Beckstead, R. M., Domesick, V. B., and Nauta, W. J. H., 1979, Efferent connections of the substantia nigra and ventral tegmental area in the rat, Brain Res., 175:191.

Beckstead, R. M., Edwards, S. B., and Frankfurter, A., 1981, A comparison of the intranigral distribution of nigrotectal neurons labeled with horseradish peroxidase in the monkey, cat and rat, J. Neurosci., 1:121.

Beckstead, R. M., and Frankfurter, A., 1982, The distribution and some morphological features of the substantia nigra neurons that project to the thalamus, superior colliculus and pedunculopontine nucleus in monkey, Neuroscience, 7:2377.

Bentivoglio, M., Van der Kooy, D., and Kuypers, H. G. J. M., 1979, The organization of the efferent projections of the substantia nigra in the rat. A retrograde fluorescent double labeling study, Brain Res., 174:1.

Bunney, B. S., and Aghajanian, G. K., 1976, The precise localization of nigral afferents in the rat as determined by a retrograde tracing technique, Brain Res., 117:423.

Carpenter, M. B., 1981, Anatomy of the corpus striatum, in: "APS Handbook of Physiology. The Nervous System," Vol. 2, Williams and Wilkins, Baltimore.

Carter, D. A., and Fibiger, H. C., 1978, The projections of the entopeduncular nucleus and globus pallidus in the rat as demonstrated by autoradiography and horseradish peroxidase histochemistry, J. Comp. Neurol., 177:113.

Coyle, J. T., Price, D. L., and DeLong, M. R., 1983, Alzheimer's disease: a disorder of cortical cholinergic innervation, Science, 219:1184.

Deniau, J. M., Hammond, C., Rizk, A., and Féger, J., 1978, Electrophysiological properties of identified output neurones of the rat substantia nigra (pars compacta and pars reticulata): evidence for the existence of branched neurons, Exp. Brain Res., 32:409.

Emmers, R., and Akert, K., 1963, "A Stereotaxic Atlas of the Brain of the Squirrel Monkey (Saimiri sciureus)," The University of Wisconsin Press, Madison.

Faull, R. L. M., and Mehler, W. R., 1978, The cells of origin of nigrotectal, nigrothalamic and nigrostriatal projections in the rat, Neuroscience, 3:989.

Grofova, I., 1975, The identification of striatal and pallidal neurons projecting to substantia nigra. An experimental study by means of retrograde axonal transport of horseradish peroxidase, Brain Res., 9:286.

Hattori, T., Fibiger, H. C., and McGeer, P. L., 1975, Demonstration of a pallidonigral projection innervating dopaminergic neurons, J. Comp. Neurol., 162:487.

Kanazawa, I., Marshall, G. R., and Kelly, J. S., 1976, Afferents to the rat substantia nigra studied with horseradish peroxidase, with special reference to fibers from the subthalamic nucleus, Brain Res., 115:485.

McBride, R. L., and Larsen, K. D., 1980, Projections of the feline globus pallidus, Brain Res., 189:3.

Nauta, W. J. H., and Mehler, W. R., 1966, Projections of the lentiform nucleus in the monkey, Brain Res., 1:3.

Parent, A., and De Bellefeuille, L., 1982, Organization of efferent projections from the internal segment of globus pallidus in primate as revealed by fluorescence retrograde double labeling method, Brain Res., 245:201.

Parent, A., and De Bellefeuille, L., 1983, The pallidointralaminar and pallidonigral projections in primate as studied by retrograde double labeling method, Brain Res., (in press).

Parent, A., Mackey, A., and De Bellefeuille, L., 1983, The subcortical afferents to caudate nucleus and putamen in primate: a fluorescence retrograde double labeling study, Neuroscience, (in press).

Parent, A., Mackey, A., Smith, Y., and Boucher, R., 1983, The output organization of the substantia nigra in primate as revealed by a retrograde double labeling method, Brain Res. Bull., 10:529.

Poirier, L. J., and Sourkes, T. L., 1964, Influence du locus niger sur la concentration des catecholamines du striatum, J. Physiol. (Paris), 56:426.

Rinvik, E., Grofova, I., and Ottersen, P., 1976,, Demonstration of nigrotectal and nigroreticular projections in the cat by axonal transport of proteins, Brain Res., 112:388.

Szabo, J., 1980a, Organization of the ascending striatal afferents in monkeys, J. Comp. Neurol., 189:307.

Szabo, J., 1980b, Distribution of striatal afferents from the mesencephalon in the cat, Brain Res., 188:3.

Van der Kooy, D., and Carter, D. A., 1981, The organization of the efferent projections and striatal afferents of the entopeduncular nucleus and adjacent areas in the rat, Brain Res., 211:15.

Van der Kooy, D., Kuypers, H.G.J.M., and Catsman-Berrevoets, C. E., 1978, Single mammillary cell bodies with divergent axon collaterals. Demonstration by a simple, fluorescent retrograde double labeling technique in the rat, Brain Res., 158:189.

DOPAMINE DECREASES THE AMPLITUDE OF EXCITATORY POST-SYNAPTIC POTENTIALS IN RAT STRIATAL NEURONES

G. Bernardi, P. Calabresi, N. Mercuri and
P. Stanzione

II Clinica Neurologica
Universita di Roma
Viale Della Universitá
30 Roma, Italy

INTRODUCTION

Recently it has been shown that dopamine (DA) has an inhibitory effect on action potential generation in several regions of the C.N.S. (Bernardi et al., 1978-bis; Herrling and Hull, 1980; Bernardi et al., 1982; Herrling, 1981; Benardo and Prince, 1982; Bernardi et al., unpublished observations in hippocampus). This action is associated with a decrease of amplitude of post-synaptic potentials in striatal and cortical neurones (Bernardi et al., 1978; Bernardi et al., 1982; Herrling and Hull, 1980). The purpose of this research was to investigate the mechanisms which mediate the decrease of amplitude of the excitatory post-synaptic potentials (EPSPs) by DA, considering that the inhibition of the action potentials could be related to the reduction of the EPSPs.

METHODS

Experiments were performed on 84 male Wistar rats weighing 200 $\pm$ 50 g., anaesthetised with pentothal sodium 40 mg/Kg i.p. (Abbott). The anaesthesia was maintained with i.p. injections (8 mg/Kg) every hour. All wound pressure points were infiltrated with xylocaine 2%. Animals were paralysed with succinylcholine (Wellcome) i.p. and then artificially ventilated. The animal temperature was maintained at 36-37.5^{o}C by means of a heating pad. The skull was exposed and a craniectomy was performed above the caudate according to the atlas of Fifkova and Masala (1967). A hole was drilled to allow penetration of a stainless steel concentric bipolar electrode positioned in the ipsilateral sensorimotor cortex. Electrical stimuli consisted of single square pulses (5-15 v intensity and 0.1-0.2 ms width) delivered at 0.5-1.0 Hz. The intracellular electrodes (Clark GC 100 TF) were pulled on a horizontal puller (Narishige) and were filled with 4M K-acetate. They had resistances ranging between 40-90 Mohm. The recording electrode was glued to the assembly of drug-ejecting barrels. The intertip distance between the recording electrode and the ejecting one ranged between 50-130 microns. The different barrels were filled by capillary action with the following solutions: dopamine (Sigma) 1M pH8; NaCl 4M (control current); NaCl 4M (balance barrel). A retaining current of 10nA was used between applications. Current flowing through the balance barrel was continually adjusted automatically to ensure neutralization of total current at the pipette tip. All the microiontophoretic manipulations were performed with the Neurophore BH2 (Medical System Corp.). Hydraulic microdrive permitted electrode penetration into the caudate nucleus. Intracellular current injection (inward, outward pulses) through a bridge circuit was used to evaluate the input resistance and to modify the synaptic potentials. At the end of various experiments animals were perfused with a formol saline solution and the brains removed for histological control of the position of the recording electrode.

RESULTS

In 56 striatal neurons intracellularly recorded membrane potentials were higher than 40 mV, with the amplitude of spikes greater than 40 mV and lasting less than 2 ms. The membrane resistance was measured by means of hyperpolarizing current steps (-0.2/-2 nA, 5-50 ms.) and ranged, at rest, between 15-35 Mohm. In these neurons cortical stimulation evoked EPSP-IPSP sequences (Hull et al., 1973), and among these cells, 43 received a mono-synaptic input, as already described by Vandermaelen and Kitai (1980) and Kitai et al. (1976). EPSP-IPSP sequences lasted 150-250 ms., and were modified in amplitude by current injection. Depolarizing pulses reduced the amplitude of the EPSP, and increased the IPSP. Hyperpolarizing pulses increased the EPSP and reduced the IPSP. These findings rule out a disfacilitatory or disinhibitory mechanism in the tested cells. At the peak of the EPSP the membrane resistance, measured by trains of short hyperpolarizing pulses, decreased up to 60%. Because these striatal neurons rarely fired spontaneously, usually direct and synaptically driven action potentials were tested. DA iontophoretically applied to 35 neurons depressed the "free running" spikes (40-100 nA balanced current) and the action potentials evoked by depolarizing pulses (70-200 nA balanced current) (Fig. 1). However, during DA application an increase of the current injection restored the spikes evoked by depolarizing pulses (Fig. 2), without changes in their shape (data not shown), ruling out an inactivation of the action potentials during DA application. In 20 of the neurones affected by DA, no

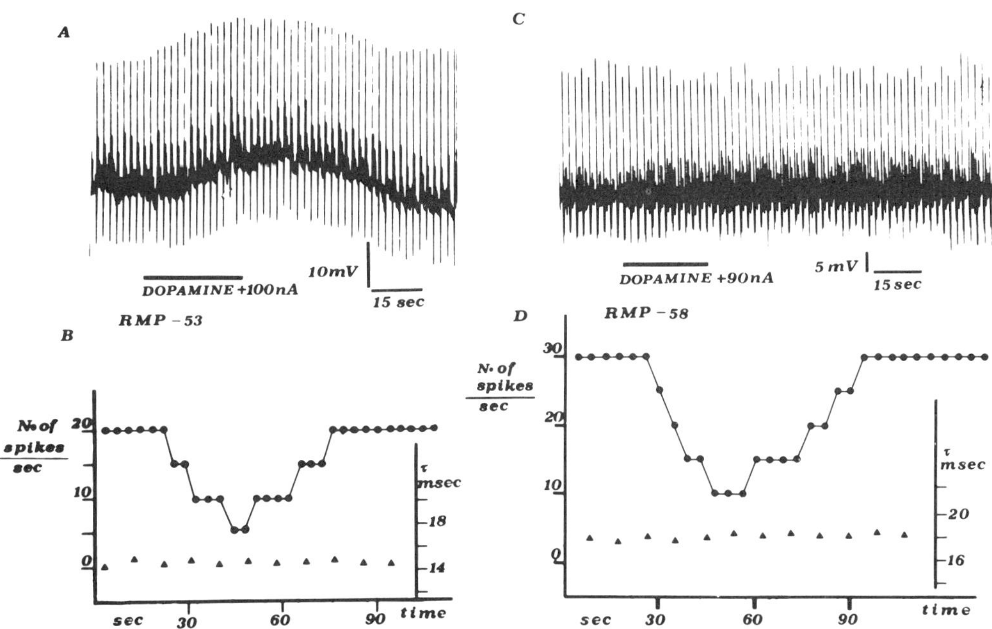

Fig. 1. DA action on two striatal neurons. In these cells action potentials were evoked by depolarizing pulses (200 ms,+1nA) and membrane resistance was measured by short hyperpolarizing pulses. Diagrams show the inhibitory action of the catecholamine on action potentials. On the left side: (A) DA (+100nA balanced current) slowly depolarizes the neuron. Action potentials, hyperpolarizing and depolarizing pulses were attenuated by slow frequency response of the pen recorder. (B) Effect of DA on spike frequency in the above neuron. Black triangles show the membrane time constant measured by short hyperpolarizing pulses (50 ms, -1 nA). The catecholamine does not change the membrane resistance. On the right side: (C) a different neuron in which the membrane potential does not change during DA ejection (+90nA balanced current). Membrane potential has been recorded with the same technical procedure as in (A). (D) Shows the decreased number of action potentials by DA in the neuron in (C). Membrane time constant is not modified in this cell by DA. In both the cells the frequency histogram is time locked with the above membrane potential.

modification of the membrane potential was observed, but in the remaining cells a slow depolarization appeared (5-15 mV) (Fig. 1). The membrane resistance, measured at rest by short hyperpolarizing pulses, was not clearly modified by DA application (Fig. 1 and Fig. 2). The amplitude of the EPSP-IPSP sequence evoked by cortical stimulation was depressed during DA induced depolarization, but the duration appeared unmodified (Fig. 3). Such a reduction of the EPSP-IPSP sequence was present also in neurons not depolarized by DA (Fig. 4). This effect excludes the possibility that the decrease of EPSP amplitude could be related to the membrane depolarization induced by DA. A clear increase in input membrane resistance, measured by trains of short hyperpolarizing pulses (5-10 ms), was detected at the peak of the excitatory post-synaptic potentials

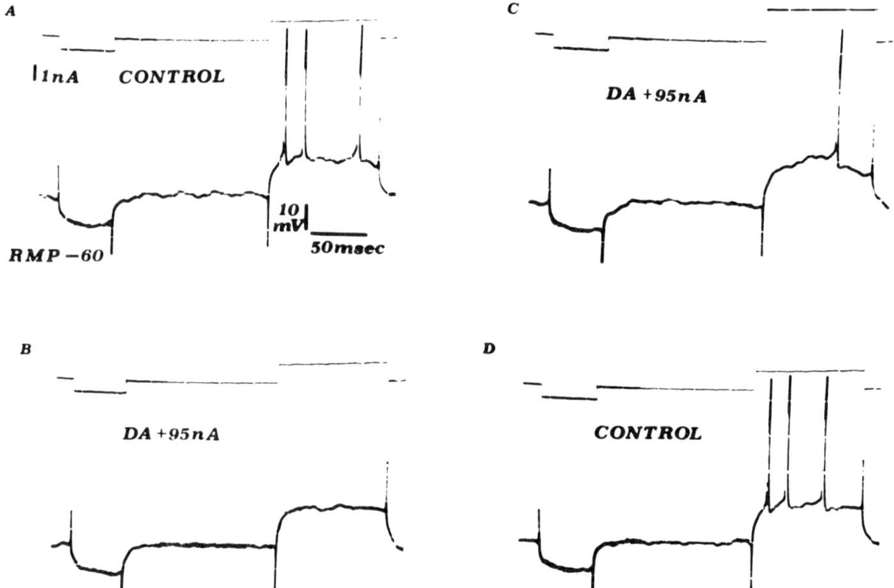

Fig. 2. Effect of DA on action potentials evoked by depolarizing pulses in a striatal neuron. Resting membrane potential: -60mV. (A) Control. In the lower trace: the hyperpolarizing pulse measures the membrane resistance, and is followed by a depolarizing pulse which evokes action potentials. The upper trace shows the amount of injected current. (B) DA (+95nA balanced current) does not change the membrane potential but inhibits the spikes. (C) The increase of the injected current restores the spike on a pulse, although DA is still ejected. This finding shows that the disappearance of the action potentials is not related to spike inactivation. (D) Control: 4 min. after the end of DA ejection.

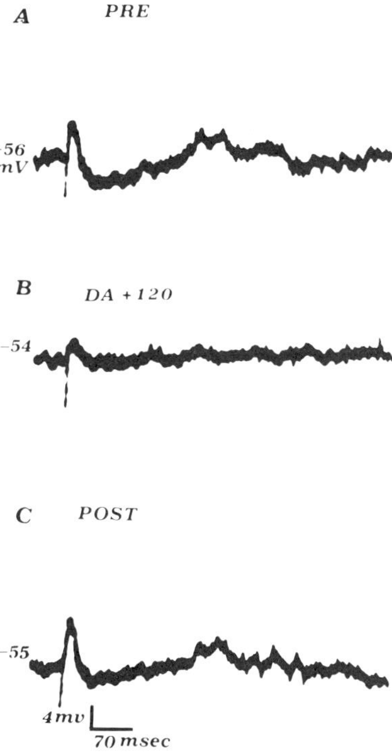

Fig. 3. The effect of DA on the EPSP-IPSP sequence evoked by cortical stimulation (13 V.). Membrane potential: -56 mV. (A) Control. (B) During DA ejection (+120 nA balanced current). Note the flattening of the EPSP-IPSP sequence. (C) Recovery.

during DA ejection (Fig. 4). This decrease of conductance was associated with the reduction of EPSP amplitude, although we cannot rule out the possibility that such a conductance change could reflect, at least in part, a concomitant decrease of the IPSP. Nevertheless these results show that DA depresses directly the EPSP, because an effect due to the flattening of the IPSP should produce an increase of the EPSP amplitude, but we never observed this response. The decrease of the EPSP amplitude indicates that this catecholamine interferes with the potency of the post-synaptic excitatory events. In order to investigate a possible post-synaptic site of action of DA, the effect of glutamic acid, considered the probable excitatory transmitter of the cortico-striatal pathway (Divac et al., 1977; Fonnum et al., 1981; Kim et al., 1977; MacGeer et al, 1979; Roberts et al., 1981) was tested before and during DA application. Glutamic acid (30/100 nA) reproducibly depolarized striatal cells, decreased the input resistance and increased the number of "spontaneous" and

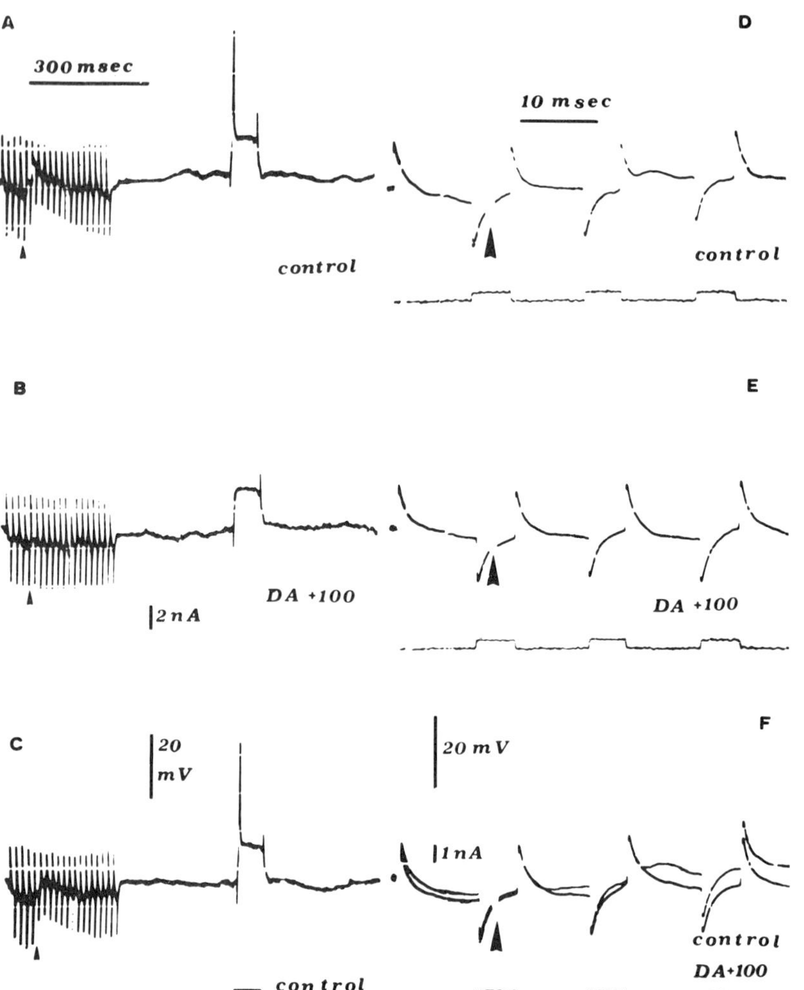

Fig. 4. DA effect on voltage and conductance of the EPSP in a striatal neuron not depolarized by DA. Membrane resistance measured by a train of short hyperpolarizing pulses delivered before and during the postsynaptic potential. Left side: (A) Control: note increase of conductance at peak of the EPSP. An action potential is evoked by a depolarizing pulse in order to test membrane excitability. RMP: -55mV. (B) EPSP flattened by DA application (+100 nA balanced current) and the related conductance change clearly decreased. The depolarizing pulse is no longer able to evoke an action potential. RMP: -55mV. (C) Recovery. RMP: -55mV.

evoked action potentials, as shown in previous reports (Bernardi et al., 1976; Herrling et al., 1983). DA was able to reduce the membrane depolarization, the increase of firing rate, and the decrease of membrane resistance produced by glutamic acid when the two substances were simultaneously applied (Fig. 5). The same amount of Na+ current ejected as control was ineffective (data not shown). These results suggest that the catecholamine modifies the potency of the excitatory aminoacid by acting at the post-synaptic level on striatal neurons.

DISCUSSION

These results show that DA has a depressant effect on the voltage of the EPSP-IPSP sequences evoked in striatal neurons by cortical stimulation. However, in this report only the action of DA on the excitatory post-synaptic potentials was studied, considering that the reduction of this potential could explain the inhibitory effect of DA on the "spontaneous" firing rate. The decrease of the EPSP was associated with an increase of membrane resistance recorded at the peak of the excitatory post-synaptic potentials, without change in the membrane conductance measured at rest. These findings suggest an action of DA on synaptic events, but the identification of a post-synaptic and/or presynaptic site of action for DA remains an open question. At this point, considering that strong evidence suggests a glutamergic transmission in the cortico-caudate projection, the interaction between DA and glutamate was investigated in a new series of experiments designed to identify the site of action of DA in the modulation of the EPSP. Our results have shown a clear interaction between DA and glutamic acid, suggesting that the catecholamine acts at the post-synaptic level by modifying the receptor affinity for the transmitter and/or by decreasing the activation of the related ionic channels. The decrease of potency of this aminoacid by DA action was not due to a negative interaction

---

Fig. 4. (cont.) Right side: Each trace in D-E represents a detail of, respectively, A-B at higher magnification, just before and after the stimulus. The lower trace indicates the amount of current injected. (D) Control. Note increase in conductance revealed by the difference in amplitude between the pulse before the stimulus and the second one after the stimulus. (E) During DA ejection. No difference appears between the pulses delivered before and after the stimulus. (F) One control trace superimposed on another oscillographic record during DA application, in order to emphasize the differences. (Upper trace: control. Lower trace: during DA ejection). Note disappearance of the increase in conductance during DA application.

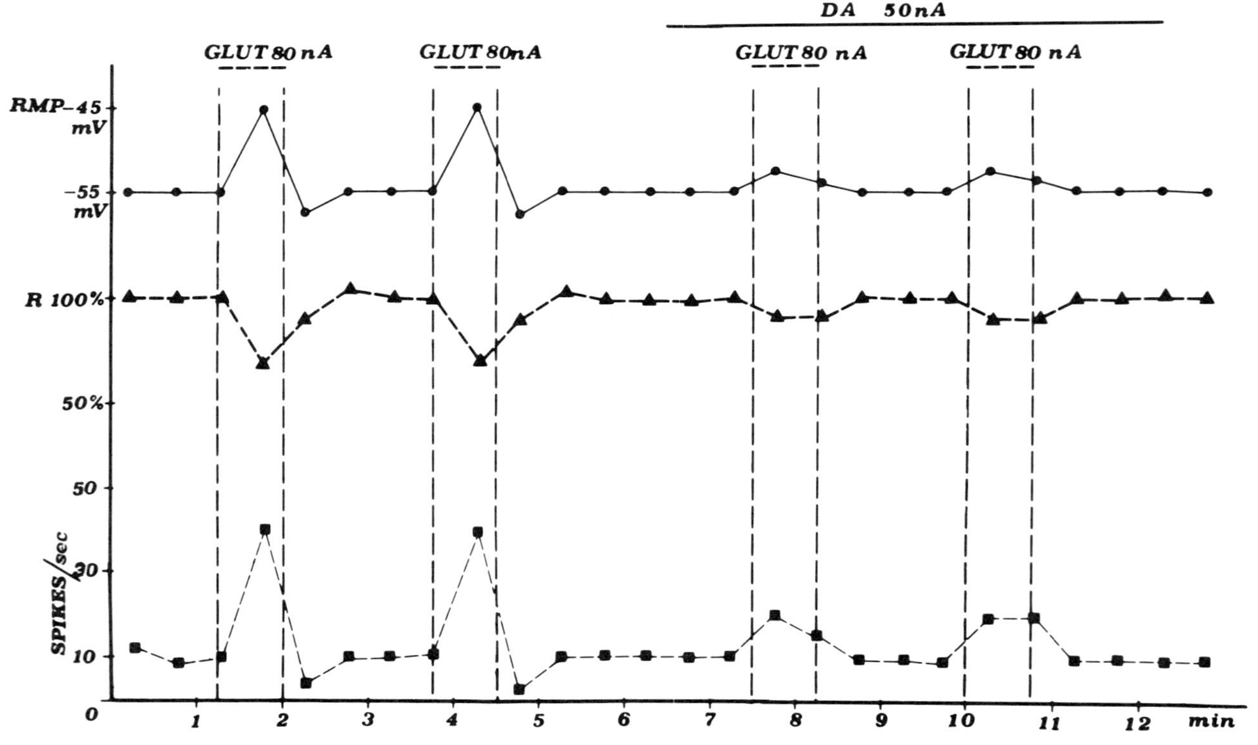

Fig. 5. Diagram summarizing the action of glutamic acid on the resting membrane potential, the membrane resistance and action potential frequency of a striatal neurone, and the interaction between this amino acid and DA.

between the positive current used for DA ejection and the negative current used for glutamate ejection, but was a pharmacological effect, because the decrease of glutamate action was not observed when Na+ current was ejected instead of DA. These results strongly support a post-synaptic effect of DA, but do not exclude a concomitant pre-synaptic action of this catecholamine. At this site DA could act by decreasing the release of the excitatory transmitter that mediates the EPSP generated by the cortico-caudate projection to striatal neurons. This possibility implies that DA interacts with specific receptors located on the terminals of this projection, which has been suggested by Schwarcz et al. (1978). In addition the decrease of H3-glutamic acid release in the striatum by DA, strengthens the possibility of a pre-synaptic site of action for DA (Mitchell and Doggett, 1980). However, considering that axo-axonic synapses have not been described in the striatum by histological studies (Kemp and Powell, 1971a; Kemp and Powell, 1971b; Hassler, 1979), the amine released from non-synaptic varicosities would have to reach specific receptors through the interstitial fluid, but at the present this mechanism has not been demonstrated.

Our results show that DA decreases the EPSP amplitude by a post-synaptic mechanism. This action could explain the decrease of the action potentials generated by the impingement of excitatory synaptic inputs on striatal neurons. However, DA was also able to depress cellular excitability, by elevating the threshold for the action potentials generated directly by depolarizing pulses. This mechanism is supported by the evidence that by increasing the current of the pulse, it was possible to restore the spikes blocked by DA. The absence of change in the rising phase of the spike suggests that the kinetics of voltage-dependent Na+ channels is not modified and that the action of DA is limited to elevating the threshold of the action potentials. A similar mechanism has been suggested for the inhibitory action of enkephalins in spinal cord cultured cells (Barker et al., 1978; Barker, 1983).

In conclusion, these results confirm the inhibitory action of DA on action potential generation in striatal neurons. This effect seems mediated by two different mechanisms: (1) depression of the EPSP, induced by DA at a post-synaptic site of action; (2) depression of intrinsic membrane excitability.

This action supports a neuromodulatory role for DA in the striatum, as previously suggested for cortical and nigral neurons (Bernardi et al., 1982; Waszczak and Walters, 1983). Such a function can play an important role in the modulation of striatal activity, and its imbalance may be involved in the pathogenesis of several extrapyramidal disorders.

REFERENCES

Barker, J. L., Gruol, D. L., Hang, L. M., MacDonald, J. F., and Smith, T. G., 1980, Peptides: pharmacological evidence for three forms of chemical excitability in cultured mouse spinal neurons, Neuropeptides, 1:62.

Barker, J. L., 1983, Peptide effects on the excitability of single nerve cells, in: "Handbook of Psychopharmacology," Vol. 16, L. L. Iversen, S. D. Iversen, S. H. Snyder, eds., Plenum Press, New York.

Benardo, L. S., and Prince, D. A., 1982, Dopamine action in hippocampal pyramidal cells, J. Neurosci., 2:415.

Bernardi, G., Floris, V., Marciani, M. G., Morocutti, C., and Stanzione, P., 1976, The action of acetylcholine and l-glutamic acid on rat caudate neurones, Brain Res., 114:134.

Bernardi, G., Marciani, M. G., Morocutti, C., and Stanzione, P., 1978a, The action of GABA and dopamine on rat caudate neurons intracellularly recorded, in: "Iontophoresis and Transmitter Mechanisms in the Mammalian Central Nervous System," R. Ryall and J. Kelly, eds., Biomedical Press, Elsevier, North Holland.

Bernardi, G., Marciani, M. G., Morocutti, C., and Stanzione, P., 1978b, The action of dopamine on rat caudate neurons intracellularly recorded, Neurosci. Lett., 8:235.

Bernardi, G., Cherubini, E., Marciani, M. G., Mercuri, N., and Stanzione, P., 1978a, Responses of intracellularly recorded caudate neurons to the iontophoretic application of dopamine, Brain Res., 245:267.

Divac, I., Fonnum, F., and Storm-Mathisen, J., 1977, High affinity uptake of glutamate in terminals of corticostriatal axons, Nature (Lond)., 266:377.

Fifkova, E., and Masala, J., 1967, Stereotaxic atlas for the rat, in: "Electrophysiological methods in biological research," J. Bures, M. Petran and J. Zachar, eds., Academic Press, New York.

Fonnum, F., Storm-Mathisen, J., and Divac, J., 1981, Biochemical evidence for glutamate as neurotransmitter in corticostriatal and corticothalamic fibres in the rat brain, Neuroscience, 6:863.

Hassler, R., 1979, Electronmicroscopic differentiation of the extrinsic and intrinsic types of nerve cells and synapses in the striatum and their putative transmitters, Adv. Neurol., 24:93.

Herrling, P. L., and Hull, C. D., 1980, Iontophoretically applied dopamine depolarizes and hyperpolarizes the membrane of cat caudate neurons, Brain Res., 192:441.

Herrling, P. L., 1981, The membrane potential of cat hippocampal neurons recorded in vivo displays four different reaction-mechanisms to iontophoretically applied transmitter agonists, Brain Res., 212:331.

Herrling, P. L., Morris, R., and Salt, T. E., 1983, Effects of excitatory amino acids on membrane and action potentials of cat caudate neurons, J. Physiol., 339:207.

Hull, C. D., Bernardi, G., Price, D. D., and Buchwald, N. A., 1973, Intracellular responses of caudate neurones to temporally and spatially combined stimuli, Exp. Neurol., 38:324.

Kemp, J. M., and Powell, T. P. S., 1971a, The structure of the caudate nucleus of the cat: light and electron microscopy, Phil. Trans. Roy. Soc. London, B,262:383.

Kemp, J. M., and Powell, T. P. S., 1971b, The site of termination of afferent fibers in the caudate nucleus, Phil. Trans. Roy. Soc. London, B,262:413.

Kim, J. -S., Hassler, R., Hang, P., and Paik, K. S., 1977, Effect of frontal cortex ablation on striatal glutamic acid level in rat, Brain Res., 132:37.

Kitai, S. T., Kocsis, J. D., Preston, R. J., and Sugimori, M., 1976, Monosynaptic inputs to caudate neurons identified by intracellular injection of horseradish peroxidase, Brain Res., 109:601.

McGeer, P. L., McGeer, E. B., Scherer, U., and Singh, K., 1977, A glutamergic corticostriatal pathway, Brain Res., 128:369.

McGeer, E. G., McGeer, P. L., and Hattori, T., 1979, Glutamate in the striatum, in: "Glutamic Acid: Advances in Biochemistry and Physiology," L. J. Filer, Jr., et al., eds., Raven Press, New York.

Mitchell, P. R., and Doggett, N. S., 1980, Modulation of striatal (H3) glutamic acid release by dopaminergic drugs, Life Sci., 176:185.

Roberts, P. J., McBean, G. J., Sharif, N. A., and Thomas, E. N., 1982, Striatal glutamergic functions: modifications following specific lesions, Brain Res., 235:83.

Schwarcz., R. I., Creese, I., Coyle, J. T., and Snyder, S. H., 1978, Dopamine receptors localized on cerebral cortical afferents to ratr corpus striatum, Nature (Lond.)., 271:76.

Vandermaelen, C. P., and Kitai, S. T., 1980, Intracellular analysis of synaptic potentials in rat neostriatum following stimulation of the cerebral cortex, thalamus and substantia nigra, Brain Res., 5:725.

Waszczak, B. L., and Walters, J. R., 1983, Dopamine modulation of the effects of gamma-aminobutyric acid on substantia nigra pars reticulata neurons, Science, 220:218.

PRESYNAPTIC ACTIONS AND DOPAMINE IN THE NEOSTRIATUM

G. W. Arbuthnott, J. R. Brown*, V. Kapoor+, and D. Whale

MRC Brain Metabolism Unit
University Department of Pharmacology
1 George Square, Edinburgh

*Glaxo Group Research, Ware, England

+Department of Medicine,
Flinders University, Adelaide, South Australia

INTRODUCTION

There are now very detailed anatomical descriptions of the structure of the neostriatum at the electronmicroscopic level (see e.g. Pasik et al., 1978) and it seems that axo-axonic synapses are rare. It should be said at the outset then, that the term presynaptic as used here does not necessarily imply any such anatomical substrate. The examples which we will discuss in detail employ very different methodologies and therefore need to be described separately in the sections below. There are three illustrations of interactions of neostriatal neurotransmitters which each warrant the use of the term "presynaptic". In the case of the action of dopamine on cortico-striatal synapses (section 1) the action is thought to be presynaptic on physiological evidence. In the other two cases, the actions are identified in slices of striatal tissue which do not contain dopamine (DA) cell bodies. Since changes in DA release are observed they must be thought of as the results of actions on DA terminals in the slices, although perhaps not due to morphologically defined synapses on them.

DA RECEPTORS ON CORTICO-STRIATE SYNAPSES

Kebabian and Calne (1979) suggested that the naming of DA receptors could be restricted. They called the "receptor" associated with adenylate cyclase D1 and the binding site D2. This simpler scheme has at least the merit of allowing the statement that D1 receptors seem to be restricted to the membranes of the striatal cells since destruction of these cells with kainic acid (KA) virtually abolishes the action of DA on adenylate cyclase (Fields et al., 1978; Leff et al., 1981). Although there are several sites with which binding is associated, it is also clear that on the cortico-striate neurones only D2 receptors are found, since only binding is influenced by decortication (Schwartz et al., 1978; Theodorou et al., 1981).

In the striatum of male Albino Wistar rats lightly anaesthetised with halothane very few spontaneous action potentials can be recorded at rates above 1 hz (Arbuthnott, 1974). In the experiments on which this report is based three methods were employed to cause these normally silent cells to fire. The cortex was stimulated and responsive cells identified after the manner of Schultz and Ungerstedt (1978). Recording electrodes glued alongside multibarrel iontophoresis electrodes (Brown et al. 1980) with a tip separation of approximately 30 μ allowed the application of pulses of glutamate with which to excite the cells. Finally, in a few experiments, an electrode implanted among the striato-nigral fibres allowed the antidromic activation of striatal output cells.

The recordings were made from adult male Albino Wistar rats anaesthetised with halothane (maintained at 0.8% in 200 ml/min of air administered through a tracheal cannula). All the recording sites were histologically verified at the end of the experiment by iontophoresis of Pontamine Skye Blue dye from the recording electrode. Under these conditions very few cells show action potentials in the absence of applied stimuli but this "spontaneous" activity is inhibited by DA (Fig. 1A). The cells can be made to fire in response to iontophoresis of glutamate (GLU). In cells responding briskly to short pulses of GLU this response was also inhibited by DA application (Fig. 1B). Finally, stimulation of the frontal cortex at a depth of about 1 mm through a concentric electrode placed approximately 1 mm anterior to the recording site evokes a response in many striatal neurones. Even at maximal stimulation strength there is very rarely more than one action potential per stimulus. Cortical stimulation is less effective during DA application close to the recording electrode (Fig. 1C). The responses to DA illustrated were the only kind of responses obtained and we could find no parameters of iontophoretic application which would differentiate them. That is, DA applications sufficient for the effect on the response to cortical stimulation, were always adequate for the demonstration of the effects on the response to GLU application (or on spontaneous activity if it could be demonstrated at all) in the cell being studied.

We saw no excitation by DA although it has been previously reported (Bevan et al., 1975), perhaps because of the tip separation

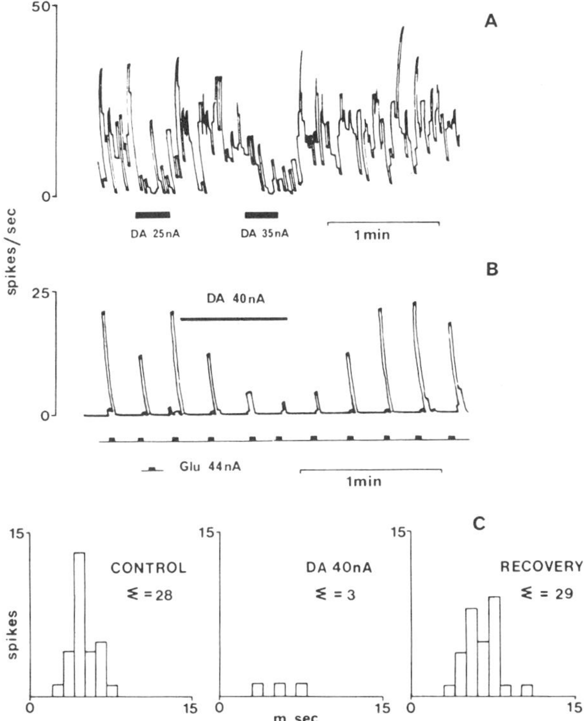

Fig. 1. The effects of iontophoresis of DA (at the current values indicated in each figure) on spontaneous activity (A) and on activity evoked by GLU application to a normally silent cell (B). In addition to these methods of recording activity from striatal units it was also possible to study the effect of DA iontophoresis in the region of striatal cells which responded to cortical stimulation. Three post stimulus histograms are shown in C. The stimulus to cortex was delivered at time zero. Σ is the total number of responses the cell made to 50 stimuli applied to the cortex. During iontophoresis of DA the response was inhibited and then recovered when the DA ejecting current was turned off. We found no excitatory effects of DA and no current which could produce the action shown in C which did not also inhibit cells as shown in A and B.

in our experiments; Herling and Hull (1980) have suggested that only an action of DA on the axon hillock will result in increased firing.

The neuroleptic fluphenazine (50-75 nA) applied 2-3 min before the iontophoresis of DA antagonised the depressant effects of the latter on 6 of the 8 neurones on which it was tried, whether the cells were spontaneously active or were responding to cortical stimulation. The antagonism was reversible in 4-5 min. and no action of the drug on spontaneous firing or on cortical response was seen. This clear cut action of the iontophoresis of fluphenazine contrasts with the lack of effect of haloperidol applied intravenously. No antagonism of the effect of DA was seen with this treatment even in doses of 0.5 mg/kg which certainly has profound effects in the behaving animal. This lack of effect of neuroleptics applied peripherally, against iontophoretically applied DA, has been reported by others (Ben-Ari et al., 1975; Zarzecki et al., 1977).

The more selective D2 antagonist (-) sulpiride (Jenner at al., 1979) did not antagonise the effect of DA on either the glutamate induced firing of striatal cells or on their spontaneous activity in all 15 cells tested. No action of the antagonist alone was seen on either spontaneous firing (6 cells) or on the excitation of the cells by GLU (27 cells). When sulpiride was tested for effects on the response of striatal cells to cortical stimulation, however, a

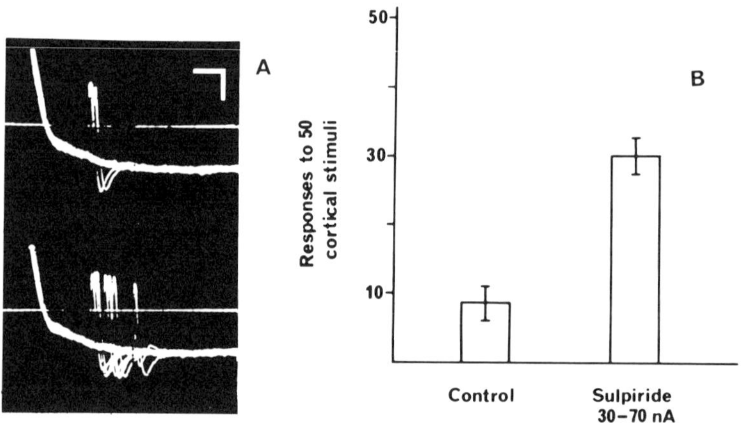

Fig. 2. Effect of sulpiride microiontophoresis on cortical responses of a striatal cell is recorded as in Fig. 1C. (A) Response of one cell is shown (responses to 50 stimuli). Calibration 0.2 mV and 2 msec. Upper trace shows control responses. Lower trace shows responses during sulpiride iontophoresis. (30 nA). (B) The effect of sulpiride on 18 cells is summarised by plotting Σ (total number of responses to 50 cortical stimuli). Columns are mean values $\pm$ S.E.M. Although iontophoresis of DA was capable of reducing this effect on 6 of these cells it never did so at doses which did not themselves inhibit the responsiveness of the cells. (From Brown et al. (1983), courtesy Pergamon, Oxford)

dramatic effect was seen in 19 of 20 cells so tested. The application of sulpiride clearly enhanced the response of the cells although they could not be made to respond more than once to each stimulus (Fig. 2). In 6 of these cells it was possible to reduce the action of sulpiride by iontophoretic application of DA; but the DA current required also affected the spontaneous activity of the cells. The (+) isomer of sulpiride was ineffective in 3 of the 5 cells on which it was possible to test it. The stimulant action on cortico-striate transmission thus appears to be specific to the effective form of the neuroleptic but the evidence that it results from activation of a DA receptor is rather less convincing.

One short series of experiments suggests that the cortico-striate pathway may be normally under the influence of DA. The responses to cortical stimulation in single units, from the two striata of 6 animals unilaterally lesioned with 6-OH-DA, were surveyed and the threshold for responding was significantly less on the side of the lesion (Table 1). Although there are other explanations of this finding, it is tempting to speculate that the cortico-striate pathway is normally under a tonic inhibitory drive from the action of the nigro-striatal DA system upon it. The removal of this tone facilitates the action of cortical stimulation upon the striatum whether it is achieved by the surgical removal of the DA neurones (with 6-OH-DA) or by the action of a DA antagonist (sulpiride).

Extracellular recordings preclude the confident assertion that the action of sulpiride is on the cortico-striatal terminals themselves, since subtle postsynaptic membrane changes would not have been detected. No difference in the action of sulpiride was seen between antidromically identified striato-nigral cells and cells not so identified. An action on small intrinsic neurones interposed between the cortical terminals and the neurones from which the recordings were made cannot be confidently excluded but monosynaptic connections between cortical afferents and striato-nigral cells do exist anatomically (Somogyi et al., 1981). This possible action on cortical terminals, along with the biochemical evidence for sulpiride binding on structures that degenerate after cortical ablation (Theodorou et al., 1981), strongly suggests that the physiological role of these binding sites is to modulate cortico-striate transmission. Since GLU may be the transmitter in this pathway (Fonnum et al., 1981) the changes in GLU release from striatal slices "in vitro" after application of DA agonists and antagonists (Mitchell and Doggett, 1980; Rowlands and Roberts, 1980) may reflect actions at these receptors, but whether they are responsible for the Parkinsonian side effects, or the therapeutic actions of neuroleptics, is at present uncertain.

Table 1. Neurophysiological data from 6-OHDA-lesioned and unlesioned striata.

Treatment	Total cells	% cells with spontaneous activity	Mean interburst frequency (Hz) (± SE)	Mean Mean firing frequency (Hz) (± SE)	Spontaneously active cells/cm	Mean stimulation current mA (± SE)	Latency of cortical stimulation (msec)	Range (msec)
lesion	37	73	117.5 ± 14.5	8.79* ± 1.33	4.50*	0.467* ± 0.052	6.19 ± 1.70	5-10.0
unlesioned	19	26	96.9 ± 51.8	0.64 ± 0.22	0.66	0.866 ± 0.004	6.58 1.40	5- 9.0

* $P < 0.01$ Mann Whitney Test

6-OH-DA (8 μg/μl) was injected in the posterior lateral hypothalamus of 4 rats under halothane anaesthesia. 6-10 days later during a neurophysiological experiment similar to those described in Fig. 1 the parameters of cell firing and responses shown were recorded from both the lesioned and unlesioned sides of the animals. Between lesion and recording all the animals were tested with 0.3 mg/kg apomorphine administered i.p. All the animals used for neurophysiology turned more than 200 times towards the intact side in the 30 min following the injection which indicates a massive depletion of dopamine on the lesioned side. The increase in mean firing rate and in the number of cells encountered confirms previous work (Schultz et al., 1978). The decrease in threshold for cortical stimulation in the absence of a change in the other parameters of the response complements the similar effect of sulpiride application and suggests that the nigro-striatal terminals may normally act on receptors on the cortical input to the striata.

STUDIES ON THE PHARMACOLOGY OF DA RELEASE

A great deal of controversy surrounds almost every aspect of the control of DA release by presynaptic receptors (Nowycky and Roth, 1978; Langer, 1981; Starke, 1981). While the effects of drugs on DA turnover are generally agreed upon, effects on release remain ambiguous.

Despite ample evidence suggesting an increase in DA turnover and utilization by muscarinic agents "in vivo", the effects of these agents on the "in vitro" release of 3H-DA remains confused. Much of the confusion arises because the experiments show nicotinic or muscarinic effects on basal release only, the physiological significance and mechanism of which remains obscure. This confusion is compounded by the fact that during these experiments (Giorguieff et al., 1976, 1977b, 1979a; de Belleroche and Bradford, 1978) the effect of drugs was examined on the basal release of newly synthesised tritiated DA (which is preferentially released), hence the possibility of an increased specific activity of a constant basal release (secondary to an increase in synthesis) was not excluded.

Using a modification of the electrochemical detector (designed by Kissinger et al., 1973 and Keller et al., 1976), and reverse-phase ion-pair chromatography, a protocol was established for studying the release of endogenous DA from small prisms (0.3 x 0.3 x 1 mm) of striatal tissue, "in vitro".

During incubation with Krebs solution (at 37°C) at a flow rate of 300-330 ul/min, the initial basal overflow of DA was 0.08 $\pm$ 0.02 ng/mg protein/2 min (mean $\pm$ S.E.M., n = 8). The initial basal overflow of dihydroxyphenyl acetate (DOPAC) was found to be much higher (0.65 $\pm$ 0.05 ng/mg protein/2 min). The basal release of DOPAC, 30 min after start of superfusion, was found to be significantly higher, i.e. 0.96 $\pm$ 0.08 ng/mg protein/2 min ($p < 0.05$, two tailed, paired student "t" test, see Fig. 3B). As shown in Fig. 3A the evoked overflow of DA, induced by a 2 min pulse of high K^+ Krebs, was complete (>95%) within 6 min of the onset of stimulation, i.e. 3 sample times. The K^+ induced overflow of DA was estimated as the total increase (above basal overflow) in DA overflow during these 6 minutes, and found to be 6.3 $\pm$ 0.4 ng/mg protein during the first stimulation (S1). A second 2 min pulse of 25 mM K^+, S2, consistently induced higher overflow of DA, i.e. 7.7 $\pm$ 0.6 ng/mg protein.

DOPAC overflow after K^+ stimulation showed a slower rate of increase and decline. The increase in DOPAC, during the 6 minutes after the onset of stimulation, was used to compare changes in the evoked release of DOPAC. Hence the increase in DOPAC release during 25 mM K^+ stimulation was taken as 2.24 $\pm$ 0.14 ng/mg protein during S1 and 3.69 $\pm$ 0.28 ng/mg protein at S2. The difference was

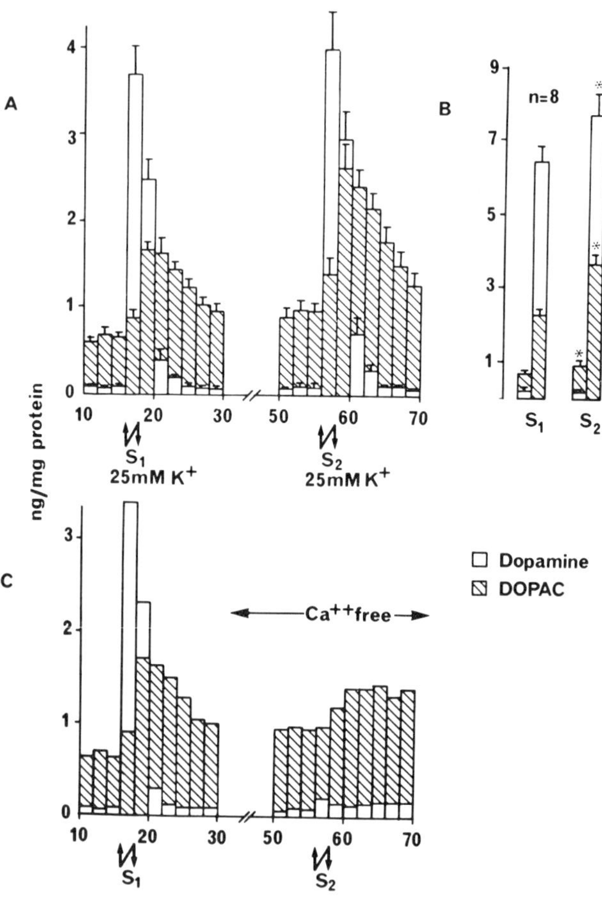

Fig. 3. Striatal slices superfused with Krebs solution were stimulated twice, at S1 and S2, with a two-minute pulse of 25 mM K^+. Two-minute fractions of the superfusate were collected and analysed for their DA and DOPAC content by HPLC-ECD. The results are shown in A expressed as ng DA or DOPAC/mg protein in each fraction (vertical bars show S.E.M., n = 8). A summary of the results is shown in B, where each pair of histograms represent the basal and the evoked (25 mM K^+) overflow of DA and DOPAC in that order (vertical bars = S.E.M.). The basal overflow of DOPAC and the evoked overflow of both DA and DOPAC were significantly higher at S2 when compared to the internal control S1 (* = $p < 0.05$ two-tailed, paired Student "t" test).
For this and subsequent diagrams the basal overflow was calculated as the average overflow of DA/DOPAC found during

significant at the $p < 0.05$ level by a two tailed, paired student "t" test (see Fig. 3B).

Substituting 0.1 mM EGTA for calcium chloride in the Krebs solution during the second stimulation, S2, with 25 mM K^+ Krebs, resulted in the complete inhibition of the stimulation evoked increase in the overflow of DA and DOPAC as shown in Fig. 3C.

The effect of electrical stimulation (biphasic, square wave pulses of 8-10 mA amplitude 2 msec duration, at 20 Hz for 30 sec) on striatal slices superfused with Krebs solution is shown in Fig. 4A, B. The results of evoked overflow were, as with K^+ stimulation, expressed as the increase above basal overflow during the 6 min after the onset of stimulation. Two stimulations, at 16 min (S1) and 56 min. (S2) from the start of superfusion were applied. The increase in DA overflow at S1, was found to be 0./44 $\pm$ 0.08 ng/mg protein (mean $\pm$ S.E.M., n = 8), the increase in DOPAC overflow was 2.75 $\pm$ 0.20 ng/mg protein. Hence while there was a much smaller increase in the evoked DA overflow, DOPAC overflow was similar to that observed during 25 mM K^+ stimulation.

At S2 there was significantly larger overflow of DA, 0.74 $\pm$ 0.08 ng/mg protein giving an S2/S1 ratio of 1.80 $\pm$ 0.20. The evoked overflow of DOPAC as S2, 2.93 $\pm$ 0.18 ng/mg protein, did not differ significantly from that at S1.

Addition of the uptake inhibitor LY5953A, 1 µm, increased the basal overflow of DA by two-fold from that seen in the absence of drugs. There was, however, no significant change in the basal overflow of DOPAC. The evoked overflow of DA at S1 was increased by about 4-fold, without any significant change in the evoked overflow

---

Fig. 3. (cont.) the 6 min before the onset of stimulation, i.e. between 10-16 mins for S1 and 50-56 mins for S2 from the onset of superfusion. Note that the columns representing DA and DOPAC are superimposed; baseline and scale are common to both columns.
In C the histograms show the DA and DOPAC content of the superfusate from striatal tissue, superfused with normal Krebs solution and stimulated with a 2 min pulse of 25 mM K^+ Krebs at S1 and then superfused with Ca^{++} free Krebs (starting 30 mins after the onset of superfusion) and again stimulated with 25 mM K^+ (Ca^{++} free) Krebs at S2 (n = 3). Omission of CA^{++} from the Krebs solution abolished the 25 mM K^+ induced increase of DA and DOPAC overflow. A small ongoing increase in the basal overflow of both DA and DOPAC was noted in the absence of Ca^{++}.

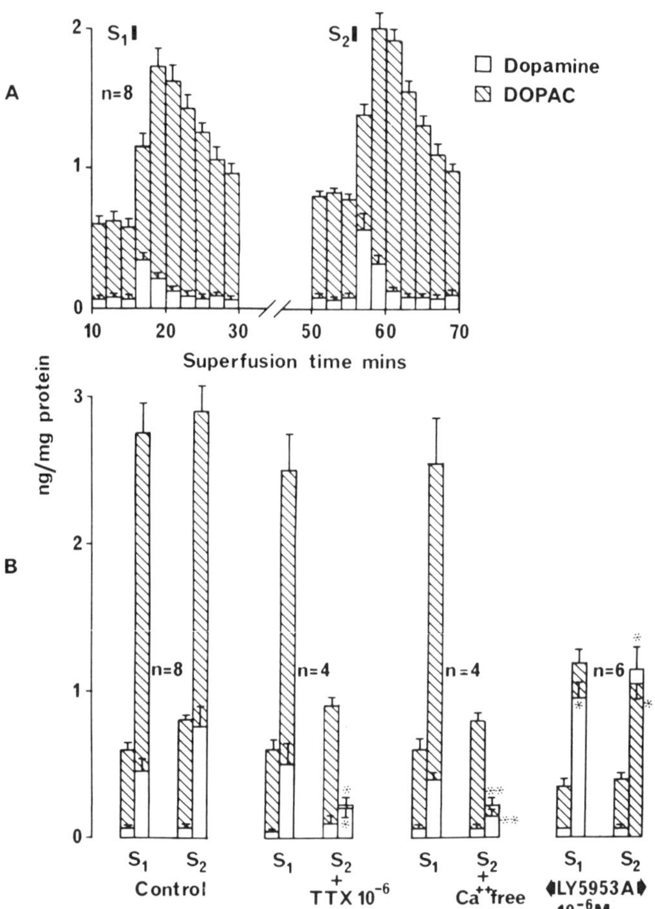

Fig. 4. A. Striatal slices superfused with Krebs solution were stimulated twice, at S1 and S2, electrically. The stimulation parameters were (as with all the other experiments, unless otherwise stated): 2 msec pulse duration, 8-10 mA amplitude, 20 Hz for 30 sec.
The superfusate was collected as 2 min fractions into tubes containing an internal standard and a protective solution. 20-50 µl of the superfusate was used for an HPLC-ECD determination of DA and DOPAC.
B. Summary of the control electrically evoked overflow of DA and DOPAC. Each pair of histograms at S1 and S2 shows the basal and evoked overflow of DA and DOPAC, for this and all subsequent figures. The basal overflow of DOPAC and the evoked overflow of DA at S2 were significantly higher than at S1.
Tetrodotoxin (1 µM) was added to the Krebs solution

of DOPAC (see Fig. 4B). During S2, the evoked overflow of DA was again increased giving an S2/S1 ratio of 1.23 $\pm$ 0.06. At S2, the evoked overflow of DOPAC, 2.09 $\pm$ 0.23 ng/mg protein, was significantly lower than the control values giving an S2/S1 ratio of 0.91 $\pm$ 0.08.

The evoked overflow of both DA and DOPAC at S2 was reduced by more than 75% in the presence of Tetrodotoxin (TTX, added to Krebs solution 20 min. before S2). Addition of TTX, 1 μm, did not have any significant effect on the basal overflow of either DA or DOPAC. Similar effects were seen when the tissues were superfused with Ca^{++} free (+ 0.1 mM EGTA) Krebs at S2 (Fig 4B).

The evoked overflow of 3H-DA, previously preincubated with the tissue, co-eluted with endogenous DA (Fig. 5B) and the peaks eluting at the retention time of authentic DA and DOPAC from superfusate samples were alumina extractable. While the DA and DOPAC overflow remained unaltered after KA lesions of the striatum, it was completely abolished by 6-OHDA lesions (Fig. 5A). Pargyline abolished the DOPAC peak selectively, while α-methyl-p-tyrosine (AMPT) inhibited the overflow of both DA and DOPAC.

CHOLINERGIC RECEPTORS AND DA RELEASE

Striatal slices superfused with Krebs solution were stimulated twice (S1 and S2), either with 25 mM K^+ (Fig. 6B) or electrically at 20 Hz for 30 sec. (Fig. 6A). Neostigmine, 1 μM, even added along with tubocurarine, 1 μM, 30 mins after the onset of superfusion, increased the evoked overflow of DA. The evoked overflow of DOPAC

Fig. 4. (cont.) superfusing striatal slices, 30 min after the onset of superfusion, i.e. after S1. The presence of TTX did not affect the basal overflow of either DA or DOPAC before S2, however, the evoked overflow of DA and DOPAC was reduced by more than 75%.
Superfusion with Ca^{++} free (0.1 mM EGTA) Krebs did not change the basal overflow of DA and DOPAC before S2, but the evoked overflow of DA and DOPAC at S2 was reduced by more than 75%. When striatal tissue was superfused with Krebs solution containing 1 μM LY5953A from the start of superfusion, the basal overflow of DA and DOPAC at S1 and S2 was not different from control, but the evoked overflow of DA was markedly enhanced. DOPAC evoked overflow at S1 did not differ from control, but at S2 it was significantly lower (*$p < 0.05$, **$p < 0.01$, two-tailed student "t" test).

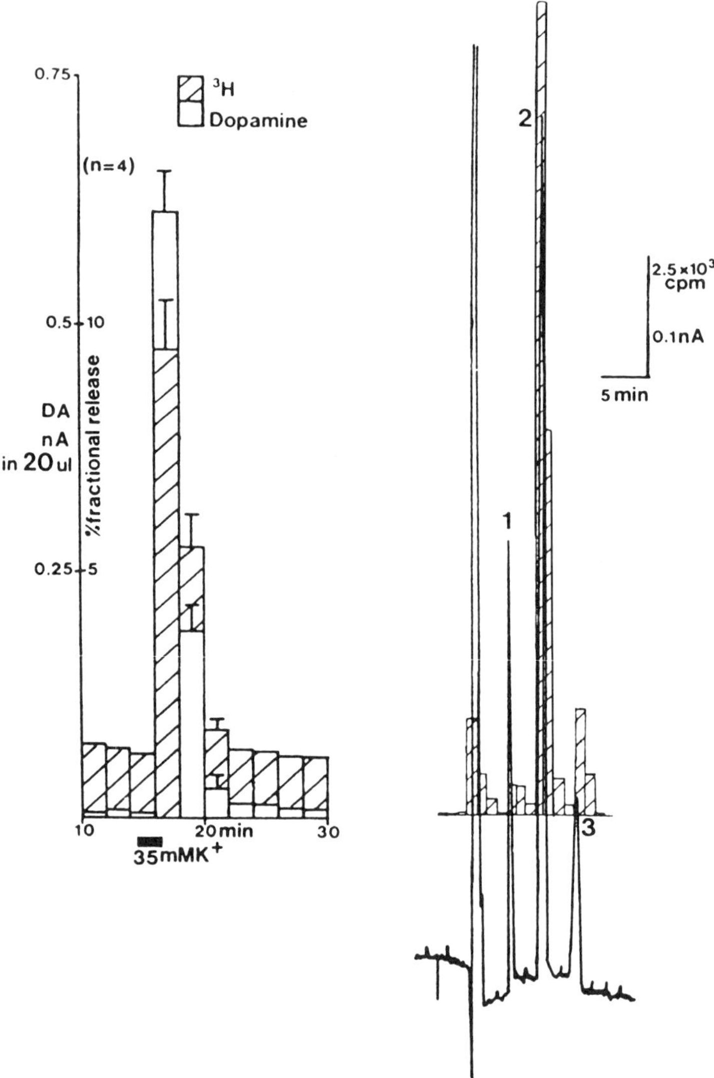

Fig. 5. (A) Comparison of the HPLC-ECD traces of superfusate samples (collected during S2, i.e. at 56 mins) from (left) the lesioned, (right) the unlesioned striatal slices of a unilaterally 6-OHDA lesioned rat. The DA/DOPAC content of superfusate samples collected during the second electrical stimulation, S2, in the presence of 1 μM LY5953A, are shown above.
1 = Dihydroxybenzylamine, internal standard. 2 = Dopamine.
3 = Dihydroxyphenylacetate (DOPAC).
No detectable DA or DOPAC was found in the lesioned side, even during electrical stimulation (left), while the overflow of DA and DOPAC from the unlesioned side (right) did not differ from control.

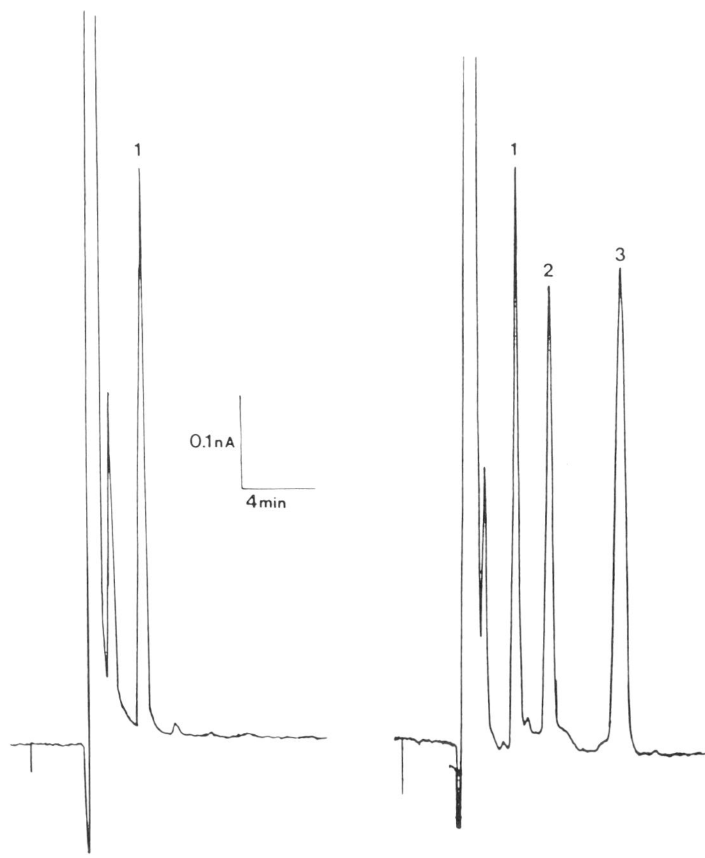

Fig. 5. (B) Striatal tissue preincubated with 3H-DA was stimulated once with a 2 min pulse of 35 mM K^+. The superfusate was analysed for its DA content by HPLC-ECD and its 3H content by liquid scintillation counting. To the left is a comparison of the DA overflow (expressed as nA in 20 ul of superfusate) and total 3H overflow (expressed as % fractional release). Vertical bars show S.E.M. (only S.E.M. >15% are shown). The overflow of both DA and 3H is simultaneously and rapidly increased by 35 mM K^+ Krebs. On the right the radioactive (3H) profile of an HPLC separation of the superfusate; sample taken during superfusion with 35 mM K^+. More than 70% of the 3H present in the sample co-elutes with DA.

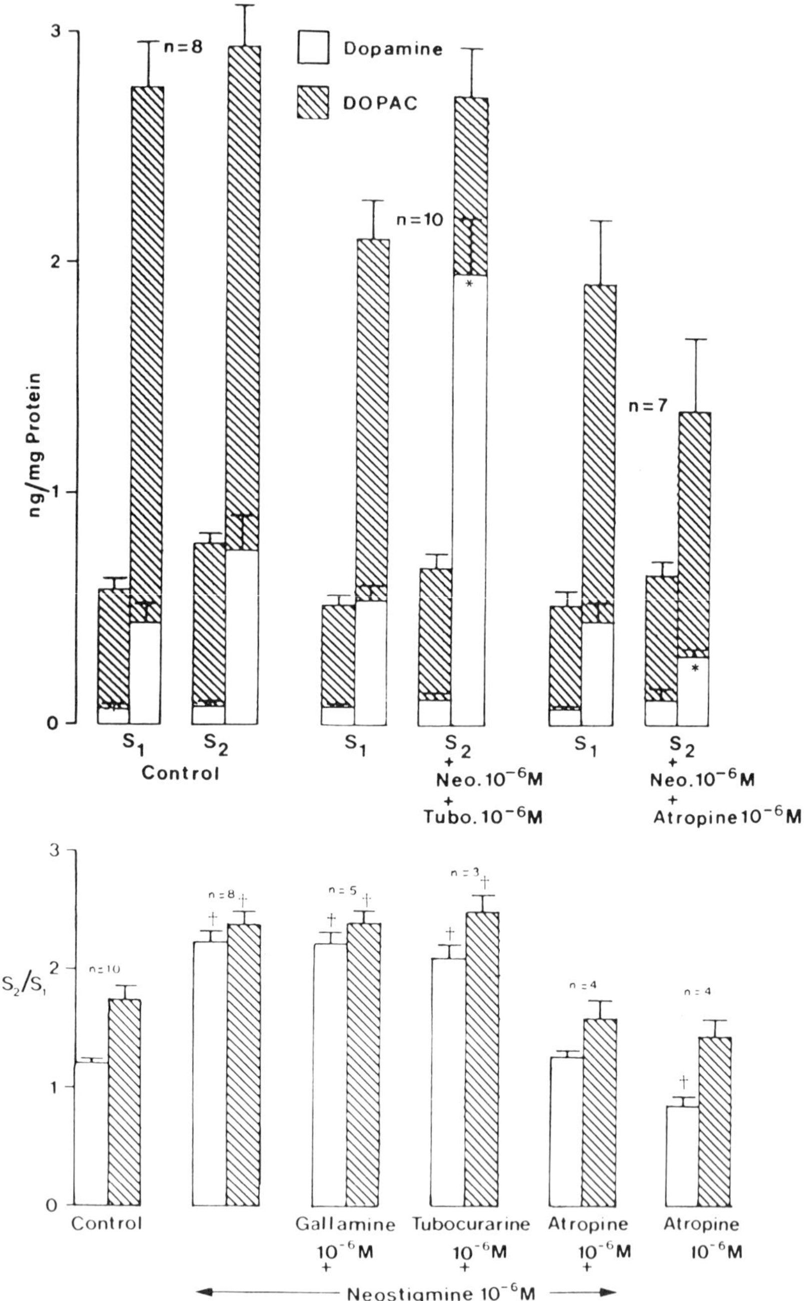

Dopamine
DOPAC
ng/mg Protein
n=8
n=10
n=7
S1
S2
Control
S2 + Neo. 10^-6M + Tubo. 10^-6M
S2 + Neo. 10^-6M + Atropine 10^-6M
S2/S1
Control
Gallamine 10^-6M +
Tubocurarine 10^-6M +
Atropine 10^-6M +
Atropine 10^-6M
Neostigmine 10^-6M

Fig. 6. ⟵ (A) After the internal control electrical stimulation S1, neostigmine (1 μM) and tubocurarine (1 μM) were added to the superfusing Krebs solution (at 30 min). The basal overflow of DA and DOPAC before S2 remained unaltered. The evoked overflow of DA and DOPAC was markedly enhanced at S2.
After the internal control stimulation, S1, neostigmine (1 μM) and atropine (1 μM) were added to the superfusing Krebs solution (at 30 min). While the basal overflow of DA and DOPAC remained unaltered, a reduction in the evoked overflow of DA and DOPAC was found. (*$p < 0.05$ two-tailed, Student "t" test).
(B) The effect of cholinergic antagonists on the facilitation of DA and DOPAC evoked overflow induced by neostigmine.
Neostigmine (1 μM) added 30 min after the onset of superfusion, enhanced the 25 mM K^+ evoked overflow of DA and DOPAC, indicated by the increases in the S2/S1 ratios. The nicotinic antagonists, gallamine (1 μM) or tubocurarine (1 μM) added with the neostigmine, failed to alter the response to neostigmine. The muscarinic agonist, atropine (1 μM) added with neostigmine completely abolished its effects.
Atropine (1 μm) added alone reduced slightly the overflow of DA. This indicates that a stimulation of muscarinic presynaptic receptors facilitates the release and synthesis of DA. Values significantly different from control, $p < 0.05$, Wilcoxon Rank test (as with all diagrams, vertical bars = S.E.M.

was also increased compared to the internal control (S1). The S2/S1 ratio for DA overflow was found to be significantly different from control by non-parametric analysis. The basal overflow of DA and DOPAC remained unaffected by the drugs.

Addition of atropine, 1 μM, with neostigmine, 1 μM, before S2 resulted in a significantly smaller evoked overflow of DA and DOPAC (Fig. 6).

Oxotremorine, 5 μM, added to the Krebs solution 30 min after the onset of superfusion caused an almost threefold increase in the evoked overflow of DA at S2 (Fig. 7). DOPAC overflow at S2 was 30% higher than at S1. The basal overflow of both DA and DOPAC remained at control levels.

Nicotine, 5 μM, added to the superfusing Krebs solution 30 min after the onset of superfusion did not significantly alter the evoked

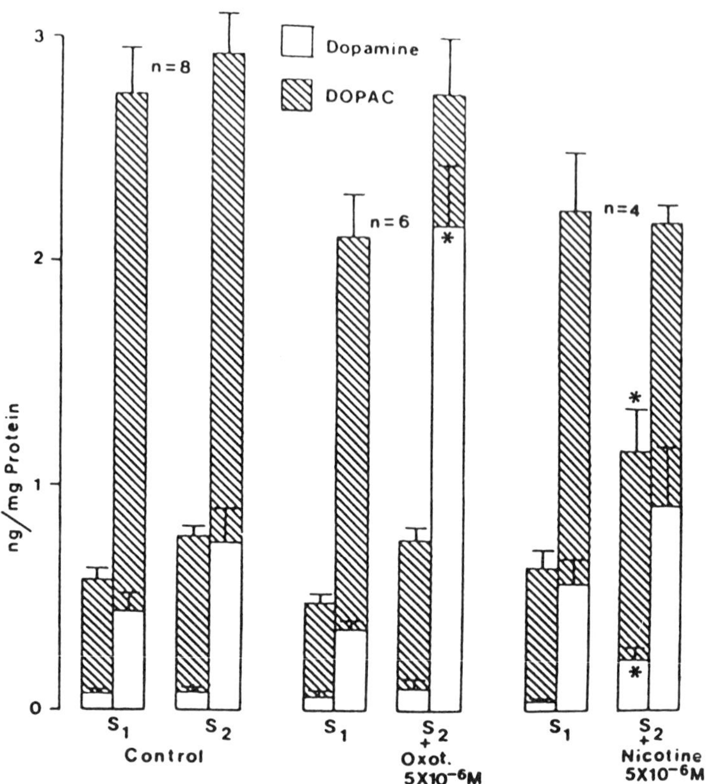

Fig. 7. Oxotremorine (Oxot.) (1 μM) enhanced the evoked overflow of DA and DOPAC at S2 (added again 30 min after start of superfusion). The basal overflow remained unaltered. Nicotine (1 μM) enhanced only the basal overflow of DA and DOPAC before S2 (added at 30 min), while the evoked overflows remained unchanged. (*$p < 0.05$, two-tailed Student "t" test).

overflow of either DA or DOPAC (Fig. 7). The basal overflow of DA was increased as was the basal overflow of DOPAC.

In order to determine whether the effect of the acetylcholine esterase inhibitor was due to an increase in the overflow or in the release of DA, neostigmine (1 μM) was added to the Krebs solution 30 min after the onset of superfusion in the presence of an uptake inhibitor. LY5953A, 1 uM, was present from the start of superfusion. The addition of neostigmine, along with the uptake inhibitor LY5953A, further enhanced the overflow of DA at S2. The evoked rise in DOPAC overflow was similarly increased (Fig. 8). The basal overflow did not differ from that in the presence of LY5953A alone.

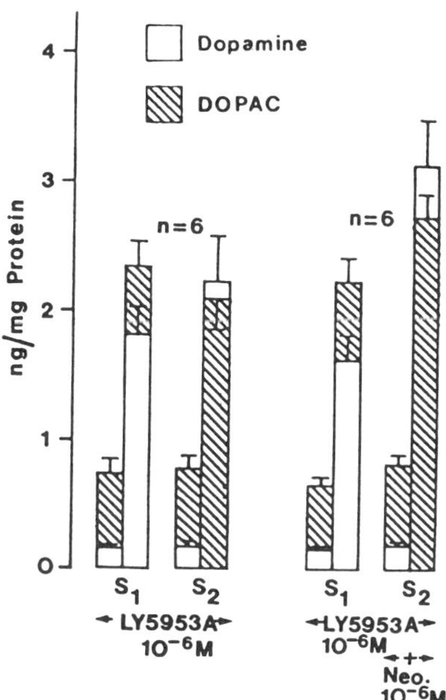

Fig. 8. Neostigmine (1 μM) added 30 min after the onset of superfusion in the presence of 1 μM LY5953A, further enhanced the evoked overflow of DA and increased the evoked overflow of DOPAC. The basal overflow remained unaltered.

Dopamine Overflow

Few data on the overflow of endogenous DOPAC "in vitro" are available. However, recent investigations by Ungerstedt and co-workers (Ungerstedt et al., 1982; Bennet et al., 1982) on the "in vivo" overflow of DA and DOPAC, mainly from the striatum, show a similar pattern to that found in the present investigation during electrical stimulation (i.e. a large overflow of DOPAC accompanied by more than an order of magnitude smaller overflow of DA) despite differences in methodology. The basal overflow of DA was found to be very low under control conditions (less than 0.1% of tissue DA content). In the absence of applied depolarizing stimuli, an increased basal overflow of DA with time was found during the present study. Cheramy et al. (1977), while examining the overflow of 3H-DA newly synthesised from 3H-tyrosine in vivo after impulse flow inhibition with γ-hydroxybutyrate, also found an increasing basal overflow of 3H-DA in the absence of impulse flow.

The large increase in the electrically evoked overflow of DA in the presence of 5 μM LY5953A (a specific DA uptake blocker, Wedley et al., 1978), suggested that up to 90% of the released DA is taken back up by nerve terminals.

While changes in current strength were not tested, the evoked overflow of DA and DOPAC was Ca^{++} and TTX dependent at 8-10 mA. Aceves and Cuello (1981) showed that while the overflow of 3H was Ca^{++} and TTX dependent up to 9 mA stimulation amplitude, the larger current strengths commonly used (20 mA) resulted in a Ca^{++}- and TTX-independent overflow of 3H.

The overflow of DA increased with increasing frequency (constant number of pulses) to a maximum around 20 Hz (Kapoor and Arbuthnott, 1982). Similar values have been reported by various authors for a maximal increase in tissue metabolite levels after in vivo stimulation of the median fore-brain bundle (Korf et al., 1976; Roth et al., 1976).

Cholinergic Receptors

Cholinergic agents - direct acting muscarinic agonists or acetylcholinesterase inhibitors (elevating the levels of evoked endogenous ACh overflow) - markedly potentiated the K^+ (Kapoor, 1982) and electrically evoked DA and DOPAC overflow. These effects could be inhibited by atropine, but not by tubocurarine or gallamine, indicating the involvement of muscarinic type cholinergic receptors. The effect of the muscarinic agents persisted in the presence of the DA uptake inhibitor, LY5953A, suggesting that muscarinic presynaptic receptor stimulation increases the release and rate of synthesis of DA.

Since all the evidence presented above was obtained from striatal slices containing nerve terminals in the absence of cell bodies, a presynaptic site of action is implied, although the existence of a "short loop feedback" by cholinergic interneurons was not completely ruled out.

The present findings support those of Lloyd and Bartolini (1975) who showed using a push-pull cannula "in vivo" that oxotremorine markedly potentiated the overflow of endogenous DA. Although Giorguieff et al. (1977) reported a muscarinic presynaptic receptor-mediated increase in the basal overflow of 3H-DA (newly-synthesised from 3H-tyrosine), interpretation of their results is complicated by the fact that an increased specific activity of the overflow of DA during a constant basal release was not ruled out.

A muscarinic agonist-induced increase of DA turnover has been reported by several authors. Paalzov and Paalzov (1975) found an increased AMPT-induced disappearance of brain DA. Muscarinic agonists have also been shown to raise DA metabolite (HVA and DOPAC) levels (Laverty and Sharman, 1965; Nose and Takemoto, 1974) in the striatum, while O'Keefe et al. (1970), found a decrease in striatal HVA after atropine. Interestingly, HVA has been shown to be derived mainly (80%) from DOPAC (Westerink and Spaan, 1982a) and so most likely reflects changes in the rate of DA synthesis. 3-MT on the other hand, is thought to be a better index of DA overflow (Westerink and Spaan, 1982a and b; Westerink and Korf, 1976; Kehr, 1976).

Javoy et al. (1975) showed an increased L-DOPA formation in vivo after oxotremorine administration. The same authors also reported that although the specific activity of tissue DA (3H-tyrosine administered I.V.) was also enhanced, the total DA content of the striatal tissue remained unaltered, suggesting a muscarinic receptor-mediated facilitation of DA synthesis and utilization simultaneously. Nicotine enhanced the basal overflow of both DA and DOPAC confirming the findings of Giorguieff et al. (1976 and 1979). Nicotine did not, however, have any significant effect on the evoked overflow of either DA or DOPAC during the current investigations.

THE EFFECT OF DOPAMINERGIC AGENTS ON THE EVOKED OVERFLOW OF DA AND DOPAC

The feedback regulation of DA turnover has been known since the 1960's (Carlsson and Lindqvist, 1963; Sharman, 1966). The early suggestion that the changes in DA turnover were mediated by a neuronal feedback loop (the striatonigral pathway, see Wright et al., 1977) still has some support (Gale et al., 1978; Gale, 1979). However, evidence gathered after inhibition of impulse flow in the nigrostriatal pathway (Walters and Roth, 1976a, and Kehr et al., 1972), after lesioning the striatonigral pathway, (Garcia-Munoz et al., 1977, DiChiara et al., 1977), and from "in vitro" work (Westfall et al., 1976, 1979), showed that despite effective isolation of the nigrostriatal terminals from effects mediated via the output pathway from neostriatum, the actions of DA agonists and antagonists on DA turnover persist; suggesting that the effects are mediated, at least in part, by presynaptic dopaminergic receptors (see reviews by Nowycky and Roth, 1978; Westerink, 1979).

After the initial observation of Farnebo and Hamberger (1971) that the neuroleptic, chlorpromazine, facilitates the evoked release of 3H-DA, several authors have reported either no effect or an inhibition of release of 3H-DA with classical neuroleptics (Seeman and Lee, 1975; Dismukes and Mulder, 1977; de Belleroche and Bradford, 1981). It was usually found that DA agonists were ineffective in modulating the overflow of 3H-DA (Dismukes and Mulder, 1977; Raiteri et al., 1978a, 1978b, 1979). Miller and Friedhoff (1979) report a biphasic effect of haloperidol on 3H-DA overflow, being stimulatory at low doses but inhibitory at high doses. There is now increasing evidence, mainly from the laboratories of Langer and Starke (see reviews Langer, (1981), Starke, (1981), for dopaminergic presynaptic receptors, inhibiting the release of 3H-DA. A likely explanation for the discrepancies in the literature, is that 3H-DA overflow is a poor index of endogenous DA release. During experiments in which

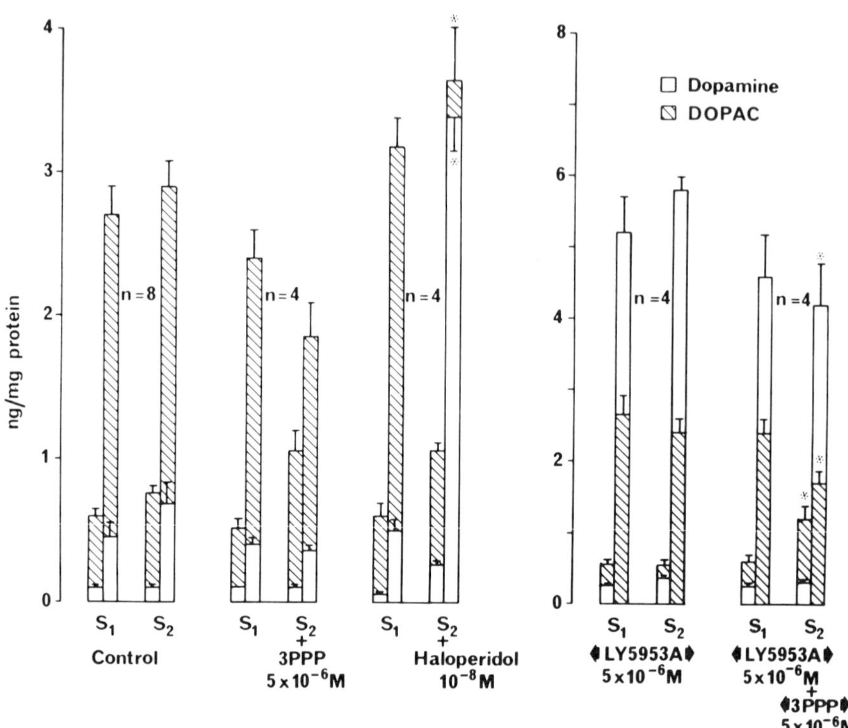

Fig. 9. The D2 agonist 3PPP (5 μM) added at 30 min of superfusion decreased the evoked overflow of DA and DOPAC. The basal overflow of DOPAC was also enhanced in the presence of 3PPP. Haloperidol (0.01 μM) enhanced the basal overflow of both DA and DOPAC. The evoked overflow of DA and DOPAC at S2 were also larger. (*$p < 0.05$, two-tailed Student "t" test). The basal and evoked overflow of DA and DOPAC in the presence of 5 μM LY5953A. The basal and evoked overflow of DA, both at S1 and S2 are significantly higher ($p < 0.05$, Student "t" test, two-tailed) than in the presence of 1 μM LY5953A. 3PPP (5 μM), added at 30 min, enhanced the basal overflow of DOPAC, in the presence of the uptake inhibitor LY5953A (5 uM).
The evoked overflow of DA and DOPAC at S2 was significantly reduced. (*$p < 0.05$, two-tailed Student "t" test).

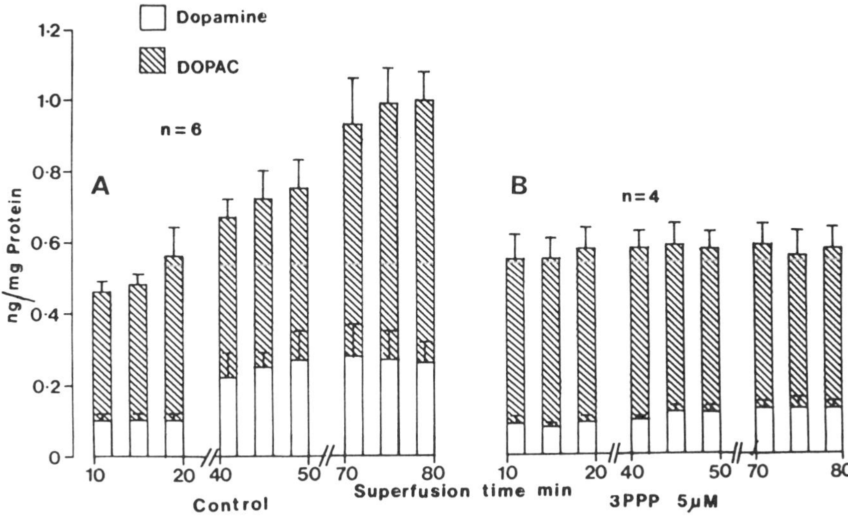

Fig. 10. A. The basal overflow of DA and DOPAC was measured from three superfusate samples between each of 10-20, 40-50 and 70-80 minutes of superfusion. No depolarizing stimuli were applied. As shown above, the basal overflow of both DA and DOPAC rose with time.
B. In the presence of 3PPP (5 μM), however, the increase in the basal overflow of DA and DOPAC with time, noted above, was abolished.

endogenous DA release ("in vivo", Lloyd and Bartholini, 1975; or "in vitro", Plotsky et al., 1977) or the release of newly-synthesized DA (Westfall et al., 1976) is measured, neuroleptics have been found to be facilitatory to DA release.

In our slices of striatum the dopaminergic agonist 3PPP [3-(3-hydroxyphenyl)-N-n-propylpiperidin], at a dose of 5 μM added to the superfusing Krebs solution 30 min after the onset of superfusion, decreased the evoked overflow at S2, of both DA, and DOPAC (Fig. 9). The S2/S1 ratios for both evoked DA and DOPAC in the presence of 3PPP were significantly different from control. The basal overflow of DA at S2 was unaffected by 3PPP, but the basal overflow of DOPAC at S2 was about 30% higher than control levels. The S2/S1 ratio for basal DOPAC was significantly higher than control. LY5953A, 5 μM, present from the start of superfusion greatly enhanced the basal overflow of DA and the evoked overflow of DA, both at S1 and at S2 compared to that seen in the presence of 1 μM LY5953A. The basal and evoked overflow of DOPAC remained unaltered.

The addition of 5 μM 3PPP at S2, in the presence of 5 μM LY5953A, reduced the evoked overflow of DA and DOPAC (Fig. 9).

The addition of haloperidol, 0.01 µM before S2 greatly enhanced both the basal and the evoked overflow of DA at S2 (Fig. 9). The evoked overflow of DA was increased by more than four-fold. Both the basal and evoked overflow of DOPAC were also increased 25% above control.

In order to observe the variation of basal overflow over longer periods of time, in the absence of applied depolarizing stimuli, striatal slices were superfused with Krebs solution for 80 min. Three superfusate samples from between 10-20, 40-50 and 70-80 min of superfusion were analysed for their DA and DOPAC content.

In the absence of drugs (Fig. 10), the average DA and DOPAC basal overflow rose from means of 0.10 ± 0.01 ng DA/mg protein/2 min and 0.50 ± 0.02 ng DOPAC/mg protein/2 min during 10-20 min of superfusion to 0.27 ± 0.05 ng DA/mg protein/2 min and 0.97 ± 0.10 ng DOPAC/mg protein/2 min during 70-80 min of superfusion.

During preliminary experiments with haloperidol (0.5 µM), present from the start of superfusion, none of the above mentioned values appeared altered.

3PPP, 5 µM, added to the superfusing Krebs solution from the start of superfusion, abolished the increase of the DA and DOPAC basal overflow with time.

Striatal slices preincubated with 3H-DA (4 µl of [3H-7,8]-dopamine, 47 Ci/mM) were superfused with oxygenated Krebs and stimulated twice (as above). The superfusate from each 2 min collection was analysed as follows:

1. 20 µl was injected into the HPLC-ECD to determine the DA/DOPAC content.
2. 500 µl was added to 8 ml of scintillation fluid and its 3H content analysed by liquid scintillation counting.

While the basal overflow of both DA and DOPAC was not different from control, the evoked overflow of DA was markedly enhanced. The rise in the evoked overflow of DOPAC did not significantly differ from control levels (Fig. 11).

The percent fractional basal overflow of 3H at S1 did not differ from that at S2 and was found to be 0.64 ± 0.08%. The evoked fractional overflow at S1 was 4.56 ± 0.29% (n = 4) with a slight decrease at S2 to 3.30 ± 0.47%

Addition of 3 PPP, 5 µM, before S2 of tissue preincubated with 3H-DA, enhanced the basal overflow of DOPAC at S2 while not affecting the DA basal overflow. The evoked overflow of both DA and DOPAC at S2 were reduced (Fig. 11).

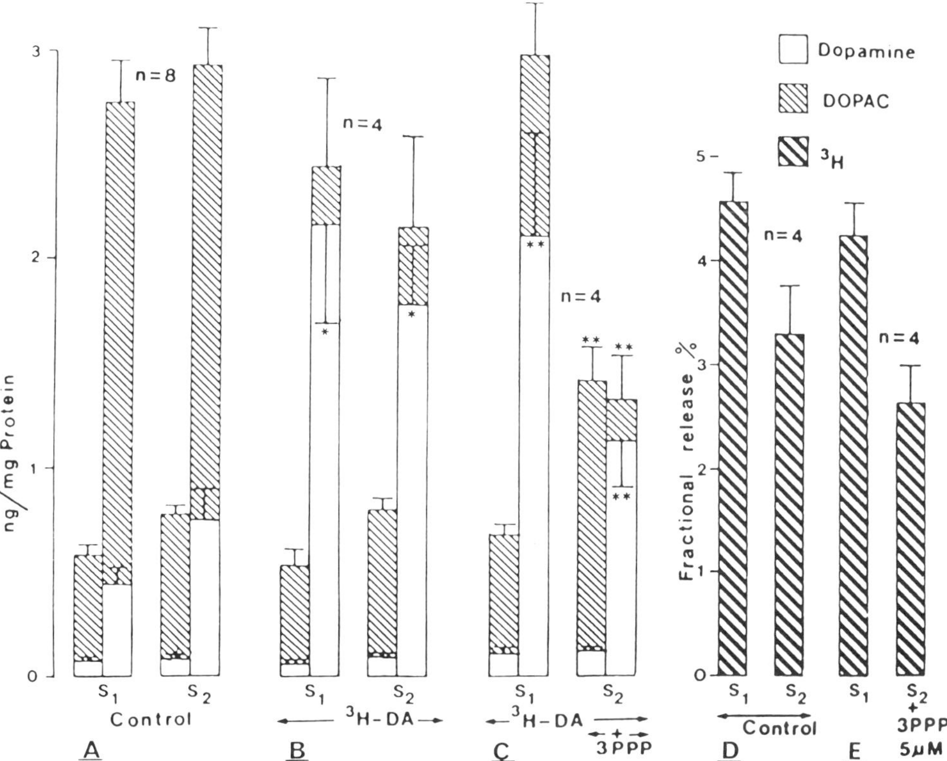

Fig. 11. A. The basal and evoked overflow of DA and DOPAC under control conditions.
B. The basal and evoked overflow of DA and DOPAC from tissue preincubated with 3H-DA. While the basal overflow of DA and DOPAC remained unaltered both at S1 and S2, the evoked overflow of DA was greatly enhanced. The DOPAC evoked overflow at S1 and S2 was not significantly different from control.
C. 3PPP (5 µM) added to the superfusing solution of tissue preincubated with 3H-DA at 30 min, enhanced the basal DOPAC overflow while decreasing the evoked overflow of both DA and DOPAC.
D. Under control conditions with 3H DA preincubation the evoked overflow of 3H at S1 was slightly but not significantly higher than that found at S2.
E. The simultaneous measurement of 3H overflow showed that the evoked overflow of 3H at S1 was larger than that at S2, in the presence of 3PPP (5 µM). However, the presence of 3PPP did not significantly alter the overflow 3H from that seen under control conditions.

The dopaminergic agonist 3PPP decreased the evoked overflow of both DA and DOPAC, while haloperidol had the opposite effect. The effects of these dopaminergic agents persisted in the presence of the uptake inhibitor, LY5953A, indicating that release rather than overflow of DA was altered. Alterations of the overflow of DOPAC suggest an inhibitory effect of dopaminergic presynaptic receptors on the rate of synthesis of DA during stimulation.

The controversy in the literature about the actions of neuroleptics on the overflow of 3H-DA remains unresolved; Lehmann et al. (1981) have recently ruled out species difference (rabbit, cat, rat) and the inhibitory effects of ascorbic acid on DA receptor mediated actions (see also Thomas and Zemp, 1977; Kayaalp and Neff, 1980) as possible causes of the differences in the literature. The most likely explanation for the lack of effect or inhibition of 3H-DA release with neuroleptics is perhaps small differences in experimental protocol, especially the presence of monoamine oxidase inhibitors. This view is supported by the finding of Zumstein et al. (1981), that the inhibitory effects of apomorphine on the overflow of 3H are abolished by pargyline.

After the original description of 3H-DA overflow inhibiting dopaminergic presynaptic receptors by Farnebo and Hamberger (1971), further supportive evidence came from "in vivo" studies of endogenous DA overflow (Lloyd and Bartholini, 1975) and "in vitro" studies on the overflow of endogenous (Plotsky et al., 1977) and newly-synthesised (Westfall et al., 1976) DA. Starke et al., (1978)) subsequently confirmed the findings of Farnebo and Hamberger (1971), and showed a presynaptic DA receptor-mediated depression of 3H-DA overflow "in vitro", using dopaminergic agonists and antagonists.

In the above mentioned studies, the neuroleptic-induced facilitation of 3H-DA release was usually found to be no more than about 50-60% above control levels (Starke et al., 1978; Miller and Friedhoff, 1979); the facilitation of endogenous or newly-synthesised DA with neuroleptics was, however, reported to be much higher, i.e. about 150-300% (in agreement with our present studies). This suggests that dopaminergic antagonists facilitate the release of newly synthesised DA (Fig. 11). This suggestion is supported by the finding that although there was a marked facilitatory effect of haloperidol on endogenous DA and DOPAC overflow and a significant inhibitory effect of 3PPP, simultaneous determination of 3H overflow failed to follow these changes accurately.

During experiments with tissue preincubated with 3H-DA, the overflow of endogenous DA and DOPAC (as measured by HPLC-ECD) differed from that found during control experiments in two major ways: (1)Whereas the overflow of DA from tissue preincubated with 3H-DA did not differ markedly from control in the presence of uptake inhibition, i.e. during high K^+ stimulation or during electrical stimulation in the presence of the uptake inhibitor LY5953A at 5 μM, there was, however, a much greater overflow of DA from tissued preincubated with 3H-DA compared to control during electrical stimulation in the absence of uptake inhibitors. Initial studies, measuring the overflow of DA from tissue preincubated with "cold" DA (results not presented) indicated that part of the apparent uptake inhibition with 3H-DA may be due to an inhibitory action of exogenous DA on uptake. (2)The evoked overflow of DA from control slices in the presence of LY5953A was roughly the same at both S1 and S2 (S2/S1 = 1.12). However, there was a significantly lower release at S2, from striata preincubated in with 3H-DA and superfused in the presence of 5 uM LY5953A (S2/S1 = 0.65), and a similar reduction of DOPAC overflow was found at S2, which was not seen in control striata. The inhibitory actions of 3PPP on DA and DOPAC overflow in control but not 3H-DA preincubated slices, (despite the presence of LY5953A) and the effectiveness of haloperidol in facilitating the overflow of DA from both tissues under the same conditions, suggests that part of the inhibition of DA and DOPAC at S2 may have been due to an action of 3H-DA at the presynaptic dopaminergic receptors.

Interestingly, the increase and the rate of DA and of DOPAC overflow found "in vitro", are similar to those found "in vivo" by Broxterman et al. (1980), after HA-966 (a gamma-butyrolactone (GBL)-like drug) administration. The DA content of the tissue at the end of "in vitro" non-stimulated experiments was also similar to that described by the above mentioned authors "in vivo".

Hence it appears that non-stimulated striatal slices "in vitro", behave similarly to the striatum "in vivo" under conditions of impulse flow inhibition.

The D2 agonist 3PPP (Hjorth et al., 1981; Watling and Williams, 1982) abolished the "in vitro" increase in DOPAC overflow seen in the absence of applied depolarizing stimuli, suggesting that in the absence of DA release, dopaminergic presynaptic receptors which decrease DA turnover are not stimulated.

Similarly, Walters and Roth (1976b) showed that the "in vivo" increase in DA turnover after GBL administration could be abolished by dopaminergic agonists, but remains unaffected by dopaminergic antagonists (Nowycky and Roth, 1978).

Thus we have demonstrated that isolation of DA terminals "in vitro" results in a steady increase in DA turnover which can be altered by the addition of DA or DA agonist (3PPP) and that when given neuroleptic the terminals increase the output of DA even more. So DA terminals respond to their own transmitter, and "feed back" control of DA release is present "in vitro". The release of DA is also subject to an action of muscarinic receptors which facilitate release by depolarisation - with a possible action of nicotinic receptors seen only in the absence of depolarising stimuli in our system.

Finally, the DA which is released seems to act on other terminals in the striatum to inhibit the efficiency of cortico-striate terminals in exciting striatal cells.

REFERENCES

Aceves, J., and Cuello, A. C., 1981, Dopamine release induced by electrical stimulation of microdissected caudate-putamen and substantia nigra of the rat brain, Neurosci., 6:2069.

Arbuthnott, G. W., 1974, Spontaneous activity of single units in the striatum after unilateral destruction of the dopamine input, J. Physiol., 399:121.

Ben-Ari, Y., and Kelly, J. S., 1975, Dopamine evoked inhibition of single cells of the feline putamen and basolateral amygdala, J. Physiol., 256:1.

Bennet, M. R., Marsden, D. A., Sharp, T., Ungerstedt, U., and Zetterstrom, T., 1982, "In vivo" measurement of dopamine and its metabolites by intra-cerebral dialysis: changes after alpha-amphetamine, Brit. J. Pharmacol, 77:355P.

Bevan, P., Bradshaw, C. M., and Szabadi, E., 1975, Effects of desipramine on neuronal responses to dopamine, noradrenaline, 5-hydroxy-tryptamine and acetylcholine in the caudate nucleus of the rat, Brit. J. Pharmacol., 54:285.

Brown, J. R., Mayer, M. L., and Arbuthnott, G. W., 1980, The use of ultra-violet setting glue for microelectrode fabrication, Journal of Neuroscience Methods, 3: 203.

Broxterman, H. J., Van Valkenburg, C. F. M., and Noach, E. L., 1980, HA-966 effects on striatal dopamine metabolism: implications for dopamine compartmentalization, J. Pharm. Pharmacol., 32: 67.

Carlsson, A., and Lindquist, M., 1963, Effect of chlorpromazine or haloperidol on formation of 3-methoxytyramine and normetanephrine in mouse brain, Acta. Pharmacol. Tox., 20:140.

Chéramy, A., Nieoullon, A., Glowinski, J., 1978, Stimulating effects of gamma-hydroxybutrate on dopamine release from the caudate nucleus and the substantia nigra of the cat, J. Pharmacol. Exp. Ther. 203:283.

De Belleroche, J. S., and Bradford, H. F., 1978, Compartmentation of synaptosomal dopamine, in: "Adv. Biochem. Pharmacol. Dopamine," P. J. Roberts, G. N. Woodruff and L. L. Iversen, eds., Raven Press, New York.

De Belleroche, J. S., and Bradford, H. F., 1981, Evidence for an inhibitory component of neuroleptic drug action, Brit. J. Pharmacol., 72:427.

Di Chiara, G., Porceddu, M. L., Fratta, W. and Gessa, G. L., 1977, Postsynaptic receptors are not essential for dopaminergic feedback regulation, Nature, 267:270.

Dismukes, K., and Mulder, A. H., 1977, Effects of neuroleptics on release of 3H-dopamine from slices of rat corpus striatum, N-S. Arch. Pharmacol., 297:23.

Farnebo, L. O., and Hamberger, B., 1971, Drug induced changes in the release of 3H-monoamines from field stimulated rat brain slices, Acta. Physiol. Scand. Suppl., 371:35.

Fields, J. A., Reisine, T. D., and Yammamura, H., 1978, Loss of striatal dopaminergic receptors after intrastriatal kainic acid injection, Life Sci. 23:569.

Fonnum, F., Storm-Mathisen, J., and Divac, I., 1981, Biochemical evidence for glutamate as neurotransmitter in corticostriatal and corticothalamic fibres in rat brain, Neuroscience, 6:863.

Gale, K., 1979, Pre- vs post-synaptic receptor control of tyrosine hydroxylase activity: the "long and short" of nigrostriatal feedback loops, in: "Presynaptic Receptors," eds. S. Z. Langer, K. Starke, M. L. Dubocovich, Pergamon, Oxford and New York.

Gale, K., Costa, E., Toffano, G., Hong, J. S., and Guidotti, A., 1978, Evidence for a role of nigral gamma-aminobutyric acid and substance P in the haloperidol-induced activation of striatal tyrosine hydroxylase, J. Pharmacol. Exp. Thera., 206:29.

Garcia-Munoz, M., Nicolau, N. M., Tulloch, I. F., Wright, A. K., and Arbuthnott, G. W., 1977, Feedback loop or output pathway in striato-nigral fibres, Nature, 265:363.

Giorguieff, M. F., Le Floc'h, M. L., Westfall, T. C., Glowinski, J., and Besson, M. J., 1976, Nicotinic effect of acetylcholine on the release of newly synthesised 3H-dopamine on rat striatal slice and cat nucleus, Brain Res., 106:117.

Giorguieff, M. F., Le Floc'h, M. L., Glowinski, J., and Besson, M. J., 1977, Involvement of cholinergic presynaptic receptors of nicotinic and muscarinic types in the control of the spontaneous release of dopamine from striatal dopaminergic terminals in the rat, J. Pharmacol. Exp. Ther., 200:535.

Giorguieff-Chesselt, M. F., Kemel, M. L., Wandscheer, D., and Glowinski, J., 1979, Regulation of dopamine release by presynaptic nicotinic receptors in rat striatal slices: Effect of nicotine in a low concentration, Life Sci., 25:1257.

Herrling, P. L. and Hull, C. D., 1980, Iontophoretically applied dopamine depolarizes and hyperpolarizes the membrane of cat caudate neurones, Brain Res., 192:442.

Hjorth, S., Carlsson, A., Wikstrom, H., Lindberg, P., Sanchez, D., Hacksell, U., Anidsson, L. E., Svensson, U., and Nilsson, J. L. G., 1981, 3-PPP, a new centrally acting DA receptor agonist with selectivity for autoreceptors, Life Sci., 28: 1225.

Javoy, F., Agid., Y., and Glowinski, J., 1975, Oxotremorine- and atropine-induced changes of dopamine metabolism in the rat striatum, J. Pharm. Pharmacol., 27:677.

Jenner, P., and Marsden, C. D., 1979, Minireview: The substituted benzamides - a novel class of dopamine antagonists, Life Sciences, 25:475.

Kapoor, V., 1982, Presynaptic cholinergic modulation of endogenous dopamine and dihydroxyphenylactic acid release from superfused rat striatal slices, J. Physiol., 322:51P.
Kapoor, V., and Arbuthnott, G. W., 1983, Release of endogenous dopamine from striatal slices "in vitro". Comparison of release following high K^+ and electrical stimulation, Brit. J. Pharmacol, 76:235P.
Kayaalp, S. O., and Neff, N. H., 1980, Differentiation by ascorbic acid of dopaminergic agonist and antagonist binding sites in the striatum, Life Sci., 26:1837.
Kebabian, J. W., and Calne, D. B., 1979, Multiple receptors for dopamine, Nature, 277:93.
Kehr, W., 1976, 3 methoxytyramine as an indicator of impulse-induced dopamine release in rat brain in vivo, N-S. Arch. Pharmacol., 293:209.
Kehr, W., Carlsson, A., Lindquist, M., Magnusson, T., and Atack, C. V., 1972, Evidence for a receptor-mediated feedback control of striatal tyrosine hydroxylase activity, J. Phar. Pharmacol. 24: 744.
Keller, R., Oke, A., Mefford, I., and Adams, R. N., 1976, Liquid chromatographic analysis of catecholamines. Routine assay for regional brain mapping, Life Sci., 19:995.
Kissinger, P. T., Reshauge, C., Dreiling, R., and Adams, R. N., 1973, An electrochemical detector for liquid chromatography with picogram sensitivity. Anal. Lett., 6:465.
Korf, J., Grasdijk, L., and Westerink, B. H. C., 1976, Effects of electrical stimulation of the nigro striatal pathway of the rat on dopamine metabolism, J. Neurochem., 26:579-584.
Langer, S. Z., 1981, Presynaptic regulation of the release of catecholamines, Pharmacol. Rev. 32:337.
Laverty, R. and Sharman, D. F., 1965, Modification by drugs of the metabolism of 3,4-dihydroxyphenyl-ethylamine, noradrenaline and 5-hydroxtryptamine in the brain, Brit. J. Pharmacol., 24:759.
Leff, S., Adams, J., Hyttel, J. and Creese, I., 1981, Kainate lesion dissociates striatal dopamine receptor radioligand binding sites. Eur. J. Pharmacol., 70:71.
Lehmann, J., Arbilla, S. and Langer, S. Z., 1981, Dopamine receptor mediated inhibition by pergolide of electrically-evoked 3H-DA release from striatal slices of cat and rat: slight effect of ascorbate, N-S. Arch. Pharmacol., 317:31.
Lloyd, K. G., and Bartholini, G., 1975, The effect of drugs on the release of endogenous catecholamines into the perfusate of discrete brain areas of the cat in vivo, Experentia, 31:560.
Miller, J. C., and Friedhoff, A. J., 1979b, Effects of haloperidol and apomorphine on the K^+-depolarized overflow of [3H]dopamine from rat striatal slices, Biochem. Pharmacol., 28:688.
Mitchell, P. R. and Doggett, N. S., 1980, Modulation of striatal [3H]-glutamic acid release by dopaminergic drugs, Life Sciences, 26:2073.

Nose, T. and Takemoto, H., 1974, Effect of oxotremorine on HVA concentration in the striatum of the rat, Eur. J. Pharmacol., 25:51.
Nowycky, M. C. and Roth, R. H., 1978, Dopaminergic neurons: role of presynaptic receptors in the regulation of transmitter biosynthesis, Prog. Neuro-psychopharmacol., 2:139.
O'Keefe, R., Sharman, D. F., and Vogt, M., 1970, Effects of drugs used in psychoses on central dopamine metabolism, Brit. J. Pharmacol., 38:287.
Paalzov, G. and Paalzov, L., 1975, Antinociceptive action of oxotremorine and regional turnover of rat brain noradrenaline, dopamine and 5-HT, Eur. J. Pharmacol., 31:261.
Pasik, P., Pasik, T., and DiFiglia, M., 1979, The internal organization of the neostriatum in mammals in: "The Neostriatum," I. Divac and R. G. E. Oberg, eds., Pergamon Press, Oxford.
Plotsky, P. M., Wightman, R. M., Chey, W., and Adams, R. N., 1977, Liquid chromato-graphic analysis of endogenous catecholamine release from brain slices, Science, 197:904.
Raiteri, M., Cerrito, F., Cervoni, A. M., del Carmine, R., Ribera, M. T. and Levi, G., 1978a, Studies on dopamine uptake and release in synaptosomes, in: Adv. Biochem. Psychopharmacol. Vol. 19. "Dopamine," P. J. Roberts, G. N. Wooidruff., and Iversen, L. L., eds., Raven, New York.
Raiteri, M., Cervoni, A. M., and del Carmine, R., 1978b, Do presynaptic autoreceptors control dopamine release, Nature, 274:706.
Raiteri, M., Cervoni, A. M., del Carmine, R., and Levi, G., 1979, Lack of presynaptic autoreceptors controlling dopamine release in striatal synaptosomes, in: "Presynaptic Receptors", S. Z. Langer, K. Starke, and M. L. Dobocovich, eds., Pergamon, Oxford and New York.
Roth, R. H., Miurrins, L. C., and Walters, J. R., 1976, Central dopaminergic neurones: effects of alterations of impulse flow on the accumulation of dihydroxyphenylacetic acid, Eur. J. Pharmacol., 36:163.
Rowlands, G. J. and Roberts, P. J., 1980, Activation of dopamine receptors inhibits calcium-dependent glutamate release from cortico-striatal terminals "in vitro", Eur. J. Pharmacol., 62:241.
Schultz, W. and Ungerstedt, U., 1978, A method to detect and record from striatal cells of low spontaneous activity by stimulating the cortico-striatal pathway, Brain. Res., 142:357.
Schwartz, R., Creese, I., Coyle, J. T., and Snyder, S. H., 1978, Dopamine receptors localized on cerebral cortical afferents to rat corpus striatum, Nature, 271:766.
Seeman, P. and Lee, T., 1975, Antipsychotic drugs: direct correlation between clinical potency and presynaptic action on dopamine neurones, Science 188: 1217.
Seeman, P., 1981, Brain dopamine receptors, Pharmacol. Rev., 32:230.

Sharman, D. F., 1966, Changes in the metabolism of 3,4-dihydroxyphenyl-ethylamine (dopamine) in the striatum of the mouse induced by drugs, Brit. J. Pharmacol., 28:153.

Somogyi, P., Bolam, J. P. Smith, A. D., 1981, Monosynaptic cortical input and local axon collaterals of identified striatonigral neurones. A light and electron microscopic study using the Golgi-peroxidase transport-degeneration procedure, J. Comp. Neurol., 195:567.

Starke, K., 1981, Presynaptic receptors, Ann. Rev. Pharmacol. Toxicol., 21:7.

Starke, K., Reimann, W., Zumstein, A., and Hertting, G., 1978, Effect of dopamine receptor agonists and antagonists on release of dopamine in the rabbit caudate nucleus in vitro, N-S. Arch. Pharmacol., 305:27.

Theodorou, A., Reavill, C., Jenner, P., and Marsden, C. D., 1981, Kainic acid lesions of striatum and decortication reduce specific [3H] sulpiride binding in rats, so D2 receptors exist post-synaptically on corticostriate afferents and striatal neurones, J. Pharm. Pharmacol., 33:439.

Thomas, T. N., and Zemp, J. W., 1977, Inhibition of dopamine sensitive adenylate cyclase from rat brain homogenates by ascorbic acid, J. Neurochem., 28:663.

Ungerstedt, U., Herrera-Marschitz, M., Jungnelius, U., Stahle, L., Tossman, U., and Zetterstrom, T., 1982, Dopamine synaptic mechanisms reflected in studies combining behavioural recordings and brain dialysis, in: Advances in the Biosciences. Vol. 37. "Advances in Dopamine Research", eds. M. Kohasaka et al., Pergamon, Oxford and New York.

Walters, J. R. and Roth, R. H., 1976a, Dopaminergic neurons: an in vivo system for measuring drug interactions with presynaptic receptors, N-S. Arch. Pharmacol., 296:5.

Walters, J. R. and Roth, R. H., 1976b, Dopaminergic neurons: Alterations in the sensitivity of tyrosine hydroxylase to inhibition by endogenous dopamine after cessation of impulse flow, Biochem. Pharmacol., 25:649.

Watling, K. J. and Williams, M., 1982, Interaction of the putative dopamine autoreceptor agonists, 3-PPP and TL-99, with the dopamine sensitive adenylate cyclase of carp retina, Eur. J. Pharmacol., 77:321.

Wedley, S., Howard, J. L., Large, B. T., and Pullar, I. A., 1978, The inhibition of monoamine uptake into rat brain synaptosomes by selected bicyclo-octanes and an analogous bicyclo-octene, Biochem. Pharmacol. 27:2907.

Westerink, B. H. C., 1979, The effects of drugs on dopamine biosynthesis and metabolism in the brain, in: "The Neurobiology of Dopamine", eds. A. S. Horn, J. Korf, B. H. C. Westerink, Academic, New York and London.

Westerink, B. H. C. and Korf, J., 1976, Turnover of acid dopamine metabolites in striatal and mesolimbic tissues of the rat brain, Eur. J. Pharmacol., 37:249.

Westerink, B. H. C. and Spaan, S. J., 1982a, Estimation of the turnover of (3MT) in the rat striatum by HPLC with (ECD): Implications for the sequence in the cerebral metabolism of dopamine, J. Neurochem. 38:342.

Westerink, B. H. C. and Spaan, S. J., 1982b, On the significance of endogenous 3-methoxytyramine for the effects of centrally acting drugs on dopamine release in the rat brain, J. Neurochem. 38:680.

Westfall, T. C., Besson, M. J., Giorguieff, M. F. and Glowinski, J., 1976, The role of presynaptic receptors in the release and synthesis of 3H-dopamine by slices of rat striatum, N-S. Arch. Pharmacol., 292:279-287.

Westfall, T.C., Perkins, N.A., and Paul, C., 1979, Role of presynaptic receptors in the synthesis and release of dopamine in the mammalian central nervous system, in: "Presynaptic Receptors", S. Z. Langer, K. Starke, M. L. Dubocovich, eds., Pergamon, Oxford and New York.

Wright, A. G., Arbuthnott, G. W., Tulloch, I. F., Garcia-Munoz, M., and Nicolaou, N. M., 1977, Are the striatonigral fibres the feedback pathway, in "Psychobiology of the striatum", A. R. Cools, A. H. M. Lohman, J. H. L. van der Berken, North Holland Biomedical, Elsevier.

Zarzecki, P., Blake, D. J., and Somjen, G. C., 1977, Neurological disturbances, nigrostriate synapses, and iontophoretic dopamine and apomorphine after haloperidol., Exp. Neurol., 57:956.

Zumstein, A., Karduck, W., and Starke, K., 1981, Pathways of dopamine metabolism in the rabbit caudate nucleus in vitro, N-S. Arch. Pharmacol., 316:205.

ANATOMY AND NEUROPHYSIOLOGY OF THE SUBTHALAMIC EFFERENT NEURONS

B. Rouzaire-Dubois, C. Hammond, J. Yelnik* and J. Féger

91 Bd de l'Hôpital, 75634 Paris, Cedex 13 and
*INSERM U3, 47 Bd de l'Hôpital, 75013 Paris, France

A. MORPHOLOGY AND ELECTROPHYSIOLOGY OF SUBTHALAMIC NEURONS

The subthalamic nucleus, part of the so-called basal ganglia, is a well-defined nucleus surrounded by fibers, the internal capsule, the ansa and fasciculus lenticularis, and in the rat the cerebral peduncle (Fig. 1). In primates its vascular or experimental lesioning gives rise to a specific motor syndrome characterized by involuntary, large, violent and stereotyped movements of the limbs (Whittier and Mettler, 1949a; Carpenter et al., 1950; Hammond et al., 1979). If the lesion is unilateral, the entire syndrome is restricted to the contralateral side. It represents a rare example

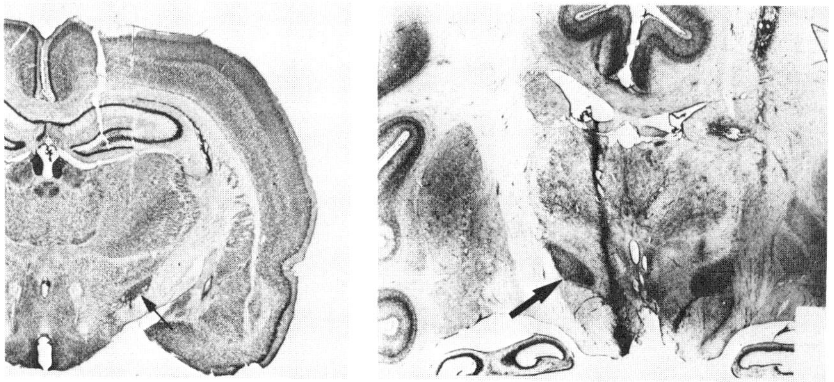

Fig. 1. Frontal sections in the rat (left) and monkey (right) brains at the level of the STN (arrow) showing its location with reference to internal capsule, peduncle and SN.

of close correlation between a restricted central lesion and the sudden development of a specific motor disorder. A changed activity of the subthalamic nucleus efferent neurons seems to correspond to the appearance of the involuntary movements, since a concomitant or secondary lesion of the internal pallidum or its projection fibers (ansa and fasciculus lenticularis) prevents or abolishes the dyskinesia (Carpenter et al., 1950). On this background we conducted a series of experiments to study the morphological and electrophysiological characteristics of the subthalamic nucleus efferent neurons. The experiments were mostly performed in rats but they will

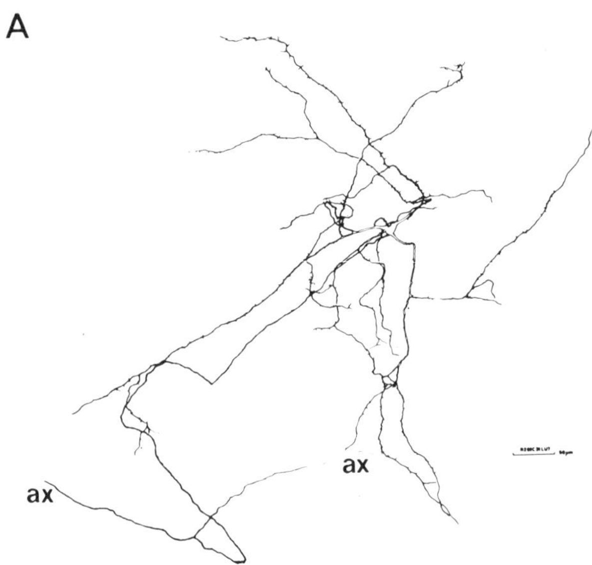

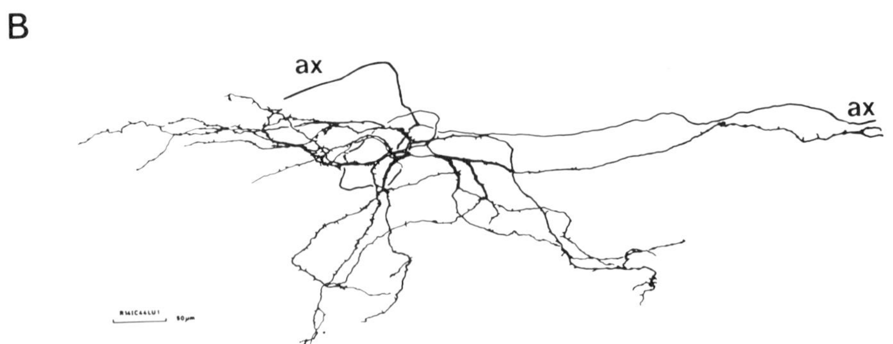

Fig. 2.

be compared to results obtained in primates either by us or by others. We will see that the old concept of the subthalamic nucleus (STN) as a satellite nucleus of the pallidum is no longer tenable. Its particular position in relation to the striato-pallido-nigro system and its effects on the two output nuclei of the basal ganglia (internal pallidum and substantia nigra) will be discussed.

Somato-dendritic and Axonal Morphology

STN neurons were intracellularly labeled with horseradish peroxidase (HRP). All the STN stained neurons studied were projection neurons since their axons were followed beyond the limits of the nucleus. They presented thin and flexuous dendrites bearing few pedunculated spines (Figs. 2 and 3). On the basis of a quantitative study including topological (Percheron, 1979) and three-dimensional parameters (Yelnik, et al., 1981), they were assumed to belong to a single neuronal population of Golgi type I neurons (Hammond and Yelnik, 1983). These parameters included the number of dendritic stems (a mean of 4, 3.76 ± 0.70) and of dendritic tips (a mean of 27, 26.6 ± 4.7). Computer reconstruction of the

Fig. 2. Camera lucida drawings of three different STN efferent neurons reconstructed from serial sagittal sections. The soma is polygonal (A and C) or fusiform (B) and the flexuous dendrites bear few spines. The parent axon gives off two (A and B) or three (C) collaterals (ax). The neuron C presents also a recurrent collateral (arrow). (Hammond & Yelnick, 1983, courtesy Pergamon, Oxford)

dendritic fields showed them to have a flat ellipsoid shape (100 µm x 600 µm x 300 µm), their principal plane being parallel to that of the nucleus (Fig. 4). The dimension of a dendritic field is very close to that of the nucleus, one STN neuron occupying nearly the whole nucleus (Fig. 3 bottom). Comparing these results with the Golgi study of Yelnik and Percheron (1979) in cat and primate, the characteristics and dimensions of the somato-dendritic tree appear quite similar in all these species though the nucleus varies greatly in size. Therefore, while a STN neuron covers nearly the whole nucleus in the rat, it occupies only one-fifth to one-ninth of its volume in primates. While intrinsic spatial organization in the rat STN seems unlikely, the situation is different in monkey and man where a topographic distribution of pallidal (Carpenter et al., 1981) and cortical (Hartmann-von-Monakow et al., 1978) afferent fibers can at leastly partly be maintained.

The axonal morphology was similar for each neuron studied. The parent axon gave off, inside or near the nucleus, two (very rarely three) branches, one running rostrally towards the entopeduncular nucleus and the globus pallidus, and the other caudally to the substantia nigra (Figs. 2 and 3). This latter branch was seen in two cases to enter the substantia nigra and divide into several thin collaterals. The caudal branch was always thinner (0.4 ± 0.1 µm) than the rostral one (0.7 ± 0.2 µm) or the parent axon (0.8 ± 0.2 µm). In a previous study (Deniau et al., 1978), we estimated that at least 78% of rat STN neurons project to both the pallidum and substantia nigra, while Van der Kooy and Hattori (1980) with double retrograde tracing technique estimated 96%. No data are at present available in the monkey.

Electrophysiological Study

The synaptic responses evoked in its two major target nuclei, the entopeduncular nucleus and the substantia nigra (SN), by STN stimulation, were studied in rats bearing chronic lesions to suppress the possible activatiion of passing fibers (Hammond et al., 1978; Hammond et al., 1983a). We also studied the minor projection to the pedunculopontine tegmental nucleus (Hammond et al., 1983b) and its possible role as an output nucleus of basal ganglia.

The subthalamo-entopeduncular pathway. In rats bearing chronic lesions of the striatum and SN for suppressing striato-entopeduncular fibers, stimulation of STN evoked a short latency suppression of spontaneous entopeduncular activity (10-60 ms duration). This was the major response recorded (89% of entopeduncular cells tested). Of these cells, 75% were identified as projecting to lateral habenula (Fig. 5) and 10% to ventral nuclei of the thalamus. The short latency suppresion of spontaneous activity was strong enough to delay the antidromic somato-dendritic spike of entopeduncular cells evoked by stimulation of lateral habenula (Fig. 5), and to reduce or

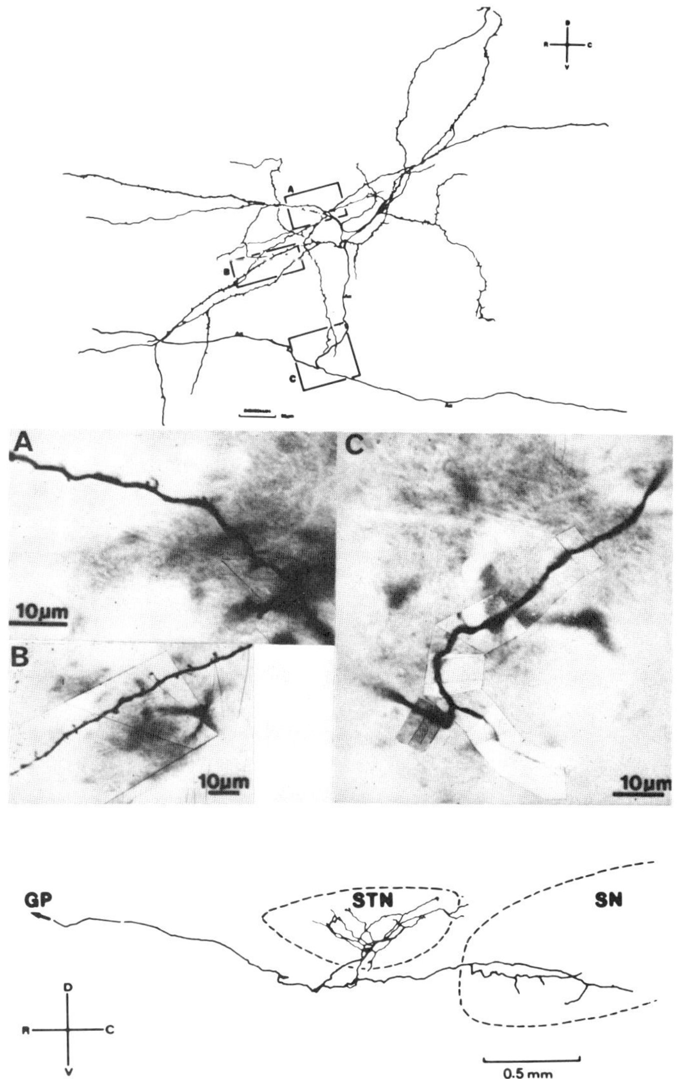

Fig. 3. Camera lucida drawing of reconstructed STN neuron. Rectangles refer to microphotographic montages showing spiny dendrites (A & B) and branching point of axon (C). Note different diameters of two axonal branches. Bottom: computer reconstruction of axonal arborization of STN neuron. Some dendrites exceed limits of nucleus (STN), and caudal axonal branch enters SN, where it divides in 2 branches forming 3 collaterals coursing dorso-ventrally in pars reticulata. Computer drawing does not represent different dendritic and axonal diameters. GP: globus pallidus. (From Hammond & Yelnick, 1983, courtesy Pergamon, Oxford)

A

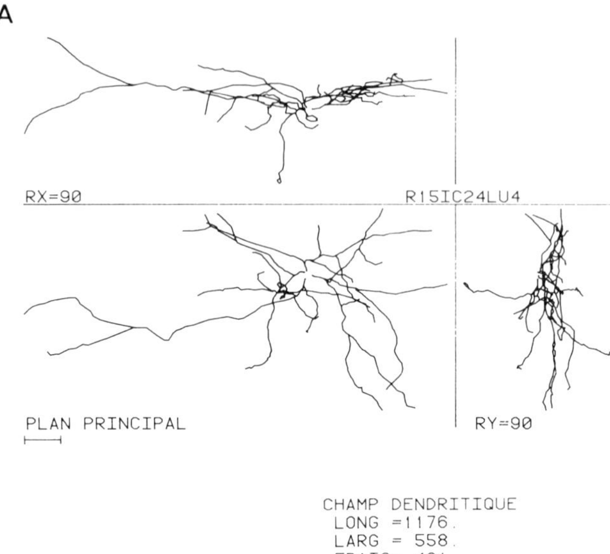

B

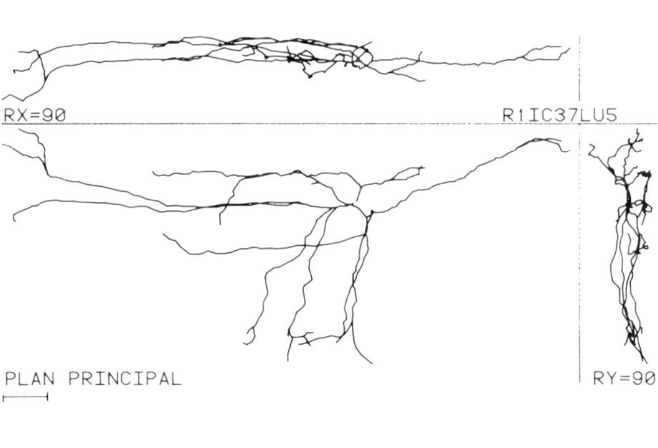

Fig. 4. Computer reconstruction of the dendritic arborizations of two different STN neurons (A and B) in the principal plane and after 90° rotations around the horizontal (RX 90) and the vertical (RY 90) axis. The respective dimensions of the dendritic trees are given (champ dendritique) under each drawing. Neuron A is also shown in Fig. 3 top and neuron B in Fig. 3 bottom.

suppress their orthodromic excitation on stimulation of the pedunculopontine nucleus (Hammond et al., 1983a). On the basis of anatomical and morphological studies, the inhibitory response of entopeduncular cells to STN stimulation is most likely due to a direct postsynaptic effect. Disfacilitation through local inhibitory interneurones is unlikely, since no Golgi type II neurons have so far been described in the pallidal complex (Fox et al., 1974; Iwahori and Mizumo, 1981). Recurrent inhibition is also unlikely since very few entopeduncular cells project to STN (Carter and Fibiger, 1978; Rinvik et al., 1979; Carpenter et al., 1981). Moreover, the mean latency of the inhibitory response (0 to 8 ms) is in agreement with the mean latency of antidromic invasion by STN stimulation (Deniau et al., 1978). A similar study in awake monkeys (Féger et al., 1975) showed that STN stimulation evokes the same type of response in the external pallidum, a suppression of spontaneous activity (Fig. 5 top).

The subthalamo-nigral pathway. Using two types of rat preparation, we showed that STN stimulation evokes an orthodromic excitatory response of SN cells (pars reticulata and compacta cells) (Fig. 6). These preparations consisted of a total deafferentation of nigral cells from rostral structures such as striatum, accumbens and globus pallidus, by a chronic hemitransection rostral to STN. This was performed either alone (Hammond et al., 1983a) or together with a chronic lesion of the pedunculopontine nucleus (Hammond et al., 1978) which is known to project to the SN and STN (Rinvik et al., 1979; Moon Edley and Graybiel, 1980; Nomura et al., 1980; Carpenter et al., 1981). The excitatory response of SN cells had a short latency (mean latency 4.16 ms) and was the major response recorded (85.7% of the tested cells). This response does not seem to represent a very powerful input since it was usually evoked only with double shock stimulation. Its possible monosynaptic mechanism is suggested by its following high frequency stimulation (200 to 700/s) (Fig. 6). However we still lack an ultrastructural study of the sites of termination of STN fibers in SN.

The subthalamo-pedunculopontine pathway. The experiments were performed in normal and chronically decorticated rats in order to reduce the possible activation, at the level of the peduncle, of cortical afferent fibers to the pedunculopontine nucleus (PPN) (Kuypers and Lawrence, 1967). PPN cells were identified by antidromic activation either from globus pallidus or from STN (Fig. 7). As these cells were silent, they were activated by continuous iontophoretic ejection of glutamate. In such preparations, STN stimulation evoked a short latency suppression of PPN activity (mean duration 24 ms, no detectable latency). A concomitant anatomical study using retrograde transport of the fluorescent compound Fast blue (Hammond et al., 1983b) confirmed the existence of a direct projection from STN to PPN. It also appeared that this projection involves a minimum of 1% of STN cells. Is this projection a collateral of the subthalamo-nigral projection? Since double

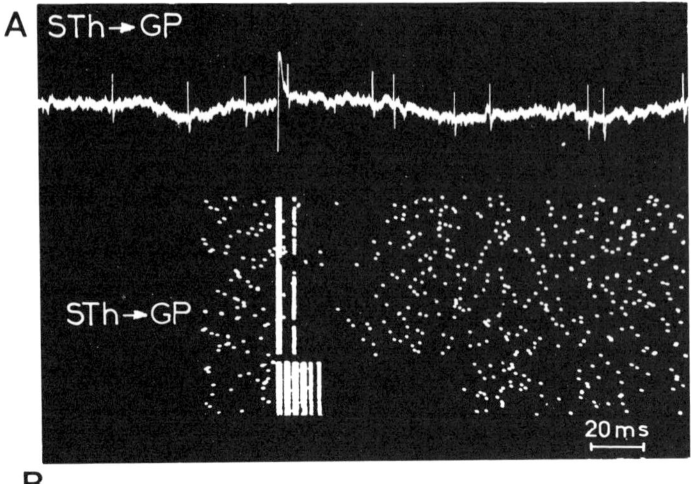
A
STh→GP
STh→GP
20 ms

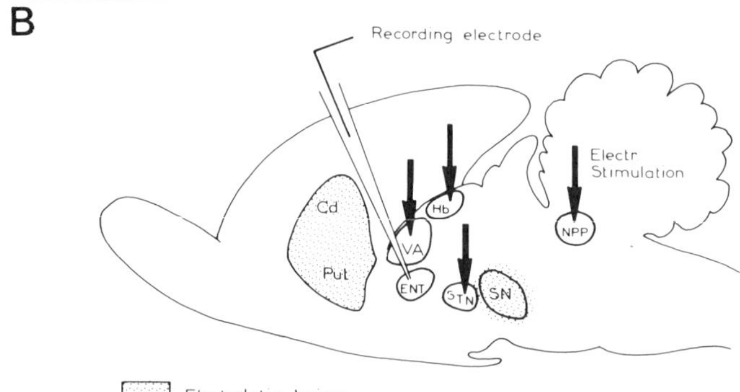
B
Recording electrode
Electr
Stimulation
Cd
Put
Hb
VA
ENT
STN
SN
NPP
Electrolytic lesions

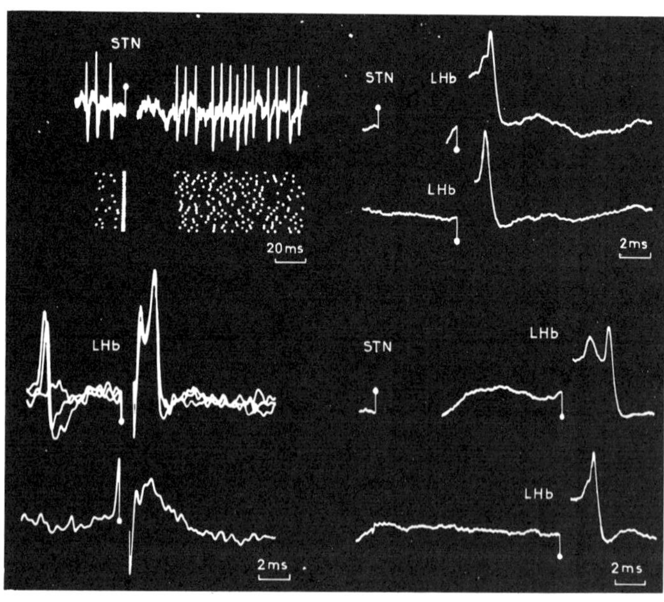
STN
20 ms
STN
LHb
LHb
2 ms
LHb
STN
LHb
LHb
2 ms
2 ms

antidromic activation of STN cells from PPN and SN could not be obtained because of a lack of antidromic activation from PPN stimulation, we cannot answer this question.

The PPN has sometimes been presented as a possible output nucleus of the basal ganglia to caudal structures such as n. reticularis pontis or spinal cord (Beckstead et al., 1979; Sakai et al., 1979; Moon Edley and Graybiel, 1980; Nomura et al., 1980). However, in view of the small percentage of STN cells projecting to PPN in the rat and monkey (Nauta and Cole, 1978; Carpenter et al., 1981) this would not seem to be its principal role, at least concerning STN.

Discussion

On the basis of quantitative morphological studies it has been shown that STN efferent neurons in the rat and primate belong to a single neuronal population of Golgi type I neurons (Rafols and Fox, 1976; Iwahori, 1978; Sakai et al., 1979; Yelnik and Percheron, 1979; Hammond and Yelnik, 1983). In the rat, we suggested along with other authors that all or nearly all STN efferent neurons project to both the entopeduncular nucleus (and/or globus pallidus) and SN (Deniau et al., 1978; Van der Kooy and Hattori, 1980; Hammond and Yelnik, 1983). A bifurcated axon with branches projecting rostrally and caudally seems to be a consistent characteristic of STN efferent neurons.

Fig. 5. Electrophysiological study of subthalamo-pallidal (STn-GP) and subthalamo-entopeduncular (STn-EP) pathways in monkey (A) and rat (B). A: spike and raster displays show suppression of spontaneous activity after subthalamic (STN) stimulation of a GP (external segment) neuron antidromically activated from STN (antidromicity of first spike is not demonstrated here). B Upper: method for study of STn-EP pathway in chronically lesioned rats. B Lower: (read left to right and top to bottom) suppression of spontaneous entopeduncular activity evoked by STN stimulation (duration 30 ms) presented in spike and raster displays. This entopeduncular cell was antidromically activated from lateral habenula (LHb). Stable latency of antidromic spike (Lat=1.4 ms, 3 superimposed traces) and collision with a spontaneous spike (collision time=0.6 ms). Timing of antidromic invasion from LHb with and without STN-evoked suppression of spontaneous activity. When LHb stimulus is 5.2 ms after STN conditioning shock, somato-dendritic invasion is delayed by 0.6 ms. It is delayed by 1.00 ms when the LHb stimulus is 12 ms after STN. All recordings from the same entopenduncular neuron. (From Hammond et al., 1983, courtesy Pergamon, Oxford)

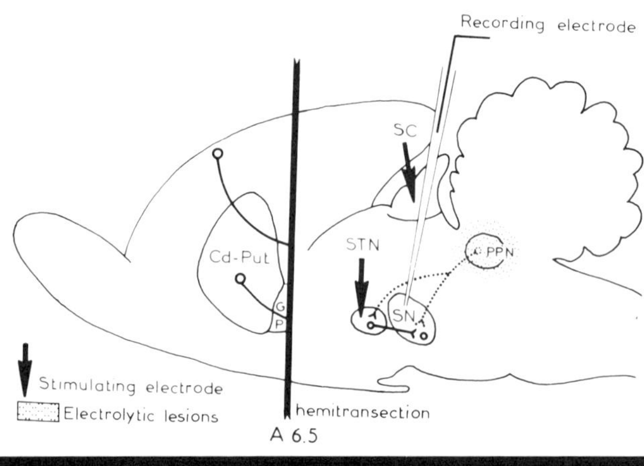

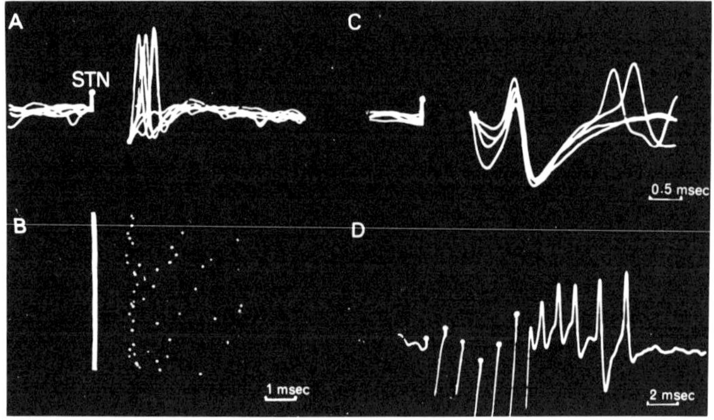

Fig. 6. Top: methodological procedure for study of the subthalamonigral pathway. Dotted line indicates possible branched projection from pedunculopontine nucleus (PPN) to STN and SN. Stimulation of such fibers could induce an excitatory response of SN cells. Nigral pars reticulata cells were identified by antidromic activation from superior colliculus (SC). Pars compacta cells were identified on the basis of their spontaneous activity. Bottom, short latency excitatory response of a pars reticulata cell. This response presented a non-stable latency (1.5 -2.5 ms) as shown in spike traces (5 superimposed traces, A) and raster display traces (B). It followed high frequency stimulation at 200/s (C, 4 superimposed traces) and 715/s (D). (Hammond et al., 1983, courtesy Pergamon, Oxford)

It therefore appears that a single STN neuronal population could modulate the activity of entopeduncular and SN output cells. However this modulation could be different in the two target nuclei. Firstly, due to the branching which represents a point of low safety factor for spike transmission (Krnjević and Miledi, 1959; Iwahori,

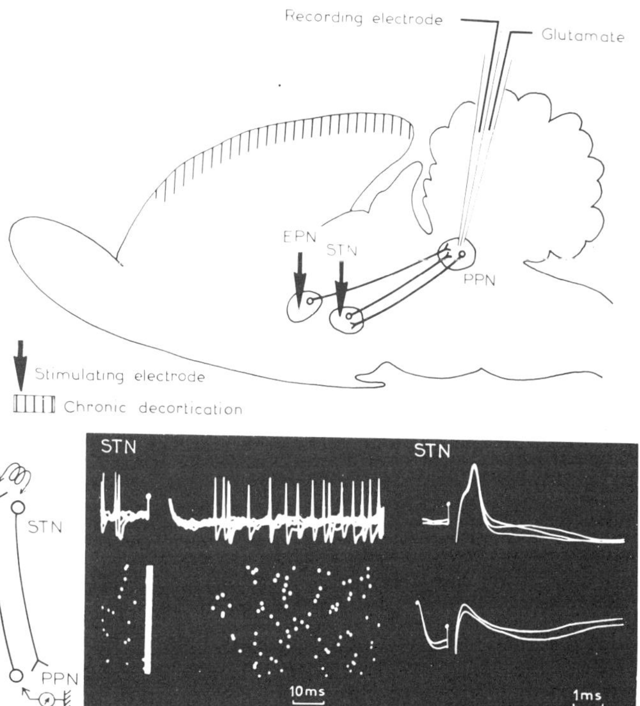

Fig. 7. Top: experimental design for study of the subthalamo-pedunculopontine pathway. PPN cells were identified by antidromic activation from entopeduncular nucleus (EPN) or STN stimulation. Bottom left: suppression of glutamate-evoked activity of a PPN cell after STN stimulation (duration 20 ms) shown in spike and raster displays. Right: the same PPN cell antidromically activated from STN. Stability of latency of the antidromic spike (0.7 ms, three superimposed traces) and its collision with a glutamate-evoked spike (collision time =1.7 ms, 2 superimposed traces). (From Hammond et al., 1983, courtesy Pergamon, Oxford)

1978; Smith, 1980), and to the differing diameters of the branches, the temporal pattern of spikes in them is likely to be different (Grossman et al., 1973). Moreover, intermittent failure of spike transmission in such circumstances occurs particularly with high frequency bursts of spikes (Waxman, 1972). STN efferent cells have been shown to respond to peripheral stimulation (Fig. 8) with bursts of spikes (3 to 7 spikes). This response, which needs an intact sensori-motor cortex, has also been obtained with cortical stimulation (Hammond et al., 1978). We hence suggest that cortical activation of STN efferent neurons could have different effects on entopeduncular and SN cells.

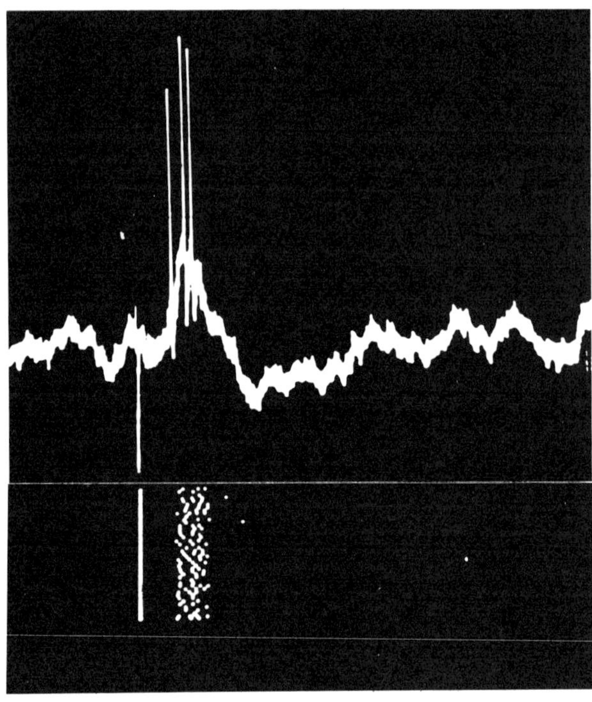

Fig. 8. Excitatory response (3-4 spikes) of rat subthalamic neuron to contralateral vibrissae stimulation. This response (lat=11 ms) is followed by suppression of spontaneous activity for 100 ms. This response disappears during cooling of the sensori-motor cortex.

Do all STN efferent neurons in primates project to both the pallidal complex and the substantia nigra? On the basis of Nauta and Cole's (1978) findings, the subthalamo-nigral projection appears less massive than the subthalamo-pallidal. This suggests that a separate subthalamo-pallidal neuronal population might exist in the monkey.

It is certainly of importance that the major projection sites of STN efferent neurons, the pallidal complex and the substantia nigra (Fig. 9) are also the targets of striatal efferent fibers. The STN appears therefore in a position to modulate the entire outflow of the striato-pallido-nigral system. In order to specify the interactions between striatal and STN efferent fibers at the level of the pallidum and substantia nigra, the ultra-structural synaptic organization should be known. Di Figlia et al. (1982), in their study of the primate pallidum, describe several types of afferent fibers and

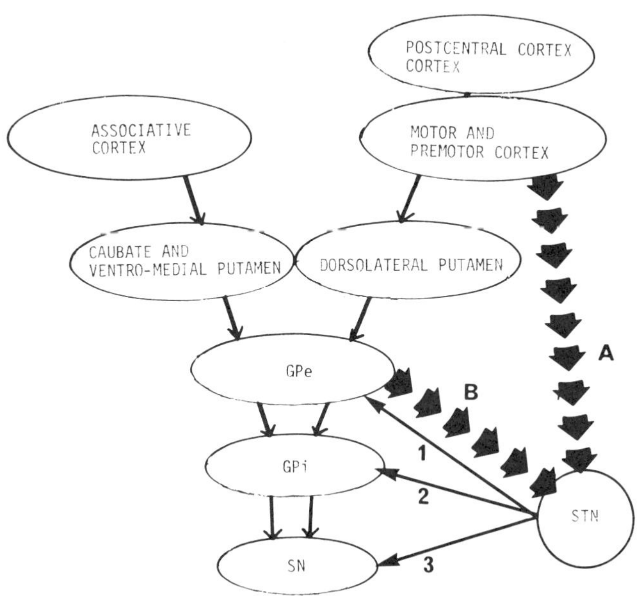

Fig. 9. Diagrammatic representation of the major connexions of STN. A, direct cortical afferent fibers. B, the massive afferent projection from the external pallidal segment (GPe) which conveys cortical information transformed in the caudate-putamen and in GPe. 1 and 2 are STN efferent projections to both pallidal segments and 3 to substantia nigra (SN).

terminations. One type corresponds to the radial fibers of striatal origin. They emit thin collaterals which follow the long pallidal dendrites and give multiple contacts along a single dendrite. Another type is attributed to subthalamic fibers and consists of thin ramifications which form symmetrical contacts with the somata or the proximal dendrites of pallidal neurons. Pallidal dendrites are completely covered with synaptic endings, many of striatal origin and some, located proximally, of subthalamic origin.

Having noted that the STN belongs to a side loop of the basal ganglia at the level of the output nuclei, the internal pallidum (or entopeduncular nucleus) and the SN, the following question arises: what does the STN tell the rest of basal ganglia? The STN receives two major afferents: direct cortical fibers and a massive projection from external pallidum (Fig. 9). The STN and the caudate-putamen are the only basal ganglia nuclei known to receive direct cortical projections. The projection from motor cortical areas has been particularly studied (Hartmann-von-Monakow et al., 1978), but other areas also project to STN (Rouzaire-Dubois - personal observation). The massive pallido-subthalamic projection (Rinvik et al., 1979; Carpenter et al., 1981), conveys cortical information transformed at the levels of the caudate-putamen and the external pallidum. In the STN, a comparison between these two types of cortical information is likely to occur, and so conditionally influence the activity of the output nuclei of basal ganglia, the internal pallidum and the SN.

B. RECIPROCAL CONNECTIONS WITH THE PALLIDAL COMPLEX

Numerous anatomical studies performed in recent years in various animal species clearly indicate the main connections of the subthalamic nucleus (STN). It receives afferents from cortex (e.g. Kitai and Deniau, 1981) and from the external segment of the pallidum (GPe) (Carpenter et al., 1968, 1981; Grofovà, 1969; Rinvik et al., 1979), and sends efferents to both parts of the pallidal complex (the external pallidum GPe, and the internal GPI, or its feline and rodent equivalent, the entopeduncular nucleus EPN) and to the SN (Carpenter and Strominger, 1967; Nauta and Cole, 1978; Ricardo, 1980; Carpenter et al., 1981). The importance of the pallido-subthalamo-entopeduncular pathway was emphasized in one of the first experimental studies of the STN (Whittier and Mettler, 1949).

The present study summarizes the nature of pallido-subthalamic and subthalamo-entopeduncular synaptic effects, and the identity of the transmitter involved. Our experimental conditions (extracellular recordings and microiontophoresis) were arranged in order to eliminate the familiar problem of fibers of passage, these nuclei being embedded in fibers of the internal capsule. The results are discussed in relation to the input-output organization of the STN.

Pallido-Subthalamic Pathway

The pallidal synaptic effect on subthalamic neurons. GP stimulation induced a depression of spontaneous firing in 36% of the STN cells recorded (Fig. 10) of mean duration 10-20 ms. This was never observed on control stimulation located either in the internal capsule (short latency excitation) or in striatum (short latency excitatory-inhibitory response); we therefore concluded that the pallidum exerts inhibitory control over STN cells.

Topographical arrangement of pallido-subthalamic fibers. Considering the small proportion of STN neurons inhibited, we looked for topography in the pallido-subthalamic projection (Fig. 11). STN neurons inhibited by rostro-medial GP stimulation were located in the most medial part of the STN, and those inhibited by caudo-medial GP stimulation were mostly in its central part. Lateral GP stimulation

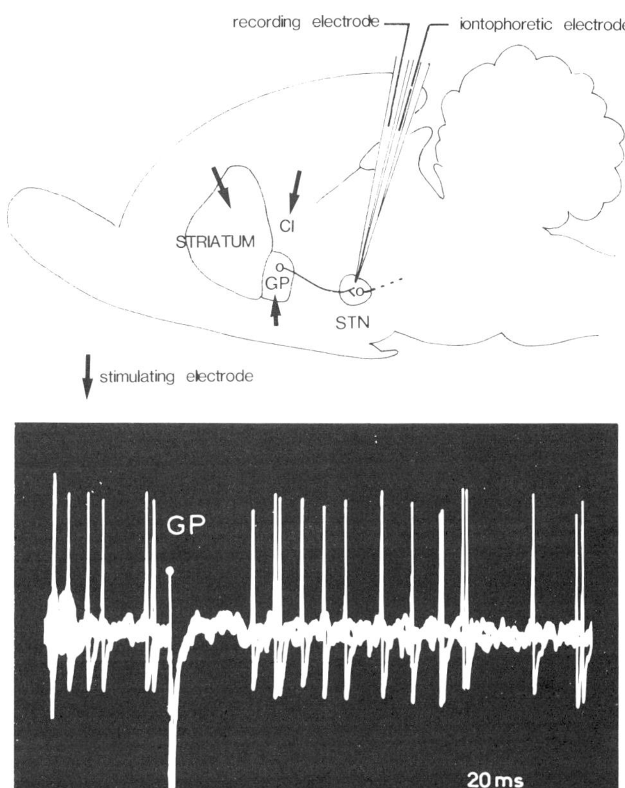

Fig.10. From top to bottom:
Experimental procedure to determine the nature of the pallido-subthalamic (GP-STN) inhibitory transmitter. Effects of GP stimulation on STN cells were compared with the effects of control stimulation located either in the striatum (St) or in the internal capsule (CI). Extracellular unit response of STN neuron to GP stimulation (3 superimposed traces). GP stimulation (0.2 ms duration, 270 μA current intensity) induced inhibition of spontaneous firing for 30 ms. (From Rouzaire-Dubois et al., 1980, courtesy Elsevier, Amsterdam)

either rostrally or caudally, induced inhibition of cells located in the most lateral part of STN.

Pharmacological antagonism of GP-induced inhibition of STN neurons. All the STN cells recorded were inhibited by iontophoresis of low doses of GABA (40 nA) without any tachyphylaxis (Fig. 12). In most experiments, the GP-induced inhibitory response of STN cells was either totally abolished or decreased in duration by

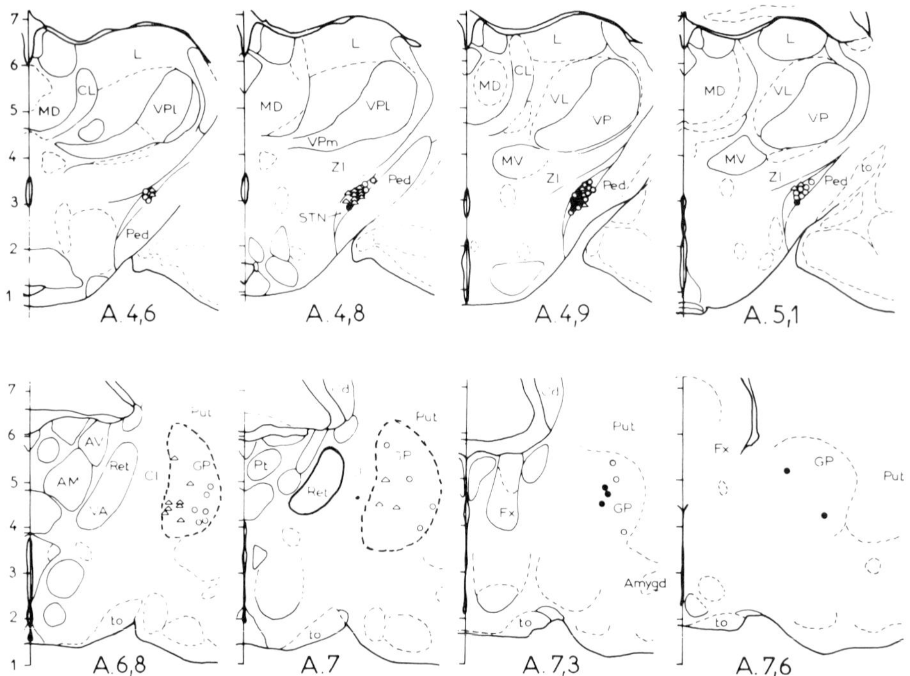

Fig.11. Drawings of frontal sections (A 4.6 to A 7.6) showing the topographical organization of GP - STN projections.
Upper: STN neurons inhibited by GP stimulation.
Lower: stimulated sites in GP. Symbols represent: dots, rostro-medial GP stimulation; triangles, caudo-medial GP stimulation;open circles, lateral GP stimulation.
Rostro-medial GP stimulation inhibited the most medial STN neurons, though lateral GP stimulation inhibited those located more laterally in STN.
Abbreviations: CI, internal capsule; GP, globus pallidus; Ped, peduncles; Put, putamen; STN, subthalamic nucleus; to, optic tract; ZI, zona incerta.

microiontophoretically applied bicuculline or picrotoxin (Figs. 12 and 13). In all cases, the spontaneous firing level of STN cells increased from 50% to 270% with bicuculline, and from 20% to 200% with picrotoxin. The antagonism of GP-induced inhibition by bicuculline was not simply due to an increase in spontaneous activity, since microiontophoretic application of glutamate, which increased the spontaneous activity to a comparable level, was not effective against the GP-induced inhibition. The inhibition was not affected by atropine, eliminating any role of acetylcholine (ACh) in the GP-STN projection.

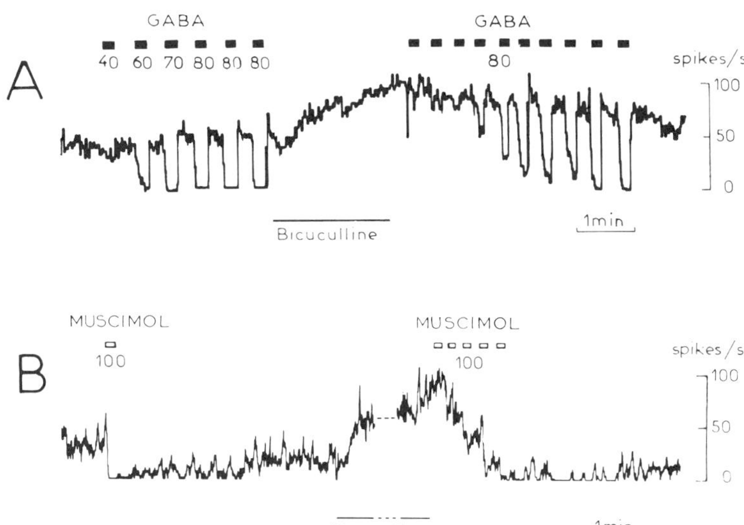

Fig.12. Inhibition of firing of STN neuron by GABA (A) and Muscimol (B). (A), note dose dependant response, and absence of desensitization when GABA applications repeated. (B), note normal level of firing reappeared only 4 min. after end of muscimol application. Bicuculline blocked GABA or muscimol-induced inhibition. In this and later figures, duration of the current ejections is indicated by horizontal bars, and current values are given in nA. (From Rouzaire-Dubois et al., 1980, courtesy Elsevier, Amsterdam)

In some cases strychnine modified the synaptic inhibitory response, but unfortunately its specificity was not tested here. The hypothesis of a GP-STN glycinergic pathway will be discussed below.

Subthalamo-entopeduncular Pathway

Subthalamic synaptic effect on EPN neurons. STN stimulation induced an inhibition of firing in EPN cells, whose electrophysiological characteristics have been discussed elsewhere (Hammond et al., 1983). Our experiments were performed on EPN cells identified as projecting to ventral anterior thalamic nucleus (VA) or lateral habenular nucleus (LHb), recorded in rats with chronic lesions of the striato-nigral pathway in order to eliminate fibers of passage (Fig. 14).

Pharmacological characteristics of EPN neurons. All EPN cells were inhibited by low doses of microiontophoretically applied GABA without any desensitization. Muscarinic cholinergic drugs (ACh and carbamylcholine) had either a small excitatory effect at long latency

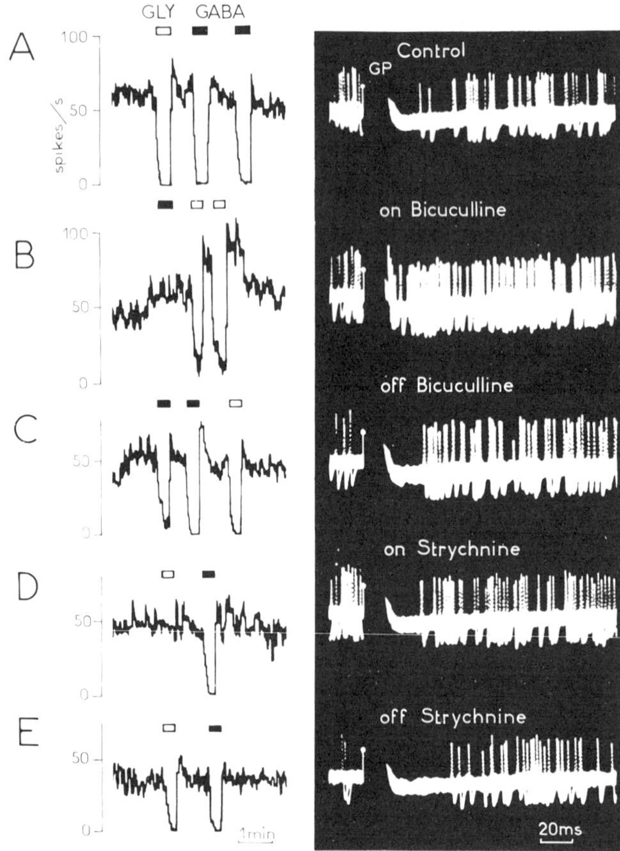

Fig.13. Recordings from a single STN neuron.
Left: ratemeter records of discharge showing effect of microiontophoretically applied GABA or glycine.
Right: extracellular inhibitory response of same STN neuron to GP stimulation (0.2 ms duration, 320 uA).
A: Before applied bicuculline, GABA (70 nA for 20 s), glycine (70 nA for 20 s) and GP (53 superimposed traces) induce inhibitory responses. B: After 6 min application of bicuculline (60 nA). Bicuculline selectively blocks GABA- and GP-induced inhibition (20 traces). C: Recovery of GABA (90 nA for 20 s), glycine (100 nA for 20 s) and GP stimulation-induced (30 traces) inhibitory responses 5 min after bicuculline discontinued. D: Strychnine (60 nA for 5 min) selectively blocks glycine-induced inhibitory response (100 nA for 20 s). E: Recovery of glycine-induced response 7 min after strychnine discontinued. (From Rouzaire-Dubois et al., 1980, courtesy Elsevier, Amsterdam)

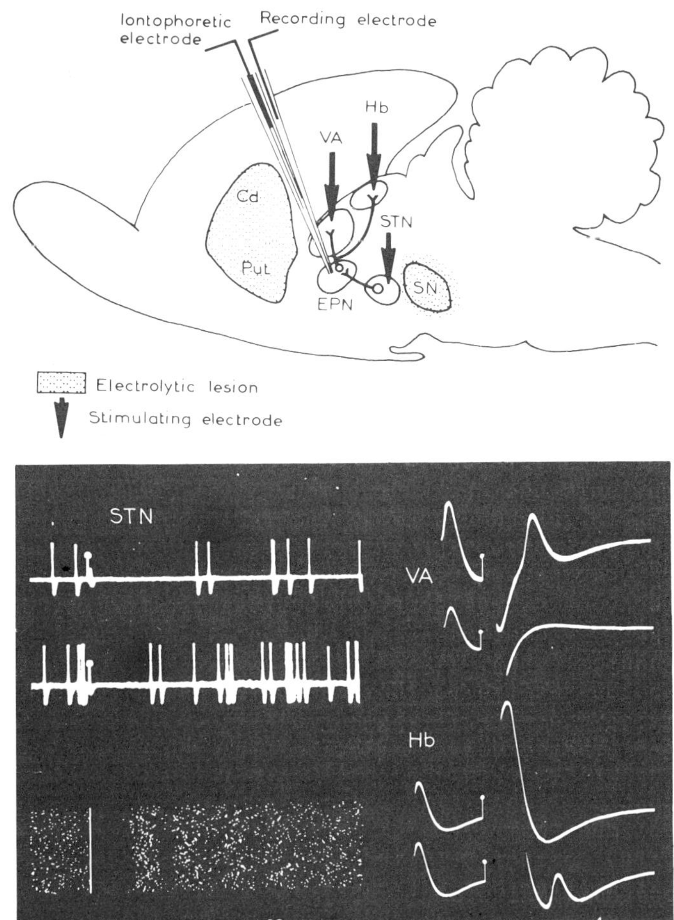

Fig.14. Upper: experimental procedure used to identify the STN-EPN inhibitory transmitter. Top avoid stimulation of passing fibers (collaterals to EPN of inhibitory caudato-nigral pathway), experiments performed on rats with chronic electrolytic lesions of striatum (St) and of substantia nigra (SN). EPN recorded neurons identified by antidromic responses to stimulation of their projection areas (lateral habenula LHb, ventral anterior thalamic nucleus VA). Lower: extracellular unit response of EPN neuron to subthalamic nucleus stimulation. Left: STN stimulation (140 µA, 0.2 ms) induced inhibition of firing for 20 ms (from top to bottom: 4 superimposed traces, 15 superimposed traces, raster display). Right: this EPN cell was antidromically activated from VA and LHb. (Hammond et al., 1983, Pergamon, Oxford)

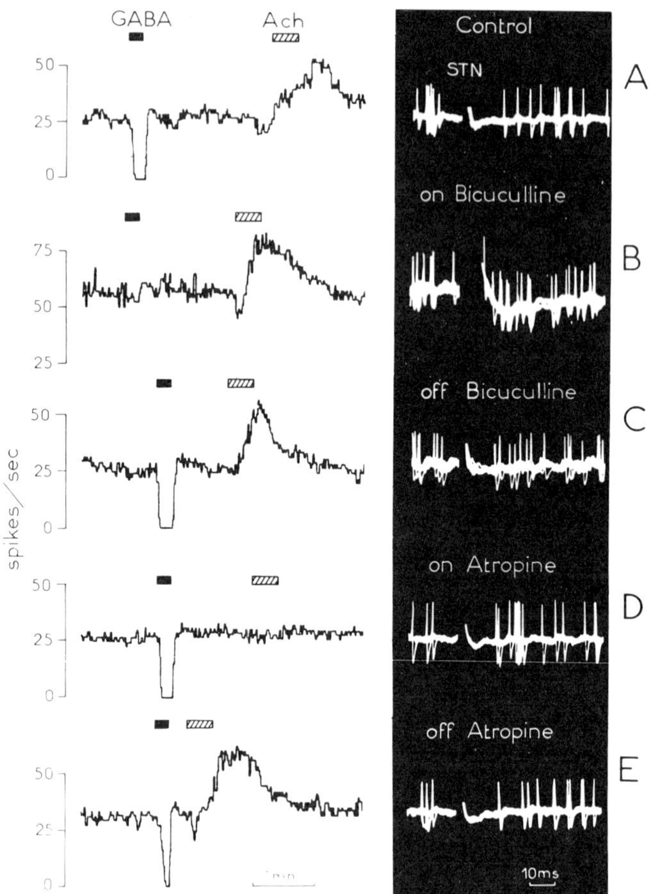

Fig.15. Recordings from single EPN neuron.
Left: Ratemeter records of firing frequency showing effect of microiontophoretically applied GABA and acetylcholine (ACh). Right: extracellular unit inhibitory response of the same EPN neuron to STN stimulation (0.2 ms duration, 175 µA).
A: before bicuculline application, GABA (150 nA for 15 s) and STN stimulation (11 superimposed traces) induce inhibitory responses and ACh (700 nA for 20 s) induces excitation.
B: after 8 min bicuculline application (250 nA). Bicuculline selectively blocks GABA and STN-induced inhibitory responses (10 superimposed traces). C: recovery of GABA and STN-induced inhibitory responses 24 min after bicuculline application discontinued (11 superimposed traces).
D: atropine (300 nA for 5 min) selectively blocks ACh-induced excitation without any action on STN-evoked inhibition.
E: recovery of ACh-induced response 2 min after atropine application discontinued. (From Rouzaire-Dubois et al., 1983, courtesy Elsevier, Amsterdam)

(44% of the cells) or no effect (56% of the cells). Another effect of local application of ACh to EPN cells, whether or not excited by it, was modification of the spontaneous discharge pattern in 20% of neurons, from very regular to irregular with bursts of spikes.

Pharmacological antagonism of STN-induced inhibition. The effects of microiontophoretically applied bicuculline, strychnine and atropine were tested on the STN-inhibition of EPN neurons. Great care was taken to use these antagonists in specific amounts, and this was always verified during our experiments using crossed tests. For instance, the dose of bicuculline tested against STN-induced inhibition was always verified as antagonizing only GABA-induced inhibition, without modifying responses to either ACh or glycine. STN-evoked inhibition was affected only by microiontophoretic specific doses of GABA antagonists, not by either atropine or strychnine (Figs. 15 and 16). Furthermore, experiments using simultaneous application of bicuculline and glycine, in order to maintain the control level of spontaneous activity, led us to conclude that bicuculline blockade is specific, and to suggest that GABA is probably involved in the subthalamo-entopeduncular pathway.

Discussion

Pallido-subthalamic pathway. The GP-evoked inhibition obtained in the present study began immediately after the stimulation artefact, without detectable latency in our extracellular recordings, thus suggesting a direct effect of GP on STN neurons. These data are in good agreement with a recent intracellular study performed in decorticated rats (Kita et al., 1983) where GP stimulation evoked a short duration and short latency hyperpolarizing potential verified as a monosynaptic IPSP. Similar results were previously described in awake monkeys (Ohye et al., 1976) and in the cat with extracellular (Tsubokawa and Sutin, 1972) or intracellular (Frigyesi and Rabin, 1971) techniques. These authors also recorded short latency excitatory responses in locally anesthetized or "encéphale isolé" cats, but no control stimulation was either tested or taken into account. In particular, the existence of a direct excitatory cortico-subthalamic pathway, partly running in the internal capsule, has been demonstrated and can explain the short latency excitation recorded after stimulation in the capsular region.

The present data confirm a medio-lateral organization of the pallidal projection onto medial to lateral parts of the STN. The topographic relationships present a certain overlap in a rostrocaudal direction. These findings are more in agreement with anatomical and biochemical data obtained in cat and rat (Grofova, 1969; Fonnum et al., 1978; Vincent et al., 1982), than in monkey, where the topographical organization of the pallido-subthalamic projection seems a little different, although undoubtedly more precise (Carpenter and Strominger, 1967; Carpenter et al., 1968; Carpenter et

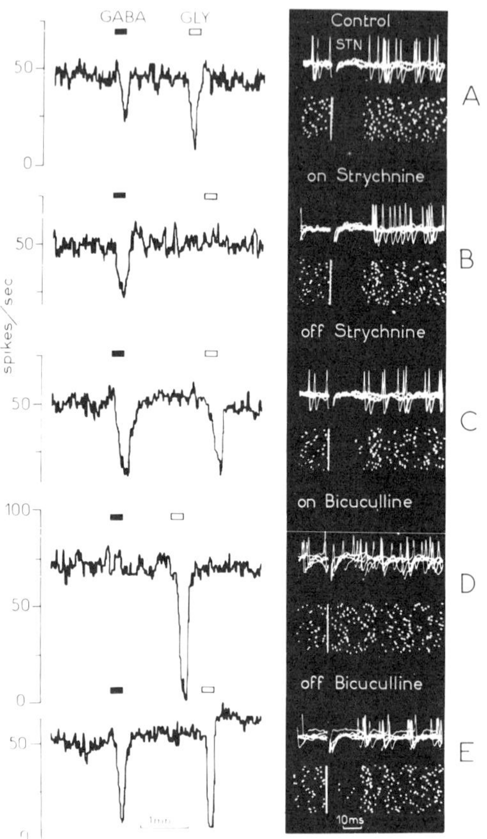

Fig.16. Effect of strychnine on STN-induced inhibitory response. Strychnine failed to suppress STN-induced inhibition of all EPN cells tested. Recording from single EPN neuron antidromically activated from LHb stimulation.
Left: Ratemeter recording of firing frequency showing effect of microiontophoretically applied GABA and glycine.
Right: extracellular unit inhibitory response of same EPN neuron to STN stimulation (0.2 ms duration, 137 µA).
A: before strychnine application, GABA (250 nA for 17 s), glycine (250 nA for 17 s) and STN (5 superimposed traces) induce inhibitory responses.
B: after 4 min strychnine application, (150 nA). Strychnine selectively blocks glycine-induced inhibitory response. GABA- and STN-evoked inhibition (6 superimposed traces) not affected.

al., 1981a,b). The significance of a topographically organized pallido-subthalamic projection in the rat is unclear, since, as described in Part A, the dendritic field of a single STN neuron covers the whole STN nucleus in the rat (Hammond and Yelnik, 1983).

The pharmacological study of STN neurons demonstrated a very powerful GABAergic inhibition, which correlates well with the relatively high GAD activity measured in this nucleus (Van der Kooy et al., 1981), as well as with the heavy staining for GABA-transaminase (the enzyme responsible for GABA degradation) revealed by immunocytochemistry (Vincent et al., 1982).

The blockade of GP-induced inhibition in STN neurons by local application of GABA antagonists, and in a few cases by a glycine antagonist, raises the problem of the specificity of these drugs. Bicuculline and picrotoxin are now well accepted as being selective antagonists of GABA (Johnston, 1976). Their irritative effects, usually attributed to an action on GABA or ACh content (for details see Rouzaire-Dubois et al., 1980 and 1983) was verified to have no interaction with antagonism of the synaptic effect. Strychnine has been commonly accepted as a specific glycine antagonist, although large doses can be less specific and have been shown to partially block GABA action in the spinal cord (Davidoff et al., 1969). However, if a specific action of strychnine on a glycinergic system in STN were subsequently demonstrated, a direct pathway coming from GP and involving glycine would have to be considered, as well as recurrent axon collaterals of the "hypothetic glycinergic" subthalamo-pallidal projection suggested by Yoshida (1974). Although such intranuclear axon collaterals have recently been shown (Iwahori, 1978; Hammond and Yelnik, 1983; Kita et al., 1983), they arise from a pathway probably using GABA as neurotransmitter (Nauta and Cuenod, 1982; Rouzaire-Dubois et al., 1983).

The present data strongly suggest GABA to be the transmitter of the pallido-subthalamic pathway, in agreement with biochemical investigations of Fonnum et al (1978) in the cat. Following lesions of GP, glutamate decarboxylase (GAD) activity fell in certain regions of the STN, in proportion to the extent of GP damage. Opposing results were obtained by Van der Kooy et al. (1981) in their similar

---

Fig. 16 (contd.) C: recovery of glycine-induced inhibitory response 7 min after strychnine application discontinued.
D: GABA- and STN-induced (5 superimposed traces) inhibitions reversed by bicuculline application (100 nA for 18 min).
E: return to control situation 9 min after bicuculline application discontinued. (From Rouzaire-Dubois et al., 1983, courtesy Elsevier, Amsterdam)

biochemical experiments performed in the rat after kainic acid lesions of GP. They found no GAD variations at the STN level for short survival times (3 days), but a fall of GAD activity seemed to appear after longer survival periods, which they attributed to a beginning of recovery. A recent histochemical study of GABA-transaminase (the enzyme catabolizing GABA) attempted to provide information regarding the distribution of GABA terminals within the STN (Vincent et al., 1982). Kainic acid lesions of the GP reduced the histochemically detectable GABA-transaminase activity in ipsilateral STN, which is consistent with our hypothesis that the pallido-subthalamic pathway contains GABA.

Subthalamo-Entopeduncular pathway. The high sensitivity of EPN cells to GABA is consistent with the relatively high level of GAD measured in the rat pallidal complex (Van der Kooy et al., 1981). Likewise the low effectiveness of ACh correlates with the absence of acetycholinesterase staining of EPN neurons in normal (Jacobowitz and Palkovitz, 1974) or DFP-treated rats (Lehmann and Fibiger, 1979). Another recent paper attempted to visualize choline acetyl transferase by immunocytochemistry in the cat (Kimura et al., 1981). Some cholinoceptive somata and sparse cholinergic neurons were shown embedded in the grey matter of EPN, but the precise characterization of these cells as belonging to EPN or to ansa lenticularis nucleus is difficult because of the absence of definite boundary between the cholinoceptive cell masses in EPN and in ansa lenticularis.

Since bicuculline and picrotoxin were the only drugs that blocked the STN-evoked inhibitory response of EPN cells at specific doses, we proposed that GABA might be the transmitter released by STN terminals onto EPN neurons. This disagrees with Yoshida's hypothesis suggesting glycine as the transmitter of the subthalamo-pallidal pathway; and glycine-specific doses of strychnine never antagonized the STN-induced inhibitory response. A recent report (Vincent et al., 1982) on basal ganglia neurons containing GABA transaminase was not conclusive about its presence in the STN, but in any case the relationship of this enzyme to the use of GABA as neurotransmitter is not clear. A preliminary study by Perkins and Stone (1981) suggested a probable role of GABA in the subthalamic projection to the GP proper (equivalent to the external pallidal segment of primates). By Dale's hypothesis, the same neurotransmitter would be released from the terminals of subthalamic collateral projections to both the lateral and medial (EPN) segment of the pallidal complex (Deniau et al., 1978; Van de Kooy and Hattori, 1980; Hammond and Yelnik, 1983; Kita et al., 1983 a,b). Nauta and Cuenod (1982) confirmed the probable intervention of GABA in the subthalamo-entopeduncular pathway, since 3H-GABA injection in one or the other segment of the pallidal complex consistently resulted in perikaryal labeling in the STN. Data on pallidal GAD levels as affected by STN lesions, and immunocytochemical evidence for GAD in STN neurones, are needed to support our iontophoretic results.

GENERAL DISCUSSION

Our findings, together with other reports, make it increasingly evident that the GP control of STN is inhibitory with GABA as neurotransmitter. This inhibitory projection is usually considered as the main STN input. The high level of spontaneous activity recorded from STN cells can be explained only by the excitatory afferents to STN, coming from PPN (Hammond et al., 1983a) and from cortex (Kitai and Deniau, 1981). The excitatory cortical input seems the more significant in arising from a region larger (personal observation) than precentral cortex as described in the literature (Hartmann-von-Monakow et al., 1978).

The GP-STN projection has a range of conduction velocity from 2.8 m/s (Deniau et al., 1978) to 3.8 m/s (Kita et al., 1983) in a system of thin myelinated fibers, about as fast as the cortico-STN excitatory system (Kitai and Deniau, 1981), but with different loci of termination on STN cells. The massive pathway from external GP terminates on more proximal parts of STN dendrites than cortico-STN fibers (Romansky et al., 1979), either on the proximal dendrites (Nakamura and Sutin, 1972) or soma (Van der Kooy et al., 1981) of neurons projecting back to GP (Van der Kooy et al., 1981; Kita et al., 1983). The homogeneous population in STN of Golgi type I cells projects to both parts of the pallidal complex and SN by way of a branched axon arising from the one neuron (See Part A, above). Consequently, all inputs arriving on this single STN neuronal population, in particular the inhibitory input from GP and the excitatory inputs coming from cortex and PPN, can potentially influence both the pallidal complex and SN.

It has been suggested that, because of these branching axons, STN must influence simultaneously both the pallidum and nigra in the same direction. But as discussed in Part A above, the low safety factor for conduction at a branch-point and differences in secondary branch diameters may produce differing impulse patterns in the two branches. Moreover, though originating from the same cells, STN fibers are inhibitory (and GABAergic) to entopeduncular output cells, and excitatory to nigro-collicular neurons. Golgi studies have not described local interneurons in the pallidal complex (Iwahori and Mizuno, 1981a,b), suggesting that the STN-evoked inhibition recorded in EPN is a monosynaptic response. At the SN level, the STN-induced excitatory response follows high frequency stimulation (Hammond et al., 1983) so that the hypothesis of a disinhibitory mechanism via local interneurons (Francois et al., 1979) is unlikely.

Since GABA is demonstrated to be the transmitter of the subthalamo-entopeduncular inhibitory pathway, by Dale's hypothesis it would also be involved in the subthalamo-nigral excitatory projection. Only a new hypothesis which takes into account unusual GABA actions (Andersen et al., 1980; Mayer et al., 1983) could explain both these responses. With intracellular recordings from hippocampal pyramidal cells, Andersen et al (1980) showed that microiontophoretically applied GABA induced hyperpolarization on somatic application and depolarization on dendritic application. Both actions could inhibit cell discharge, although in some cases the depolarizing response could facilitate other excitatory influences, of by itself cause cell firing. Such opposite effects of GABA suggest that two different postsynaptic receptors (affinity, location on the soma and dendritic tree, physico-chemical characteristics) could be localized to EPN and SN.

As noted in Part A, the two output targets of STN (the pallidal complex and SN) are also the output targets of the striatum, so that the STN seems to be in a strategic position to influence the main outputs of the basal ganglia.

ACKNOWLEDGEMENTS

Figures 2, 3, 5, 6 and 7, previously published in Neuroscience are reprinted with permission from authors and Pergamon Press Ltd. Figures 10 (bottom), 12, 13, 14, 15, 16, previously published in Brain Research, are reprinted with permission from authors and Elsevier Science Publishers.
The authors gratefully acknowledge financial support from Inserm (CRL 80/6/003), Université R. Descartes (UER de Psychologie) and Fondation pour le Recherche Médicale.

REFERENCES

Andersen, P., Dingledine, R., Gjerstad, L., Langmoen, I. A., and Mosfeldt Laursen A., 1980, Two different responses of hippocampal cells to application of gamma-amino butyric acid, J. Physiol., 305: 279.

Beckstead, R. M., Domesick, V. B., and W. J. H. Nauta, 1979, Efferent connections of the substantia nigra and ventral tegmental area in the rat, Brain Research, 175:191.

Carpenter, M. B., Batton, R. R. III, Carleton, J. C., and Keller, J. T., 1981, Interconnections and organization of pallidal and subthalamic nucleus in the monkey, J. Comp. Neurol., 197:579.

Carpenter, M. B., Carleton, S. C., Keller, J. T., and Conte, P., 1981, Connections of the subthalamic nucleus in the monkey, Brain Res., 224:1.

Carpenter, M. B., Fraser, R. A. R., and Shriver, J. E., 1968, The organization of pallido-subthalamic fibers in the monkey, Brain Res., 11:522.

Carpenter, M. B., and Strominger, N. L., 1967, Efferent fibers of the subthalamic nucleus in the monkey, Am. J. Anat., 121:47.

Carpenter, M. B., Whittier, J. R., and F. A. Mettler, 1950, Analysis of choreoid hyperkinesia in the rhesus monkey. Surgical and pharmacological analysis of hyperkinesia resulting from lesions of the subthalamic nucleus of Luys, J. Comp. Neurol., 92:293.

Carter, D. A., and Fibiger, H. C., 1978, The projections of the entopeduncular nucleus and globus pallidus in rat as demonstrated by autoradiographic and horseradish peroxidase histochemistry, J. Comp. Neurol., 177:113.

Davidoff, R. A., Aprison, M. H., and Werman, R., 1969, The effects of strychnine on the inhibition of interneurons by glycine and gamma-aminobutyric acid, Int. J. Neuropharmacol., 8:191.

Deniau, J. M., Hammond, C., Chevalier, G., and Féger, J., 1978, Evidence for branched subthalamic nucleus projections to substantia nigra, entopeduncular nucleus and globus pallidus, Neurosc. Lett., 9, 117.

DiFiglia, M., Pasik, P., and Pasik, T., 1982, A golgi and ultrastructural study of the monkey globus pallidus, J. Comp. Neurol., 212:53.

Féger, J., Ohye, C., Gallouin, F., and Albe-Fessard, D., 1975, Stereotaxic technique for stimulation and recording in nonanesthetized monkeys: application to the determination of connections between caudate nucleus and substantia nigra, in: "Advances in Neurology," Vol. 10, B. S. Meldrum and C. D. Marsden, eds., Raven Press, New York.

Fonnum, F., Grofova, I., and Rinvik, E., 1978, Origin and distribution of glutamate decarboxylase in the nucleus subthalamicus of the cat, Brain Res., 153:370.

Fox, C. A., Andrade, A. N., Lu Qui, I. J., and Rafols, J. A., 1974, The primate globus pallidus: a golgi and electron microscopic study, J. fur. Hirnforschung, 15:75.

Fox, C. A., and Rafols, J. A., 1975, The radial fibers in globus pallidus, J. Comp. Neurol., 159:177.

Francois, C., Percheron, G., Yelnik, J., and Heyner, S., 1979, Demonstration of the existence of small local circuit neurons in the Golgi-stained primate substantia nigra, Brain Res., 172:160.

Frigyesi, T., and Rabin, A., 1971, Basal ganglia-diencephalic synaptic relation in the cat. III. An intracellular study of ansa lenticularis, lenticular fasciculus and pallido-subthalamic projection activities, Brain Res.35:67.

Grofova, I., 1969, Experimental demonstration of a topical arrangement of the pallido-subthalamic fibers in the cat, Psychiat. Neurol. Neurochir., 72:53.

Grossman, Y., Spira, M. E., and I. Parnas, 1973, Differential flow into branches of a single axon, Brain Research, 64:379.

Hammond, C., Deniau, J. M., Rizk, A., and Féger, J., 1978, Electrophysiological demonstration of an excitatory subthalamo-nigral pathway in the rat, Brain Research, 151:235.

Hammond, C., Deniau, J. M., Rouzaire-Dubois, B., and Féger, J., 1978, Peripheral input to the rat subthalamic nucleus, an electrophysiological study, Neurosci. Lett., 9: 171.
Hammond, C., Féger, J., Bioulac, B., and Souteyrand, J. P., 1979, Experimental hemiballism in the monkey produced by unilateral kainic acid lesion in corpus luysii, Brain Research, 171:577.
Hammond, C., Rouzaire-Dubois, B., Féger, J., Jackson, J., and Crossman, R. A., 1983, Anatomical and electrophysiological studies on the reciprocal projections between the subthalamic nucleus and nucleus tegmenti pedunculopontinus in the rat, Neurosci., 9:41.
Hammond, C., Shibazaki, T., and Rouzaire-Dubois, B., 1983, Branched output neurons of the rat subthalamic nucleus: electrophysiological study of the synaptic effects on identified cells on the two main target nuclei, the entopeduncular nucleus and the substantia nigra, Neurosci. 9:511.
Hammond, C., and Yelnik, J., 1983, Intracellular labelling of rat subthalamic neurones with horseradish peroxydase: computer analysis of dendrites and characterization of axon arborization, Neurosci., 8:781.
Hartmann-Von Monakow, K., Akert, K., and Kunzle, H., 1978, Projections of the precentral motor cortex and other cortical areas of the frontal lobe to the subthalamic nucleus in the monkey, Exp. Brain Res., 33:395.
Iwahori, N., 1978, A Golgi study on the subthalamic nucleus of the cat, J. Comp. Neurol., 182:383.
Iwahori, N., and Mizumo, N., 1981, A Golgi study of the globus pallidus of the mouse, J. Comp. Neurol., 197:29.
Iwahori, N., and Mizumo, N., 1981, Entopeduncular nucleus of the cat: a Golgi study, Exp. Neurol., 72:654.
Jacobowitz, D. M., and Palkovitz, M., 1974, Topographic atlas of catecholamines and acetylcholinesterase containing neurons in the rat brain. I. Forebrain, J. Comp. Neurol., 157:13.
Johnston, G. A. R., 1976, Physiologic pharmacology of GABA and its antagonists in the vertebrate nervous system, in: "GABA in nervous system function," E. Roberts, T. N. Chase, and D. E. Tower, eds., Raven Press, New York.
Kimura, H., McGeer, P. L., Peng, J. H., and McGeer, E. G., 1981, The central cholinergic system studied by choline acetyltransferase immunocytochemistry in the cat, J. Comp. Neurol., 200:151.
Kita, H., Chang, H. T., and Kitai, S. T., 1983, The morphology of intracellularly labelled rat subthalamic neurons: a light microscopic analysis, J. Comp. Neurol, 215:245.
Kita, H., Chang, H. T., and Kitai, S. T., 1983, Pallidal input to subthalamus: intracellular analysis, Brain Res., 264:255.
Kitai, S. T., and Deniau, J. M., 1981, Cortical inputs to the subthalamus: intracellular analysis, Brain Res., 214:411.
Krnjević, K., and Miledi, R., 1959, Presynaptic failure of neuromuscular propagation in rats, J. Physiol., 149:1.

Kuypers, H. G. J. M., and Lawrence, D. G., 1967, Cortical projections to the red nucleus and the brain stem in the rhesus monkey, Brain Research, 4:151.
Lehmann, J., and Fibiger, H. C., 1979, Acetylcholinesterase and the cholinergic neuron, Life Sciences, 25:1939.
Mayer, M. L., Higashi, H., Gallagher, J. P., and Shinnick-Gallagher, 1983, On the mechanism of action of GABA in pelvic vesical ganglia: biphasic responses evoked by two opposing actions on membrane conductance, Brain Res., 260:233.
Moon Edley, S., and Graybiel, A. M., 1980, Connections of the nucleus tegmenti pedunculopontinus, pars compacta (TPi) in cat, Anat. Rec., 196:129A.
Nakamura, S., and Sutin, J., 1972, The pattern of termination of pallidal axons upon cells of the subthalamic nucleus, Exp. Neurol., 35:254.
Nauta, H. J. W., and Cole, M., 1978, Efferent projection of the subthalamic nucleus, an autoradiographic study in monkey and cat, J. Comp. Neurol., 180:1.
Nauta, H. J. W., and Cuenod, M., 1982, Perikaryal cell labelling in the subthalamic nucleus following the injection of 3H-gamma-aminobutyric acid into the pallidal complex: an autoradiographic study in the cat, Neurosci., 7:2725.
Nomura, S., Mizuno, N., and Sugimoto, T., 1980, Direct projections from the pedunculopontine tegmental nucleus to the subthalamic nucleus in the cat, Brain Research, 196:223.
Ohye, C., Le Guyader, C., and Féger, J., 1976, Responses of subthalamic and pallidal neurons to striatal stimulation: an extracellular study on awake monkeys, Brain Res., 111:241.
Percheron, G., 1979, Quantitative analysis of dendritic branching I. II., Neurosci. Lett., 14:287.
Perkins, M. N., and Stone, T. W., 1981, Iontophoretic studies on pallidal neurones and the projection from the subthalamic nucleus, Quarterly J. Exp. Physiol., 66:225.
Rafols, J. A., and Fox, C. A., 1976, The neurons in the primate subthalamic nucleus: a golgi and electron microscopic study, J. Comp. Neurol., 168:75.
Ricardo, J. A., 1980, Efferent connections of the subthalamic region in the rat. I. The subthalamic nucleus of Luys, Brain Res., 202:257.
Rinvik, E., Grofova, I., Hammond, C., Féger, J., and Deniau, J. M., 1979, Afferent connections in the monkey and the cat studied with the HRP technique, in: "The extrapyramidal system and its disorders," L. J. Poirier, and T. L. Sourkes, eds., Raven Press, New York.
Romansky, K. V., Usunoff, K. G., Ivanov, D. P., and Galabov, G. P., 1979, Cortico-subthalamic projection in the cat, an electron microscopic study, Brain Res., 163:319.
Rouzaire-Dubois, B., Hammond, C., Hamon, B., and Féger, J., 1980, Pharmacological blockade of the globus pallidus-induced inhibitory response of subthalamic cells in the rat, Brain Res., 200:320.

Rouzaire-Dubois, B., Scarnati, E., Hammond, C., Crossman, A. R., and Shibazaki, T., 1983, Microiontophoretic studies on the nature of the neurotransmitter in the subthalamo-entopeduncular pathway of the rat, Brain Res., 271:11.

Sakai, K., Sastre, J. P., Salvert, D., Mouret, M., Tohjama, M., and Jouvet, M., 1979, Tegmentoreticular projections with special reference to the muscular atonia during paradoxical sleep in the cat: an HRP study, Brain Research, 176:233.

Smith, D. O., 1980, Mechanisms of action potential propagation failure at sites of axon branching in the crayfish, J. Physiol., 301:243.

Tsubokawa, T., and Sutin, J., 1972, Pallidal and tegmental inhibition of oscillatory slow waves and unit activity in the subthalamic nucleus, Brain Res., 41:101.

Van der Kooy, D., and Hattori, T., 1980, Single subthalamic nucleus neurons project to both the globus pallidus and substantia nigra in the rat, J. Comp. Neurol., 192:751.

Van der Kooy, D., Hattori, T., Shannak, K., and Hornikiewcz, O., 1981, The pallido-subthalamic projection in rat: anatomical and biochemical studies, Brain Res., 204:253.

Vincent, S. R., Kimura, H., and McGeer, E. G., 1982, GABA transaminase in the basal ganglia: A pharmacohistochemical study, Brain Res., 251:93.

Vincent, S. R., Kimura, H., and McGeer, E. G., 1982, A histochemical study of GABA transaminase in the afferents of the pallidum, Brain Res., 241:162.

Waxman, S. G., 1972, Regional differentiation of the axon, a review with special reference to the concept of the multiplex neuron, Brain Res., 47:269.

Whittier, J. R., and Mettler, F. A., 1949, Studies on the subthalamus of the rhesus monkey. I. Anatomy and fiber connections of the subthalamic nucleus of Luys, J. Comp. Neurol., 90:315.

Whittier, J. R., and Mettler, F. A., 1949, Studies on the subthalamus of the rhesus monkey. II. Hyperkinesia and other physiologic effects of subthalamic lesions, with special reference to the subthalamic nucleus of Luys, J. Comp. Neurol., 90:319.

Yelnik, J., and Percheron, G., 1979, Subthalamic nucleus in primates: a quantitative and comparative analysis, Neurosci., 4:1717.

Yelnik, J., Percheron, G., Perbos, J., and Francois, C., 1981, A computer-aided method for the quantitative analysis of dendritic arborizations reconstructed for serial sections, J. Neurosci. Methods, 4:347.

Yoshida, M., 1974, Functional aspects of, and role of transmitter in the basal ganglia, Confin. Neurol., 36:282.

NEURAL ACTIVITY IN BASAL GANGLIA OUTPUT NUCLEI

AND INDUCED HYPERMOTILITY

P. W. Everett, R. E. Kemm and J. S. McKenzie

Department of Physiology
University of Melbourne
Parkville 3052 Australia

INTRODUCTION

Systemic administration of low doses of dopamine (DA) agonists produces a hypermotility response, which at higher doses becomes a pattern of stereotyped behaviour variably composed of licking and gnawing with suppression of locomotor activity (Iversen, 1977). The neural mechanisms underlying these behavioural responses respectively involve the action of DA-ergic receptors in the nucleus accumbens (NAcc) and neostriatum (NS), as the behaviour is reproduced by local injections of DA agonists into NAcc and NS (Costall et al., 1977) and abolished by lesions which destroy the DA-ergic afferents to them (Kelly et al., 1975).

Neostriatal efferents project principally to the globus pallidus (GP), entopeduncular nucleus (EP) and substantia nigra pars reticulata (SNr) (Graybiel and Ragsdale, 1979; Carpenter, this volume). Hence it is assumed that one or more of these structures are important in mediating these responses. The principal transmitter used by the NS efferent system is the inhibitory gamma-amino-butyric acid (GABA) (Kitai, 1981). Local injections of GABA antagonists into GP produce hypermotility similar to that seen following injections of DA agonists into NAcc (Jones and Mogenson, 1980), while injection of GABA agonists into GP attenuates the hypermotility elicited by injection of DA agonists into NAcc (Jones and Mogenson, 1980; Pycock et al., 1976; Slater et al., 1982). Local injections of GABA antagonists into SNr produce catalepsy (Di Chiara et al., 1978); injections of GABA agonists do not produce hypermotility, but rather posturing and head-turning when given unilaterally (Arnt and Scheel-Kruger, 1979; Scheel-Kruger et al., 1977), and stereotyped behaviour which includes licking and biting when injected bilaterally (Iversen, 1977; Scheel-Kruger et al., 1977). This implies that GP is important in mediating the hypermotility response seen with low doses of DA agonists, while SNr is important in mediating the stereotyped behaviour seen at higher doses. Further support is given to this proposition by the finding that bilateral kainic acid lesions of SNr abolish the stereotyped behaviour pattern seen at high doses of DA agonists but do not affect the hypermotility response seen at lower doses (Morelli et al., 1980).

From these studies it might be predicted that the hypermotility following systemic administration of low doses of DA agonists is associated with an elevated firing of neurons in GP but not SNr. We report here some preliminary results of recording unit activity in GP and SNr of awake behaving rats during the hypermotility induced by systemic administration of the DA agonists amphetamine and apomorphine.

METHODS

Male hooded rats weighing 250 to 350 g were used. Stereotaxic co-ordinates were derived from the atlas of Albe-Fessard et al., (1966). Ketamine anaesthesia was used for surgery (150 mg/kg with supplementary doses of 50 mg/kg as required).

Implantation of Recording Electrodes

Animals were implanted with 8 fine wire recording electrodes (Medwire No 316 SS 3t; 76 μm stainless steel wire with an insulating coat of Teflon giving a total diameter of 110 μm), 2 in GP and 2 in SNr on each side of the brain.

The position of each electrode was marked on the skull at stereotaxic co-ordinates: AP:7.6, LR:2.5 and AP:6.8, LR:3.5 for GP, and AP:3.5, LR:2.0 and AP:2.6, LR:3.0 for SNr, and burr holes drilled to allow insertion of the eight recording electrodes, anchoring screws and a large-surface earth electrode. Neural activity on each electrode was monitored as it was inserted under stereotaxic guidance. The electrode was cemented in place with a drop of dental acrylic when a single unit, with signal to noise ratio of at least 3 to 1, was located in the target structure. When all eight electrodes were implanted, they were crimped onto gold plated female sockets which were inserted into a connector which was in turn cemented to the skull. This head piece connected to a multiple plug containing a head-stage amplifier constructed from 8 FET source followers (McKenzie et al., 1983). At least 4 days were allowed for recovery from surgery before recording commenced.

Extracellular Recording

Up to 6 electrodes were recorded simultaneously. Neural activity was amplified 1000X and filtered above 10KHz and below 100Hz (roll off: 12 db/octave) before storage on magnetic tape. After habituating the animal to the recording situation, a 15 minute control recording was made prior to the administration of amphetamine (2 mg/kg i.p.), apomorphine (1 mg/kg s.c.) or saline (0.2 ml i.p.). Recording continued for a further 45 minutes. At least 2 days were allowed before another drug trial was conducted.

At the completion of recording, a 10 µA anodal current was passed between each recording electrode and earth for 10 seconds and the brain perfused via the left ventricle with 10% potassium ferrocyanide in 10% formalin to form a Prussian Blue spot at the tip of each electrode. Frozen sections are cut at 60 µm and stained with Cresyl Violet to verify recording sites.

Analysis of Extracellular Recordings

Recordings were played back on an oscilloscope at a fast sweep speed, and the shape and amplitude of each spike drawn on graph paper to verify identity as a single unit. Well-isolated single units were then played back through a signal conditioner and the time between successive events stored on computer floppy disc. Firing rate was displayed with a digital X-Y plotter. Interspike interval histograms were also constructed from within the pre-drug and post-drug periods to examine any changes in firing pattern. A change in firing rate of more than 30% of baseline was considered significant.

RESULTS

Both amphetamine and apomorphine elicited increases in locomotor activity. Prior to drug administration the rats sat quietly with rare periods of grooming and locomotion. Following amphetamine and apomorphine, they spent most of their time walking about the cage, sniffing.

A total of 24 extracellular recordings were made from GP during amphetamine and apomorphine induced hypermotility. Of the 14 unit recordings following amphetamine, 12 showed increased firing rate and 2 did not change. All 10 recordings made following apomorphine showed an elevated firing rate. None of 6 recordings made following i.p. saline injection showed any change in firing rate.

Two discharge patterns could be distinguished in GP prior to drug administration. The first was characterized by burst discharges of 7 - 12 spikes, separated by intervals of several hundred ms, during which single spikes were seen once or twice. The interspike interval histograms of these cells showed a single narrow peak at very short intervals (2 to 3 ms) corresponding to the high frequency discharge within the bursts (Fig. 1). The second pattern displayed regular firing which was interrupted by brief (80 - 200 ms) periods of silence and frequent short bursts of 3 to 5 spikes. The interspike interval histograms of these cells showed two peaks. The first was narrow at short intervals (1.5 to 2.5 ms) corresponding to the frequency of firing within the bursts, the second a broader peak at longer intervals (10 to 35 ms) corresponding to the longer periods when the cell was firing regularly (Fig. 2). Amphetamine and apomorphine administration abolished the bursts in both cases, leaving a more regular pattern of firing. Thus in both cases the narrow histogram peaks at short intervals disappeared along with the bursting, and a single broad peak at longer modal interval remained.

A total of 9 recordings were made in SNr during amphetamine- and apomorphine-induced hypermotility. The 4 unit recordings following amphetamine and the 5 recordings following apomorphine all showed increased firing rate. The regularity of firing pattern observed during control periods did not change following drug administration. None of the 4 unit recordings made following saline showed any change in firing rate or pattern.

DISCUSSION

Mechanisms of DA-ergic Action on GP and SNr

Dopaminergic agonists could influence neural discharge levels in the basal ganglia targets of NS, i.e. GP and SNr, by various mechanisms when administered systemically to awake mobile animals.

Effects on DA receptors in SNr, of DA released by amphetamine from nearby SNc dendrites (Chéramy, 1981), or of directly acting DA agonists such as apomorphine, are likely. Nigrostriatal axons give off DA-ergic collaterals to form a sparse terminal plexus in GP (Lindvall and Björklund, 1979) which suggests the possiblity of DA-ergic action also on GP neurons.

Indirect effects on GP and SNr would be derived from direct action on NS, with consequent effects on GABA-ergic efferents and probably on efferents using Substance P or other peptides. Further, since in mobile animals systemic DA-ergic stimulants produce increased locomotion or other movements, the afferent input thereby generated might be capable of affecting neurons in GP and SNr (Féger et al., 1978; Matsunami and Cohen, 1975; Schneider et al., 1982). In support of this possibility is our observation that during pre-drug control periods, the usually quiet and immobile rats occasionally engaged in exploratory or grooming activity which was correlated with increased firing rates in GP and SNr.

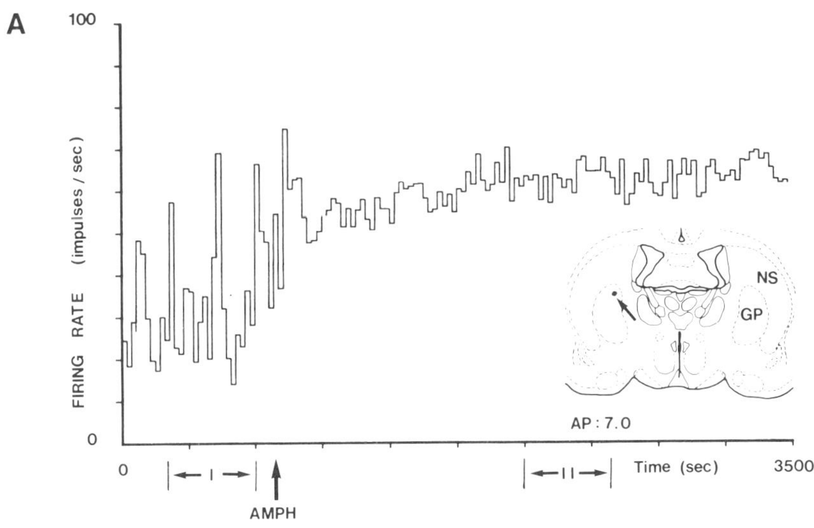

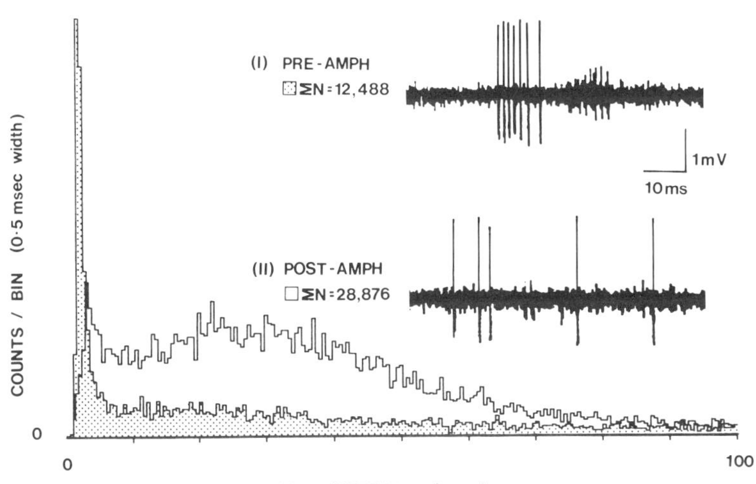

Fig. 1. Single unit recorded from GP of an intact rat before and after systemic amphetamine (2 mg/kg i.p.). (A) Firing rate of unit before and after amphetamine (injected at arrow). Recording site in inset. (B) Interspike interval histogram with inset unit traces showing the change in firing pattern following amphetamine. The pre-drug histogram (shaded) corresponds to the pattern in the upper trace recorded during period I in A. The post-drug histogram (open) corresponds to the lower trace recorded during period II in A.

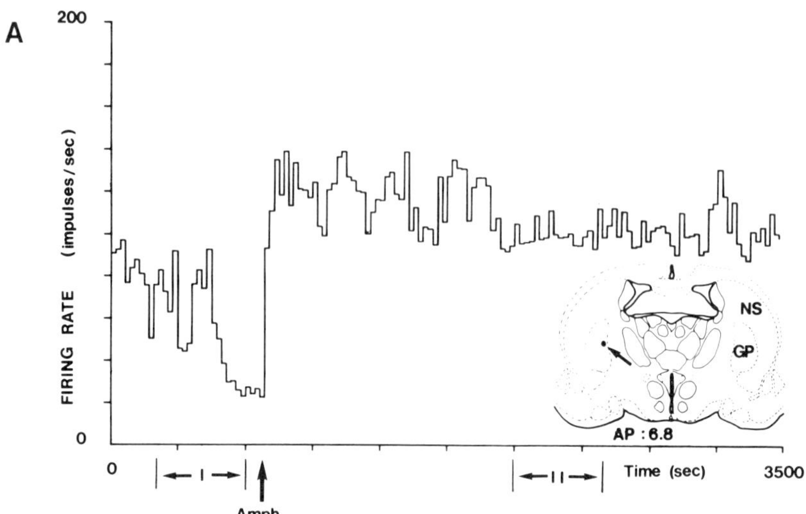

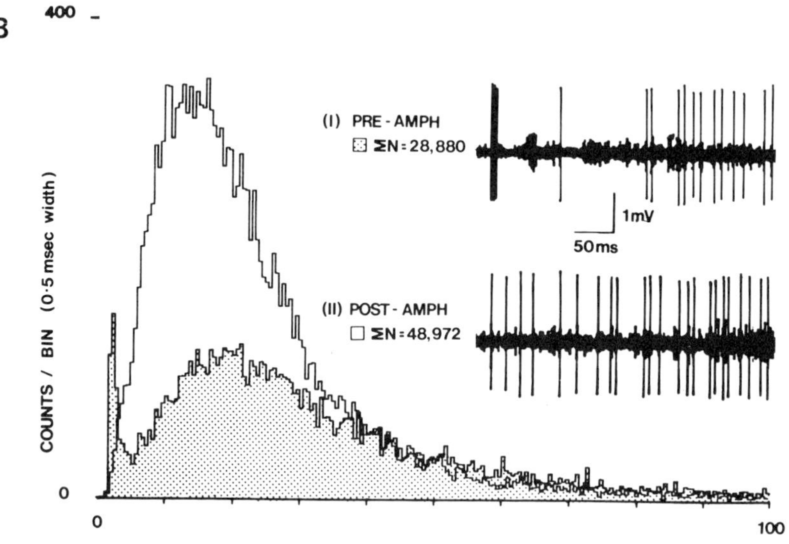

Fig. 2. Single unit recorded from GP of an intact rat before and after systemic amphetamine (2 mg/kg i.p.). (A) Firing rate of unit before and after amphetamine (injected at arrow). Recording site in inset. (B) Interspike interval histogram with inset traces of the unit showing change in firing pattern following amphetamine. Shaded and open histograms as in Fig. 1.

The increased discharge produced in GP and SNr by systemic amphetamine and apomorphine in our mobile rats corresponds to effects reported in immobilised animals, either under general anaesthesia or paralysed and locally anaesthetised (Bergstrom et al., 1982; Bergstrom and Walters, 1981; Rebec and Groves, 1975). For this similarity of response between mobile and immobilized animals to be the consequence of DA-ergic action on NS neurons, that action would need to influence NS firing rates in the same direction in each type of preparation. But, on the contrary, systemic DA stimulants produce a decrease in discharge rate of NS cells in anaesthetised or paralysed animals, particularly at low to moderate doses (Bashore et al., 1978; Rebec and Segal, 1978), whereas they increase NS firing in awake and mobile animals (Hansen and McKenzie, 1979; McKenzie and Hansen, 1980; Trulson and Jacobs, 1974). Therefore other factors or modes of action must be considered.

It is possible that the differing NS responses of mobile and immobilised animals to DA-ergic agents result from a sampling of different neuronal populations between animal preparations, depending on the recording methods and possibly the dose levels used. These factors might disguise an underlying similarity of response direction in a small contingent of NS output neurons.

Another likely source of difference for NS activity is that the increased locomotor activity induced by DA-ergic stimulants in mobile animals gives rise to afferent activity capable of exciting NS neurons. However this factor does not explain the similarity of response direction in NS targets, GP and SNr, between preparation types. One explanation might be that movement-generated afferent activity feeding into GP and SNr could counterbalance the inhibitory effects of GABA-ergic inputs derived from NS output activity in mobile animals. Furthermore, a direct excitatory action on SNr neurons of DA applied iontophoretically has been described (Ruffieux and Schultz, 1980) and is capable of overriding the inhibitory action of GABA (Waszczak and Walters, 1983). Direct DA action might also be a factor in the vicinity of nigrostriatal terminals in GP, mentioned above.

Further investigations are required to decide among these various possibilities.

Our finding that systemic amphetamine and apomorphine abolish bursting of neurons in GP is in agreement with Filion (1979) who showed that decreased DA receptor stimulation, produced either by lesions of the substantia nigra pars compacta or by systemic administration of the DA antagonist haloperidol, increased bursting activity in both the internal and external segments of GP in awake monkeys. The abolition of GP burst discharge may be due to either the action of DA agonists per se, or changes in the behavioural repertoire induced by DA agonists. Hull et al., (1974) showed that

NS neurons display more bursting activity following the destruction of DA-ergic afferents to NS. Thus the increased bursting in GP following DA depleting lesions seen by Filion (1979) may have resulted from increased bursting activity in NS projections to GP. It may be that the decreased bursting found by us following systemic DA agonists is the reverse of this, i.e. increased DA receptor stimulation in NS reduces the bursting activity in NS output neurons resulting in decreased bursting in GP. Given the similarity between NS outputs to GP and SNr, the fact that no change in discharge pattern was seen in neurons recorded in SNr argues against the proposal that the loss of bursting in GP arises from changes in discharge pattern in NS outputs.

The question of the involvement of GP and SNr in hypermotility is complex and a consensus has not yet been reached. The finding that bilateral kainic acid lesions of SNr do not affect hypermotility induced by low doses of DA agonists (Morelli et al., 1980) suggests that SNr is not critically involved in this behaviour (and points to GP). Despite this, Di Chiara and co-workers have put forward a model (Di Chiara et al., 1981) based on the SNr projection to the ventromedial thalamic nucleus (VM) and the fact that interference with VM can influence the motility component of induced rotation, in which hypermotility is mediated by SNr output neurons. An alternative viewpoint is that of Mogenson and co-workers (Mogenson and Yim, 1981) who believe that hypermotility is mediated by GP or possibly by the ventral pallidum which is known to receive extensive projections from NAcc (Nauta et al., 1978).

Our findings provide direct support for the postulate of Jones and Mogenson (1980) that hypermotlity following low doses of DA agonists is associated with elevated rates of firing in GP. The results are unable to decide between the relative contribution of GP and SNr outputs to the behavioural response.

REFERENCES

Arnt, J. and Scheel-Kruger J., 1979, Behavioural differences induced by muscimol selectively injected into pars compacta and pars reticulata of substantia nigra. Naunyn-Schmiedeberg's Arch. Pharmacol., 310:43.

Bashore, T. R., Rebec, G. V. and Groves, P. M., 1978, Alterations of spontaneous neuronal activity in the caudate-putamen, nucleus accumbens and amygdaloid complex of rats produced by d-amphetamine. Pharmac. Biochem. Behav., 8:467.

Bergstrom, D. A., Bromley, S. D. and Walters, J. R., 1982, Apomorphine increases activity of rat globus pallidus neurons. Brain Res., 238:266.

Bergstrom, D. A. and Walters, J. R., 1981, Neuronal responses of the globus pallidus to systemic administration of d-amphetamine: investigation of the involvement of dopamine, norepinephrine and serotonin. J. Neuroscience, 1:292.

Chéramy, A., Leviel, V. and Glowinski, J., 1981, Dendritic release of dopamine in the substantia nigra. Nature (Lond.), 289:537.

Costall, B., Naylor, R. J., Cannon, J. G. and Lee, T., 1977, Differentiation of the dopamine mechanisms mediating stereotyped behaviour and hyperactivity in the nucleus accumbens and caudate-putamen. J. Pharm. Pharmac., 29, 337.

Di Chiara, G., Morelli, M., Porceddu, M. L. and Gessa, G. L., 1978, Evidence that nigral GABA mediates behavioural responses elicited by striatal dopamine receptor stimulation. Life Sciences, 23:2045.

Di Chiara, G., Porceddu, M. L., Imperato, A. and Morelli, M., 1981, Role of GABA neurons in the expression of striatal motor functions. In "GABA and the Basal Ganglia," Advances in Biochemical Psychopharmacology Vol. 30, G. Di Chiara and G. L. Gessa, eds., Raven Press, New York.

Féger, J., Jacquemin, J. and Ohye, C., 1978, Peripheral excitatory input to substantia nigra. Exp. Neurol., 59:352.

Filion, M., 1979, Efects of interruption of the nigrostriatal pathway and of dopaminergic agents on the spontaneous activity of globus pallidus neurons in the awake monkey. Brain Res., 178:425.

Graybiel, A. M. and Ragsdale, D. W., 1979, Fibre connections of the basal ganglia. In "Development and Chemical Specificity of Neurons" Progress in Brain Research, Vol. 51, M. Cuenod, G. W. Kreutzberg and F. E. Bloom, eds., Elsevier, Amsterdam.

Hansen, E. L. and McKenzie, G. M., 1979, Dexamphetamine increases striatal neuronal firing in freely moving rats. Neuropharmacol. 18:547.

Hull, C. D., Levine, M. S., Buchwald, N. A., Heller, A. and Browning, R. A., 1974, The spontaneous firing pattern of forebrain neurons: I The effects or dopamine and non-dopamine depleting lesions on caudate unit firing patterns. Brain Res., 73:241.

Iversen, S. D., 1977, Striatal function and stereotyped behaviour. In "Psychobiology of the Striatum", A. R. Cools, A. H. M. Lohman & J. H. L. van den Bercken, eds., North Holland, Amsterdam.

Jones, D. L. and Mogenson, G. J., 1980, Nucleus accumbens to globus pallidus GABA projection subserving ambulatory activity. Am. J. Physiol., 238:R65.

Kelly, P. H., Serviour, P. W. and Iversen, S. D., 1975, Amphetamine and apomorphine response in the rat following 6-OHDA lesions of the nucleus accumbens septi and corpus striatum. Brain Res., 94:507.

Kitai, S. T., 1981, Electrophysiology of the corpus striatum and brain stem integrating systems. In "Handbook of Physiology", Section 1, vol. 2, V. B. Brooks, ed., American Physiological Society, Bethesda.

Lindvall, O. and Bjorklund, A., 1979, Dopaminergic innervation of the globus pallidus by collaterals from the nigrostriatal pathway. Brain Res., 172:169.

McKenzie, G. M. and Hansen, E. L., 1980, GABA agonists dissociate striatal unit activity from drug induced stereotyped behaviour. Neuropharmacol., 19:957.

McKenzie, J. S., Everett, P. W. and Dally, L. J., 1983, A method for simultaneously recording neural activity and rotation in the rat. Physiol. Behav., 30:653.

Matsunami, K. and Cohen, B., 1975, Afferent modulation of unit activity in globus pallidus and caudate nucleus: changes induced by vestibular nucleus and pyramial tract stimulation. Brain Res., 91:140.

Mogenson, G. J. and Yim, C. H., 1981, Electrophysiological and neuropharmacological-behavioural studies of the nucleus accumbens: Implications for its role as a limbic-motor interface. In "The Neurobiology of the Nucleus Accumbens", R. B. Chronister and J. F. DeFrance, eds., Haer Institute, Maine.

Morelli, M., Porceddu, M. L. and Di Chiara, G., 1980, Lesions of the substantia nigra by kainic acid: effects on apomorphine induced stereotyped behaviour. Brain Res., 191:67.

Nauta, W. J. H., Smith, G. P., Domesick, V. B. and Faull, R. L. M., 1978, Efferent connections and nigral afferents of the nucleus accumbens septi in the rat. Neuroscience, 3:385.

Pycock, C., Horton, R. W. and Marsden, C. D., 1976, The behavioural effects of manipulating GABA function in the globus pallidus. Brain Res., 116:353.

Rebec, G. V. and Segal, D. S., 1978, Dose-dependent biphasic alterations in the spontaneous activity of neurons in the rat striatum produced by d-amphetamine and methylphenidate. Brain Res., 150:353.

Rebec, G. V. and Groves, P. M., 1975, Apparent feedback from the caudate nucleus to the substantia nigra following amphetamine administration. Neuropharmacology, 14:275.

Ruffieux, A. and Schultz, W., 1980, Dopaminergic activation of reticulata neurons in the substantia nigra. Nature, 285:240.

Scheel-Kruger, J., Arnt, J. and Magelund, G., 1977, Behavioural stimulation induced by muscimol and other GABA agonists injected into the substantia nigra. Neuroscience Letters, 4:351.

Schneider, J. S., Morse, J. R. and Lidsky, T. I., 1982, Somatosensory properties of globus pallidus neurons in awake cats. Exp. Brain Res., 46:311.

Slater, P., Longman, D. A. and Dickinson, S. L., 1982, Effects of intrapallidal drugs on hyperactivity induced by nucleus accumbens dopamine receptor stimulation. Naunyn-Schmiedeberg's Arch. Pharmacol., 321:201.

Trulson, M. E. and Jacobs, B. L., 1979, Effects of d-amphetamine on striatal unit activity in freely moving cats. Neuropharmacol., 18:735.

Waszczak, B. L. and Walters, J. R., 1983, Dopamine modulation of the effects of gamma-aminobutyric acid on substantia nigra pars reticulata neurons. Science, 220:218.

TONIC NIGRAL CONTROL OF TECTO SPINAL/TECTO DIENCEPHALIC BRANCHED NEURONS: A POSSIBLE IMPLICATION OF BASAL GANGLIA IN ORIENTING BEHAVIOR

G. Chevalier, S. Vacher, J. M. Deniau, and
D. Able-Fessard

Laboratoire de Physiologie des Centres Nerveux
Université Pierre et Marie Curie
4 Place Jussieu
F-7523300 Paris Cedex 05, France

The observation in man that a dysfunction of the basal ganglia is responsible for the occurrence of motor disorders has given considerable impetus to experimental studies on this neuronal system. In this respect, the use of animal models offers unique advantages for analysing the physiological processes by which the basal ganglia participate in the elaboration of motor behavior. In the rat for instance the alteration of striatal or nigral activity induces a postural asymmetry (compulsive head turning, circling behavior) which is associated with a sensory neglect for stimuli occurring in the contralateral field (Di Chiara et al., 1979a; Dray, 1980; Dunnett and Iversen, 1982; Feeney and Wier, 1979; Ljunberg and Ungerstedt, 1976; Marshall et al., 1980; Pycock, 1980; Siegfried and Bures, 1978). Considering the marked axio-cephalic component of these basal ganglia induced disorders in rodents, particular attention has been paid to the anatomical relationships between the basal ganglia and midbrain structures and particularly to the superior colliculus (SC). Indeed the nigrotectal pathway which originates from the gabaergic neurons of the substantia nigra-pars reticulata (SNr) is involved in relaying striatal information towards the deep strata of the SC. Experimental evidence that this projection is important for the expression of basal ganglia influence on head movements has been assessed by recent studies. By lesioning the lateral part of the SC, Di Chiara et al. (1982), and Kilpatrick et al. (1982) succeeded in reducing or blocking basal ganglia induced compulsive head turning. Interestingly, the neurons projecting to cervical segments of the spinal cord lie, in the rat, in the lateral part of the SC (Murray and Coulter, 1982). Considering the involvement of the tecto-spinal projection in head-orienting processes, their control by the SN could be hypothesized (York and Faber, 1977).

The recent morphological description of tecto spinal neurons in the cat by Grantyn and Grantyn (1982) has strengthened the functional importance of such a putative basal ganglia influence on these cells. As revealed by these authors, each tectospinal neuron gives off, as well as an ascending collateral directed towards the diencephalon, an impressive network of axon collaterals to several rhombencephalic ocular and cephalic premotor circuits. From these data, the concept that these multibifurcated "tecto-spinal" cells represent a major output system for the SC has arisen. It is within this conceptual framework that we have undertaken in the rat the identification of this tectal efferent system and analysed its putative control by the SN.

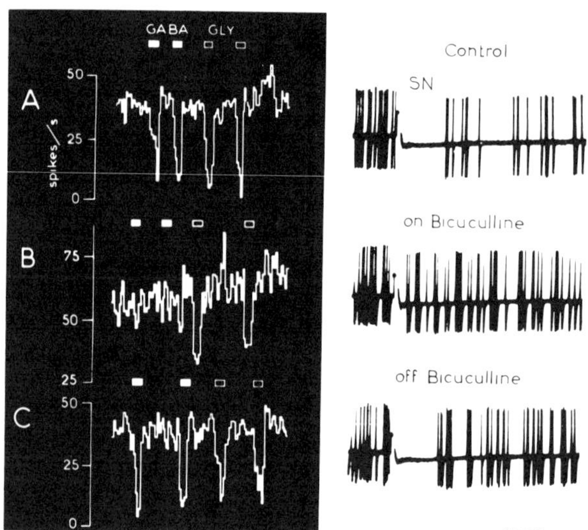

Fig. 1. Specific blockade of the SN evoked inhibition of a SC neuron by microiontophoretic application of a GABA receptor antagonist (bicuculline). Left: time frequency histogram showing the sensitivity of a SC cell to application of GABA (60 nA for 5 sec) and glycine (70 nA for 5 sec). Right: response to SNr stimulation (10 superimposed sweeps). A: before application of bicuculline. Note the depressant action of GABA and glycine. B: during the application of bicuculline (70 nA); while the sensitivity of the collicular cell is specifically abolished for GABA, the nigral induced inhibition is suppressed. C: 12 min after bicuculline application, both GABA sensitivity and nigral effect recovered.
Calibration bar: 20 msec.

THE SYNAPTIC INFLUENCE EXERTED BY SNr ON SC CELLS

An earlier study (York and Faber, 1977) devoted to this question indicated that in the rat electrical stimulation of SN induced an excitatory effect in collicular cells. However biochemical evidence (Di Chiara et al., 1979b; Vincent et al., 1978) that nigrotectal neurons might use GABA as their neurotransmitter questioned the excitatory nature of this pathway. The SN being partly crossed and bordered by numerous fiber bundles running towards the tectum (cortico- and retino-tectal axons), electrical stimulation of SN may also activate these fibers of passage. Thus the origin of the nigral evoked effects had to be determined. We have therefore re-analysed nigral evoked responses in the tectum of the rat.

As well as some excitatory responses, the majority of collicular cells of the intermediate layers responded to SN stimulation with pure inhibition (Deniau et al., 1978a). By comparing the results obtained before and after lesioning peduncular and optic tract fibers, we concluded that the specific effect of nigrotectal cells is a short latency (1,3 msec mean) short duration (12 msec mean) inhibition (Chevalier et al., 1981a). This effect is powerful enough to completely suppress spontaneous as well as glutamate-evoked activity of tectal cells. In accordance with biochemical data, this nigral induced inhibition resulted from the release of GABA since it was specifically supressed by microiontophoretic application of bicuculline (Chevalier et al., 1981b), Fig. 1.

THE TECTOSPINAL TECTODIENCEPHALIC BRANCHED SYSTEM IN THE RAT

Anatomical Identification of Tectospinal Cells

Following unilateral injection of horse radish peroxidase (HRP) in the cervical segments of the rat spinal cord, about 300 cells were found to be retrogradely labelled in the controlateral SC. As depicted on Fig. 2 these cells are concentrated in the lateral half of the structure. They occupy most of the rostro-caudal extent of the SC, intermingled with the fiber fascicles of the stratum album intermedium. As described by others, these tectospinal cells mainly innervate the upper cervical cord. The number of retrogradely labelled neurons considerably decreases as the tracer is deposited in spinal segments caudal to C3.

Ascending Branches of Tectospinal Neurons

To determine if the tectospinal cells, in rats as in cats, give an ascending axonal collateral, and in order to reveal some of their targets, we made use of the antidromic activation method. While this technique discloses the branching pattern of long projecting axons its application is considerably facilitated when preliminary anatomical data indicate the target structures that might be

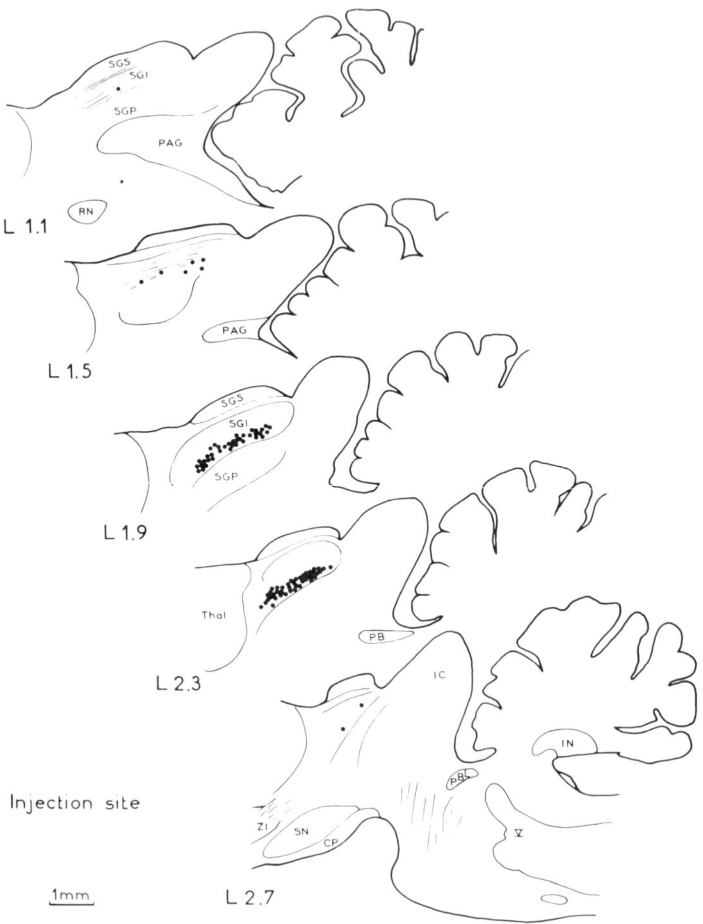

Fig. 2. Distribution of neuronal cell bodies retrogradely labelled in the SC after injection of HRP in the contralateral cervical spinal cord (C1 to C5). Note that the population of tectospinal neurons constitutes a unilamellar cell layer occupying most of the rostrocaudal extent of the lateral SC.

innervated by these collaterals. Since the tectospinal neurons in the rat lie in the deep strata of the lateral SC, we initially deetermined the diencephalic areas innervated by this tectal locus using the anterograde axonal transport of WGA conjugated with HRP. Next, we analysed if a single tectal cell could be antidromically driven by stimulation of both the cervical spinal cord and some of the diencephalic nuclei where labelled terminals had been previously observed (Chevalier et Deniau, in press).

Injection of WGA-HRP in the lateral deep SC results in a dense anterograde labelling in intralaminar nuclei (Parafascicular, Central lateral, paralamellar portion of Mediodorsal, Paracentral, Centromedial), in the Ventromedial nucleus of the thalamus, and in the subthalamic area (Zona incerta and Forel's field). These colliculodiencephalic projections are predominanatly ipsilateral to the injection site.

In the electrophysiological approach, the controlateral cervical spinal cord was stimulated in combination with some of the diencephalic structures onto which lateral SC projects. In the present study our investigation was restricted to the ipsilateral medio dorsal/central lateral nuclei (MD/CL), the paracentral region (Pc), the ventro medial thalamic nucleus (VM), and the zona incerta (ZI). The antidromic nature of action potentials elicited from each stimulation site was determined on the basis of the three classical criteria: fixed latency at threshold; ability to follow a high frequency train of stimulation pulses; and collision within an appropriate time interval with a spontaneous or synaptically evoked spike potential.

Since somata of tectospinal neurons lie in a fibrous lamina, particular attention has been paid to distinguish fiber activity from soma discharge. The latter is characterized on the basis of shape: diphasic with IS-SD breaks. When a single neuron was antidromically driven from different loci, a reciprocal collision test was systematically applied to distinguish between stimulation of divergent axon collaterals and activation of a single axon at various sites along its course. For more details on this procedure see Chevalier et al., in press; Deniau et al., 1978b.

Eighty-one SC neurons were identified as projecting to the spinal cord. 90% of them could also be antidromically activated from at least one of the diencephalic areas studied. As a rule single tectospinal neurons give off an ascending branch which secondarily splits to innervate the subthalamic area and the intralaminar nuclei.

The present findings in the rat and those obtained in the cat are indicative of the high degree of collateralization exhibited by tectospinal neurons in the two species. As reported for the cat the neuronal circuits connected by the rhombencephalic collaterals of tectospinal neurons are all concerned with eye and head movements. Assuming the excitatory nature of these neurons, it has been suggested that they may provide "a spatiotemporal pattern of facilitation which promotes rapid orientation of eye, head and body towards the controlateral hemifield" (Grantyn and Grantyn, 1982). Interestingly, such a view might be extended to the diencephalic arborization of tectospinal cells whose targets include MD/CL, Pc, and VM. Each of these nuclei innervate motor and premotor cortices which include the oculo-cephalomotor areas (Bentivoglio et al., 1981; Glenn et al., 1982; Herkenham, 1979; Jones and Leavitt, 1974; Krettek and Price, 1977; Moran et al., 1982; Orem and Schlag, 1973). Furthermore, ZI, another diencephalic target for the tectospinal cells, connects several brainstem structures also related to the control of eye and head movements (Ricardo, 1981).

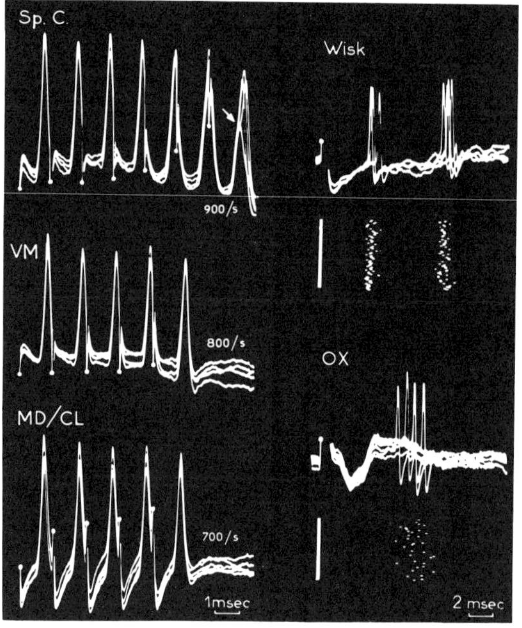

Fig. 3. Responses of an identified TSD cell to somatosensory and optic chiasm stimulation. Left: multiple antidromic activation of a single TSD cell elicited by high frequency stimulation pulses delivered to spinal cord (SPc), ventro medial thalamic nucleus (VM) and medio dorsal/central lateral (MD/CL) thalamic area stimulation. The dissociation of the antidromic potential in its IS-SD components (arrow in upper trace) is indicative of soma activity. Right: response of this TSD cell to stimulation of the snout (Whisk) and of the optic chiasm (OX), shown by superimposing sweeps (upper) and by successive dot displays (lower) at the same time course.

Traditionally, SC is considered to be involved in the orientation of the eyes and head towards sensory cues in the environment (see review from Wurtz and Albano, 1980). Since the tectospinal cells influence through their profuse rhombencephalic and diencephalic collaterals most of the ocular and cephalic motor and premotor circuits, it was of interest to test their responsiveness to peripheral stimulation.

Peripheral Input to Tectospinal-tectodiencephalic Cells (TSD cell)

Seventy TSD cells were tested for their response to somatic, visual and auditory stimuli.

Somatosensory stimulation consisted of light displacement of individual whiskers or brushing the snout with a paint brush. In some cases electrical stimulation was also applied through a pair of needles inserted into the tissue of the snout. Visual stimuli consisted of either a flash of light, or moving a luminous bar produced with an opthalmoscope. Acoustic stimuli were multifrequency sounds, such as finger snaps.

Sixty-two of the 70 recorded cells were driven by stimulation of the snout. Usually the TSD cells, which were rather silent, gave a high frequency discharge in response to a brisk air puff which moved a limited group of whiskers (two or three) from their resting position. When electrical stimuli were applied, TSD cells exhibited a characteristic two-component excitatory response, whose latencies were 3-4 msec and 7-10 msec respectively (Fig. 3). In accordance with the general somatotopic plan described in the rat SC (Kassell, 1982), a topographic representation of the whiskers has been observed in the TSD neuronal population. The more laterally located TSD cells responded to stimulation of the lower rows of whiskers; whereas the more rostral ones were fired by moving the anterior whiskers. In contrast to somatic stimulation, very few TSD cells (n=7), could be influenced by application of visual or auditory stimuli. However,TSD cells probably receive visual input since 40% of them were activated following direct stimulation of the retinofugal fibers (see Fig. 3). The predominance of somatosensory input observed in TSD cells compared to visual or acoustic input might be related to our experimental conditions (anesthetized and paralyzed animals), but is also probably relevant to the importance played by the whisker apparatus during exploratory behavior. From this input-output relationship it appears that the TSD cells might be considered as one of the sensorimotor links through which SC participates in the elaboration of eye and head orienting movements.

THE NIGRAL CONTROL OF TSD CELLS

To see if TSD cells receive an inhibitory nigral influence, we identified these neurons by the antidromic activation method and tested their response to SN stimulation (Chevalier et al., in press). Out of 37 units identified, 50% of them exhibited the typical short latency and short duration nigral-induced inhibition (Fig. 4). During the course of the nigral effect, the excitability of the TSD cells was considerably reduced. In fact the nigral input caused a phasic suppression of their spontaneous as well as their peripherally evoked activity (Fig. 5). Moreover the temporal coincidence between the nigral evoked inhibition and the antidromic activation resulted in the dissociation of the antidromic spike into its IS-SD components.

In a recent experimental analysis using monkeys performing ocular orienting tasks, Hikosaka and Wurtz (1983) reported that SNr and SC neuronal activities are organized in a mirror fashion. While the SC neurons increase their firing rate before targeting movements, SNr cells and particularly identified nigro-tectal neurons exhibit a phasic arrest of their tonic spontaneous discharge. Similar changes in SNr activity preceding eye-head orienting movements have also been observed in the cat (Joseph and Boussaoud, 1982). Considering the high level of spontaneous discharge of SNr neurons currently observed in all studied mammals, it has been suspected that their function is to exert a tonic inhibitory influence on their target cells. Then through this arrest of activity they would participate in the elaboration of SC presaccadic discharges via a disinhibitory mechanism.

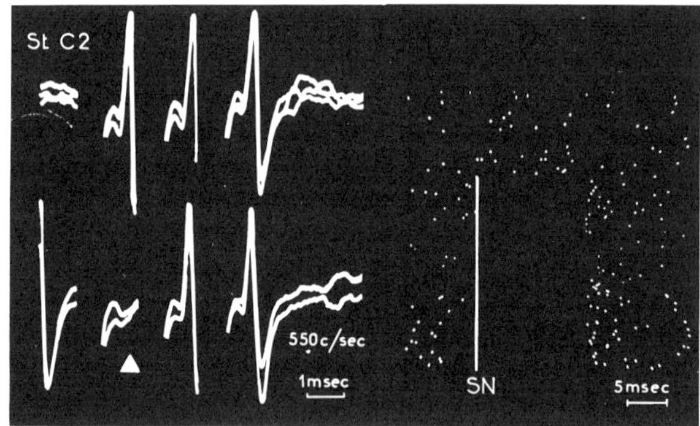

Fig. 4. Nigral evoked inhibition on a TSD cell. Left: identification of a TSD cell. Its antidromic invasion following stimulation of the cervical spinal cord (C2). The antidromic response is characterized by fixed latency, persistance on high frequency stimulation, and collision with spontaneous spikes (triangle). Right: nigral stimulation produced in this cell a typical brisk and short-lasting arrest of discharge.

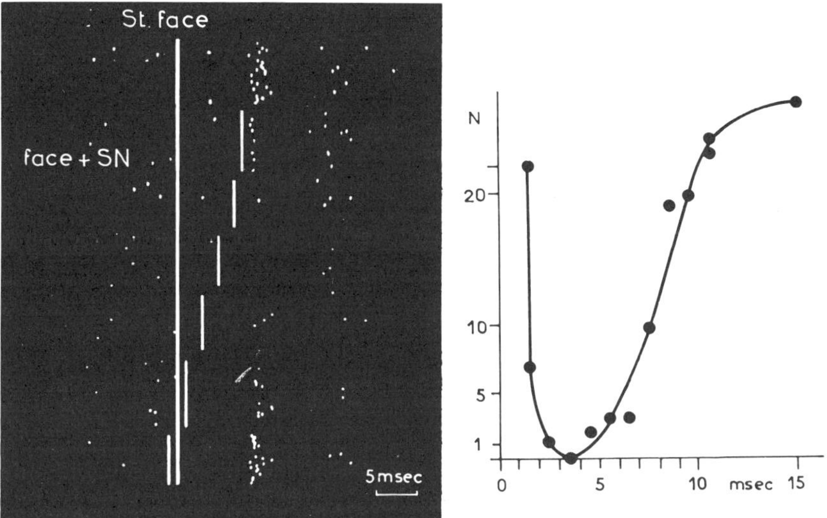

Fig. 5. Inhibitory effect of nigral stimulation on the somatosensory evoked discharge of a TSD cell. Stimulation of the snout (st face) is presented at the same delay (continuous vertical bar). Stimulation of the SN is given at various delay intervals subsequent to peripheral stimulation. Note the inhibitory influence of the conditioning SN stimulation. On the right, the number of spikes evoked by peripheral stimulation is plotted as a function of the interval separating the onset of the peripheral evoked responses from that of the conditioning stimulation. Note the short latency and short duration of the inhibitory nigral influence, and its effectiveness in blocking the sensory input.

To test the accuracy of such a disinhibitory mechanism, we have examined the consequences of a transitory blockade of nigral discharge on the activity of electrophysiologically identified TSD cells. For this purpose SNr and TSD neurons were simultaneously recorded while 50 to 100 nl of a 1M GABA solution was injected in the SNr. Concomitant with the arrest of the SNr cells' activity, TSD neurons exhibited a brisk and vigorous increase in firing (Fig. 6). The time course of these changes in firing rate were closely related to each other. While the nigral firing rate was returning to the control level, TSD activity slowed down. Thus it can be concluded that SNr does indeed exert a tonic inhibitory influence whose arrest is able to provoke a discharge in its collicular target cells.

In conclusion, from our current knowledge of nigro-collicular relationship, it appears: firstly, that the role of SNr cells seems to be a sustained and potent inhibitory action on their collicular

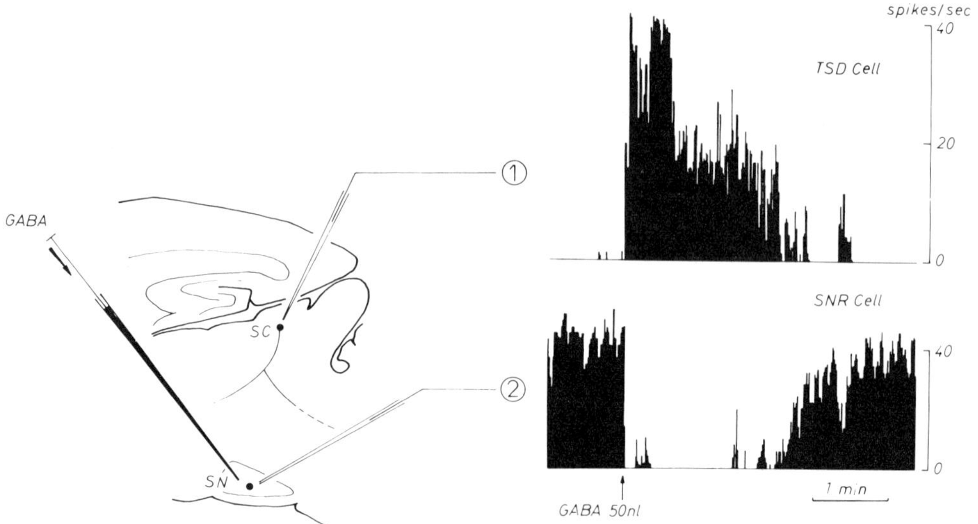

Fig. 6. Effect of a transient arrest of SNr neuronal firing on the activity of a TSD cell. Activity of an identified TSD neuron (1) and of a SNr cell (2) simultaneously recorded during intranigral application of GABA. While SNr discharge is blocked by GABA, the TSD cell is active.

target cells; secondly, by acting on the branched tecto-spinal neurons, SNr can exert a coordinated control of most of the diencephalic, mesencephalic, rhombencephalic and spinal areas which cooperate in the elaboration of eye and head movement.

Since a previous demonstration that nigro-collicular neurons are themselves under powerful inhibitory effect of the striatum (Deniau et al., 1978b), one could question the functional meaning of such a double striato-nigro-collicular inhibitory chain (see Fig. 7). The present finding that the transient arrest of tonic nigro-collicular inhibition, induced by intranigral application of GABA, is able to provoke a phasic release of activity of TSD cells, leads to the hypothesis that, in behavioral conditions, the GABAergic striato-nigral neurons would also be able to generate such firing in TSD cells. This basal ganglia evoked discharge might participate in the set of activity required for triggering eye and head movements. Furthermore, considering that SNr activity depresses the responsiveness of TSD neurons to sensory input, we propose that the striato-nigro-collicular circuit acts on TSD cells, through a disinhibitory mechanism, by switching on an appropriate sensorimotor link which could allow stimuli to be taken as a target for orienting

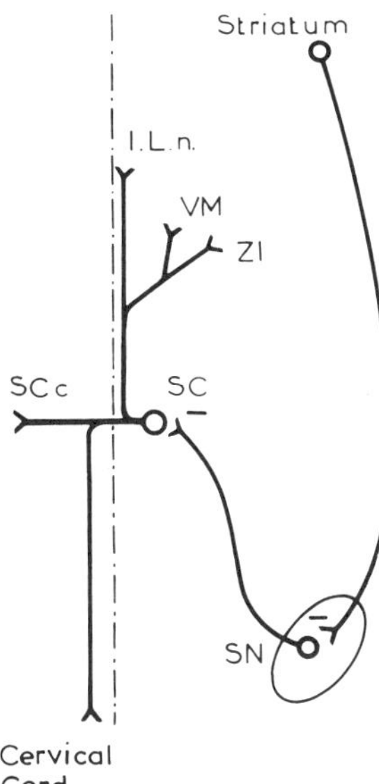

Fig. 7. Diagrammatic representation of the double inhibitory neuronal chain linking striatum to TSD cells. ILn: Intralaminar nuclei; VM: Ventro medial thalamic nucleus; ZI: Zona incerta; SC: Superior colliculus; SCc: Contralateral superior colliculus; SN: Substantia nigra.

eye/head movements. Disruption of this functional link, following striatal or nigral lesion, might result in a persistent inability of TSD cells to be driven by sensory inputs. Such a pathological process might partly underlie the sensory neglect syndrome which readily accompanies basal ganglia impairment in rodents.

REFERENCES

Bentivoglio, M., Macchi, C., and Albanese, A., 1981, The cortical projections of the thalamic intralaminar nuclei, as studied in cat and rat with the multiple fluorescent retrograde tracing technique, Neurosci. Lett., 26:5-10.

Chevalier, G., Deniau, J. M., Thierry, A. M., and Féger, J., 1981a, The nigrotectal pathway. An electrophysiological reinvestigatation in the rat, Brain Res., 213:253-263.

Chevalier, G., Thierry, A. M., Shibazaki, T., and Feger, J., 1981b, Evidence for a GABAergic inhibitory pathway in the rat, Neuroscience Lett., 21:67-70.

Chevalier, G., and Deniau, J. M., Spatio-temporal organization of a branched tecto-spinal/tecto-diencephalic neuronal system, Neuroscience (in press).

Chevalier, G., Vacher, S., and Deniau, J. M., Inhibitory nigral influence on tecto-spinal neurons. A possible implication of basal ganglia in orienting behavior, Exp. Brain Res., (in press).

Deniau, J. M., Chevalier, G., and Féger, J., 1978a, Electrophysiological study of the nigro tectal pathway in the rat, Neurosci. Lett., 10:215-220.

Deniau, J. M., Hammond, C., Riszk, A., and Féger, J., 1978b, Electrophysiological properties of identified output neurons of the rat substantia nigra (pars compacta and pars reticulata). Evidences for the existence of branched neurons, Exp. Brain Res., 32:409-422.

Di Chiara, G., Porceddu, M. L., Gessa, G. L., 1979a, Substantia nigra as an output station for striatal dopaminergic responses: role of a GABA-mediated inhibition of pars reticulata neurons, Naunyn-Schmiedeberg's Arch. Pharmacol., 306:153-159.

Di Chiara, G., Porceddu, M. L., Morelli, M., Mulas, M. L., and Gessa, G. L., 1979b, Evidence for a GABAergic projection from the substantia nigra to the ventromedial thalamus and to the superior colliculus of the rat, Brain Res., 176:173-184.

Di Chiara, G., Morelli, M., Imperato, A., and Porceddu, M. L., 1982. A reevaluation of the role of superior colliculus in turning behaviour, Brain Res., 237:61-77.

Dray, A., 1980, The physiology and pharmacology of mammalian basal ganglia, Prog. Neurobiol., 14:221-335.

Dunnett, S. P.., and Iversen, S. D., 1982, Sensorimotor impairments following localized kainic acid and 6-hydroxydopamine lesions of the neostriatum, Brain Res., 248:121-127.

Feeney, D. M., and Wier, C. S., 1979, Sensory neglect after lesions of substantia nigra or lateral hypothalamus: differential severity and recovery of function, Brain Res., 178:329-346.

Glenn, L. L., Hada, J., Roy, J. P., Deschenes, M., and Steriade. M., 1982, Anterograde tracer and field potential analysis of the neocortical layer I projection from nucleus ventralis medialis of the thalamus in cat, Neuroscience, 7, 1861-1878.

Grantyn, A., and Grantyn, R., 1981, Axonaal patterns and sites of termination of cat superior colliculus neurons projecting in the tecto-bulbo-spinal tract, Exp. Brain Res., 46:243-256.

Herkenham, M., 1979, The afferent and efferent connections of the ventromedial thalamic nucleus in the rat, J. Comp. Neur., 183:487-518.

Hikosaka, O., and Wurtz, R. H., 1983, Visual and oculomotor functions of monkey substantia nigra pars reticulata. Relation of substantia nigra to superior colliculus, J. Neurophysiol., 49:1285-1301.

Jones, E. G., and Leavitt, R. Y., 1974, Retrograde axonal transport and the demonstration of non-specific projections to the cerebral cortex and striatum from thalamic intralaminar nuclei in the rat, cat and monkey, J. Comp. Neur., 154:349-378.

Joseph, J. P., and Boussaoud, D., 1982, Involvement of the substantia nigra pars reticulata in eye and head movements in the cat, Neurosci. Lett., S 10:257.

Kassel, J., 1982, Somatotopic organization of SI corticotectal projections in rats, Brain Res., 231:247-255.

Kilpatrick, I. C., Collingridge, G. L., and Starr, M. S., 1982, Evidence for the participation of nigrotectal aminobutyrate containing neurones in striatal and nigral derived circling in the rat, Neuroscience, 7:207-222.

Krettek, J. E., and Price, J. L., 1977, The cortical projections of the mediodorsal nucleus and adjacent thalamic nuclei in the rat, J. Comp. Neur., 171: 157-192.

Ljunberg, T., and Ungerstedt, U., 1976, Sensory inattention produced by 6-hydroxydopamine induced degeneration of ascending dopamine neurons in the brain, Exp. Neurol., 53:585-600.

Marshall, J. F., Berrios, N., and Sawyers, S., 1980, Neostriatal dopamine and sensory inattention, J. Comp. Physiol. Psychol., 94:833-846.

Moran, A., Avandano, C., and Reinoso-Suarez, F., 1982, Thalamic afferents to the motor cortex in the cat. A horseradish peroxidase study, Neurosci. Lett., 33:229-233.

Murray, E. A., and Coulter, J. D., 1982, Organization of tecto-spinal neurons in the cat and rat superior colliculus, Brain Res., 243:201-214.

Orem, J., and Schlag, J., 1973, Relations between thalamic and corticofrontal sites of oculomotor control in the cat, Brain Res., 60:503-507.

Pycock, C. J., 1980, Turning behaviour in animals, Neuroscience, 5:461-514.

Ricardo, J. A., 1981, Efferent connections of the subthalamic region in the rat. II. The zona incerta, Brain Res., 21:43-60.

Siegfried, B.., and Bures, J., 1978, Asymmetry of EEG arousal in rats with unilateral 6-hydroxydopamine lesion of the substantia nigra: quantification of neglect, Exp. Neurol., 62:173-190.

Vincent, S. R., Hattori, T., and McGeer, E. G., 1978, The nigrotectal projection: a biochemical and ultrastructural characterisation, Brain Res., 151:159-164.

Wurtz, R. H., and Albano, J. E., 1980, Visual-motor function of the primate superior colliculus, Ann. Rev. Neurosci., 3:189-226.

York, D. H., and Faber, J. E., 1977, An electrophysiological study of nigrotectal relationships: a possible role in turning behaviour, Brain Res., 130:383-386.

TRANSMITTERS AND RECEPTORS IN THE BASAL GANGLIA

Philip M. Beart

University of Melbourne
Clinical Pharmacology and Therapeutics
Austin Hospital
Heidelberg, Victoria 3084

This chapter surveys the neurotransmitters and their receptors within the neostriatum (caudate-putamen, NS), globus pallidus (GP) and substantia nigra (SN), but is deliberately biased towards those transmitters involved in the principal input and output pathways of the basal ganglia in an attempt to relate neuronal circuitry to function. Where appropriate, behavioural data have also been presented and reference has been made to the abnormal movement disorders affecting basal ganglia function. Morphological evidence for the relevant pathways and connections can be found elsewhere within this volume. No attempt has been made to consider the peptide transmitters, neurotensin, somatostatin, TRH, angiotensin, cholecystokinin and the opiates, which are found in the basal ganglia, but whose involvement in motor function has not been well established. Whereas species differences may be of considerable importance in any anatomical review, there is insufficient evidence pertaining to the variation across species of pharmacological and neurochemical data, and thus attempts have generally not been made to nominate the animal species employed. The present article will not discuss the ventral striatum (nucleus accumbens septi), whose connections are less well defined, but these have been dealt with elsewhere (Fonnum and Walaas, 1979; Heimer and Van Hoesen, 1979).

DOPAMINE

The monoamine, dopamine (DA), has long been considered a key neostriatal transmitter, particularly since neuropathological studies have implicated DA in a number of disease states (Parkinson's disease, Huntington's chorea and tardive dyskinesia; Hornykiewicz,

1979; Bird and Iversen, 1974). More recent histofluorescence studies reveal that neostriatal dopaminergic neurones arise not only from the SN, but also from the adjacent ventral tegmental area, and thus these neurones are probably better described by the more collective terminology "mesostriatal" (Moore and Bloom, 1978; Lindvall and Bjorklund, 1978). The organization and morphology of the mesostriatal dopaminergic pathways have been the subject of many reviews (Lindvall and Bjorklund, 1978; Moore and Bloom, 1978). Mesostriatal dopaminergic neurones appear essential for the expression of a number of striatal-associated overt behaviours, most of which can be elicited by the administration of dopaminergic drugs. Electrophysiological studies indicate that DA has both inhibitory and excitatory effects within the NS, with the former effect apparently being mediated by collaterals and neostriatal interneurones (York, 1979). DA plays a key role in neurotransmission within the basal ganglia in that it can mediate the balance between inhibitory and excitatory outflow from the NS : a balance which can apparently be altered by pathological changes and drug treatments (Alloway and Rebec, 1983; Groves, 1983; Kanazawa and Yoshida, 1980; Neve et al., 1982).

D-1 and D-2 Receptors

The concept of multiple DA receptors was first suggested by Spano (1978), subsequently elaborated by Kebabian and Calne (1979), and has been extensively reviewed (Beart, 1982, Creese et al., 1983). As initially postulated, the two DA receptor subtypes (designated D-1 and D-2) were differentiated on the basis of the linkage of the D-1 subtype to the dopamine-stimulated adenylate cyclase. Although the actual division of DA receptors into D-1 and D-2 subtypes is widely accepted, and in fact strengthened by the design of subtype-specific drugs, a division based purely on adenylate cyclase is no longer acceptable. Firstly, there is a paucity of evidence supporting a role for the dopamine-stimulated adenylate cyclase and implicating a physiological involvement of cyclic AMP in neostriatal dopaminergic transmission. Secondly, guanine nucleotides (Leff and Creese, 1983; Gundlach et al., 1983) and alterations in adenylate cyclase activity may be associated with the activation of D-2 receptors. The corollary to these objections is that the definition of D-1 and D-2 receptors should be based purely on the pharmacological actions of agonists and antagonists at these sites : a status that is coming about with the advent of new, subtype drugs. An attempt has been made to classify DA receptors as DAi and DAe receptors according to whether they are inhibition-or excitation-mediating within the NS, but this more functional classification has not met with general acceptance (Cools, 1982). Two further DA receptor subtypes, D-3 and D-4, have been postulated mainly from evidence obtained in radioligand ligand binding - very little evidence supporting their existence has been forthcoming, although the D-3 receptor has some properties consistent with DA autoreceptors (Sokoloff et al., 1980; Leff and Creese, 1983).

The subtype of receptor designated D-2 is the one at which ergots used to treat Parkinson's disease (e.g. bromocryptine, pergolide) act as high affinity agonists (Gundlach et al., 1983; Kebabian and Calne, 1979). With advances in structure-activity relationships it has been possible to design partial ergolines with high pharmacological specificity for D-2 receptors (e.g. LY 141865; Cannon, 1983; McPherson and Beart, 1983). The butyrophenone (e.g. haloperidol) and diphenylbutylpiperidine (e.g. pimozide) classes of neuroleptic drugs are effective D-2 antagonists (Beart, 1982; Creese et al., 1983; Kebabian and Calne, 1979). Appreciable evidence supports a functional role for D-2 receptors within the central nervous system, both in motor control and in hormone secretion (Schachter et al., 1980). Indeed D-2 receptors are altered in the basal ganglia of post-mortem brains from Parkinsonian patients (Lee et al., 1978) and in the NS of hyperactive rodents (Helmeste and Seeman, 1982). Our understanding of the importance of D-1 receptors has been hindered by a lack of specific pharmacological agents. Until recently the phenylbenzazepine, SKF 38393, was the only available D-1 specific drug (Setler et al., 1978). (3H) thiaxanthenes (cis-flupenthixol and piflutixol) have been used to label D-1 receptors (Cross and Owen, 1980; Hyttel, 1980), but both ligands also have appreciable affinity for D-2 receptors. Recently new D-1 agonists and an antagonist have been reported (Hahn et al., 1982; Hyttel, 1983), and we can look forward to some of the mysteries of the D-1 receptor being unravelled. Indeed, since SKF 38393 produces circling behaviour in the rat model of Ungerstedt (unilateral nigral 6-hydroxydopamine lesion), D-1 receptors are also likely to play an important role in motor function (Setler et al., 1978).

Although the terminology D-1 and D-2 infers _per se_ no particular localization on a specific neurone within the basal ganglia, attempts have been made to analyse the distribution of neostriatal DA receptors. Such studies have employed a variety of lesioning techniques (Kainate, 6-hydroxydopamine and cortical ablation) and much of the data has recently been summarized (Hall et al., 1983). The very large loss of neostriatal D-1 receptors after a striatal kainate lesion contrasts with the less extensive loss of D-2 receptors (Cross and Waddington, 1981; Leff et al., 1981). Moreover, the decrease in D-1 receptors correlates well with the loss of neostriatal GABAergic neurones (as shown by reduced glutamate decarboxylase activity; Cross and Waddington, 1981), which represents the main striatal projection pathway in the form of the striato-pallido-nigral system. Thus, D-1 receptors may be relatively more important than D-2 receptors in regulating neostriatal outflow. Receptors of the D-2 subtype appear to be divided between striatal (kainate-sensitive) neurones and corticostriatal afferent fibres (Garau et al., 1978; Schwarcz et al., 1978). D-2 receptors are certainly localized on neostriatal cholinergic interneurones, since

D-2 agonists increase striatal levels of acetylcholine - an effect which is antagonized by neuroleptic drugs (Sethy, 1979; Scatton, 1982). In vitro the specific D-2 agonist, LY 141865, inhibits the evoked-release of 3H-acetylcholine from slices of rat neostriatum (Scatton, 1982). The D-1 agonist, SKF 38393, is ineffective in both of these biochemical models.

In the SN autoradiographic and lesion studies suggest that D-2 receptors are mainly located on the cell bodies and processes of dopaminergic neurones (Murrin et al., 1979; Quik et al., 1979). In fact intrastriatal kainic acid and striatonigral lesions have no effect on the distribution of D-2 receptors (Murrin et al., 1979). Receptors of the D-1 subtype are also present in the SN (Cross and Owen, 1980), and, if the dopamine-stimulated adenylate cyclase can be taken as a guide, are likely to be localized on non-dopaminergic elements. Nigral 6-hydroxydopamine lesions have no effect on the dopamine-stimulated adenylate cyclase and thus D-1 receptors are restricted to the pars reticulata (Jessell, 1978; Reubi et al., 1978), apparently mainly on GABAergic (and not substance P) neurones. Thus D-1 receptors probably also play an important role in modulating output from the SN.

Autoreceptors

Since the demonstration that receptors sensitive to DA were present on dopaminergic nerve terminals (autoreceptors), considerable effort has been devoted to the design of specific autoreceptor drugs. These autoreceptors are now known to be also located on dopaminergic somata and dendrites (Aghajanian and Bunney, 1973), and their activation reduces the firing rate of mesostriatal dopaminergic neurones, decreases the synthesis and turnover of DA, and suppresses locomotor activity in rodents (Strombom, 1976; Wachtel et al., 1979). Indeed, in electrophysiological experiments, nigral autoreceptors for DA were six to ten times more sensitive to the inhibotory effects of DA agonists than postsynaptic receptors of the NS (Skirboll et al., 1979). Clinical reports have suggested that low doses of DA agonists (which preferentially activate autoreceptors) have beneficial antipsychotic properties (Tamminga et al., 1978; Corsini et al., 1981), and DA autoreceptor agonists may thus represent a novel approach for the treatment of disorders characterized by hyperactive dopaminergic neurones (e.g. Huntington's chorea, tardive dyskinesia and schizophrenia).

Although the properties of DA autoreceptors remain to be fully elucidated, it seems that they are not associated with DA-stimulated adenylate cyclase (Carlsson, 1977; Iversen et al., 1976) and that many neuroleptic drugs are effective DA antagonists at these sites (Walters and Roth, 1976). Although DA agonists such as apomorphine, ergots and tetralin-derivatives (e.g. TL-99 in Fig. 1) have the

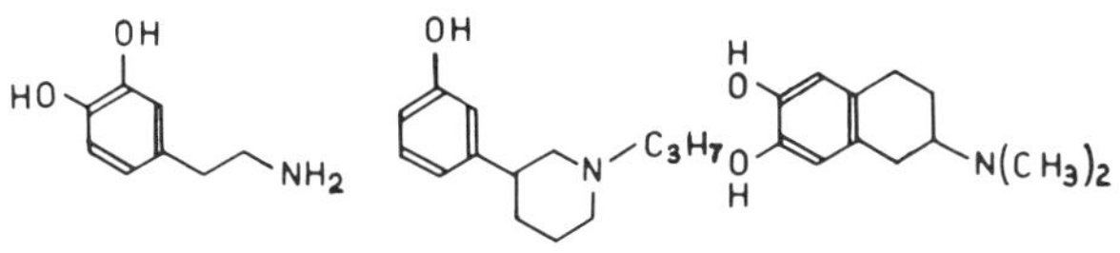

DOPAMINE 3-PPP TL-99

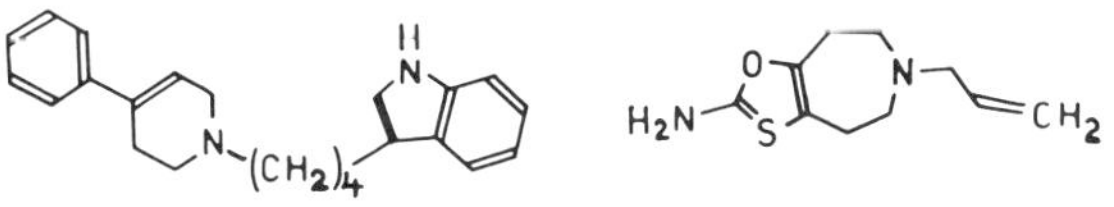

EMD 23 448 B-HT 920

Fig. 1. Dopamine autoreceptor agonists

capacity to activate both D-2 and autoreceptors (Gundlach et al., 1983; Walters and Roth, 1976; Williams and Totaro, 1982), recent advances in structure-activity relationships have allowed the synthesis of DA agonists with specificity for DA autoreceptors. All of the drugs shown in Figure 1 possess actions at DA autoreceptors - TL-99 was the first compound proposed as a selective autoreceptor agonist (Goodale et al., 1980), but subsequent investigations have shown that it also possesses actions as a postsynaptic D-2 agonist (Gundlach et al., 1983). Certainly, from the diversity of chemical structures depicted in Figure 1, much remains to be learnt about the structure activity constraints that confer autoreceptor activity upon a molecule. Both BHT 920 and EMD 23 448 have been comparatively recently studied for autoreceptor properties and further investigations are needed to establish their specificity and potency (Anden et al., 1982; Seyfried and Fuxe, 1982). 3-(3-Hydroxphenyl)-N-n-propylpiperidine (3-PPP, Fig. 1) should probably be regarded as the prototype for DA autoreceptor agonists. The drug has almost negligible postsynaptic activity, decreases neostriatal DA turnover after peripheral administration (Hjorth et al., 1981) and suppresses spontaneous- and d-amphetamine-induced locomotor activity in rats. Congeners of 3-PPP are now available that possess even greater affinity for DA autoreceptors (Nilsson and Carlsson, 1982), and we should expect this to be an area of appreciable progress in the next decade.

Somatodendritic Mechanisms

Cajal was the first to describe the rather prominent dendrites of nigral dopaminergic neurones and drew attention to their very long, shaggy and discretely divided arborization. With refinements in the histofluorescence methodology for catechomamines, Bjorklund and Lindvall (1975) were able to demonstrate the presence of DA in these dendritic processes. These dopaminergic dendrites often run in bundles that spread out into the pars reticulata, and at a distance of several hundred microns from their cell bodies they ramify into fluorescent varicose terminals (Bjorklund and Lindvall, 1975; Wilson et al., 1977). Both axons and dendrites form synapses *en passage* in the pars compacta, and dendrodendritic profiles are found in the pars reticulata (Wilson et al., 1977). Complementary electrophysiological and biochemical evidence has indicated that DA is released from dendrites of nigral dopaminergic neurones and contributes to the modulation of neuronal activity in the substantia nigra. The released DA may regulate the activity of dopaminergic neurones and the release of transmitters from nigral afferent fibres (Groves et al., 1975; Cheramy et al., 1981; Waszcak and Walters, 1983). DA can be synthesized, stored and inactivated in dopaminergic dendrites, although some of these mechanisms may differ from those of axonal nerve terminals. The release of dopamine from dendrites has been demonstrated both *in vitro* and *in vivo* (see Cheramy et al., 1981).

More recent elegant work from Glowinski's laboratory has shown that *in vivo* changes in the activity of neurones with functional connections to the SN, or involved in sensory motor processes, affected the dendritic release of newly synthesized (3H)DA. Thus, for example, (3H)DA release could be altered by electrical stimulation of motor and visual cortices (Cheramy et al., 1981). The regulation of the dendritic release of DA appears to be very complex, involving additional pathways contralateral to SN under study, with thalamic nuclei seemingly playing a central role (Cheramy et al., 1983). Release studies have also provided an indication of the tramittters likely to influence the dendritic release of DA, and these include GABA, glycine, substance P, 5-hydroxytryptamine and acetylcholine (Reubi et al., 1978; Starr, 1978; Cheramy et al., 1981). At the present time it remains unclear whether such regulatory processes are functionally important in regulating dopaminergic activity, since they take place at dendrites located some appreciable distance from the dopaminergic somata. Their importance will probably only be revealed in experiments performed in unanaesthetised, freely moving animals. However, it is interesting to note that synaptic circuits through dendrites have now been found in the retina, cerebral cortex, thalamus and olfactory bulb.

GABA

Abnormalities of the γ-aminobutyric acid (GABA) system occur in extrapyramidal disorders, including Huntington's chorea and Parkinson's disease (Bird and Iversen, 1974; Marsden, 1979; Lloyd, 1980). Thus is it of some interest that the highest brain concentrations of GABA and its synthesising enzyme, glutamate decarboxylase (GAD), are found within some of the structures of the basal ganglia (Fahn, 1976). Immunocyotochemical studies have revealed GAD-positive reaction product associated with axon terminals in the NS, GP, entopeduncular nucleus and SN (Ribak, 1981).

Within the NS, medium-sized spiny projection neurones give rise to a large number of intrinsic GABAergic synapses which contain GAD-positive reaction product (Ribak, 1981). These neurones also give rise to numerous local axon collaterals which terminate in the vicinity of their cell bodies. Axons arising from these GABAergic neurones also innervate the GP and entopeduncular nucleus (Fonnum et al., 1978; Ribak, 1981). GAD activity appears to be higher in the GP and entopeduncular nucleus than in the NS (Fahn, 1976; Ribak, 1981), where the enzyme is concentrated in the ventrocaudal region (Fonnum et al., 1978). In fact the striatopeduncular and striatonigral GABAergic pathways arise preferentially from the posterior NS. The GP also receive a GABAergic input from the nucleus accumbens which terminates in the ventromedial pallidum (Jones and Mogenson, 1980), whereas neostriatal projections seem more associated with the dorsal region of the GP. GABA is also the transmitter of a number of extrinsic pathways arising in this region of the basal ganglia - the striatonigral, pallidonigral and pallido(entopeduncular)-subthalamic pathways. Pallidal lesions decrease GAD activity in the subthalamus (Fonnum et al., 1978) and the inhibition of subthalamic neurones evoked by pallidal stimulation is attenuated by the GABA antagonist, bicuculline (Rouzaire-Dubois et al., 1980). GABAergic afferents to the SN from the NS and GP have been the centre of appreciable interest, and there is general agreement that rostrally situated GABA-containing neurones of the NS project almost exclusively to the pars reticulata, while caudal neostriatal and pallidal neurones innervate both the reticulata and compacta (Brownstein et al., 1977; Fonnum et al., 1978; Jessell et al., 1978). The pallidonigral pathway may be a relatively minor one. Divergent branched axons from the nucleus accumbens, which innervate the GP (see above), also provide a further GABAergic input to the SN (Dray and Oakley, 1978).

Within the SN, somata containing GAD-positive reaction product are located in both pars compacta and reticulata (Fonnum et al., 1978; Ribak, 1981). Further evidence for nigral GABAergic neurones is the persistence of GAD-positive terminals within the ventral pars reticulata after hemisections interrupting striato- and pallidonigral GABAergic pathways (Ribak, 1981). The GAD-positive neurones of SN are small and medium-sized local circuit neurones, some having projections throughout the SN and others giving rise to local axon collaterals (Juraska et al., 1977). These neurones also appear to send collaterals to the dopaminergic neurones of the pars compacta. Nigral GABAergic neurones also project to diverse brain regions including the ventromedial thalamus, nucleus parafascicularis, superior colliculus, periaqueductal grey, formatio reticularis and pedunculopontine nucleus (Beckstead et al., 1979; Herkenham, 1979).

Moreover, many of these nigral GABAergic neurones have branched axons and thus possess the capacity to influence neuronal activity in several outflow systems (Anderson and Yoshida, 1980; Deniau et al., 1978).

Many areas of the basal ganglia are able to support behavioural responses after the local application of GABAergic drugs - behaviours which can be either DA-independent or DA-dependent. In the rat, the intrastriatal administration of drugs which interfere with GABAergic transmission induces dyskinesias (Standefer and Dill, 1977; Tarsy et al., 1978), while intrastriatal GABA abolishes DA-dependent peri-oral (stereotyped) movements (Costall et al., 1978). High doses of GABA agonists upon microinjection into the ventral striatum and GP elicit initially short-lasting catalepsy and then long-lasting stimulation (Scheel-Kruger et al., 1981). GABAergic drugs produce prominent behavioural effects upon infusion into the SN, and it is clear that these distinct GABA-mediated behaviours depend upon topographical differences in neuronal organization. A wide variety of GABAergic drugs elicit circling behaviour upon unilateral microinjection into the SN of unlesioned rats. The actual direction of circling is extremely dependent upon the precise location of the microinjection and the injection volume, and opposite effects have been routinely observed after injections into the reticulata and compacta (Kilpatrick and Starr, 1981). Circling produced by GABAergic drugs is generally described as DA-dependent or -independent (i.e. unaffected by 6-hydroxydopamine, neuroleptics and alpha-methyl-p-tyrosine). High doses of the GABA agonist, muscimol, produce stereotypy with reduced frequency of circling, while bilateral intranigral muscimol is even more effective at inducing stereotyped behaviour which is DA-independent (Scheel-Kruger et al., 1978). Kainic acid administered into the pars reticulata induces a behavioural, DA-independent syndrome analagous to that produced by GABA-mimetics, viz. circling after unilateral application and sterotypy upon bilateral microinjection. These behaviours probably result from the destruction of nigral efferent pathways and/or nigral interneurones (Olianas et al., 1978). Moreover, hemi-transections rostral to the SN appear not to attenuate or abolish GABA-mediated circling, stereotypy and catalepsy (Gale, 1981; Reavill et al., 1981). Recent evidence would suggest that nigral GABAergic efferents to the ventromedial thalamus, parafascicular nucleus, superior colliculus and mesencephalic reticular formation are all involved in motor-associated behaviours (Di Chiara et al., 1981 and this volume). In interpreting these data, much of which pertains to circling behaviour, it should be remembered that there are both postural and locomotor components of circling (Kelly and Moore, 1976).

GABA Receptors

Largely as a result of microelectrode studies, the existence of discrete receptor subtypes for GABA has been considered likely for many years (Curtis and Johnston, 1975; Andrews and Johnston, 1979; Enna and Gallagher, 1983). Such experiments led to the terminology bicuculline-sensitive and -insensitive to delineate two groups of analogues, one with inhibitory actions on the firing of central neurones which could be blocked by the GABA antagonist, bicuculline (Curtis and Johnston, 1975; Johnston, 1981). Included in the group of bicuculline-insensitive GABA analogues are a number of amino acids of folded conformation. Recent attention has centred on one member of this group, baclofen (p-chlorophenyl GABA, lioresal), and in ligand binding studies ^3H-baclofen (and ^3H-GABA) has been shown to label a unique population of sites in brain homogenates (Hill and Bowery, 1981). These sites possess pharmacological characteristics different from those of classical GABA receptors (GABA-A; see below) and both bicuculline and picrotoxin are ineffective at competing for the binding sites. These GABA-B receptors appear to be associated with Ca^{2+} and guanyl nucleotides (Bowery et al., 1982), and independent of sites affected by barbiturates and benzodiazepines. Baclofen-sensitive GABA receptors are widely distributed throughout the central nervous system (Bowery, 1982) and may mediate presynaptic inhibition (Curtis et al., 1981). Further progress in this area requires the development of selective GABA-B antagonists. Within the NS, GABA-B receptors are apparently localized on dopaminergic nerve terminals, where they regulate the release of transmitter (Bowery et al., 1980). Baclofen elicits circling upon intranigral administration and thus GABA-B receptors also occur within the SN (Waddington and Cross, 1979).

Much more is known about the properties of GABA-A receptors, mainly because the characteristics of the GABA receptor-complex have been investigated for 10 years. The GABA receptor-ionophore complexes appear to be composed of a number of interacting domains, the characteristics of which vary greatly with membrane preparation. Barbiturates and benzodiazepines stimulate GABA binding by different mechanisms (Karobath et al., 1979; Skolnick et al., 1980; Johnston, 1981), and it is possible that these classes of drugs act upon different populations of GABA receptors. Other evidence suggests that barbiturates and picrotoxin both bind to the GABA-linked chloride ionophore (Olsen et al., 1979) without altering the binding of GABA to its recognition site. Bicuculline labels only a single population of GABA-sensitive sites of relatively low affinity (Mohler and Okada, 1978). The kinetics and characteristics of GABA binding are very dependent upon a range of substances that are normally incorporated into the membrane associated with the GABA-receptor complex. These substances have been collectively called GABARINS (GABA Receptor Inhibitors) and include occluded GABA, phosphatidylethanolamine, peptides, GRIF (GABA Receptor Inhibitory Factor) and purines (Johnston, 1981). GABA binding is affected by various treatments including the chronic administration of phenobarbitone and diazepam (Mohler et al., 1978), stress (Skerritt et al., 1981) and striatal lesions (Waddington and Cross, 1978).

Within the SN, GABA receptors are located on nigral dopaminergic neurones, possibly on dendrites (Starr, 1978; Cheramy et al., 1978a and b). Moreover, electrophoretically applied muscimol depresses the firing rate of nigral pars compacta cells (Grace and Bunney, 1979; Waszczak et al., 1980). Nigral benzodiazepine receptors associated with the striatonigral GABAergic pathway are located postsynaptically and on axonal terminals (Lo et al., 1983). Striatal GABA receptors are believed to be localized on dopaminergic nerve terminals, cholinergic interneurones (Bartholini, 19800, and on glutamate-releasing corticostriatal terminals (Mitchell, 1980).

SUBSTANCE P

Substance P is an undecapeptide, and only one of a class of closely related peptides, the tachykinins. Since the characterization of substance P it has been possible to map the distribution of the peptide in brain by radioimmunoassay techniques and immunocytochemistry (see reviews by Nicoll et al., 1980; Iversen et al., 1980; Iversen, 1982). The SN contains the highest concentration of substance P immunoreactivity and the peptide is found in punctate structures throughout the nigra (Cuello and Kanazawa, 1978). The substance P-containing pathway originates in the most rostal part of the NS, and there are separate projections from GP to SN, and from NS to entopeduncular nucleus (Jessell et al., 1978). There is a partial topographic separation of striato- and pallidonigral substance P pathways from GABA pathways projecting to the SN (see above). Knife cuts in the frontal plane, which separate the NS from the GP, markedly decrease substance P levels in pars reticulata, whereas more caudal cuts effect maximal decreases in glutamate decarobxylase activity (Brownstein et al., 1977; Jessell et al., 1978). Substance P levels are decreased in GP and SN relative to controls in post-mortem brain tissue from Huntingtonian patients (Kanazawa et al., 1979). The substance P deficiencies in Huntington's disease are consistent with the loss of striatonigral and striatopallidal substance P neurones. Substance P is generally believed to be an excitatory transmitter and excites cells in the dopaminergic SN when applied locally by iontophoresis (Davies and Dray, 1976; Guyenet and Aghajanian, 1979). However, the effects of caudate stimulation on SN are predominantly reflected as inhibition and not excitation - the excitation observed after picrotoxin (causes GABA blockade) may be attributable to substance P pathways and to release from GABA-mediated inhibition (Kanazawa and Yoshida, 1980).

Substance P Receptors

The calcium-dependent release of substance P immunoreactivity has been demonstrated *in vitro* from nigral preparations in response to depolarizing stimuli (Iversen et al., 1980). Ljungdahl et al. (1978) localized substance P in fibres synapsing on dopamine somata in the pars compacta of SN. Substance P receptors may be localized on dopaminergic neurones within the SN for the peptide has direct effects on the release of (3H) dopamine both in vivo (Michelot et al., 1979) and in vitro (Reubi et al., 1978). However, unlike GABA, substance P-containing afferents to the SN seem to lack DA receptors as the release of substance P from nigral slices was not influenced by DA and dopaminergic agonists (Jessell, 1978). The intranigral application of substance P elevates (3H)DA release from the neostriatum in vivo (Cheramy et al., 1977) and elicits circling behaviour (James and Starr, 1977; Olpe and Koella, 1977). Ligand binding techniques have recently been utilized to label substance P receptors on brain membranes (Hanley et al., 1980; Beaujouan et al., 1982), and reveal a single population of saturable, high affinity binding sites. The autoradiographic distribution of (3H)substance P receptors has recently been analysed using in vitro slide-mounted sections of rat brain (Quirion et al., 1983). Moderate levels of (3H)substance P receptors exist in the NS (and nucleus accumbens), but particularly surprising was the absence of substance P binding sites from the SN. In fact the distribution of ^{3}H-substance P receptors gave only a reasonable correlation with that of endogenous substance P. The sites described in the autoradiographic study of Quirion et al. (1983) are probably of the substance P-P receptor subtype. Those receptors at which all tachykinins exert potent actions, with some preference for physalaemin (an amphibian peptide structurally related to substance P), have been described as substance P-P receptors, while the term substance P-E receptors has been employed to denote a subtype at which eledoisin (and kassinin; two other tachykinins) are the preferred agonists (Erspamer et al., 1980; Lee et al., 1982; Iversen, 1982). It should be noted that these subtypes have been described mainly on the basis of bioassays in peripheral tissues, and that at this early stage little information is available about the identity and characteristics of central substance P receptors. In view of the forementioned evidence as to the direct pharmacological actions of substance P in the substantia nigra, the nigral sites may differ from the striatal receptors labelled by (3H)substance P and perhaps be of the substance P-E subtype. Further developments in substance P pharmacology can be expected as information becomes available on the different receptor subtypes, the properties of substance P antagonists (Leander et al., 1981), and the coexistence of substance P and classical transmitters in neurones (Schultzberg et al., 1982).

SEROTONIN

Serotonin (5-hydroxtryptamine, 5-HT) has been implicated in a range of overt behaviour including locomotor activity (Geyer et al., 1976b), circling behaviour (James and Starr, 1980; Porceddu et al., 1983) and stereotypy (Carter and Pycock, 1981). Dysfunction of neostriatal serotonergic neurones occurs in Parkinson's disease along with the commonly recognized disruption of dopaminergic transmission (Chase, 1974; Hornykiewicz, 1979). The levels of 5-HT and its metabolite, 5-hydroxyindoleacetic acid, are high throughout the basal ganglia with 5-HT being distributed such that its concentration in GP > SN > NS (see e.g. Mefford et al., 1982). Serotonergic afferents to the NS and GP arise mainly from the dorsal raphe nucleus (Miller et al., 1975; Geyer et al., 1976a; Van der Kooy and Hattori, 1980), with fibres innervating the NS by way of the internal capsule pathway, while those reaching the GP course by way of the ansa lenticularis (Azmitia, 1978). There has also been some suggestion of a serotonergic projection to the anteromedial NS originating from the median raphe nucleus (Geyer et al., 1976a; Azmitia, 1978), but this has not been generally confirmed. Biochemical evidence suggests that serotonergic terminals are concentrated mainly in the ventrocaudal part of the NS and decrease in the rostral direction (Ternaux et al., 1977). While the serotonergic pathway to the NS has been widely studied, comparatively little is known about that to the GP. Serotonergic terminals are found in both the pars compacta and reticulata of the SN, and the results of histofluorescence, lesion, autoradiographic and electrophysiological studies reveal that these projections arise from both the dorsal and median raphe nuclei (Dray et al., 1976; Fibiger and Miller, 1977). The majority of dorsal raphe cells projecting to the SN also send collateral projections to the NS : these cells seem to be located in the dorsal part of the dorsal raphe nucleus (Van der Kooy and Hattori, 1980). Of course, many cells throughout the dorsal raphe innervate the NS, but not the SN. The serotonergic projections to the NS and SN produce predominantly inhibition of cell firing (Miller et al., 1975; Dray et al., 1976; Fibiger and Miller, 1977).

Serotonin Receptors

Although the existence of multiple receptor subtypes for 5-HT has been considered likely for many years, it is only recently that Peroutka et al. (1979) have postulated the concept of 5-HT1 (or S1) and 5-HT2 (or S2) receptors on the basis of receptor binding studies. Receptors designated 5-HT1 are labelled by (3H)-5-hydroxytryptamine, regulated by guanine nucleotides and linked to the 5-HT-sensitive adenylate cyclase in brain membranes (Peroutka et al., 1979 and 1981). The last point is still somewhat contentious (Hamon et al., 1980), perhaps because (3H)-5-HT may label multiple sites (Pedigo et al., 1981). 5-HT2 receptors, labelled by (3H)spiroperidol and (3H)mianserin, mediate behavioural responses associated with central 5-HT receptor stimulation (Jacobs, 1976) and there is an excellent correlation between the *in vivo* blockade of 5-hydroxytryptophan-induced head twitches and drug affinity for 5-HT2 binding sites (Peroutka et al., 1981; Leysen et al., 1982). There are no good functional correlates for the 5-HT1 recognition site, although a number of *in vitro* preparations afford opportunities to investigate the two 5-HT receptor subtypes (Fozard, 1983). Recent evidence suggests that a population of 5-HT1 receptors may be synonymous with those autoreceptors present on serotonergic neurones (Martin and Sanders-Bush, 1982; Gothert and Schliker, 1983) which regulate transmitter release, but as with most studies of 5-HT receptor subtypes there are pharmacological inconsistencies (Fozard, 1983).

There have been relatively few attempts to localize 5-HT receptors in the basal ganglia, although there certainly is evidence for 5-HT1 receptors located postsynaptically. Neostriatal kainate lesions reduce (3H)-5-HT binding by 60% of control values, suggesting a localization on intrinsic neurones, while 5-HT terminals are left intact (Schwarcz et al., 1977). This kainate-induced decrease in 5-HT1 receptor binding is consistent with the 50% reduction in (3H)-5-HT binding observed in post-mortem NS from Huntington's patients (Enna et al., 1976). At least some of these 5-HT1 receptors are located on striatal cholinergic interneurones because 5-HT modulates acetylcholine release (Euvrard et al., 1977; Vizi et al., 1981). The remaining neostriatal (3H)-5-HT binding is not on intrinsic neurones and seems not to correspond with 5-HT2 or autoreceptors (Nelson et al., 1980). Some of these 5-HT1 sites may be located postsynaptically on neostriatal dopaminergic nerve terminals, where they mediate DA release (de Belleroche and Bradford, 1980; Ennis et al., 1981). However, the appropriate lesion studies have not been performed to substantiate such a localization of 5-HT1 sites. As might be expected, a range of receptors appear to be localized presynaptically on serotonergic nerve terminals and Glowinski's group has reported that DA (Hery et al., 1980), glutamate and substance P (Reisine et al., 1982) modulate 5-HT release in both the NS and SN. GABA receptors are located on serotonergic afferent terminals in the SN (Soubrie et al., 1981; Gale, 1982).

ACETYLCHOLINE

Cholinergic terminals are rather unevenly distributed rostocaudally in the NS, being somewhat concentrated in the rostral portion (Fonnum et al., 1978; Scally et al., 1978). While the NS has high levels of choline acetyltransferase, acetylcholinesterase, (3H)choline uptake, and (3H)quinuclidinyl benzilate binding (muscarinic receptors), the corresponding values for GP are less than 10% of neostriatal levels (Yamamura et ala., 1974) - the SN has still lower values (Kobayashi et al., 1975). There has been considerable interest in striatal cholinergic neurones ever since it was appreciated that dopaminergic-cholinergic interplay was important in Parkinson's disease, and it is now well established that mesostriatal dopaminergic neurones exert a direct inhibitory influence on the activity of neostriatal cholinergic neurones (Stadler et al., 1973); Guyenet et al., 1975; Costa et al., 1978). Virtually all neostriatal cholinergic neurones are intrinsic (Fonnum et al., 1978; Kimura et al., 1980), and as might be expected the excitotoxin, kainic acid, causes large decreases in neostriatal choline acetyltransferase and in muscarinic receptor binding (Schwarcz and Coyle, 1977). These kainate-induced changes parallel quite closely the loss of cholinergic parameters observed in post-mortem brains from Huntingtonian patients (Bird and Iversen, 1974; Enna et al., 1976). There may also be a cholinergic thalamostriatal pathway, although its existence is still somewhat controversial (Fonnum and Walaas, 1979). Simke and Saelens (1977) found that a lesion of the parafascicular nucleus decreased neostriatal choline acetyltransferase activity and this pathway deserves further investigation, particularly since thalamostriatal neurones appear to have profound effects on neostriatal DA (Cheramy et al., 1983). The role of acetylcholine in the SN has also not been widely studied, but this region contains acetylcholine, and cholinergic terminals and somata (Olivier et al., 1970; Palkovits and Jacobowitz, 1974). Some of these terminals may be associated with cholinergic striatonigral afferents (Olivier et al., 1970), but the fact that intranigral kainic acid reduces choline acetyltransferase activity and (3H)quinuclidinyl benzilate binding in pars reticulata (de Montis et al., 1979) certainly favours the existence of cholinergic cell bodies within the SN. Cholinergic drugs have profound behavioural effects (non-dopamine dependent circling and catalepsy) following intranigral administration, and nigral GABAergic-cholinergic balance could be essential to the control of posture (de Montis et al., 1979).

Considerable recent interest has centred on the discovery of cholinergic neurones in the ventral pallidum, especially since these neurones undergo profound (> 75%) and selective degeneration in Alzheimer's disease (Whitehouse et al., 1982). In primates this region coincides with the nucleus basalis magnocellularis of Meynert (nbM), and has as its counterpart in the rat the ventral GP and substantia innominata. The nbM is in a critical position to integrate information from a variety of subcortical influences and to directly influence the cerebral cortex, as well as the hippocampus, thalamus, amygdala and brainstem. Thus the nbM should perhaps be regarded as a key output station of the basal ganglia, and a special locus for interfacing motor and cognitive functions. Additionally nbM pathways to the hippocampus may provide a biochemical basis for the memory disorder of Alzheimer's disease (Bartus et al., 1982). One can only speculate as to the involvement of these cholinergic neurones in the frequently observed cognitive deficits in Parkinson's disease (see e.g. Gottfies et al., 1980). The cholinergic neurones of nbM form a number of loosely connected cell clusters which have characteristic topographical projections to motor, parietal and occipital cortices (Divac, 1975; Lehmann et al., 1980). Both electrolytic and kainic acid lesions of the nbM cause appropriate reductions of cholinergic parameters in neocortex and nbM (Johnson et al., 1979; Lehmann et al., 1980; McKinney et al., 1983). These changes correspond quite closely to the 60% decrease in choline acetyltransferase activity described for neocortical post-operative and post-mortem specimens from Alzheimer's patients (Davies and Maloney, 1976; Spillane et al., 1977). Very little information is available on how other basal ganglia neurones interact with cholinergic neurones of the nbM, but Costa et al. (1978) has reported that DA receptor agonists decrease acetylcholine turnover in GP and concomitantly elevate pallidal GABA turnover. Together with other pharmacological data (Wood and Richard, 1982), these studies implicate a role for GABAergic neurones in regulating cholinergic neurones of mbM.

By contrast, there has been a whole host of studies examining the modulation of neostriatal cholinergic function. The concept of dopaminergic/cholinergic imbalance within the basal ganglia is largely based on pharmacological evidence (see Stadler et al., 1973; Guyenet et al., 1975; Costa et al., 1978; Bartholini, 1980). With subsequent advances in our understanding of the pharmacology of DA receptor subtypes it has become apparent that receptors of the D-2 subtype mediate dopaminergic effects on the target cholinergic neurones (see above; Sethy, 1979; Scatton, 1982). Other pharmacological evidence is consistent with this localization of DA receptors on neostriatal cholinergic neurones, in that lesion of cortico-striatal afferents does not affect the ability of apomorphine to increase striatal acetylcholine, but almost totally abolishes the effects of GABA agonists (Scatton and Bartholini, 1980). This study also provides evidence for a GABA-glutamate-acetylcholine link in the NS. Data from release studies further implicates the existence of glutamate receptors on neostriatal cholinergic neurones (Lehmann and Scatton, 1982). As mentioned in the preceding section serotonergic afferents to the NS also modulate cholinergic activity. Both muscarinic and nicotinic cholinergic receptors are located presynaptically on striatal dopaminergic neurones where they act to control DA release (Westfall, 1974; Giorguieff et al., 1977). Cholinergic receptors within the NS may also regulate the release of GABA (Besson et al., 1982) and 5-HT (Westfall et al., 1983).

GLUTAMATE

While it has been known for many years that cortical projections to the NS are topographically well organized so that each cortical region exerts its influence mainly on a special sector of the NS (although there is some overlap), it is less than 10 years since the evidence from electrophysiological, biochemical and lesion studies suggested glutamate as the transmitter released by the terminals of corticostriatal afferents. Both in primates and rodents the frontal cortex projects to rostral NS (including nucleus accumbens), while the posterior cortical areas project more progressively to more caudal parts. Classical degeneration studies, anterograde autoradiographic fibre tracing and horseradish peroxidase methods all yield quite similar results (Webster, 1961; Kemp and Powell, 1971; see also this volume). A number of studies employing cortical ablations have shown that the high affinity uptake of glutamate (associated with glutamate-releasing nerve terminals) by slices or synaptosomes of NS and the levels of endogenous glutamate are decreased by some 30-60% of control values 5-14 days post-lesion (Divac et al., 1977; Kim et al., 1977; McGeer et al., 1977; Walaas, 1981). Perhaps one criticism that can be leveled at these investigations is that the nature of the ablations was rather gross and that further efforts in this direction might well benefit from use of the second generation excitotoxins (ibotenate and N-methyl-D-asparate; Nadler et al., 1981). Other evidence implicating glutamate as the transmitter of corticostriatal fibres includes the antagonism of cortical excitation by glutamic acid diethyl ester, a glutamate antagonist (Spencer, 1976), and the release of glutamate both _in vivo_ and _in vitro_ (Reubi and Cuenod, 1979; Godukhin et al., 1980). Recent autoradiographic studies of the receptor binding sites for (3H)kainic acid, a conformationally restricted analogue of glutamate, demonstrated high uniform labelling over NS, no specific grains localized to GP and legligible labelling of the SN (Unnerstall and Walmsely, 1983). While these studies might suggest that there is no corticopallidal pathway, there certainly is some evidence for a corticonigral projection. Glutamate may be the transmitter of this corticonigral pathway (Afifi et al., 1974; Beckstead, 1979), since cortical lesions reduce the high affinity uptake of glutamate (Carter, 1980), and glutamate receptors appear to be present in the SN (Torrens et al., 1981).

Glutamate Receptors

Electrophysiological and biochemical evidence indicates that there are multiple receptors for glutamate in the central nervous system (Watkins, 1981; Johnston, 1979; Krogsgaard-Larsen and Honore, 1983). Much of this evidence comes from microelectrode studies of natural and synthetic amino-acids structurally related to glutamate and aspartate, which are powerful excitants of central neurones (Watkins, 1981; Curtis, 1981). Radioligand binding studies also favour the existence of multiple receptors for glutamate (Johnston, 1979; Krogsgaard-Larsen and Honore, 1983). Based largely on the relative sensitivity of central neurones to L-glutamate and L-aspartate, and the actions of certain antagonists, there seem to be at least three classes of receptors for excitatory amino acids. (1) Receptors for quisqualate (a cyclic glutamate analogue) where glutamate diethyl ester is an antagonist, (2) receptors for N-methyl-D-aspartate where 2-amino-5-phosphonovalerate and α-aminoadipate are antagonists, and (3) kainate receptors. Multiple receptors for glutamate are present in the NS, where they exist postsynaptically on neurones utilizing DA, acetylcholine, substance P and GABA as their transmitters. Good evidence for glutamate receptors on the last three types of intrinsic neostriatal neurones is the toxic effect of intrastriatal kainate (Coyle et al., 1981). Moreover, cortical lesions modulate the turnover rates of acetylcholine and GABA (Wood et al., 1979). Receptors of the N-methyl-D-aspartate type mediate the release of newly synthesised acetylcholine from slices of NS (Lehmann and Scatton, 1982). Other release studies support the hypothesis of a direct control by glutamatergic neurones of striatal DA release (Giorguieff et al., 1977; Roberts and Sharif, 1978; Roberts and Anderson, 1979) - these receptors may be of the quisqualate subtype and are destroyed by intranigral 6-hydroxy-dopamine (Roberts et al., 1982). Earlier evidence has been given for a population of DA receptors located on corticostriatal terminals which inhibit glutamate release (see above; Rowlands and Roberts, 1980). A novel GABA receptor may modulate glutamate release from corticostriatal terminals (Mitchell, 1980). Nigral glutamate receptors have received little or no attention, but glutamate does reduce the potassium-induced release of substance P from nigral slices - an effect that is blocked by glutamate diethyl ester (Torrens et al., 1981).

TRANSMITTER INTERACTIONS

Much of the impetus for the multitude of investigations seeking to analyse the contributions of the various transmitters to basal ganglia function has come from our knowledge (albeit limited) of the transmitter deficiencies found in abnormal movement disorders, especially Parkinson's disease and Huntington's chorea. Such evidence seems to have produced a somewhat biased view of the importance of mesostriatal dopaminergic neurones in basal ganglia function and in voluntary movements. Cetainly L-dopa replacement therapy or the use of a directly acting DA agonist (e.g. bromocriptine) markedly improves the symptoms of Parkinson's disease. Additionally, drugs which antagonize the action of DA (haloperidol, thioridazine, tetrabenzine, reserpine) employed to control the involuntary movements of Huntington's chorea, while dopaminergic agonists worsen the symptoms. While one can certainly understand how DA came to be accepted as the central transmitter regulating neostriatal neuronal activity and outflow, new evidence would suggest that the activity of mesostriatal dopaminergic neurones is remarkably imperturbable during normal body function including movement (see Groves, 1983). Thus evidence on the relative importance of transmitters and their interactions, as deduced from the pharmacological and biochemical changes occurring in abnormal movement disorders, may not be directly relevant to the operative regulatory mechanisms in the NS of a normally functioning animal because there are such gross disruptions of neurotransmission in these basal ganglia disorders. Our understanding of the basal ganglia and their normal function is incomplete, and current thinking in this area is taking a greater account of all of the transmitters involved in basal ganglia circuits and of their various inputs and outputs.

A considerable body of evidence has shown that neostriatal circuitry can be considered as consisting of two functional efferent systems which project to the SN. This circuitry has been the subject of several reviews (see e.g. Dray, 1979; Moore and Wuerthele, 1979; Bartholini, 1980; Jones et al., 1981; Groves, 1983). By far the most important of these circuits is that involving Spiny type I neostriatal neurones, which utilize GABA as their transmitter and which form a matrix where collaterals provide recurrent self-inhibition (see above; Pasik et al., 1979; Groves, 1983). These neurones receive direct cortical and thalamic inputs, and DA probably exerts a weak excitatory influence on them. Their firing pattern is characteristically low, with phasic bursts, and seems to parallel quite closely the spontaneous activity of single neurons in the NS (Wilson and Groves, 1980 and 1981). The GABAergic neurones are numerically the major source of outflow from the NS to the GP and SN. Substance P seems to represent the other feedforward pathway, which is apparently less important, especially since inhibition (GABA-mediated) and not excitation (substance P-mediated) is predominantly observed in the SN (Kanazawa and Yoshida, 1980). As described above, GABAergic afferents to the SN are directed mainly at the non-dopaminergic pars reticulata and make very little contact with dopaminergic dendrites (Rinvik and Grofova, 1970; Gulley and Smithberg, 1971). The second circuit involves substance P (probably the transmitter of Spiny type II neurones; Pasik et al., 1979; Groves, 1983), and the well described cholinergic interneurone (Aspiny type II), which provides excitatory drive to the substance P efferents (Dray, 1980). GABAergic interneurones are likely to regulate the activity of this second circuit, which also provides the basis for dopaminergic/cholinergic balance via a powerful direct inhibitory influence on the cholinergic cell (Groves, 1983). This dopaminergic-cholinergic balance has been described above, but the relatively small number of Spiny type II substance P neurones, and the importance of GABA-mediated inhibition in SN, might suggest that dopaminergic-cholinergic interaction becomes of pharmacological importance only during abnormal neurotransmission (e.g. Parkinson's disease). Then, the loss of inhibitory influence due to the degeneration of dopaminergic neurones results in overactivity of cholinergic neurones and hence substance P-mediated excitatory outflow - the beneficial effects of anticholinergics in Parkinson's disease would be attributable to a diminution of cholinergic activity.

Several other mechanisms may act to regulate the activity of basal ganglia neurones. Many neurones possess autoreceptors on their presynaptic nerve terminals (including DA, 5-HT and GABA) which can act to diminish transmitter release, but their contribution to the modulation of transmitter release remains unclear, particularly as experiments designed to analyse their role do not appear to have been performed in unanaesthetized, freely moving animals. In the peripheral nervous system autoreceptors seem to play a functional role only when neurotransmission is stressed or abnormal as in disease states (Langer, 1981). Postsynaptic receptors seem to exert much more powerful actions on neurotransmission : thus postsynaptic blockade of central dopaminergic transmission by neuroleptic drugs causes catalepsy in rodents, whereas interruption by activation of autoreceptors produces reduced locomotor activity. Similar questions also apply to the physiological function of dendritically released DA in the SN (Groves et al., 1975; Cheramy et al., 1981). A most interesting observation relevant to these nigral dendrites in pars reticulata is that DA may exert a neuromodulatory role on nigral efferent neurones (mainly GABAergic), which project to the thalamus, superior colliculus and reticular formation (Waszczak and Walters, 1983). A further mechanism, which has received only sparse attention, is that of tolerance-plasticity. Although perhaps really two separate phenomena, the terminology as utilized here is meant to refer to the ability of basal ganglia neurones to adapt in response to alterations to their normal milieu. Mesostriatal dopaminergic neurones have been shown to become tolerant to drug treatments (e.g. Asper et al., 1973), but other neostriatal neurones may also exhibit adaptive changes after injury (Neve et al., 1982) and drug treatment (Alloway and Rebec, 1983). Such phenomena are worthy of further attention, particularly because of their direct relevance to diseases of the basal ganglia.

There has been some discussion in the last few years about the role played by GABAergic striato- and pallidonigral pathways. Initially these GABAergic neurones were believed to provide an inhibitory, regulatory input to the DA cell bodies present in the zona compacta (Groves et al, 1975; James and Starr, 1978; Dray, 1979). More recent evidence does not favour this "feedback hypothesis" because the majority are feedforward; neostriatal output (mainly GABAergic) is in fact directed towards the non-dopaminergic pars reticulata. Additionally, hemisections that would leave nigral circuitry untouched while disrupting mesostriatal dopaminergic transmission, leave most overt behaviours intact (Childs and Gale, 1983). Further discussion of these two issues has been presented above in the section devoted to GABA. Perhaps GABA, and not DA, should be viewed as the key transmitter in mediating basal ganglia function. Thus the "output hypothesis", where striato- and pallidonigral GABAergic pathways represent just the first link in a chain transmitting motor signals via output neurones of zona reticulata, has gained greater current acceptance (Garcia-Munoz et al., 1977; Marshall and Ungerstedt, 1977; James and Starr, 1978; Porceddu et al., 1983; Starr et al., 1983; see Di Chiara this volume). There has been an appreciable upsurge of interest in the role of thalamic nuclei in regulating basal ganglia function. The thalamus is ideally situated to integrate basal ganglia function, receiving inputs from the SN and GP and sending efferents to the NS (Herkenham, 1979; Di Chiara et al., 1979; Cheramy et al., 1981; Garcia-Munoz et al., 1983). The transmitters of the thalamostriatal and thalamonigral pathways have not been identified, and an improved understanding of thalamic function would advance our knowledge of the mechanisms regulating the basal ganglia.

ACKNOWLEDGEMENTS

Philip Beart is a Senior Research Fellow of the National Health and Medical Research Council of Australia and wishes to thank Marion Cincotta, Jenni Caulfield and Andrew Gundlach for their assistance in the preparation of this article.

REFERENCES

Afifi, A. K. A., Bahuth, N. B., Kaelber, W. W., Michael, E., and Nassar, S., 1974, The corticonigral fibre tract. An experimental Fink-Heimer study in cats, J. Anat., 118:469.

Aghajanian, G. K., and Bunney, B. S., 1973, Central dopaminergic neurons: Neurophysiological identification and responses to drugs, in: "Frontiers in Catecholamine Research," S. H. Snyder and E. Usdin, eds., Pergamon Press, New York.

Alloway, K. D., and Rebec, G. V., 1983, Shift from inhibition to excitation in the neostriatum but not in the nucleus accumbens following long-term amphetamine, Brain Res., 273:71.

Anden, N. E., Golembiowsak-Nikitin, K., and Thornstrom, U., 1982, Selective stimulation of dopamine and noradrenaline autoreceptors by B-HT 920 and B-HT 933 respectiovely, Naunyn-Schmiedeberg's Arch. Pharmacol., 321:100.

Anderson, M. E., and Yoshida, M., 1980, Axonal branching patterns and location of nigrothalamic and nigrocollicular neurones in the cat, J. Neurophysiol., 43:883.

Andrews, P. R., and Johnston, G. A., 1979, GABA agonists and antagonists, Biochem. Pharmacol., 28:2967.

Asper, H., Baggiolini, M., Burki, H. R., Laeuner, H. and Stille, G., 1973, Tolerance phenomena with neuroleptics, catalepsy, apomorphine stereotypies and striatal dopamine metabolism in the rat after single and repeated administration of loxapine and haloperidol, Eur. J. Pharmacol., 22:287.

Azmitia, E. C., 1978, The serotonin-producing neurones of the midbrain median and dorsal raphe nuclei, in: "Handbook of Psychopharmacology 9," L. L. Iversen, S. D. Iversen and S. H. Snyder, eds., Plenum Press, New York.

Bartholini, G., 1980, Interaction of striatal dopaminergic, cholinergic, and GABAergic neurons: relation to extrapyramidal function, Trends Pharmacol. Sci., 1:138.

Bartus, R. T., Dean, R. L., Beer, B., and Lippa, A. S., 1982, The cholinergic hypothesis of geriatric memory dysfunction, Science, 217:408.

Beart, P. M., 1982, Multiple dopamine receptors - new vistas, Trends Pharmacol. Sci., 3:100.

Beaujouan, J. C., Torrens, Y., Herbet, A., Daguet, M. C., Glowinski, J., and Prochiantz, A., 1982, Specific binding of an immunoreactive and biologically active 125I-labelled substance P derivative to mouse mesencephalic cells in culture, Mol. Pharmacol., 22:48.
Beckstead, R. M., 1979, An autoradiographic examination of corticortical and subcortical projections of the mediodorsal-projection (prefrontal) cortex in the rat, J. Comp. Neurol., 184:43.
Beckstead, R. M., Domesick, V. M., and Nauta, W. J. H., 1979, Efferent connections of the substantia nigra and ventral tegmental area in the rat, Brain Res., 175:191.
Besson, M. J., Keriel, M. L., Gauchy, C., and Glowinski, J., 1982, Bilateral asymmetrical changes in the nigral release of [3H]GABA induced by unilateral application of acetylcholine in the cat caudate nucleus, Brain Res., 241:241.
Bird, E. D., and Iversen, L. L., 1974, Huntington's chorea-post-mortem measurement of glutamic acid decarboxylase, choline acetyltransferase in basal ganglia, Brain, 97:457.
Bjorklund, A., and Lindvall, O., 1975, Dopamine in dendrites of substantia nigra neurones: suggestions for a role in dendritic terminals, Brain Res., 83:531.
Bowery, N. G., 1982, Baclofen: 10 years on, Trends Pharmacol. Sci., 3:400.
Bowery, N. G., Hill, D. R., and Hudson, A. L., 1982, Guanyl nucleotides decrease the binding affinity of GABA-B receptors in the mammalian C.N.S., Br. J. Pharmacol., 75:86P.
Bowery, N. G., Hill, D. R., Hudson, A. L., Doble, A., Middlemiss, D. N., Shaw, J., and Turnbull, M., 1980, (-) Baclofen decreases neurotransmitter release in the mammalian CNS by an action at a novel GABA receptor, Nature, 283:92.
Brownstein, M. J., Mroz, E. A., Tappaz, M. L., and Leeman, S., 1977, On the origin of substance P and glutamic acid decarboxylase (GAD) in the substantia nigra, Brain Res., 135:315.
Cannon, J. G., 1983, Structure-activity relationships of dopamine agonists, Ann. Rev. Pharmacol. Toxicol., 23:103.
Carlsson, A., 1977, Dopamine autoreceptors: background and implications, Adv. Biochem. Psychopharmacol., 16:439.
Carter, J., 1980, Evidence for glutamatergic projections from the prefrontal cortex to the striatum, nucleus accumbens, amygdaloid complex and substantia nigra in the rat, Neurosci. Letts., Supplement 5:580.
Carter, C. J., and Pycock, J., 1981, The role of 5-hydroxytryptamine in dopamine-dependent stereotyped behaviour, Neuropharmacology, 20:261.
Chase, T. N., 1974, Serotonergic-dopaminergic interactions and extrapyramidal function, Adv. Biochem, Psychopharmacol., 11:377.

Cheramy, A., Chesselet, M. F., Romo, R., Leviel, V., and Glowinski, J., 1983, Effects of unilateral electrical stimulation of various thalamic nuclei on the release of dopamine from dendrites and nerve terminals of neurons of the two nigrostriatal dopaminergic pathways, Neurosci., 8:767.

Cheramy, A., Leviel, V., and Glowinski, J., 1981, Dendritic release of dopamine in the substantia nigra, Nature, 289:537.

Cheramy, A., Nieoullon, A., and Glowinski, J., 1978a, GABAergic processes involkved in the control of dopamine release from nigrostriatal dopaminergic neurons in the cat, Eur. J. Pharmacol., 48:281.

Cheramy, A., Nieoullon, A., and Glowinski, J., 1978b, Effects of unilateral nigral application of baclofen on dopamine release in the two caudate nuclei of the cat, Eur. J. Pharmacol, 58:133.

Cheramy, A., Nieoullen, A., Michelot, R., and Glowinski, J., 1977, Effects of intranigral applications of dopamine and substance P on the in vivo release of newly synthesized 3H-dopamine in the ipsilateral caudate nucleus of the cat, Neurosci. Letts., 4:105.

Childs, J. A., and Gale, K., 1983, Neurochemical evidence for a nigrotegmental GABAergic projection, Brain Res., 258:109.

Cools, A. R., 1982, The puzzling 'cascade' of multiple receptors for dopamine, Trends Pharmacol. Sci., 2:178.

Corsini, G. U., Pitzalis, G. F., Bernardi, F., Bochetta, A., and Del Zompo, M., 1981, The use of dopamine agonists in the treatment schizophrenia, Neuropharmacology, 20:1309.

Costa, E., Cheney, D. L., Mao, C. C., and Moroni, F., 1978, Action of antischizophrenic drugs on the metabolism of gamma-aminobutyric acid and acetylcholine in globus pallidus, striatum and n. accumbens, Fed. Proc., 37:2408.

Costall, B., Naylor, R. J., and Owen, R. T., 1978, GABAminergic and serotonergic modulation of the antidyskinetic effects of tiapride and oxiperomide in the model using 2-(N,N-dipropyl)-amino-5,6-dihydroxy tetralin, Eur. J. Pharmacol., 49:407.

Coyle, J. T., Bird, S. J., Evans, R. H., Gulley, R. L., Nadler, J. V., Nicklas, W. J., and Olney, J. W., 1981, Excitatory amino acid neurotoxins: selectivity, specificity and mechanisms of action, Neurosci. Res. Program Bull., 19:1.

Creese, I., Sibley, D. R., Hamblin, M. W., and Leff, S. E., 1983, The classification of dopamine receptors: relationship to radioligand binding, Ann. Rev. Neurosci., 6:43.

Cross, A. J., and Owen, F., 1980, Characteristics of 3H-cis-flupenthixol binding to calf brain membranes, Eur. J. Pharmacol., 65:341.

Cross, A. J., and Waddington, J. L., 1981, Kainic acid lesions dissociate [3H]cis-flupenthixol binding sites in rat striatum, Eur. J. Pharmacol., 71:327.

Cuello, A. C., and Kanazawa, I., 1978, The distribution of substance P immunoreactive fibres in the rat central nervous system, J. Comp. Neurol., 178:129.

Curtis, D. R., 1981, Problems in the evaluation of glutamate as a central nervous system transmitter, in: "Glutamic Acid: Advances in Biochemistry and Physiology," L. J. Filer, S. Garattini, M. R. Kare, W. A. Reynolds, and R. J. Wurtman, eds., Raven Press, New York.

Curtis, D. R., and Johnston, G. A. R., 1975, Amino acid transmitters in the mammalian central nervous system, Rev. Physiol., 69:97.

Curtis, D. R., Lodge, D., Bornstein, J. C., and Peet, M. J., 1981, Selective effects of (-) baclofen on spinal synaptic transmission in the cat, Exp. Brain Res., 42:158.

Davies, J., anmd Dray, A., 1976, Substance P in the substantia nigra, Brain Res., 107:623.

Davies, P., and Maloney, A. J. F., 1976, Selective loss of central cholinergic neurones in Alzheimer's disease, Lancet, ii:1403.

de Belleroche, J. S., and Bradford, H. F., 1980, Presynaptic control of the synthesis and release of dopamine from striatal synaptosomes: a comparison between the effects of 5-hydroxytryptamine, acetylcholine and glutamate, J. Neurochem., 35:1227.

de Montis, G. M., Olianas, M. C., Serra, G., Tagliamonte, A., and Scheel-Kruger, J., 1979, Evidence that a nigral GABAergic-cholinergic balance controls posture, Eur. J. Pharmacol., 53:181.

Deniau, J. M., Hammond, C., Riszk, A., and Féger, J., 1978, Electrophysiological properties of identified output neurons of the rat substantia nigra (pars compacta and pars reticulata): evidences for the existence of branched neurons, Exp. Brain Res., 32:409.

Di Chiara, G., Morelli, M., Porceddu, M. L., and Gessa, G. L., 1978, Evidence that nigral GABA mediates behavioural responses elicited by striatal dopamine receptor stimulation, Life Sci., 23:2045.

Di Chiara, G., Morelli, M., Porceddu, M. L., and Gessa, G. L., 1979, Role of thalamic gamma-aminobutyrate in motor functions: catalepsy and ipsiversive turning after intrathalamic muscimol, Neurosci., 4:1453.

Di Chiara, G., Porceddu, M. L., Imperato, A., and Morelli, M., 1981, Role of GABA neurons in the expression of striatal motor functions, Adv. Biochem. Psychopharmacol., 30:129.

Divac, I., 1975, Magnocellular nuclei of the basal-forebrain project to neocortex, brainstem, and olfactory bulb. Review of some functional correlates, Brain Res., 93:385.

Divac, I., Fonnum, F., and Storm-Mathisen, J., 1977, High affinity uptake of glutamate in terminals of corticostriatal axons, Nature, 266:377.

Dray, A., 1979, The striatum and substantia nigra: A commentary on their relationships, Neurosci., 4:1407.

Dray, A., 1980, The physiology and pharmacology of mammalian basal ganglia, Prog. Neurobiol., 14:221.

Dray, A., Gonye, T. J., Oakley, N. R., and Tanner, T., 1976, Evidence for the existence of a raphe projection to the substantia nigra in rat, Brain Res., 113:45.

Dray, A., and Oakley, N. R., 1978, Projections from nucleus accumbens to globus pallidus and substantia nigra in the rat, Experientia, 34:68.

Enna, S. J., Bird, E. D., and Bennett, J. P., Jr., 1976, Huntington's chorea, Changes in neurotransmitter receptors in the brain, N. Engl. J. Med., 294:1305.

Enna, S. J. and Gallagher, J. P., 1983, Biochemical and electrophysiological characteristics of mammalian GABA receptors, Int. Rev. Neurobiol., 24:181.

Ennis, C., Kemp, J. D., Cox, B., 1981, Characterisation of inhibitory 5-hydroxytryptamine receptors that modulate dopamine release in the striatum, J. Neurochem., 36:1515.

Erspamer, G. F., Erspamer, V., and Piccinnelli, D., 1980, Parallel bio-assay of physalaemin and kassinin, a tachykinin dodecapeptide from the skin of the African frog kassina senegalensis, Naunyn Schmiedeberg's Arch. Pharmacol., 311:61.

Euvrard, C., Javoy, F., Herbet, A., and Glowinski, J., 1977, Effect of quipazine, a serotinin-like drug, on striatal cholinergic interneurones, Eur. J. Pharmacol., 41:281.

Fahn, S., 1976, Regional distribution studies of GABA and other putative neurotransmitters and their enzymes, in: "GABA in Nervous System Function," E. Roberts, T. N. Chase, and D. B. Tower, esq., Raven Press, New York.

Fibiger, H. C., and Miller, J. J., 1977, An anatomical and electrophysiological investigation of the serotoninergic projection from the dorsal raphe nucleus to the substantia nigra in the rat, Neuroscience, 2:975.

Fonnum, F., Gottesfeld, Z., and Grofova, I., 1978, Distribution of glutamate decarboxylase, choline acetyltransferase and aromatic acid decarboxylase in the basal ganglia of normal and operated rats. Evidence for striatopallidal, striatoentopeduncular and striatonigral GABAergic fibres, Brain Res., 143:125.

Fonnum, F., and Walaas, I., 1979, Localization of neurotransmitter candidates in the neostriatum, in: "The Neostriatum," I. Divac, and R. G. E. Oberg, eds., Pergamon Press, Sydney.

Fozard, J. R., 1983, Functional correlates of 5-HT1 recognition sites, Trends Pharmacol. Sci., 4:288.

Gale, K., 1981, Relationship between the presence of dopaminergic neurons and GABA receptors in substantia nigra: effects of lesions, Brain Res., 210:410.

Gale, K., 1982, Evidence for GABA receptors on serotonergic afferent terminals in the substantia nigra and on nigral efferent projections to the caudal mesencephalon, Brain Res., 231:209.

Garau, L., Govoni, S., Stefanini, E., Trabucchi, M., and Spano, P. F., 1978, Dopamine receptors: pharmacological and anatomical evidence indicate that two distinct dopamine receptor populations are present in rat striatum, Life Sci., 23:1745.

Garcia-Munoz, M., Nicolaou, N. M., Tulloch, I. F., Wright, A. K., and Arbuthnott, G. W., 1977, Feedback loop or output pathway in striatonigral fibres, Nature, 265:363.

Garcia-Munoz, M., Patino, P., Wright, A. J., and Arbuthnott, G. W., 1983, The anatomical substrate of the turning behaviour seen after lesions in the nigrostriatal dopamine system, Neuroscience, 8:87.

Geyer, M. A., Puerto, A., Dawsey, W. J., Knappi, S., Bullard, W. P., and Mandell, A. J., 1976a, Histologic and enzymatic studies of the mesolimbic and mesostriatal serotonergic pathways, Brain Res., 106:241.

Geyer, M. A., Puerto, A., Menkes, D. B., Segal, D. S., and Mandell, A. J., 1976b, Behavioural studies following lesions of the mesolimbic and mesostriatal serotonergic pathways, Brain Res., 106:257.

Giorguieff, M. F., Kemel, M. L., and Glowinski, J., 1977, Presynaptic effects of L-glutamic acid on the release of dopamine in rat striatal slices, Neurosci. Letts., 6:73.

Godukhin, O. V., Zharikova, A. D., and Novoselor, V. I., 1980, The release of labelled L-glutamic acid from rat neostriatum in vivo following stimulation of frontal cortex, Neuroscience, 5:2151.

Goodale, D. B., Rusterholz, D. B., Long, J. P., Flynn, J. R., Walsh, B., Cannon, J. G., and Lee, T., 1980, Neurochemical and behavioural evidence for a selective presynaptic dopamine receptor agonist, Science, 210:1141.

Göthert, M., and Schlicker, E., 1983, Autoreceptor-mediated inhibition of 3H-5-hydroxytryptamine release from rat brain cortex slices by analogues of 5-hydroxytryptamine, Life Sci., 32:1183.

Gottfries, C. G., Adlfson, R., Aquilonius, S. M., Carlsson, A., Oreland, L., Svennerholm, L., and Winblad, B., 1980, Parkinsonism and dementia disorders of Alzheimer type: Similarities and differences in Parkinson's disease, in: "Current Progress, Problems and Management," U. K. Rinne, M. Klinger, and G. Stamm, eds., Elsevier, Amsterdam.

Grace, A. A., and Bunney, B. S., 1979, Paradoxical GABA excitation of nigral dopaminergic cells, indirect mediation through reticulata inhibitory neurons, Eur. J. Pharmacol., 59:211.

Groves, P. M., 1983, A theory of the functional organization of the neostriatum and the neostriatal control of voluntary movement, Brain Res., 286:109.

Groves, P. M., Wilson, C. J., Young, S. J., and Rebec, G. V., Self-inhibition by dopaminergic neurons, Science, 190:522.

Gulley, R. L., and Smithberg, M., 1971, Synapses in the cat substantia nigra, Tissue Cell, 3:691.

Gundlach, A. L., Krstich, M., and Beart, P. M., 1983, Guanine nucleotides reveal differential actions of ergot derivatives at D-2 receptors labelled by [3H]spiperone in striatal homogenates, Brain Res., 278:155.

Guyenet, P. G., Agid, Y., Javoy, F., Beaujouan, J. C., Boissier, J., and Glowinski, J., 1975, Effects of dopaminergic receptor agonists and antagonists on the affinity of the neostriatal cholinergic system, Brain Res., 84:277.

Guyenet, P. G., and Aghajanian, G. K., 1979, Antidromic identification of dopaminergic and other output neurons of the rat substantia nigra, Brain Res., 150:69.

Hahn, R. A., Wardell, J. R., Sarau, H. M., and Ridley, P. T., 1982, Characterisation of the peripheral and central effects of SK- and F-82526, a novel dopamine receptor agonist, J. Pharmacol. Exp. Ther., 223:305.

Hall, M. D., Jenner, P., Kelly, E., and Marsden, C. D., 1983, Differential anatomical location of [3H]-spiperone binding sites in the striatum and substantia nigra of the rat, Brit. J. Pharmacol., 79:599.

Hamon, M., Nelson, D. L., Herbert, A., and Glowinski, J., 1980, Multiple receptors for serotonin in the rat brain, in: "Receptors for Neurotransmitters and Peptide Hormones," G. Pepeu, M. J. Kuhar, and S. J. Enna, eds., Raven Press, New York.

Hanley, M. R., Sandberg, B. E. B., Lee, C. M., Iversen, L. L., Brundish, D. E., and Wade, R., 1980, Specific binding of 3H-substance P to rat brain membranes, Nature, 286:810.

Hedreen, J. C., 1977, Corticostriatal cells identified by the peroxidase method, Neurosci. Letts., 4:1.

Heimer, L., and van Hoesen, G., 1979, Ventral striatum, in: "The Neostriatum." I. Divac and R. G. E. Oberg, eds., Pergamon Press, Sydney.

Helmeste, D. M., and Seeman, P., 1982, Amphetamine-induced hypolocomotion in mice with more brain D2 dopamine receptors, Psychiatry Res., 7:351.

Herkenham, M., 1979, The afferent and efferent connections of the ventromedial thalamic nucleus in the rat, J. Comp. Neurol., 183:679.

Hery, F., Soubrie, P., Bourgoin, S., Motastruc, J. L., Artaud, F., and Glowinski, J., 1980, Dopamine released from dendrites in the substantia nigra controls the nigral and striatal release of serotonin, Brain Res., 193:143.

Hill, D. R., and Bowery, N. G., 1981, 3H-Baclofen and 3H-GABA bind to bicuculline-insensitive GABA-B sites in rat brain, Nature, 290:149.

Hjorth, S., Carlsson, A., Wikstrom, H., Lindberg, P., Sanchez, D., Hacksell, U., Arvidsson, L. -E., Svensson, U., and Nilsson, J. L. G., 1981, 3-PPP, a new centrally acting DA-receptor agonist with selectivity for autoreceptors, Life Sci., 28:1225.

Hornykiewicz., O., 1979, Brain dopamine in Parkinson's disease and other neurological disturbances, in: "The Neurobiology of Dopamine," A. S. Horn, J. Korf and B. H. C. Westerink, eds., Academic Press, London.

Hyttel, J., 1980, Further evidence that 3H-cis(Z)flupenthixol binds to the adenylate cyclase-associated dopamine receptor (D-1) in rat corpus striatum, Psychopharmacology (Berlin), 67:107.
Hyttel, J., 1983, SCH 23390 - The first slective dopamine D-1 antagonist, Eur. J. Pharmacol., 91:153.
Iversen, L. L., 1982, Substance P., Br. Med. Bull., 38:277.
Iversen, L. L., Lee, C. M., Gilbert, R. F., Hunt, S., and Emson, P. C., 1980, Regulation of neuropeptide release, Proc. R. Soc. Lond. [Biol]., 210:91.
Iversen, L. L., Rogawski, M. A., and Miller, R. J., 1976, Comparison of the effects of neuroleptic drugs on pre- and postsynaptic dopaminergic mechanisms in the rat striatum, Mol. Pharmacol., 12:251.
Jacobs, B. L., 1976, An animal behaviour model for studying central serotonergic synapses, Life Sci., 19:777.
James, T. A., and Starr, M. S., 1977, Behavioural and biochemical effects of substance P injected into the substantia nigra of the rat, J. Pharm. Pharmacol., 29:181.
James, T. A., and Starr, M. S., 1978, The role of GABA in the substantia nigra, Nature, 275:229.
James, T. A., and Starr, M. S., 1980, Rotational behaviour elicited by 5-HT in the rat: evidence for an inhibitory role of 5-HT in the substantia nigra and corpus striatum, J. Pharm. Pharmacol., 32:196.
Jessell, T. M., 1978, Substance P release from the rat substantia nigra, Brain Res., 151:469.
Jessell, T. M., Emson, P. C., Paxinos, G., and Cuello, A. C., 1978, Topographic projections of substance P and GABA pathways in the striato- and pallido-nigral system: a biochemical and immunohistochemical study, Brain Res., 152:487.
Johnston, G. A. R., 1979, Central nervous system receptors for glutamic acid, in: "Glutamic Acid: Advances in Biochemistry and Physiology," L. J. Filer, Jr., ed., Raven Press, New York.
Johnston, G. A. R., 1981, GABA Receptors, in: "The role of Peptides and Amino Acids as Neurotransmitters," J. B. Lombardini, ed., Liss Inc., New York.
Johnston, M. V., McKinney, M., and Coyle, J. T., 1979, Evidence for a cholinergic projection to neocortex from neurones in basal forebrain, Proc. Natl. Acad. Sci. (USA), 76:5392.
Jones, D. L., and Mogenson, G. J., 1980, Nucleus accumbens to globus pallidus GABA projection: electrophysiological and iontophoretic investigations, Brain Res., 188:93.
Jones, D. L., Mogenson, G. J., and Wu, M., 1981, Injections of dopaminergic, cholinergic, serotonergic and GABAergic drugs into the nucleus accumbens: effects on locomotor activity in the rat, Neuropharmacol., 20:29.
Juraska, J. M., Wilson, C. J., and Groves, P. M., 1977, The substantia nigra of the rat: a Golgi study, J. Comp. Neurol., 172:585.

Kanazawa, I., Bird, E. D., Gale, J. S., Iversen, L. L., Jessell, T. M., Muramoto, O., Spokes, E. G., and Sutoo, D., 1979, Substance P: decrease in substantia nigra and globus pallidus in Huntington's disease, in: "Huntington's Disease, Advances in Neurology," Raven Press, New York.

Kanazawa, I., and Yoshida, M., 1980, Electrophysiological evidence for the existence of excitatory fibres in the caudato-nigral pathway in the cat, Neurosci. Letts., 20:301.

Karobath, M., Placheta, P., Lippitsch, M., and Krogsgaard-Larsen, P., 1979, Is stimulation of benzodiazepine receptor binding mediated by a novel GABA receptor? Nature, 278:748.

Kebabian, J. W., and Calne, D. B., 1979, Multiple receptors for dopamine, Nature, 277:93.

Kelly, P., and Moore, K. E., 1976, Mesolimbic dopamine neurons in the rotational model of nigrostriatal function, Nature, 263:695.

Kemp, J. M., and Powell, T. P. S., 1971, The site of termination of afferent fibres in the caudate nucleus, Phil. Trans. R. Soc. Lond [Biol.], 262:413.

Kilpatrick, I. C., and Starr, M. S., 1981, Involvement of dopamine in circling responses to muscimol depends on intranigral sites of injection, Eur. J. Pharmacol., 69:407.

Kim, J-S., Hassler, R., Haug, P., and Paik, K. S., 1977, Effect of frontal cortex ablation on striatal glutamic acid level in rat, Brain Res., 132:370.

Kimura, H., McGeer, P. L., Peng, F., and McGeer, E. G., 1980, Choline acetyltransferase-containing neurons in rodent brain demonstrated by immunohistochemistry, Science, 208:1057.

Kobayashi, R. M., Brownstein, M., Saavedra, J. M., and Palkovits, M., 1975, Choline acetyltransferase content in discrete regions of the brainstem, J. Neurochem., 24:637.

Krogsgaard-Larsen, P., and Honore, T., 1983, Glutamate receptors and new glutamate agonists, Trends Pharmacol. Sci., 4:31.

Langer, S. Z., 1981, Presynaptic regulation of the release of catecholamines, Pharmacol. Rev., 81:337.

Leander, S., Hakanson, R., Rosell, S., Folkers, K., Sundler, F., and Tornqvist, K., 1981, A specific substance P antagonist blocks smooth muscle contractions induced by non-cholinergic, non-adrenergic nerve stimulation, Nature, 294:467.

Lee, C. M., Iversen, L. L., Hanley, M. R., and Sandberg, B. E., 1982, The possible existence of multiple receptors for substance P, Naunyn Schmeidebergs Arch. Pharmacol., 318:281.

Lee, T., Seeman, P., Rajput, A., Farley, I. J., and Hornykiewicz, O., 1978, Receptor basis for dopaminergic supersensitivity in Parkinson's disease, Nature, 273,59.

Leff, S., Adams, L., Hyttel, J., and Creese, I., 1981, Kainate lesion dissociates striatal dopamine receptor radioligand binding sites, Eur. J. Pharmacol., 70:71.

Leff, S. E., and Creese, I., 1983, Dopamine receptors re-explained, Trends Pharmacol. Sci., 4:463.

Lehmann, J., Nagy, J. I., Atmadia, S., and Fibiger, H, C., 1980, The nucleus basalis magnocellularis: the origin of a cholinergic projection to the neocortex of the rat, Neurosci., 5:1161.
Lehmann, J., and Scatton, B., 1982, Characterization of the excitatory amino acid receptor-mediated release of [3H]acetylcholine from rat striatal slices, Brain Res., 252:77.
Leysen, J. E., Niemegeers, C. J. E., Van Nueten, J. M., and Laduron, P., 1982, [3H]Ketanserin (R 41 468), a selective 3H-ligand for serotonin 2 receptor binding sites. Binding properties, brain distribution, and functional role, Mol. Pharmacol., 21:301.
Lindvall, O., and Bjorklund, A., 1978, Anatomy of the dopaminergic neuron systems in the rat brain, Adv. Biochem. Psychopharmacol., 19:1.
Ljungdahl, A., Hokfelt, T., and Nilsson, G., 1978, Distribution of substance P-like immunoreactivity in the central nervous system of the rat. I. Cell bodies and nerve terminals, Neurosci., 3:861.
Lloyd, K. G., 1980, The neuropathology of GABA neurons in extrapyramidal disorders, J. Neural. Transm. [Suppl.], 16:217.
Lo, M. S. L., Neihoff, D. L., Kuhar, M. J., and Snyder, S. H., 1983, Differential localization of type I and type II benzodiazpeine binding sites in substantia nigra, Nature, 306:57.
Marsden, C. D., 1979, GABA in relation to extrapyramidal diseases, with particular relevance to animal models, in: "GABA Neurotransmitters," P. Krogsgaard-Larsen, J. Scheel-Kruger, and H. Kofod, eds., Munksgaard, Copenhagen.
Marshall, J. F., and Ungerstedt, U., 1977, Striatal efferent fibres play a role in maintaining rotational behaviour in the rat, Science, 198:62.
Martin, L. L., and Sanders-Bush, E., 1982, Comparison of the pharmacological characteristics of 5-HT1 and 5-HT2 sites with those of serotonin autoreceptors which modulate serotonin release, Naunyn-Schmiedelberg's Arch. Pharmacol., 321:165.
McGeer, P. L., McGeer, E. G., Scherer, U., and Singh, K., 1977, A glutamatergic corticostriatal tract? Brain Res., 128:369.
McKinney, M., Coyle, J. T., and Hedreen, J. C., 1983, Topographical analysis of the innervation of the rat neocortex and hippocamnpus by the basal forebrain cholinergic system, J. Comp. Neurol., 217:103.
McPherson, G. A., and Beart, P. M., 1983, The selectivity of some ergot derivatives for alphal and alpha2 adrenoceptors of rat cerebral cortex, Eur. J. Pharmacol., 91:363.
Mefford, I. N., Foutz, A., Noyce, N., Jurik, S. M., Handen, C., Dement, W. C., and Barchas, J. D., 1982, Distribution of norepinephrine, epinephrine, dopamine, serotonin, 3,4-dihydroxyphenylacetic acid, homovanillic acid and 5-hydroxyindole-3-acetic acid in dog brain, Brain Res., 236:339.

Michelot, R., Leviel, V., Giorguieff-Chesselet, M. F., Cheramy, A., and Glowinski, J., 1979, Effects of the unilateral nigral modulation of substance P transmission on the activity of the two nigrostriatal dopaminergic pathways, Life Sci., 24:715.

Miller, J. J., Richardson, T. L., Fibiger, H. C., and McLennan, H., 1975, Anatomical and electrophysiological identification of a projection from the mesencephalic raphe to the caudate-putamen in the rat, Brain Res., 97:133.

Mitchell, R., 1980, A novel GABA receptor modulates stimulus-induced glutamate release from cortico-striatal terminals, Eur. J. Pharmacol., 67:119.

Mohler, H. and Okada, T., 1978, Properties of gamma-aminobutyric acid receptor binding with (+)-[3H]bicuculline methiodide in rat cerebellum, Mol. Pharmacol., 14:256.

Mohler, H., Okada, T., and Enna, S. J., 1978, Benzodiazepine and neurotransmitter receptor binding in rat brain after chronic administration of diazepam or phenobarbital, Brain Res., 156:391.

Moore, R. Y., and Bloom, F. E., 1978, Central catecholamine neuron system: Anatomy and physiology of the dopamine systems, Ann. Rev. Neurosci., 1:29.

Moore, K. E., and Wuerthle, S. M., 1979, Regulation of nigrostriatal and tuberinfundibular-hypophyseal dopaminergic neurons, Prog. Neurobiol., 13:325.

Murrin, L. C., Gale, K., and Kuhar, M. J., 1979, Autoradiographic localization of neuroleptic and dopamine receptors in the caudate-putamen and substantia nigra: effects of lesions, Eur. J. Pharmacol., 60:229.

Nadler, J. V., Evenson, D. A., and Cuthbertson, G. J., 1981, Comparative toxicity of kainic acid and other acidic amino acids toward rat hippocampal neurons, Neurosci., 6:2505.

Nelson, D. L., Herbet, A., Enjaalbert, A., Bockaert, J., and Hamon, M., 1980, Serotonin-sensitive adenylate cyclase and [^{3}H]serotonin binding sites in the central nervous system of the rat, Biochem. Pharmacol., 29:2445.

Neve, K. A., Kozlowski, M. R., and Marshall, J. F., 1982, Plasticity of neostriatal dopamine receptors after nigrostriatal injury: relationship to recovery of sensorimotor functions and behavioural supersensitivity, Brain Res., 244:33.

Nicoll, R. A., Alger, B. E., and Jahr, C. E., 1980, Peptides as putative excitatory neurotransmitters: carnosine, enkephalin, substance P and TRH, Proc. R. Soc. Lond. [Biol.], 210:133.

Nilsson, J. L. G., and Carlsson, A., 1982, Dopamine-receptor agonist with apparent selectivity for autoreceptors: a new principle for anti-psychotic action? Trends Pharmacol. Sci., 3:322.

Olianas, M. C., De Montis, G., Mulas, G., and Tagliamonte, A., 1978, The striatal dopaminergic function is mediated by the inhibition of a nigral non-dopaminergic neuronal system via a strio-nigral, GABAergic pathway, Eur. J. Pharmacol., 49:233.

Olivier, A., Parent, A., Simard, H., and Poirier, L. J., 1970, Cholinesterasic striatopallidal and striatonigral efferents in the cat and monkey, Brain Res., 18:273.
Olpe, H. R., and Koella, W. P., 1977, Rotatory behaviour in rats by intranigral application of substance P and an eleodoisin fragment, Brain Res., 126:576.
Olsen, R. W., Ticku, M. K., Greenlee, D., and Van Ness, P., 1979, GABA receptor and ionophore binding sites: interaction with various drugs, in: "GABA-Neurotransmitters," P. Krogsgaard-Larsen, J. Scheel-Kruger, and H. Kofod, eds., Munksgaard, Copenhagen.
Palkovits, M., and Jacobowitz, D. M., 1974, Topographic atlas of catecholamine and acetylcholinesterase-containing neurons in the rat brain. II. Hindbrain (mesencephalon, rhombencephalon), J. Comp. Neurol., 157:29.
Pasik, P., Pasik, T., and Di Figlia, M., 1979, The internal organization of the neostriatum in mammals, in: "The Neostriatum," I. Divac and R. G. E. Oberg, eds., Pergamon Press, Oxford.
Pedigo, N. W., Yamamura, H. I., and Nelson, D. L., 1981, Discrimination of multiple [3H]5-hydroxytryptamine binding sites by the neuroleptic spiperone in rat brain, J. Neurochem., 36:220.
Peroutka, S. J., Lebovitz, R. M., and Snyder, S. H., 1979, Serotonin receptor binding sites affected differentially by guanine nucleotides, Mol. Pharmacol., 16:700.
Peroutka, S. J., Lebovitz, R. M., and Snyder, S. H., 1981, Two distinct central serotonin receptors with different physiological functions, Science, 212:827.
Porceddu, M. L., Imperato, A., Melis, M. R., and Di Chiara, G., 1983, Role of ventral mesencephalic reticular formation and related noradrenergic and serotonergic bundles in turning behaviour as investigated by means of kainate, 6-hydroxydopamine and 5,7-dihydroxytryptamine, Brain Res., 262:187.
Quik, M., Emson, P. C., and Joyce, E., 1979, Dissociation between the presynaptic dopamine-sensitive adenylate cyclase and [3H]spiperone binding sites in rat substantia nigra, Brain Res., 167:355.
Quirion, R., Shults, C. W., Moody, T. W., Pert, C. B., Chase, T. N., and O'Donohue, T. L., 1983, Autoradiographic distribution of substance P receptors in rat central nervous system, Nature, 303:714.
Reavill, C., Leigh, N., Jenner, P., and Marsden, C. D., 1981, Critical role of midbrain reticular formation in the expression of dopamine-mediated circling behaviour, Adv. Biochem. Psychopharmacol., 30:187.
Reisine, T., Soubrie, P., Artaud, F., and Glowinski, J., 1982, Sensory stimuli differentially affect in vivo nigral and striatal [3H]-serotonin release in the cat, Brain Res., 232:77.

Reubi, J., and Cuenod, M., 1979, Glutamate release in vitro from cortico-striatal terminals, Brain Res., 176:185.

Reubi, J. C., Emson, P. C., Jessell, T. M., and Iversen, L. L., 1978, Effects of GABA, dopamine and substance P on the release of newly synthesized 3H-5-hydroxytryptamine from rat substantia nigra in vitro, Naunyn Schmiedeberg's Arch. Pharmacol., 304:271.

Ribak, C. E., 1981, The GABAergic neurons of the extrapyramidal system as revealed by immunocytochemistry, Adv. Biochem. Psychopharmacol., 30:23.

Rinvik, E., and Grofova, 1970, Observations on the fine structure of the substantia nigra in the cat, Exp. Brain Res., 11:229.

Roberts, P. J., and Anderson, S. D., 1979, Stimulatory effects of L-glutamate and related amino acids on (3H)-dopamine release from rat striatum: an in vitro model for glutamate actions, J. Neurochem., 32:1539.

Roberts, P. J., McBean, G. J., Sharif, N. A., and Thomas, E. M., 1982, Striatal glutamatergic function: modifications following specific lesions, Brain Res., 235:83.

Roberts, P. J., and Sharif, N. A., 1978, Effects of L-glutamate and related amino acids upon the release of [3H]dopamine from rat striatal slices, Brain Res., 157:391.

Rouzaire-Dubois, B., Hammond, C., Hamon, B., and Feger, J., 1980, Pharmacological blockade of the globus pallidus-induced inhibitory response of subthalamic cells in the rat, Brain Res., 200:321.

Rowlands, G. J., and Roberts, P. J., 1980, Activation of dopamine receptors inhibits calcium-dependent glutamate release from cortico-striatal terminals in vitro, Eur. J. Pharmacol., 62:241.

Scally, M. C., Ulus, I. H., Wurtman, R. J., and Pettibone, D. J., 1978, Regional distribution of neurotransmitter-synthesizing enzymes and substance P within the rat corpus striatum, Brain Res., 143:556.

Scatton, B., 1982, Further evidence for the involvement of D2, but not D1 dopamine receptors in dopaminergic control of striatal cholinergic transmission, Life Sci., 31:2883.

Scatton, B., and Bartholini, G., 1980, Increase in striatal acetylcholine levels by GABAergic agents: dependence on corticostriatal neurons, Brain Res., 200:174.

Scatton, B., and Lehmann, J., 1982, N-methyl-D-aspartate-type receptors mediate striatal 3H-acetylcholine release evoked by excitatory amino acids, Nature, 297:422.

Schachter, M., Bedard, P., Debono, A. G., Jenner, P., Marsden, C. D., Price, P., Parkes, J. D., Keenan, J., Smith, B., Rosenthaler, J., Horowki, R., and Dorow, R., 1980, The role of D-1 and D-2 receptors, Nature, 286:157.

Scheel-Kruger, J., Arnt, J., Braestrup, C., Christensen, A. V., Cools, A. R., and Magelund, G., 1978, GABA-dopamine interaction in substantia nigra and nucleus accumbens - relevance to behavioural stimulation and stereotyped behaviour, Adv. Biochem. Psychopharmacol., 19:343.

Scheel-Kruger, J., Magelund, G., and Olianas, M. C., 1981, Role of GABA in striatal output system: globus pallidus, nucleus entopeduncularis, substantia nigra, and nucleus subthalamicus, Adv. Biochem. Psychopharmacol., 30:165.

Schultzberg, M., Hokfelt, and Lundberg, J. M., 1982, Coexistence of classical transmitters and peptides in the central and peripheral nervous systems, Br. Med. Bull., 38:309.

Schwarcz, R., Bennett, J. P. Jr., Coyle, J. T. Jr., 1977, Loss of striatal serotonin synaptic receptor binding induced by kainic acid lesions: correlations with Huntington's Disease, J. Neurochem., 28:867.

Schwarcz, R., and Coyle, J. T., 1977, Striatal lesions with kainic acid: neurochemical characteristics, Brain Res., 127:235.

Schwarcz, R., Creese, I., Coyle, J. T., and Snyder, S. H., 1978, Dopamine receptors localized on cerebral cortical afferents to rat corpus striatum, Nature, 271:766.

Sethy, V. H., 1979, Regulation of striatal acetylcholine concentration by D2-dopamine receptors, Eur. J. Pharmacol., 60:397.

Setler, P. E., Sarau, H. M., Sirkle, C. L., and Saunders, H. L., 1978, The central effects of a novel dopamine agonist, Eur. J. Pharmacol., 50:419.

Seyfried, C. A., and Fuxe, K., 1982, Neuropharmacological and neurochemical effects of 3-(4-phenyl-1,2,3,6-tetra hydropyridyl-1)-butyl) indole (EMD 23 498) a new, long-acting dopamine agonist, Arzneim, Forsch., 32:892.

Simke, J. P., and Saelens, J. K., 1977, Evidence for a cholinergic fibre tract connecting the thalamus with the head of the striatum of the rat, Brain Res., 126:487.

Skerritt, J. H., Trisdikoon, P., and Johnston, G. A. R., 1981, Increased GABA binding in mouse brain following acute swim stress, Brain Res., 215:398.

Skirboll, L. R., Grace, A. A., and Bunney, B. S., 1979, Dopamine auto- and post-synaptic receptors: Electrophysiological evidence for differential sensitivity to dopamine agonists, Science, 206:80.

Skolnick, P., Paul, S. M., and Barker, J. L., 1980, Pentobarbital potentiates GABA-enhanced [3H]diazepam binding to benzodiazepine receptors, Eur. J. Pharmacol., 65:125.

Sokoloff, P., Martres, M. -P., and Schwartz, J. C., 1980, Three classes of dopamine rceptors (D-2, D-3, D-4) identified by binding studies with 3H-apomorphine and 3H-dompepridone, Naunyn Schmiedeberg's Arch. Pharmacol., 325:89.

Soubrie, P., Montsatruc, J. L., Bourgoin, S., Reisine, T., Artaud, F., and Glowinski, J., 1981, In vivo evidence for GABAergic control of serotonin release in the cat substantia nigra, Eur. J. Pharmacol., 69:483.

Spano, P. F., Memo, M., Stefanini, E., Fresia, P., and Trabucchi, M., 1978, Detection of multiple receptors for dopamine, in: "Receptors for Neurotransmitters and Peptide Hormines, G. Pepeu, M. J. Kuhar, and S. J. Enna, eds., Raven Press, New York.

Spencer, H. J., 1976, Antagonism of cortical excitation of striatal neurones by glutamic acid diethyl ester: Evidence for glutamic acid as an excitatory transmitter in the rat striatum, Brain Res., 102:91.

Spillane, J., White, P., Goodhart, M., Flack, R., Bowe, D. M., and Davison, A., 1977, Selective vulnerability of neurons in organic dementia, Nature, 266:558.

Stadler, H., Lloyd, K. G., and Gadea-Ciria, M., 1973, Enhanced striatal acetylcholine release by chlorpromazine and its reversal by apomorphine, Brain Res., 55:476.

Standefer, M. J., and Dill, R. E., 1977, The role of GABA in dyskinesias induced by chemical stimulation of the striatum, Life Sci., 21:1515.

Starr, M. S., 1978, GABA potentiates potassium-stimulated 3H-dopamine release from slices of rat substantia nigra and corpus striatum, Eur. J. Pharmacol., 48:325.

Starr, M. C., Summerhayes, M., and Kilpatrick, I. C., 1983, Interactions between dopamine and gamma-aminobutyrate in the substantia nigra: implications for the striatonigral output hypothesis, Neurosci., 8:547.

Strombom, U., 1976, Catecholamine receptor agonists. Effects on motor activity and rate of tyrosine hydroxylation in mouse brain, Naunyn Schmeideberg's Arch. Pharmacol., 292:167.

Tamminga, C. A., Schaffer, M. H., Smith, R. C., and Davies, J. M., 1978, Schizophrenic symptoms improve with apomorphine, Science, 200:567.

Tarsy, D., Pycock, C. J., Meldrum, B. S., and Marsden, C. D., 1978, Focal contralateral myoclonus produced by inhibition of GABA action in the caudate nucleus of rats, Brain, 101:143.

Ternaux, J. P.,. Hery, F., Bourgoin, S., Adrien, J., Glowinski, J., and Hamon, M., 1977, The topographic distribution of serotonergic terminals in the neostriatum of the rat and the caudate nucleus of the cat, Brain Res., 121:311.

Torrens, Y., Beaujouan, J. C., Besson, M. J., Michelot, R., and Glowinski, J., 1981, Inhibitory effects of GABA, L-glutamic acid and nicotine on the potassium-evoked release of substance P in the substantia nigra slices of the rat, Eur. J. Pharmacol., 71:383.

Unnerstall, J. R., and Wamsley, J. K., 1982, Autoradiographic localization of high-affinity [3H] kainic acid binding sites in the rat forebrain, Eur. J. Pharmacol., 86:361.

Van der Kooy, D., and Hattori, T., 1980, Dorsal raphe cells with collateral projections to the caudate-putamen and substantia nigra: a fluorescent retrograde double labelling study in the rat, Brain Res., 186:1.

Vizi, E. S., Harsing, L. G., and Zsilla, G., 1981, Evidence of the modulatory role of serotonin in acetylcholine release from striatal interneurones, Brain Res., 212:89.

Wachtel, H., Ahlenius, S., and Anden, N., E., 1979, Effects of locally applied dopamine to the nucleus accumbens on the motor activity of normal rats and following alpha-methyltyrosine or reserpine, Psycho-Pharmacol., 63:203.

Waddington, J. L. and Cross, A. J., 1978, Denervation supersensitivity in the striatonigral GABA pathway, Nature, 276:618.

Waddington, J. L. and Cross, A. J., 1979, Baclofen and muscimol: behavioural and neurochemical sequelae of unilateral intranigral administration and effects on 3H-GABA receptor binding, Naunyn Schmiedebergs Arch. Pharmacol., 306:275.

Walaas, I., 1981, Biochemical evidence for overlapping neocortical and allocortical glutamate projections to the nucleus accumbens and rostral caudate-putamen in the rat brain, Neuroscience, 6:309.

Walters, J. R., and Roth, R. H., 1976, Dopaminergic neuron: An in vivo system for measuring drug interactions with presynaptic receptors, Naunyn Schmiedeberg's Arch. Pharmacol., 296:5.

Waszczak, B. L., Eng. N., and Walters, J. R., 1980, Effects of muscimol and picrotoxin on single unit activity of substantia nigra neurons, Brain Res., 188:185.

Waszczak, B. L., and Walters, J. R., 1983, Dopamine modulation of the effects of gamma-aminobutyric acid on substantia nigra pars reticulata neurones, Science, 220:218.

Watkins, J. C., 1981, Pharmacology of excitatory amino acid receptors, in: "Glutamate: Transmitter in the Central Nervous System," P. J. Roberts, J. Storm-Mathisen, and G. A. R. Johnston, eds., John Wiley and Sons, Chichester.

Webster, K. E., 1961. Cortico-striate interrelations in the albino rat, J. Anat. 95:532.

Westfall, T. C., Grant, H., and Perry, H., 1983, Release of dopamine and 5-hydroxytryptamine from striatal slices following activation of nicotinic cholinergic receptors, Gen. Pharmacol., 14:321.

Westfall, T. C., 1974, Effect of nicotine and other drugs on the release of 3H-norepinephrine and 3H-dopamine from rat brain slices, Neuropharmacology, 13:693.

Whitehouse, P. J., Price, D. L., Struble, R. G., Clarke, A., Coyle, J. T., and DeLong, M. R., 1982, Alzheimer's disease: a loss of neurons in the basal forebrain, Science, 215:1237.

Williams, M., and Totaro, J. A., 1982, Interaction of the putative dopamine autoreceptor agonists 3-PPP and TL-99 with [3H]apomorphine binding sites in rat striatal membranes, Eur. J. Pharmacol., 86:39.

Wilson, C., J., and Groves, P. M., 1980, Fine structure and synaptic connections of the common spiny neuron of the rat neostriatum: a study employing intracellular injections of horseradish peroxidase, J. Comp. Neurol., 194:599.

Wilson, C., J., and Groves, P. M., 1981, Spontaneous firing patterns of identified spiny neurones in the rat neostriatum, Brain Res., 220:67.

Wilson, C., J., Groves, P. M., and Fifkova, E., 1977, Monoaminergic synapses, including dendro-dendritic synapses in the rat substantia nigra, Exp. Brain Res., 30:161.

Wood, P. L., Moroni, F., Cheney, D. L., and Costa, E., 1979, Cortical lesions modulate rates of acetylcholine and gamma-aminobutyric acid, Neurosci. Letts., 12:349.

Wood, P. L., and Richard, J., 1982, GABAergic regulation of the substantia innominata-cortical cholinergic pathway, Neuropharmacol., 21:969.

Yamamura, H. I., and Snyder, S. H., 1974, Muscarinic cholinergic binding in rat brain, Proc. Natl. Acad. Sci., U.S.A., 71:1725.

York, D. H., 1979, The neurophysiology of dopamine receptors, in: "The Neurobiology of Dopamine," A. S. Horn, J. Korf, and B. H. C. Westerink, eds., Academic Press, London.

THE REGULATION OF STRIATAL DOPAMINE RECEPTORS:

SUBSENSITIVITY INDUCED BY HYPERTHYROIDISM OR REM SLEEP DEPRIVATION

David H. Overstreet, Marion A. Joschko, Peter F. Harris and Ann D. Crocker

Schools of Biological Sciences and Medicine and
Centre for Neuroscience
The Flinders University of South Australia
Bedford Park, South Australia 5042

INTRODUCTION

Considerable interest in the dopaminergic system(s) has developed since the recognition that dopamine (DA) might be a neurotransmitter in its own right as well as being a precursor to noradrenaline. The initial discovery of a deficiency of dopamine in post mortem brains from individuals with Parkinson's Disease led to a greater interest in the potential relevance of DA in clinical disorders. More recent studies have begun to focus on the potential involvement of changes in DA receptor sensitivity as well as presynaptic DA metabolism in particular disease states. Two examples of these recent developments are the use of direct DA receptor agonists to treat Parkinsonian patients who appear to be resistant to L-DOPA therapy (Williams and Calne, 1981) and the suggestion that an increase in DA receptors may be involved in the positive symptoms of schizophrenia (e.g. Cross et al, 1981). In the present paper we wish to review much of the recent evidence for the regulation of DA receptor sensitivity and present some of our recent findings implicating thyroid hormones and REM sleep deprivation in the modulation of DA receptor sensitivity.

Dopaminergic Systems

It is now well recognised that there are at least five clearly defined dopaminergic pathways in the central nervous system (CNS). The nigrostriatal pathway from the substantia nigra (pars compacta) to the neostriatum has been the one which has been most carefully documented. Other pathways include the mesolimbic pathway, the mesocortical pathways, the tuberoinfundibular pathway, and the amacrine cells in the retina. The reader is referred to Lindvall (1979) for a more comprehensive discussion of these. Because of the greater attention that has been given to the nigrostriatal pathway, we will devote most of our discussion about the regulation of DA receptors to those in the striatum.

It is generally recognised that stimulation of the nigrostriatal pathway will lead to a number of motor behaviours, including sterotyped sniffing, licking and if the conditions are appropriate, gnawing. If the stimulation is unilateral, then contralateral turning is frequently seen. Conversely, if the nigrostriatal pathway is lesioned or DA antagonists are given, severe akinesia and catalepsy may occur. Because of the often clear cut behavioural syndromes induced by DA agonists and antagonists, there has been an extensive literature on their use to detect possible changes in DA receptor sensitivity. Much of this earlier literature has been reviewed (Overstreet and Yamamura, 1979; Creese and Sibley, 1981) and we propose to concentrate our discussion on the more recent literature.

Heterogeneity of DA receptors

Since the publication of these earlier reviews, it has become apparent that there are a number of different types of DA receptors. The two which have been most extensively characterised are the D1 receptor which is linked to adenylate cyclase and is relatively insensitive to blockade by DA antagonists, and the D2 receptor which is sensitive to blockade by DA antagonists (see Beart, 1982). Some workers (e.g. Seeman, 1981) have proposed as many as four different classes, based upon their relative affinity for agonists and antagonists. However, the functional relevance of these various receptors is still far from clear.

In addition to the well recognised pharmacological heterogeneity of DA receptors, there is also an anatomical heterogeneity. It appears that DA receptors may be located on striatal neurons innervated by the nigrostriatal pathway (postsynaptic receptors), on the terminals of these incoming nigrostriatal neurons (presynaptic receptors), on the terminals of incoming corticostriate neurons (presynaptic receptors), or upon the cell bodies of the nigrostriatal neurons (autoreceptors)(Beart, 1982). Once again, there is some dispute in the literature on the best way of classifying these different receptors. Moreover, it is not clear how the anatomical location of the receptors relates to the pharmacological distinctions of D1, D2, etc. made above. Nevertheless, there is some headway being made toward developing more specific receptor agonists (see Chapter by Beart in this volume). Such agents will eventually help to elucidate the degree to which the different types of DA receptors may be regulated in a similar way.

Up-regulation of DA Receptors

It would appear that striatal DA neurons function according to Cannon's law of denervation, because it has been clearly demonstrated that treatments which chronically decrease the interaction of DA with its receptor result in an increase in the sensitivity of these striatal neurons. Among those treatments which have been reported to increase the sensitivity of striatal DA neurons are (1) lesions of the nigro-striatal pathway (e.g. Creese et al, 1977); (2) chronic treatment with DA antagonists (Burt et al, 1976); (3) chronic treatment with alpha methyl tyrosine, an inhibitor of tyrosine hydroxylase (Tarsy and Baldessarini, 1974). As indicated previously, these studies have been summarised in recent reviews (Overstreet and Yamamura, 1970; Creese and Sibley, 1981); consequently, only the more recent papers will be reviewed.

One shortcoming of this earlier literature, that few investigators studied both behavioural and biochemical indices of DA receptor sensitivity (Overstreet and Yamamura, 1979), has been overcome in some of the more recent studies. For example, Lai et al (1981) compared apomorphine-elicited stereotyped behaviour and striatal DA binding sites in rats given three weeks of chronic treatment with the classical neuroleptics haloperidol and thioridazine or the typical neuroleptic zotepine. They found highly significant correlations between apomorphine-elicited sterotypies and DA binding sites in control and haloperidol-treated rats, but not in rats treated with zotepine or thioridazine. It was also of interest to note that while the latter two drugs led to an up-regulation (increase) in DA receptors, there was no corresponding increase in apomorphine-elicited stereotyped behaviour. This paper clearly emphasizes the point made by earlier reviewers: it is important for workers to obtain both behavioural and neurochemical indices of changes in receptor sensitivity.

More recent studies have indicated that other factors may lead to an increase in dopamine receptor sensitivity. The finding that female rats are more sensitive to the behavioural effects of DA agonists (Savageau and Beatty, Robinson et al, 1980) suggests that DA receptors may be modulated by the sex hormones. In fact, several studies have indicated that DA receptor sensitivity may be increased by estrogen (Hruska et al, 1982). Despite this evidence for hormonal modulation of DA receptor sensitivity, it is not yet clear as to whether the modulation comes about from a direct interaction of the hormones with the binding site or is produced indirectly by inducing a decrease in presynaptic DA metabolism and release.

Very recently in our laboratory we have determined that DA receptor sensitivity may be increased in hypothyroid rats (Crocker and Overstreet, 1983; Overstreet et al, 1984). In one experiment the rats were made hypothyroid by treatment with propylthiouracil; these rats were less affected by the neuroleptic haloperidol and exhibited an increased number of DA binding sites when chronically treated with a low dose (0.1 mg/kg) of haloperidol which by itself did not increase DA binding sites in controls (Crocker and Overstreet, 1983). In a second study rats were made hypothyroid by being reared on iodine-deficient diets; these rats exhibited greater behavioural sensitivity to apomorphine and had greater numbers of DA binding sites (Overstreet et al, 1984). Thus, if thyroid deficiency is severe enough and is carried out for a sufficient period, an up-regulation of DA receptors will occur. Again, however, it is not yet clear whether thyroid hormone interacts directly with the receptor or acts indirectly via changes in presynaptic DA metabolism.

Thus, in summary, it is clear that if DA is prevented from interacting with its receptor by lesioning the DA pathways or administering receptor blockers, an up-regulation of DA receptors will occur. It is also true that hormones may modulate the sensitivity of DA receptors, but the exact mechanisms by which they do so remain to be elucidated.

Down-regulation of DA Receptors

One of the more intriguing paradoxes in the literature on dopaminergic systems is the increased sensitivity or "reverse tolerance" that develops during the chronic administration of psychostimulants, putative DA agonists (e.g. Ellinwood and Kilbey, 1980). It should be stressed that classical tolerance to stimulants has also been observed, and the factors that determine whether increased sensitivity or tolerance will develop remain unclear (Post, 1981). Recently Demellweek and Goudie (1983) have reviewed this extensive literature and presented a case for considering behavioural mechanisms. Thus, the behavioural changes associated with chronic stimulant administration are extremely complex (Demellweek and Goudie, 1983). It is not surprising, therefore, that the effects of chronic stimulant administration on DA receptor binding are similarly complex: some studies have reported increases in DA receptor binding (Borison et al, (1979) while others have reported significant decreases (Howlett and Nahorski, 1979; Hitzemann et al, 1980; Nielsen et al, 1980). Still others have reported decreases which were not statistically significant (Owen et al, 1981; Ridley et al, 1982), while some investigators have failed to detect any effect of chronic psychostimulant treatment on DA receptor binding (Burt et al, 1976; Algeri et al, 1980; Jackson et al, 1981, Finnegan et al, 1982).

Further studies have reported regionally selective increases and decreases (Akiyiama et al, 1981).

As indicated earlier, it is generally agreed that behavioural and receptor binding studies should be conducted wherever possible in the same animals (Overstreet and Yamamura, 1979; Demellweek and Goudie, 1983). However, most of the papers referred to above did not do so. In those papers that did (Ridley et al., 1982; Finnegan et al., 1982), there was usually no correlation between the changes in behaviour and receptor binding. Also, as noted recently, because tolerance and sensitization can develop concurrently to different parameters of amphetamine-induced stereotypy it is difficult to account for tolerance and/or increased sensitivity in terms of receptor changes.

Nevertheless, even more recent studies with chronic DA agonists have reported evidence for a down-regulation of DA receptors (Quirion et al., 1982; Robertson and Paterson, 1981; Nielsen et al, 1983), a finding that is consistent with several reports of reduced sensitivity to the directly acting DA agonist, apomorphine, in animals chronically treated with stimulants (Weston and Overstreet, 1976; Finnegan et al., 1982). However, in the Finnegan et al. (1982) study the animals did not show any changes in DA binding even though they were subsensitive to apomorphine and supersensitive to haloperidol. There has also been a recent report of an increased behavioural sensitivity to apomorphine developing during its chronic administration (see Riffee et al., 1982; Wilcox et al., 1981). Thus, the literature on chronic DA agonist administration is not supportive of a simple symmetrical relationship between stimulation of the DA receptors and regulation of those receptors. These studies also illustrate that behavioural indices of receptor sensitivity do not necessarily agree with biochemical indices, as seen previously with the literature on antagonists. Consequently, measurements of both behavioural and biochemical indices should always be collected when studying changes in dopaminergic systems.

Although the evidence is also somewhat controversial, there is some indication that thyroid hormones may modulate the sensitivity of DA receptors to agonists. This evidence will be reviewed in greater detail in the next section. At this time we wish only to point out that the majority of the recent literature is consistent with the hypothesis that the sensitivity of DA receptors is reduced by hyperthyroidism. If this relationship can be confirmed, it would suggest an inverse relationship between thyroid status and DA receptor sensitivity: increased thyroid function being associated with reduced DA receptor sensitivity and decreased thyroid function being associated with elevated DA receptor sensitivity to agonists.

EFFECTS OF HYPERTHYROIDISM ON DOPAMINE RECEPTOR SENSITIVITY

Introduction

Early reports indicated that behavioural responses to the direct DA agonist, apomorphine, were greater in hyperthyroid than in euthyroid guinea pigs (Klawans et al, 1974). However, more recent reports, including one from our laboratory (Joschko et al, 1982), have indicated that hyperthyroid rats and mice exhibit a reduced behavioural response to apomorphine (Strombom et al., 1977; Atterwill, 1981). These latter findings are consistent with reports on increased sensitivity of hyperthyroid animals to DA antagonists (Crocker and Overstreet, 1983; Lake and Fann, 1973). Because of the controversy about whether apomorphine has a reduced or increased behavioural effect in hyperthyroid animals, we decided to re-examine this question. In our studies we have looked at a wider range of behavioural and physiological responses to apomorphine, thereby increasing the validity of our conclusions.

Methods

The animals used were male Sprague-Dawley albino rats, approximately 90 days of age and weighing 300-350 g at the beginning of the study. They were housed in groups of eight under conditions of constant temperature (22°C), humidity (45%) and lighting. They had free access to food but water was restricted as indicated below.

Rats were trained to press a bar in an operant chamber to obtain water. The 8 operant boxes were each housed in sound-proofed chambers and were controlled by a TRS 80 micro-computer with Lehigh-Valley interfaces. After being trained to bar-press, the rats were maintained on an FR5 (5 bar-presses for one reward - 0.05 ml) schedule of reinforcement. The operant sessions were 15 mins in duration and were followed by 15 mins of water in the home cage.

Core body temperatures of the rats were recorded by inserting a thermistor probe 6-8 cm into the rectum and reading the output on a CRL digital recorder. Normally a stable temperature (no change in 15 secs) was reached for each animal within one minute. Temperatures were used both to assess the course of hyperthyroidism and the hypothermic response to apomorphine.

Stereotyped behaviour was observed by placing the rats in a rectangular open field (60 x 30 cm) made of perspex for a one-minute period and observations were carried out in a room lit only by a red fluorescent light. The degree of stereotypy was assessed according to the 6-point scale adapted from Creese and Iversen (1973).

Rats were made hyperthyroid by daily subcutaneous injections of 200 µg/kg l-thyroxine. Euthyroid rats were given daily subcutaneous injections of isotonic saline (1 ml/kg). There were 16 rats in each of these chronic treatment groups.

After the rats had been stabilized on the operant responding task and had been treated with thyroxine or saline for at least five weeks, they were challenged with either 0.3 mg/kg apomorphine hydrochloride or isotonic saline (1 ml/kg) given subcutaneously. Stereotypy ratings were taken at 11 and 25 mins after the injections, operant responding was recorded for 10 mins between 13 and 23 mins after the injection, and body temperature was recorded at 24 mins after the injection. These times and doses were selected on the basis of preliminary studies (Joschko et al, 1982).

Rats were sacrificed by decapitation 24 hours after the last dose of thyroxine or saline (approximately 50 days after the beginning of treatment). Trunk blood was collected, allowed to clot, centrifuged and serum collected and frozen at -20°C. Thyroxine and TSH were measured by radioimmunoassay techniques developed at the Queen Elizabeth Hospital, Woodville, South Australia, as described previously (Crocker and Overstreet, 1983). After sacrifice brains were removed; striata were dissected out, homogenized in 10 volumes of ice cold 50 mM Tris buffer (pH 7.4) and stored at -20°C. Aliquots containing 10 mg of striatal tissue were incubated with concentrations of 3H-spiroperidol ranging from 0.1-2.0 nM in the presence and absence of 1 µM (+) butaclamol. Binding data were analysed by Scatchard analysis to obtain K_D in nM and B_{max}, the maximum number of binding sites, expressed as pmols/100 mg protein. Protein was estimated by the Folin method (Lowry et al, 1951).

Data for operant responding and for body temperature were expressed as changes from appropriate baselines. These percentage scores and the ratings for stereotypy were analysed by the Mann-Whitney U Test (Siegel, 1957). Student's t test was used to analyse differences in baseline body temperatures and in hormone levels and receptor concentration.

Results

As early as three weeks after the initiation of treatment, the thyroxine-treated rats exhibited a significant elevation of temperature (38.3 ± 0.07 vs 38.6 ± 0.06, $t = 7.20$, $P < 0.001$); however, most rats had not yet shown a stabilised operant response rate. On the day of the apomorphine challenge two weeks later, the temperature was still significantly elevated in the hyperthyroid group (39.3 ± 0.11 vs 38.2 ± 0.08, $t = 7.68$, $P < 0.001$). As indicated in Figure 1, apomorphine produced a dramatic hypothermic response in euthyroid rats and a much reduced response in the hyperthyroid rats. Thus, using temperature as a variable, hyperthyroid rats exhibited a subsensitivity to apomorphine.

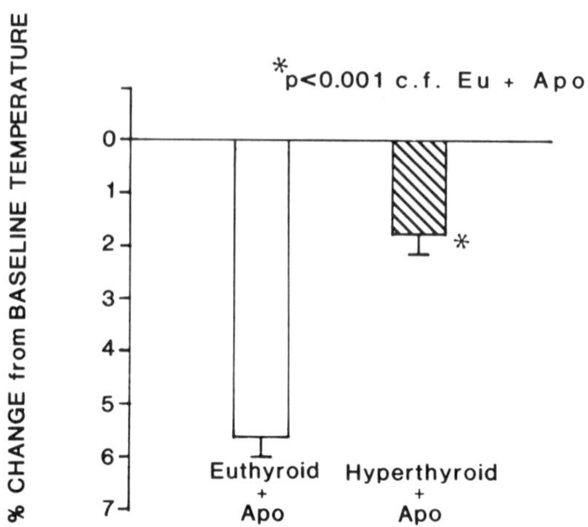

Fig. 1. The effects of apomorphine on body temperature in euthyroid (chronically treated with isotonic saline) and hyperthyroid rats (chronically treated with T_4). The data represent the mean % change from baseline 24 mins after the subcutaneous injection of 0.3 mg/kg apomorphine. There were 8 rats in each group.

The values for T_4 levels in trunk blood collected at sacrifice indicated that the T_4-treated rats were still hyperthyroid. The saline-treated rats had mean values of 54 $\pm$ 2.8 nmol/l; in contrast, the T_4-treated rats had mean values of 138 $\pm$ 14.2 nmol/l. The assay for TSH was unsatisfactory, so no reliable values could be obtained.

The effects of apomorphine on operant responding for a water reward are illustrated in Figure 2. The results for the 10-min challenge session were expressed as a percentage of the basline responses on the preceding day. Apomorphine almost completely suppressed the responding of the euthyroid rats, while the suppression induced by apomorphine in the hyperthyroid rats was significantly reduced. Saline had very little effect on the operant responding of either group, indicating that the animals had reached a stable level of responding. These data also indicate that the hyperthyroid rats are subsensitive to apomorphine.

The effects of apomorphine on stereotypy ratings at 25 min are illustrated in Figure 3. Euthyroid and hyperthyroid rats treated with saline exhibited moderate amounts of activity with intermittent sniffing, thus obtaining ratings of 1 or 2. Euthyroid rats treated with apomorphine exhibited continuous sniffing that was often in one location and nearly always in a random path, thus obtaining ratings of 3.5 or 4 whereas thyroxine-treated rats were significantly less

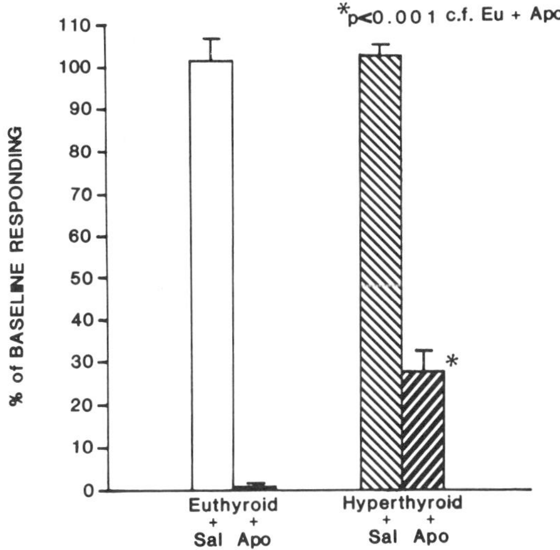

Fig. 2. The effects of apomorphine on operant responding in euthyroid and hyperthyroid rats. The data represent the mean % of baseline responding 13-23 mins after the subcutaneous injection of 0.3 mg/kg apomorphine. There were 8 rats in each group.

affected by apomorphine exhibiting either moderate activity with discontinuous sniffing or sniffing along a fixed path, thus obtaining ratings of 2 or 3.

The results from analysis of the spiroperidol binding data indicated that there was a decreased concentration of dopamine receptors in the hyperthyroid compared with the euthyroid group. However, because striata from several animals were pooled the sample sizes have been reduced substantially and there is no significant difference between the groups. Further experiments are in progress to clarify these findings.

Discussion

The data presented in this study clearly indicate that DA receptor subsensitivity is associated with hyperthyroidism. Thus, the findings confirm the conclusions of Atterwill (1981) and are at variance with the findings of Klawans et al. (1974). It is possible that species differences may account for this discrepancy. Because each of the three parameters used (core body temperature, operant responding for water reward, and stereotypy) demonstrated the same patterns of effects, it is difficult to escape the conclusion of DA subsensitivity. In particular, the measure of body temperature and operant responding are less subjective than ratings of stereotypy. Although the findings for temperature are complicated by the fact

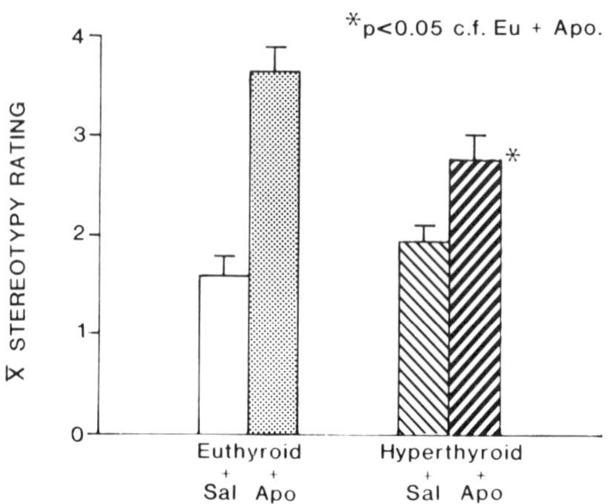

Fig. 3. The effects of apomorphine and isotonic saline on stereotypy ratings in euthyroid and hyperthyroid rats. The data represent the mean stereotypy ratings 25 mins after the subcutaneous injection of 0.3 mg/kg apomorphine or isotonic saline. There were 8 rats in each group.

that hyperthyroid rats have elevated baseline body temperature, the operant responding data are not. However, it is not yet clear whether the suppression of bar-pressing induced by apomorphine is a reflection of striatal dopamine stimulation. Studies are underway using chronically implanted cannulae to administer apomorphine into the striatum and nucleus accumbens in order to resolve this issue.

Both Atterwill (1981) and we in our earlier studies (Crocker and Overstreet, 1983) failed to detect any change in DA receptor concentration in hyperthyroid rats even though Atterwill (1981) reported a subsensitivity to apomorphine. In our previous studies the rats were sacrificed within 3 weeks after the initiation of chronic treatment, whereas the data from the present experiment came from animals treated for at least 5 weeks. Therefore, it is possible that the longer duration of treatment was sufficient to induce a down-regulation of DA receptors. A parallel finding was mentioned early with regard to hypothyroidism: Crocker and Overstreet (1983) did not detect any up-regulation of DA receptors in hypothyroid rats treated for 2-3 weeks with propylthiouracil, whereas Overstreet et al (1984) did detect an up-regulation in hypothyroid rats raised on an iodine-deficient diet throughout life. These two sets of results indicate that chronic high or low levels of thyroid hormone may produce inverse changes in DA receptor concentration if sufficiently prolonged. Nevertheless, other studies indicate that the receptor sensitivity changes may be detected by alterations in the sensitivity of apomorphine before there are biochemically demonstrable changes in receptors (cf. Atterwill, 1981; Joschko et al., 1982).

EFFECT OF REM SLEEP DEPRIVATION ON DOPAMINE RECEPTOR SENSITIVITY

Introduction

A common procedure used to deprive rodents of REM sleep in recent years has been the platform technique whereby the rat or mouse is placed on a platform surrounded by water. When muscle relaxation occurs during REM sleep, the rat falls into the water and wakes up. Depending upon the size of the platforms it is hypothesized that the rat can be selectively deprived of REM sleep. However, there is considerable controversy over the validity of this technique and two recent reviews have severely criticized earlier studies (Hicks et al., 1977; McGrath and Cohen, 1978).

Despite these reviews which criticize the use of the platform technique, two recent papers employing an inadequate platform methodology have been published which implicated the dopaminergic system in REM sleep deprivation. In fact, these studies (Tufik, 1981; Tufik et al., 1978) suggested that REM sleep-deprived rats exhibit a receptor supersensitivity when challenged with apomorphine, a result which is at variance with the reported increases in DA levels in REM sleep-deprived animals (Ghosh, et al., 1976; Hernandez-Peon et al., 1969; Mark et al., 1969). Stern and Morgane (1974), using a similar experimental design, reported that rats confined to a small platform (presumably REM sleep-deprived) exhibited a subsensitivity to the locomotor activating effects of amphetamine, an indirect DA agonist.

Because of these conflicting results and the inadequate methodology employed by the earlier workers, it was decided to re-examine the question of DA receptor sensitivity in REM sleep-deprived rats. We wish to report on a study which suggests that REM sleep deprivation may lead to a down-regulation of DA receptors.

Methods

The animals were male Sprague-Dawley rats, bred and supplied by the Flinders University School of Biological Sciences animal house. They were housed under conditions of constant temperature (122°C), humidity (45%) and lighting. They had free access to food and water.

The REM sleep deprivation apparatus consisted of a single wooden box measuring 145 x 84 x 64 cms deep. A plastic tank was placed inside the box and was divided into 8 chambers each measuring 31 x 31 x 46 cms deep. The floor of the chambers consisted of a sheet of aluminium on which the platform stems were mounted. The stems were located in the centre of each compartment and rose to 15 cms above the floor level. The circular platform tops were screwed onto the stems and were made of aluminium. Water was added to within 0.5 cm of the platform top. A wire mesh ceiling was placed 30 cms above the tops. Through this ceiling, food hoppers were suspended. On the underside of the tank lid six lights were positioned so as to illuminate each chamber. An exhaust fan was mounted in the lid with adjacent holes to ensure a continual flow of air to the tank when the lid was down.

There were five experimental groups utilized in this series of experiments: A group cage control, two groups confined to platforms in which the weight to area (W/A) ratio was 1.7 (large platform-confinement), and two groups confined to platforms in which the W/A ratio was 6.5 (small platform-confinement). Two of the platform-confined groups were naive to the platform conditions (NH) prior to being placed on the platforms and two were habituated to the platform conditions for four days (4H) one week prior to being exposed to the platforms. The five groups were designated by the following code: G.C.C., 1.7 W/A NH, 6.5 W/A NH, 1.7 W/A 4H, and 6.5 W/A 4H. Habituation to the platforms was included in this design because Harris et al. (1982) indicated that this could be an important factor.

To examine the hypothermic effects of apomorphine, rats received subcutaneous injections of 1.0, 3.5, 5.0 and 7.5 mg/kg apomorphine approximately one hour after the completion of the platform phase. These dosages were chosen on the basis of earlier reports by Cox et al. (1976) and Tufik et al. (1978) that they induced a significant hypothermic effect in normal rats. Core body temperature was measured at 0 (predrug baseline), 30 and 60 mins after drug administration. Temperatures were recorded on a CRL digital temperature unit as described previously.

To examine the stereotypic effects of apomorphine, rats received subcutaneous injections of 5.0 mg/kg apomorphine approximately one hour after the completion of the platform phase. The stereotypy rating for each rat was performed at 30, 60 and 90 mins after the administration of apomorphine, using an adaptation of the rating scale reported by Creese and Iversen (1973), as indicated previously.

Because there were baseline differences in temperature among the five groups, the results for temperature were expressed as deviations from baseline. Two-way analysis of variance was used to evaluate the significance of the temperature data. The Mann-Whitney U test was used to evaluate the data on stereotyped behaviour.

At the conclusion of behavioural testing, or approximately three hours after removal from the platforms, the rats were sacrificed by decapitation. The brains were removed and the striata were dissected out, weighed and homogenized in 10 volumes of ice cold 50 mM tris buffer (ph 7.4). The striata from each treatment group were pooled and homogenates stored at -20°C. Dopamine receptor concentration and affinity were calculated as described previously.

Results

The hypothermic effects of apomorphine at 30 mins after injection are illustrated in Figure 4. Significant drug ($F3,52$ = 8.08, $P < 0.001$) and platform ($F4,52 = 13.85$, $P < 0.001$) effects were observed. Furthermore, a significant drug x platform interaction was found ($F2,52 = 10.1$, $P < 0.001$). As illustrated in Figure 5, similar results were observed at the 60 min time point, with the drug ($F3,52 = 6.14$, $P < 0.001$), platform ($F4,52 = 24.4$, $P < 0.001$) and interaction ($F12,52 = 4.65$, $P < 0.001$) effects all being significant. Because of these results, two further analyses were performed, separately for the habituated and nonhabituated groups. For the

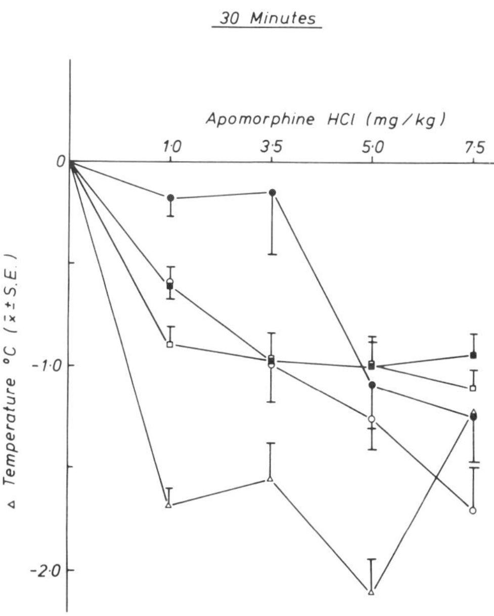

Fig. 4. The effects of apomorphine on body temperature in rats confined to large or small platforms at 30 mins after injection. The data represent the mean deviations from baseline temperature in group cage controls (open triangle) or animals housed on platforms under four conditions: W/A 1.7 NH (open squares), W/A 6.5 NH (filled squares), W/A 1.7 4H (open circles), or W/A 6.5 4H (filled circles). There were at least 6 animals in each group.

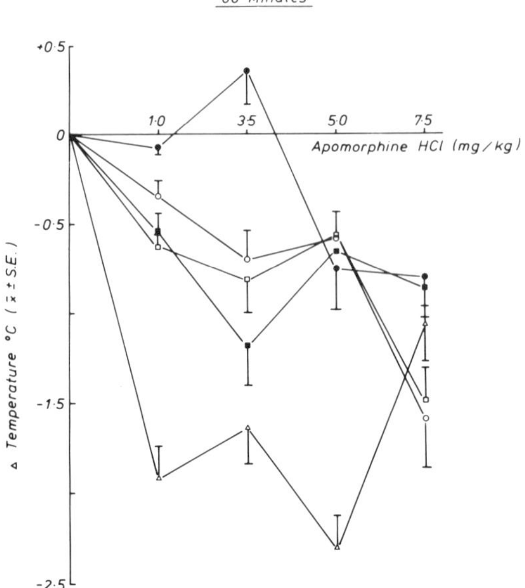

Fig. 5. The effects of apomorphine on body temperature in rats confined to large or small platforms at 60 mins after injection. See legend to Figure 4 for description of groups.

naive condition there was a significant drug effect at both 30 ($F_{3,70}$ = 6.5, P < 0.001) and 60 mins (F = 16.5, P < 0.001), however there were no significant platform or interaction effects (all P > 0.1). Thus, for the naive groups apomorphine produced a similar dose-related hypothermia. For the habituated condition there was a significant drug effect at 30 mins ($F_{3,47}$ = 4.32, P < 0.001) but not at 60 mins ($F_{3,47}$ = 1.82, P > 0.05). In contrast, there were no significant platform or interaction effects at 30 mins (both P > 0.05), but both effects were significant at 60 mins (platform: $F_{1,47}$ = 5.24, P < 0.05; interaction: $F_{3,47}$ = 3.1, P < 0.05). These latter results were primarily due to the lack of a hypothermic effect for the W/A 6.5 group after 3.5 mg/kg apomorphine (see Figure 5).

The general conclusion from these results is that all of the platform conditions exhibited a significant subsensitivity to the hypothermic effects of apomorphine when compared with group cage controls. This was clearly evident for the lower doses of apomorphine, i.e. 1.0 and 3.5 mg/kg (see Figures 4 & 5). Moreover, the greatest subsensitivity to this hypothermic effect was observed in the W/A 6.5 4H condition.

The effects of apomorphine on stereotypy ratings are summarized in Figure 6. The pattern of results is similar to those observed for temperature: the highest ratings were exhibited by the group cage control and the lowest by the W/A 6.5 4H group, indicating that the habituated group confined to the small platform was sub-sensitive to apomorphine. Mann-Whitney U tests confirmed that the W/A 6.5 4H group showed ratings that were significantly smaller than those of the group cage control at all time points. In addition, they were significantly smaller than the ratings of the W/A 1.7 4H group at 30 and 90 minutes.

Preliminary experiments in which DA receptor concentrations and affinity were calculated from Scatchard analysis suggested that there was a reduction in receptor concentration in the small platform-

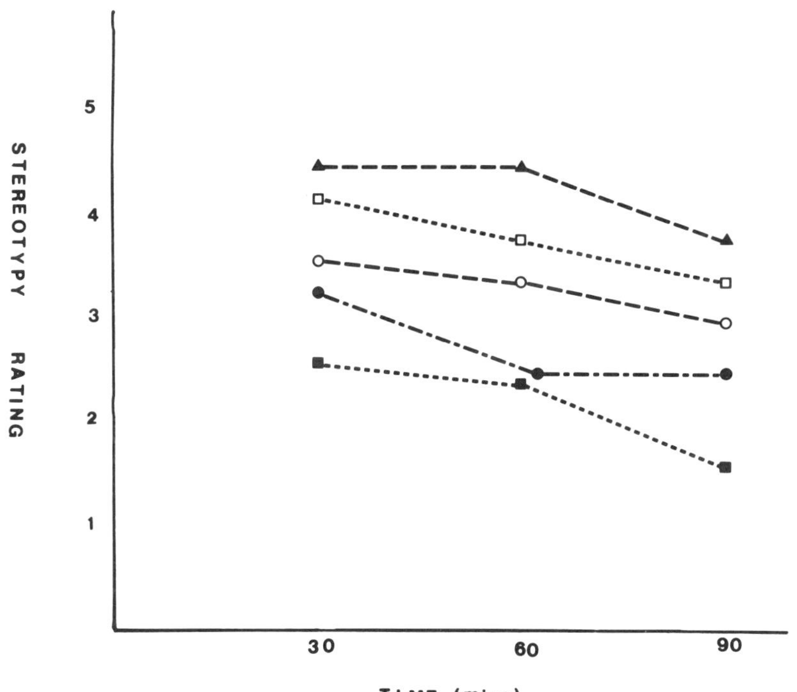

Fig. 6. The effects of apomorphine (5 mg/kg) on stereotypy ratings of rats confined to large or small platforms. The data represent the mean stereotypy ratings at various times after a subcutaneous injection of 5 mg/kg in group cage controls (filled triangles) or animals housed under four different platform conditions: W/A 1.7 NH (open circle), W/A 6.5 NH (filled circles), W/A 1.7 4H (open squares), W/A 6.5 4H (filled squares). There were at least 6 rats per group.

confined compared to the large platform-confined group and that receptor concentration was smallest in the habituated small platform-confined groups. No differences in the affinity of the DA receptors was found. Further experiments are necessary to confirm this finding.

Discussion

The findings on temperature contradict the results of Tufik et al. (1978) who reported that rats confined to small platforms exhibited a supersensitive response to the hypothermic effect of apomorphine. They suggested that REM sleep deprivation induces a supersensitivity of dopaminergic systems, whereas the present findings are consistent with a subsensitivity. However, a closer examination of these results suggests in fact that the subsensitivity may be a consequence of stress induced by the platform technique rather than REM sleep deprivation per se. Firstly, the two naive groups exhibited an equivalent hypothermic response to apomorphine at both time points. The absence of a platform or drug x platform interaction suggests that the conditions of the platform technique are responsible for the observed subsensitivity to the hypothermic effect of apomorphine, rather than REM sleep deprivation. Secondly, only a small platform effect was observed in the habituated groups and this effect reflects only a difference in the relative degree of subsensitivity to apomorphine, which may reflect either a difference in stress or a difference in the degree of REM sleep deprivation.

The data on apomorphine induced stereotypy present a general similarity to the temperature data. However, only the habituated small platform-confined rats exhibited ratings that were significantly lower than those of the group cage control. Again, these results are interpretable as a greater degree of subsensitivity, which as indicated above may reflect either a difference in stress or a difference in the degree of REM sleep deprivation. Nevertheless, the findings are still contradictory to the concept of REM sleep deprivation-induced dopamine supersensitivity (Tufik, 1981).

The down-regulation of DA receptors is generally consistent with the subsensitivity to apomorphine-induced hypothermia and stereotypy in the platform-confined groups. The overall pattern of results is therefore entirely consistent with a large body of literature which has either investigated the platform technique or stress and has uniformly reported a reduced catecholamine sensitivity (Mogilnicka et al., 1980, 1981; Stern and Morgane, 1974; Stone, 1980, 1981). In this context, the findings of Tufik et al. (1978) were inconsistent with this general trend of research.

GENERAL DISCUSSION

Both of the present series of experiments indicate that there was reduced sensitivity to apomorphine-induced behavioural and physiological responses. This subsensitivity was accompanied by a trend for a down-regulation of DA receptors. Thus, it is tempting to suggest that both hyperthyroidism and REM sleep deprivation either directly or indirectly influence the regulation of DA receptors. There is some evidence for an increased release and/or turnover of DA in hyperthyroidism (Strombom et al., 1977; Atterwill, 1981) and there is some evidence for alterations in DA levels in REM sleep-deprived animals (e.g. Ghosh et al., 1976). It appears, therefore, that the subsensitivity to DA agonists induced by hyperthyroidism or REM sleep deprivation may be a response to the effects these treatments have on DA metabolism and turnover. Further studies must be conducted to confirm this suggestion.

One possible explanation of these findings is that REM sleep deprivation may lead to an increased metabolism of apomorphine. However, Harris (1983) failed to find any difference in brain apomorphine uptake in rats confined to platforms. Preliminary studies have indicated that hyperthyroid rats are subsensitive to apomorphine directly injected into the striatum. Thus, decrease in apomorphine metabolism is unlikely to be responsible for the subsensitivity observed in hyperthyroid and REM sleep-deprived rats. The fact that there was a trend for DA receptors to decrease in both studies suggests that a down-regulation of DA receptors mediated the subsensitivity.

Although both treatment conditions clearly indicate a subsensitivity to apomorphine, there were large differences in the time required to produce these effects; hyperthyroidism must be present for rather long periods before the subsensitivity is apparent, while REM sleep deprivation produces an effect within a few days. It is possible that intermittent (in hyperthyroidism) versus continuous release of DA (in REM sleep-deprived rats) may provide an explanation. It should be remembered in this context that the demonstration of down-regulation of DA receptors in rats treated chronically with DA agonist was most easily shown in rats which received nearly continuous administration of the drugs (Nielsen et al., 1983).

A final avenue for further study is the examination of other DA receptors. For example, do DA receptors in the nucleus accumbens exhibit similar changes? It would also be interesting to see if the down-regulation is confined to either D1 or D2 receptors, or involves both subtypes. Such information may eventually lead to a better understanding of schizophrenia, an illness where the DA receptors may be pathologically up-regulated (Cross et al., 1981).

ACKNOWLEDGEMENTS

This research was supported in part by research grants from the Australian Research Grants Scheme and the Flinders University Research Budget. We express our appreciation to Robert Westphalen and Bob Frost for technical assistance.

REFERENCES

Akiyama, K., Sato, M., Kashihara, K. and Otsuki, J., 1982, Lasting changes in high affinity 3H-spiperone binding to the rat striatum and mesolimbic area after chronic methamphetamine administration: Evaluation of dopaminergic and serotonergic receptor components, Biol. Psychiatry, 17:1389-1402.

Algeri, S., Brunello, N. and Vantani, G., 1980, Different adaptive responses by rat striatal dopamine synthetic and receptor mechanisms after repeated treatment with d-amphetamine, methylphenidate and nomifersine, Pharmacol. Res. Commun. 12:675-681.

Atterwill, C. K., 1981, Effect of acute and chronic tri-iodothyroxine (T_3) administration to rats on central 5-HT and dopamine-mediated behavioural responses and related brain biochemistry, Neuropharmacology, 20:131-144.

Beart, P. M., 1982, Multiple dopamine receptors - new vistas, TIPS, March 1982, 100-102.

Borison, R. L., Hitri, A., Klawans, H. L. and Diamond, H. I., 1979, A new animal model for schizophrenia: Behavioural and receptor binding studies, in: "Catecholamines: Basic and Clinical Frontiers," Vol. 1, E. Usden, I. J. Kopin and J. Barchas, eds., Pergamon, New York.

Burt, D. R., Creese, I. and Snyder, S. H., 1977, Antischizophrenic drugs: chronic treatment elevates dopamine receptor binding in brain, Science, 196:326-328.

Cox, B., Ary, M., and Lomax, P., 1976. Dopaminergic involvement in withdrawal hypothermia and thermoregulatory behaviour in morphine-dependent rats, Pharmac. Biochem. Behav., 4:254-262.

Creese, I., Burt, D. R., and Snyder, S. H., 1977, Dopamine receptor binding enhancement accompanies brain-induced behavioural supersensitivity, Science, 197:596-598.

Creese, I., and Iversen, S. D., 1973, Blockage of amphetamine-induced motor stimulation and stereotypy in adult rat following neonatal treatment with 6-hydroxydopamine, Brain Res., 55:369-382.

Creese, I., and Sibley, D. R., 1981, Receptor adaptations to centrally acting drugs, Ann. Rev. Pharmacol. Toxicol., 21:357-391.

Cross, A. J., Crow, T. J., and Owen, F., 1981, 3H-flupenthixol binding in postmortem brains of schizophrenics: Evidence for a selective increase in dopamine D2 receptors, Psychopharmacology, 74:122-124.

Crocker, A. D., and Overstreet, D. H., 1983, Modification of the behavioural effects of halopridol and of dopamine receptor regulation by altered thyroid status, Psychopharmacology, (in press).
Demellweek, C., and Goudie, A. J., 1983, Behavioural tolerance to amphetamine and other psychostimulants: the case for considering behavioural mechanisms, Psychopharmacology, 80:287-307.
Ellinwood, E. K., and Kilbey, M. M., 1980, Fundamental mechanisms underlying altered behaviour following chronic administration of psychomotor stimulants, Biol. Psychiatry, 15:749-758.
Finnegan, K. T., Ricaurte, G., Seiden, L. S., and Schuster, C. R., 1982, Altered sensitivity to d-methylamphetamine, apomorphine and haloperidol in rhesus monkeys depleted of caudate dopamine by repeated administration of d-methylamphetamine, Psychopharmacology, 77:43-52.
Ghosh, P., Hrdina, P., and Ling, G., 1976, Effect of REMS deprivation on striatal dopamine and acetylcholine in rats, Pharmac. Biochem. Behav., 4:401-405.
Harris, P. F., The platform technique of REM sleep deprivation: Increase in stress and decreases in catecholamine receptor sensitivity. Unpublished doctoral dissertation, Flinders University.
Harris, P. F., Overstreet, D. H., and Orbach, J., 1982, Disruption of passive avoidance memory by REM sleep deprivation. Methodological and pharmacological considerations, Pharmac. Biochem. Behav., 17:1119-1121.
Hernandez-Peon, R., Drucker-Colin, R., Del Angel, A., Charez, B., and Seriano, P., 1969, Brain catecholamines and serotonin in rapid sleep deprivation, Physiol. Behav., 4:654-661.
Hruska, R. E., Ludmer, L. M., Pitman, K. I., Derych, M., and Silbergeld, E. K., 1982, Effects of estrogen on striatal dopamine receptor function in male and female rats, Pharmacol. Biochem. Behav., 16:285-291.
Hicks, R. A., Okuda, A., and Thomsen, D., 1977, Depriving rats of REM sleep: The identification of a methodological problem, Amer. J. Psychol., 90:95-102.
Hitzemann, R., Wu, J., Hom, D., and Loh, H., 1979, Acute and chronic amphetamine treatments modulate striatal dopamine receptor binding sites, Psychopharmacology, 72:93-101.
Howlett, D. R., and Nahorski, S. R., 1979, Acute and chronic amphetamine treatments modulate striatal dopamine receptor binding sites, Brain Res., 161: 173-178.
Jackson, D. M., Bailey, R. C., Christie, M. J., Crisp, E. A., and Skeritt, J. H., 1981, Long term d-amphetamine in rats: Lack of change in post-synaptic dopamine receptor sensitivity, Psychopharmacology, 73:276-280.
Joschko, M., Crocker, A. D., and Overstreet, D. H., 1982, The effects of haloperidol and apomorphine on behavioural responses in rats with modified thyroid status, Proceedings of Australasian Society of Clinical and Experimental Pharmacologists.

Klawans, H. L., Goetz, C. L., and Weiner, W. J., 1974, Dopamine receptor site sensitivity in hyperthyroid and hypothyroid guinea pigs, Adv. Neurol., 5:445-501.

Lai, M., Carino, A., and Horita, A., 1981, Chronic treatments with zotepine, thioridazine and haloperidol affect apomorphine-elicited stereotypic behavior and striatal 3H-spiroperidol binding sites in the rat, Psychopharmacology, 75:388-390.

Lake, C. R., and Fann, W. E., 1973, Possible potentiation of haloperidol neurotoxicity in acute hyperthyroidism, Brit. J. Psychiat., 123:523-525.

Lindvall, O., 1979, Dopamine pathways in the rat brain, in: "The Neurobiology of Dopamine," J. Korf and B. H. C. Westerink, eds., Academic Press, London.

Mark, J., Heiner, L., Mandel, P., and Godin, Y., 1969, Norepinephrine turnover in brain and stress reactions in rats during paradoxical sleep deprivation, Life Sci., 8:1085-1093.

McGrath, M., and Cohen, D. P., 1978, REM sleep facilitation of adaptive waking behaviour, Psychol. Bull., 85:24-57.

Mogilnicka, Anpilla, S., Deportee, H. and Langer, S., 1980, Rapid-eye movement sleep deprivation decreases the density of 3H-dihydroalprenalol and 3H-mepramine binding in rat cerebral cortex, Eur. J. Pharmacol., 65:289-292.

Mogilnicka, E., and Pilc, A., 1981, Rapid-eye-movement sleep deprivation inhibits chloridine-induced sedation in rats, Eur. J. Pharmacol., 71:123-126.

Nielsen, E. B., Nielsen, M., Ellison, G., and Braestrup, C., 1980, Decreased spiroperidol and LSD binding in rat brain after continuous amphetamine, Eur. J. Pharmacol., 66:149-154.

Nielsen, E. B., Nielsen, M., and Braestrup, C., 1983, Reduction of 3H-spiroperidol binding in rat striatum and frontal cortex by chronic amphetamine: Dose response time course and role of sustained dopamine release, Psychopharmacology, 81: 81-85.

Overstreet, D. H., and Yamamura, H. I., 1979, Receptor alterations and drug tolerance, Life Sci., 25:1865-1878.

Overstreet, D. H., Crocker, A.D., Evans, C. A., and McIntosh, G. H., 1984, Alterations in the dopaminergic systems and behaviour in rats reared on iodine-deficient diets, Pharmac. Biochem. Behav., (in press).

Owen, F., Baker, H. F., Ridley, R. M., Crose, A. J., and Crow, T. J., 1981, Effect of chronic amphetamine administration on central dopaminergic mechanisms in the vervet, Psychopharmacology, 74:213-216.

Post, R. M., 1981, Central stimulants. Clinical and experimental evidence on tolerance and sensitization, in: "Research advances in alcohol and drug problems," Vol. 6., Y. Israel, F. B. Glaser, H. Kalant, R. E. Popham, W. Schmidt., and R. G. Smart, eds., Plenum, New York.

Quirion, R., Bayork, M. A., Zerbe, R., and Pert, C. B., 1982, Chronic phencyclidine treatment decreases phencyclidine and dopamine receptors in rat brain, Pharmac. Biochem. Behav., 17:69-702.

Ridley, R. M., Baker, H. F., Owen, F., Cross, A. J.,, and Crow, T. J., 1982, Behavioural and biochemical effects of chronic amphetamine treatment in the vervet monkey, Psychopharmacology, 78:245-251.

Riffee, W. H., Wilcox, R. E. Vaughn, D. M., and Smith, R. V., 1982, Dopamine receptor sensitivity after chronic dopamine agonists. Striatal 3H-spiroperidol binding in mice after chronic administration of high doses of apomorphine, 1 N-n-propyl-norapomorphine and dextroamphetamine, Psychopharmacology, 77:146-149.

Robertson, H. A., and Paterson, M. R., 1982, Chronic phencyclidine or amphetamine treatment produces a decrease in 3H-spiroperidol binding in rat striatum and nucleus accumbens, Prog. Neuropsychopharmacol., 5:311-318.

Robinson, T. E., Becker, J. B., and Ramirez, V. D., 1980, Sex differences in amphetamine-elicited rotational behavior and the lateralization of striatal dopamine in rats, Brain Res. Bull., 5:539-546.

Savageau, M. J., and Beatty, W. W., 1981, Gonadectomy and sex differences in the behavioural responses to amphetamine and apomorphine in rats, Pharmac. Biochem. Behav., 14:17-21.

Seeman, P., 1981, Brain Dopamine Receptors, Pharmacological Reviews, Vol. 32, No. 3:229-271.

Siegel, S., 1957, Nonparametric Statistics for the Behavioral Sciences, McGraw-Hill, New York.

Stern, W. C., and Morgane, P. J., 1974, Theoretical view of REM sleep function. Maintenance of catecholamine systems in the central nervous system, Behav. Biol., 11:1-32.

Stone, E. A., 980, Subsensitivity to norepinephrine as a link between adaptation to stress and antidepressant therapy: an hypothesis, Res. Commun. Psychol. Psychiat. Behav., 4:241-255.

Stone, E. A., 1981, Mechanism of stress-induced subsensitivity to norepinephrine, Pharmacol. Biochem. Behav., 14:719-723.

Strombon, U., Svensson, T. H., Jackson, D. M., and Engstron, O., 1977, Hyperthyroidism: Specifically increased response to central Na( ) receptor stimulation and generally increased monoamine turnover in brain, J. Neural. Trunem., 41:73-92.

Tarsy, D., and Baldessarini, R. J., 1974, Behavioral supersensitivity to apomorphine following chronic treatment with drugs which interfere with the synaptic function of catecholamines, Neuropharmacology, 13:927-940.

Tufik, S., 1981, Changes of response to dopaminergic drugs in rats submitted to REM sleep deprivation, Psychopharmacology, 72:257-260.

Tufik, S., Lindsey, C. and Carlini, E., 1978, Does REM sleep deprivation induce a supersensitivity of dopaminergic receptors in the rat brain? Pharmacology, 16:98-105.

Weston, P. F., and Overstreet, D. H., 1976, Does tolerance to low doses of d-amphetamine and l-amphetamine on locomotor activity in rats? Pharmac. Biochem. Behav., 5:645-649.

Wilcox, R. E., Riffee, W. H., Reynolds-Vaughn, R. A., Leamons, B. J., Vaughn, D. A., and Smith, R. V., 1981, Behavioral supersensitivity but decreased striatal (3H)-spiroperidol receptor binding after chronic apomorphine, N-n-propylnorapomorphine and dextroamphetamine administration, *Soc. Neurosci. Abstr.*, 7:803.

Williams, A., and Calne, D. B., 1981, Treatment of Parkinson's Disease, *in*: "Disorders of Movement," A. Barbeau, ed., MP Press Ltd, Lancaster.

CAN ENZYMES RELEASED FROM THE NIGRO-STRIATAL PATHWAY ACT AS NEUROMODULATORS?

Susan A. Greenfield

University Laboratory of Physiology
Parks Road
Oxford OX1 3PT
England

INTRODUCTION

"Neuromodulator" is a term widely used but rarely defined. Transmitters themselves are sometimes described as having a "modulatory" action on neuronal activity: clearly therefore, the term "neuromodulation" implies a mode of intercellular information transfer different from the familiar events occurring at the axonal synapse, following propagation of a presynaptic action potential. Hence it seems that the critical issue is the way a substance modifies neuronal activity, rather than the transmitter-status or otherwise, of the substance itself.

A signal could be considered modulatory when it differs from that of classical synaptic transmission in any of the following ways: its initiation may be unrelated to pre-synaptic neuronal discharge; it may have longer latency or duration of action than synaptic transmission; the target sites of this action may be less restricted.

The substantia nigra is an area of the brain where neuromodulation in this sense is thought to occur. Dopamine (DA) is released not only from axon terminals of nigro-striatal neurons but also from dendrites of these cells, actually within the substantia nigra (Cheramy et al., 1981). Release from dendrites, unlike that from axon terminals, is resistant to blockade of nerve impulse flow by either tetrodotoxin (TTX) (Cheramy et al., 1981) or gamma-butyrolactone (Hefti and Lichtensteiger, 1978). Hence dendritic release could occur in the absence of an action potential.

There are very few dendro-dendritic synapses in the substantia nigra (Wassef et al., 1981): hence DA released from dendrites could be expected to affect many cells indiscriminately, over a relatively long period of time. In a physiological situation, therefore, the inhibitory action of DA on nigrostriatal cells (Bunney et al., 1973), may be modulatory (cf. Hefti and Lichtensteiger, 1978).

Not only DA but proteins also are released within the substantia nigra and caudate nucleus (Greenfield et al., 1983a). The question that we shall consider here is whether the release of certain neuronal proteins indicates that they too are modulators of neuronal activity in the nigro-striatal system.

RELEASE OF PROTEIN

The technique used to measure release of DA from the substantia nigra and caudate nucleus in vivo (Cheramy et al., 1981) is local tissue perfusion via push-pull cannulae (Gaddum, 1961) implanted in the appropriate brain region. Using this technique in an identical preparation (the halothane-anaesthetized cat) it has been possible to demonstrate that proteins are spontaneously released from the caudate nucleus and substantia nigra.

The release of proteins from the substantia nigra can be enhanced by local elevation of concentrations of potassium (via push-pull cannulae). This release of protein in the substantia nigra is not accompanied by any change in release of proteins from the ipsilateral caudate nucleus. In the contralateral caudate nucleus and substantia nigra, however, there is a respective increase and decrease in protein release. Furthermore, the release of at least thirty percent of proteins evoked by potassium in the substantia nigra is calcium dependent (Greenfield et al., 1983a). As yet, only four specific proteins, all enzymes, have been studied: aminopeptidase, lactate dehydrogenase (LDH) acetylcholinesterase (AChE) and non-specific cholinesterase (ChE).

Experiments have been performed on anaesthetized cats, rats and rabbits with push-pull cannulae implanted in the caudate nucleus and substantia nigra. The aim of these studies has been to see how high concentrations of potassium in the substantia nigra affect release of the enzymes from the substantia nigra and ipsilateral caudate nucleus: secondly, to discover whether release of the enzymes from the caudate nucleus and substantia nigra is associated with the dopaminergic nigro-striatal pathway. This second question has been tackled in two ways, namely by observing the effects on enzyme release of the DA agonist amphetamine and by lesioning the nigro-striatal pathway with the neurotoxin, 6-hydroxydopamine (6-OHDA).

RELEASE OF AMINOPEPTIDASE

Both the substantia nigra and caudate nucleus are rich in neuroactive peptides (Emson, 1979) and aminopeptidases (Shaw, 1978). However a high proportion of this group of enzymes is associated not with neurons but with the endothelial cells of blood vessels (Shaw, 1978). Using standard assay procedures on tissue homogenates, it is very difficult to establish which aminopeptidases are present in neurones, and which are contained in the vasculature.

However, it is conceivable that the concentration of enzymes associated with neurons could be modified by changing nervous activity, whereas the aminopeptidases in the vasculature would be unaffected. It might be possible thus to identify neuronal aminopeptidases. Hence, we have investigated the possibility that aminopeptidase might be released from the caudate nucleus and substantia nigra in vivo, whilst changing the activity of the nigro-striatal system by local application of high concentrations of potassium to the substantia nigra.

In perfusates of both the caudate nucleus and substantia nigra, an isoenzyme of aminopeptidase is present, with fast electrophoretic mobility. Although this isoenzyme has a similar mobility to that in plasma related to neuronal activity, local application of potassium to the substantia nigra led to an increase in release of this enzyme from the substantia nigra of eighty-four percent above resting levels; this effect was accompanied by a decrease in release from the caudate nucleus of twenty-six percent (Greenfield et al., 1983b).

Following a 6-OHDA lesion of the nigro-striatal pathway, the potassium-evoked decrease in aminopeptidase release from the caudate nucleus was abolished, while both the spontaneous and evoked release from the substantia nigra was reduced by approximately fifty percent (Greenfield et al., 1983b).

Hence we concluded that approximately twenty-six percent of the enzyme in the caudate nucleus perfusates originated either presynaptically from DA-containing terminals of nigro-striatal neurons, or postsynaptically from cells upon which these terminals directly synapsed. About fifty percent of the fast-mobility aminopeptidase released from nigral cells was derived from the somata and dendrites of nigro-striatal neurons.

When amphetamine was added to the perfusate of a push-pull cannula implanted in the substantia nigra, another form of aminopeptidase was observed. This isoenzyme had a much slower electrophoretic mobility, was not released spontaneously (Greenfield and Shaw, 1982) and its release was not evoked by potassium (Greenfield et al., 1983b). Following infusion of amphetamine into one substantia nigra, however, there was both local release and release from the contralateral caudate nucleus and substantia nigra.

It is possible that this isoenzyzme is stored exclusively in DA-containing nigral neurons and that its release is not directly impulse-related but rather associated with events at either the DA "autoreceptor" (Groves et al., 1975) located on the somata and dendrites of nigro-striatal neurons or at presynaptic DA receptors on afferent axon terminals in the substantia nigra (Reubi et al., 1977).

RELEASE OF LACTATE DEHYDROGENASE

Lactate dehydrogenase (LDH) is generally known as a soluble enzyme found in cytoplasm. Its activity has been initially monitored in perfusates of push-pull cannulae to serve as an index of cell damage (Greenfield et al., 1980; 1983a) or plasma contamination (Greenfield and Shaw, 1982). For example, application of amphetamine to the substantia nigra did not affect LDH levels in perfusates of either substantia nigra or caudate nucleus (Greenfield and Shaw, 1982).

However, in the anaesthetized rat, addition of potassium to the substantia nigra led to a local increase in release of LDH and a decrease in the enzyme amounts released from the caudate nucleus (Greenfield et al., 1983b).

Lesions with 6-ODHA of the nigro-striatal tract decrease the spontaneous release of LDH from the caudate nucleus: this decreased release from the caudate nucleus of lesioned animals corresponded very closely to the potassium-evoked decrease in the caudate nucleus of control animals, i.e. about sixty-five percent of control resting levels (Greenfield et al., 1983b). It would seem then that LDH is released from a subcellular compartment of the caudate nucleus that is derived from DA-containing nerve terminals, and related to impulse traffic.

In the substantia nigra, a 6-OHDA lesion of the nigro-striatal tract led to a decrease in spontaneous release of LDH and a partial reduction (about fifty percent) of the potassium-evoked increase. As seen already in the case of the fast-mobility aminopeptidase, LDH in the substantia nigra therefore appears partly derived from DA-containing nigro-striatal neurons. In fact, there was a very close correlation between LDH release and that of the fast mobility aminopeptidase. This parallel in release of the two enzymes occurred in perfusates of both substantia nigra and caudate nucleus, in both control and 6-OHDA lesioned animals (Greenfield et al., 1983b).

RELEASE OF ACETYLCHOLINESTERASE AND NON-SPECIFIC CHOLINESTERASE

Although normally considered to be membrane-bound enzymes (Silver, 1974) both acetylcholinesterase (AChE) and non-specific cholinesterase (ChE) also exist in soluble forms (Chubb and Smith, 1975a). By means of polyacrylamide gel electrophoresis, it is possible to distinguish five separate soluble forms of AChE (Chubb and Smith, 1975a) of which the isoenzyme with slowest mobility, "Isoenzyme 5", is released from both peripheral (Chubb and Smith, 1975b) and CNS (Chubb et al., 1976) tissue.

When either the caudate nucleus or substantia nigra is stimulated electrically, there is a marked increase in AChE, but not ChE, release into cisternal cerebrospinal fluid (CSF) (Greenfield and Smith 1979). Hence activity of the nigro-striatal system would appear in some way implicated in the release of AChE into CSF. It was therefore important to see whether AChE and ChE were released directly from neurons in the caudate nucleus and substantia nigra.

When these structures were perfused using push-pull cannulae, spontaneous AChE and ChE release was detectable (Greenfield et al., 1980; 1983a). Following application of potassium locally to one substantia nigra, there was no change in ChE release from either the caudate nuclei or substantiae nigrae. Release of AChe, however, was increased both locally and from the contralateral caudate nucleus: release from the ipsilateral caudate nucleus and contralateral substantia nigra was reduced (Greenfield et al., 1980).

Curiously enough, these complex potassium-induced changes in AChE release in the two nigro-striatal systems are difficult to relate to changes in the activity of possible cholinergic cells in the substantia nigra. First, there is very little acetylcholine (ACh) and choline acetyltransferase (CAT) in the substantia nigra (Silver, 1974; Fonnum et al., 1974). Secondly, although nigro-striatal neurons are cholinoceptive (Walker et al., 1976; Kimura et al., 1981) no cholinergic synapses have been identified (Kimura et al., 1981; Levey et al., 1983). Thirdly, release of AChE from the substantia nigra is not affected by either activation (Cuello et al., 1981) or blockade (Greenfield and Smith, 1979) of cholinergic receptors.

Instead, nigral AChE appears to be closely associated with DA-containing nigro-striatal neurons: Butcher and Woolf (1982) have shown that virtually every nigro-striatal neuron that contains DA, also contains AChE. Furthermore, twelve percent of AChE in striatal homogenates appears attributable to pre-synaptic dopaminergic afferent axon terminals of nigrostriatal cells (Lehmann and Fibiger, 1978). It was of interest, therefore, to see to what extent releasable AChE, and ChE, could be attributed to the DA-containing nigro-striatal system.

Following a 6-OHDA lesion of the nigro-striatal tract, there was a decrease in spontaneous release of both AChE and ChE from the substantia nigra and caudate nucleus (Greenfield et al., 1983b). In the caudate nucleus, the 6-OHDA-induced decrease in AChE release (to approximately thirty percent of control levels) corresponded to that seen in the caudate nucleus of control animals following potassium application to the substantia nigra. These observations suggest that about seventy percent of AChE released from the caudate nucleus originates presynaptically from DA-containing nigro-striatal terminals. This figure is far higher than the twelve percent estimate derived from biochemical studies (Lehmann and Fibiger, 1978), and thus illustrates that the distribution between membrane bound and soluble, releasable AChE is not necessarily parallel.

In the substantia nigra the decrease in spontaneous AChE release, following a 6-OHDA lesion, corresponded very closely to the amount of DA itself depleted. Hence, it seems that AChE released from nigral neurons is derived exclusively from DA-containing nigro-striatal cells. This idea was further corroborated by the finding that potassium application to the substantia nigra of lesioned animals no longer evoked release of AChE (Greenfield et al., 1983a).

The spontaneous release of ChE from the caudate nucleus and substantia nigra also fell markedly, following a 6-OHDA lesion. This result implied that, even though release of ChE could not be modified by depolarising concentrations of potassium, release of ChE occurred preferentially from nigro-striatal neurons (Greenfield et al., (1983b).

When amphetamine is infused into one substantia nigra, release of ChE is again unchanged in both the substantia nigra and caudate nucleus. AChE release, however, increases in the substantia nigra and decreases in the ipsilateral caudate nucleus (Greenfield and Shaw, 1982). These modifications in release of AChE from the nigro-striatal pathway are reminiscent of those seen in release of DA following nigral application of amphetamine (Leviel et al., 1979).

Further studies performed on the release of AChE from nigro-striatal cells again indicate a close parallellism between release of AChE and dendritic release of DA. Release neither of DA (Cheramy et al., 1981) nor of AChE (Greenfield and Chesselet, unpublished observations) is blocked by TTX: furthermore, potassium-evoked release of both DA (Cheramy et al., 1981) and AChE (Greenfield et al., 1983) is calcium-dependent.

SIGNIFICANCE OF RELEASE OF ENZYMES

Table I summarizes the effect of the various treatments on release of enzymes from (A) the caudate nucleus and (B) the substantia nigra. Do these results indicate a possible neuromodulatory function for released enzymes, and if so, what could it be?

Aminopeptidase

The fact that the fast mobility form of aminopeptidase is released spontaneously and can in part be modified by depolarization and lesions of the nigro-striatal pathway, suggests that at least some aminopeptidase is released in a fashion linked to the firing rate of dopaminergic nigro-striatal neurons. In the substantia nigra, there are a wealth of potential peptide substrates for aminopeptidase; indeed Substance P - reactive boutons have actually been shown to form direct contact with dendrites and somata of nigro-striatal neurons (Somogyi et al., 1982). It is easy to imagine, therefore, that an extracellular role for the fast mobility aminopeptidase might be, in the substantia nigra and indeed in the caudate nucleus, to break down peptides.

Unlike the fast mobility aminopeptidase, the release of a slow mobility form of the enzyme appears totally unrelated to neuronal activity. There is no spontaneous nor potassium-evoked release. In addition, amphetamine, which inhibits nigro-striatal cell firing (Groves et al., 1975) actually leads to release of the enzyme, not only locally in the substantia nigra, but in the contralateral caudate nucleus and substantia nigra as well. Although we do not know whether the slow mobility aminopeptidase is derived from DA-containing nigro-striatal cells, its release is in some way influenced by activation of DA receptors and/or relatively long term changes in the availability of DA: amphetamine might enhance extracellular DA levels more effectively than potassium-induced depolarisation.

Considering the above, the slow mobility form of aminopeptidase could be a neuromodulator. Its actual function in this capacity is, at the moment, impossible to imagine. Needless to say, it need not necessarily be involved in modification of peptide transmission in either the caudate nucleus or substantia nigra. As we have seen, released AChE from nigro-striatal neurons appears unrelated to cholinergic transmission. By analogy, it is possible that other released enzymes do not solely have a catalytic function.

Table I.

	Spont.	K^+SN	AMPH.,SN	6-OHDA SPONT	6-OHDA + K^+, SN
(A) Caudate Nucleus					
Aminopep. Fast	Yes	Decrease	No effect	No effect	No effect
Aminopep. Slow	No	No effect	No effect	No effect	No effect
LDH	Yes	Decrease	No effect	Decrease	No effect
AChE	Yes	Decrease	Decrease	Decrease .	No effect
ChE	Yes	No effect	No effect	Decrease	No effect
(B) Substantia Nigra					
Aminopep. Fast	Yes	Increase	No effect	Decrease	Partial block
Aminopep.	No	No effect	Evoked release	No effect	No effect
LDH	Yes	Increase	No effect	Decrease	Partial block
AChE	Yes	Increase	Increase	Decrease	No effect
ChE	Yes	No effect	No effect	Decrease	No effect

Lactate Dehydrogenase

The finding that LDH release could be modified in both the caudate nucleus and substantia nigra by high concentrations of nigral potassium, suggests that it may not merely be a marker for cell damage. The close correlation between LDH release and that of the fast mobility aminopeptidase, in all conditions, implies that the two enzymes may be co-released or even co-stored. As yet, no studies have been performed investigating the possibility of an extracellular function of LDH released from either DA- or non-DA- containing neurons in the substantia nigra and caudate nucleus.

Non-Specific Cholinesterase

The release of ChE seems to be largely derived from striatal afferent terminals and cell bodies and dendrites of DA-containing nigro-striatal cells. However, in none of the species used (cat, rabbit and rat) and following no treatment, could this release be modified. If released ChE has a function at all, therefore, it would not be a modulatory response to specific signals. In peripheral nerves, it has been proposed that protein may be released from axons as "trophic material" to maintain the integrity of the myelin sheath (Singer, 1969; Singer and Steinberg, 1972; Hines and Garwood, 1977). Perhaps extracellular ChE could perform a similar function on myelinated axons in the substantia nigra or an analogous role to maintain the integrity of the membrane of nigro-striatal neurons themselves.

Acetylcholinesterase

The significance of release of AChE from nigro-striatal terminals in the caudate nucleus is hard to interpret. The fact that a far higher proportion of releasable enzyme is present presynaptically, compared to membrane-bound, suggests that the releasable enzyme is not simply present to inactivate ACh binding to striatal presynaptic receptors (Giorguieff et al., 1977). However, it may well be the case that AChE released from striatal axons reflects the firing rate of nigro-striatal neurons. Amphetamine, locally applied to the substantia nigra, depresses the firing rate of nigro-striatal neurons (Groves et al., 1975) and indeed causes a decrease in AChE release from the ipsilateral caudate nucleus. Yet, any further understanding of a non-cholinergic role for AChE released in the caudate nucleus is hampered by the fact that this structure is very rich in cholinergic cells (Dray, 1979).

In the substantia nigra, however, there is no convincing evidence for the existence of cholinergic transmission. Furthermore, the release of AChE from the substantia nigra is straightforward to study, since it appears to originate exclusively from DA-containing nigro-striatal cells and to have characteristics of release similar to that of dendritic DA. Indeed, AChE is actually stored in the smooth endoplasmic reticulum of nigro-striatal cell dendrites (Liesli et al., 1980) as is DA (Mercer et al., 1979).

A further similarity between release of DA and AChE in the

substantia nigra is that neither appear to be related to firing rate. Not only is release of neither DA nor AChE blocked by TTX, but there appears to be no discernible relationship between the action a transmitter has on the firing of nigro-striatal cells and the effect that the same transmitter has on DA or AChE release. For example, serotonin and ACh both enhance dendritic DA release (Glowinski and Cheramy, 1981) yet the former has a depressant effect on the activity of the cells (Dray et al., 1976) whereas the latter is excitatory (Walker et al., 1976). A similar problem arises regarding AChE release from the substantia nigra. As we have already seen, amphetamine depresses nigrostriatal neuron activity, yet enhances the release of nigral AChE.

Since storage sites in, and release of AChE from, nigro-striatal cell dendrites, so closely resemble those of DA, the same criteria for considering that nigral DA could act as a neuromodulater (see Introduction) would also apply to AChE. Clearly, the first step in testing this hypothesis is to see whether extracellular AChE has any effect on the activity of nigral neurons.

ACTION OF ACETYLCHOLINESTERASE

We have attempted to mimic a raised concentration of extracellular AChE by introducing exogenous AChE into the substantia nigra. The effect of exogenous enzyme on the activity of the nigro-striatal pathway has then been monitored in two ways: by extracellular recording and by using a behavioural model for detecting a disparity between the activity of the two nigro-striatal pathways, i.e. the monitoring of circling behaviour.

Electrophysiological Studies

We have found that exogenous AChE inhibits the firing rate of identified nigro-striatal cells (Greenfield et al., 1981). Butyrylcholinesterase, even in higher concentrations, is without effect. It would appear then, that the inhibitory action of AChE is not induced indirectly by the hydrolysis of ACh. Another parallel can, however, be drawn between nigral AChE and DA, since DA also depresses nigro-striatal cell activity (Bunney et al., 1973). Although we do not as yet know the mechanisms by which AChE could inhibit nigrostriatal cell firing, the enzyme, once released, seems to have a physiological significance.

Behavioural Studies

Following a unilateral 6-OHDA lesion of the nigro-striatal pathway, animals turn in a direction away from the intact side (contraversive rotation) (Ungerstedt, 1971); hence circling behaviour can be regarded as indicative of an imbalance in the functional amount of terminal DA in the two striata (cf. Pycock, 1980).

Following a single injection of AChE into the substantia nigra of unlesioned rats, contraversive rotation occurred in the presence of a systemic amphetamine challenge. Butyrylcholinesterase was without effect (Greenfield et al., 1983c).

Two interesting observations arose from this study. First, a purified enzyme preparation (Chubb et al., 1980) containing only "Isoenzyme 5" (see above) was far more potent in eliciting circling behaviour, than commercial enzyme containing several forms of AChE and which have a far higher overall activity towards ACh. Secondly, the AChE-induced circling persisted for up to three weeks following a single injection. During this period systemic administration of apomorphine induced transient ipsiversive rotation, consistent with a reduction in DA receptor sensitivity (Ungerstedt, 1971) in the striatum ipsilateral to the AChE-treated substantia nigra (Greenfield et al., 1983c). Hence it seems that AChE released in the substantia nigra has an action independent of ACh hydrolysis: this action may be modulatory and may influence the activity of the dopaminergic nigrostriatal system.

CONCLUSIONS

In all the studies described here, enzyme release has been investigated from the caudate nucleus and substantia nigra. However, all manipulations of neuronal activity have been either locally in the substantia nigra, or in the dopaminergic nigro-striatal tract. Conclusions concerning release of enzymes from the caudate nucleus have therefore been limited to those released either pre- or post-synaptically from afferent DA terminals or their striatal target cells. To a greater or lesser extent, however, all four enzymes studied are also released from striatal cells that are not directly involved with striatal dopaminergic transmitter systems. Any understanding of these phenomena awaits experiments that manipulate, locally, striatal cell activity and that identify the non-DA cell types from which each released enzyme originates.

Nevertheless, aminopeptidase, LDH and AChE released from the caudate nucleus are all modified when the somata of nigro-striatal neurons are depolarized. Whether or not these enzymes in the extracellular space subsequently have any function, modulatory or otherwise, remains to be tested.

In the substantia nigra, the study of neuromodulation has proved easier due to the extensive information already available concerning dendritic DA release. As seen already for nigral DA release (Introduction) any substance released from nigro-striatal cells in the substantia nigra could already be regarded as a good candidate for neuromodulator, given the absence of DA axon terminals and collaterals, and the presence of dendro-dendritic synapses (Wassef et al., 1981).

Of the four enzymes studied, AChE has proved to be the most promising to pursue as a neuromodulator in the substantia nigra, since it is released exclusively from nigro-striatal cells independently of action potentials but in response to specific signals. In additon, the finding that AChE can exert an action on the activity of the nigro-striatal system strongly suggests that its release has functional significance.

In general, enzymes released into the extracellular space could be neuromodulators, due to their relatively large size and subsequent slow rates of diffusion and degradation. The studies on enzyme release reported here at least clarify the factors relevant to such a function: i.e. the part of the cell (terminals or dendrites) from which release occurs; the relationship between impulse traffic and release; the co-storage of transmitter contained in the neurons from which the enzyme is released; the particular form or forms of the enzyme which are releasable by different agents; the nature of possible targets for released enzyme. The studies on release and action of AChE suggest, in addition, that just as conventional transmitters can function unconventionally as neuromodulators, so can enzymes play a modulatory role independent of their normal enzymatic action.

REFERENCES

Bunney, B. S., Aghajanian, G. K., and Roth, R. H., 1973, Comparison of effects of L-dopa, amphetamine and apomorphine on firing rate of rat dopaminergic cells, Nature, New Biol., 24:123.

Butcher, L. L. and Woolf, N. J., 1982, Monoaminergic-cholinergic relationships in the chemical communication matrix of the substantia nigra and neostriatum, Brain Res. Bull., 9:475.

Chéramy, A., Leviel, V., and Glowinski, J., 1981, Dendritic release of dopamine in the substantia nigra, Nature, 289:537.

Chubb, I. W., and Smith, A. D., 1975a, Isoenzymes of soluble and membrane-bound acetylcholinesterase in bovine splanchnic nerve and adrenal medulla, Proc. R. Soc. B., 191:245.

Chubb, I. W., and Smith, A. D., 1975b, Release of acetyl-cholinesterase into the perfusate from the ox adrenal gland, Proc. R. B., 191:263.

Chubb, I. W., Goodman, S., and Smith, A. D., 1976, Is acetyl-cholinesterase secreted from central neurons into the cerebrospinal fluid, Neuroscience, 1:57.

Chubb, I. W., Hodgson, A. J., and White, G. H., 1980, Acetylcholin-esterase hydrolyses substance P, Neuroscience, 5:2065.

Cuello, A. C., Romero, E., and Smith, A. C., 1981, In vitro release of acetylcholinesterase from the rat substantia nigra, J. Physiol., Lond., 312:14P.

Dray, A., Gonge, T. J. Oakley, N. R., and Tanner, T., 1976, Evidence for the existence of a raphe projection to the substantia nigra in rat, Brain Res., 113:45.

Dray, A., 1979, The striatum and the substantia nigra: a commentary on their relationships, Neuroscience, 4:1407.

Emson, P. C., 1979, Peptides as neurotransmitter candidates in the mammalian CNS, Prog. Neurobiol., 13:61.

Fonnum, F., Grofova, I., Rinvik, E., Storm-Mathisen, J. and Walberg, F., 1974, Origin and distribution of glutamate decarboxylase in substantia nigra of the cat, Brain Res., 71:77.

Gaddum, J. H., 1961, Push-pull cannulae, J. Physiol. Lond., 155:1P.

Giorguieff, M. F., Le Foch, M. L., Glowinski, J., and Besson, M. J., 1977, Involvement of cholinergic presynaptic receptors of nicotinic and muscarinic types in the control of the spontaneous release of dopamine from striatal dopaminergic terminals in the rat, J. Pharmacol. Exp. Ther., 200:535.

Glowinski, J. and Chéramy, A., 981, Dendritic release of dopamine: its role in the substantia nigra, in: "Chemical Neurotransmission, 75 Years," L. Stjarne, P. Hedqvist, H. Lagercrantz, and A. Wennmalm, eds., Academic Press.

Greenfield, S. A., Chéramy, A., Leviel, V., and Glowinski, J., 1980, In vivo release of acetylcholinesterase in the cat substantiae nigrae and caudate nuclei, Nature, 284:355.

Greenfield, S. A., Stein, J. F., Hodgson, A. J., and Chubb, I. W., 1981, Depression of nigral pars compacta cell discharge by exogenous acetylcholinesterase, Neuroscience, 6:2287.

Greenfield, S. A., and Shaw, S. G., 1982, Amphetamine-evoked release of acetylcholinesterase and aminopeptidase, in vivo, Neuroscience, 7:2883.

Greenfield, S. A., Chéramy, A., and Glowinski, J., 1983a, Evoked release of proteins from central neurons in vivo, J. Neurochem., 40:1048.

Greenfield, S. A., Grunewald, R. A., Foley, P., and Shaw, S. G., 1983b, Origin of various enzymes released from the substantia nigra and caudate nucleus: effects of 6-hydroxydopamine lesions of the nigro-striatal pathway, J. Comp. Neurol., 214:87.

Greenfield, S. A., Chubb, I. W., Grunewald, R. A., Henderson, Z., May, J., Portnoy, S., Weston, J., and Wright, M. C., 1983c, A non-cholinergic function for acetylcholinesterase in the substantia nigra: behavioural evidence, Exptl. Brain Res., (In press).

Groves, P. M., Wilson, C. J., Young, G. J., and Rebec., G. V., 1975, Self-inhibition by dopaminergic neurons, Science, 190:522.

Hefti, F., and Lichtensteiger, W., 1978, Subcellular distribution of dopamine in substantia nigra of the rat brain: effects of alphabutyrolactone and destruction of noradrenergic afferents suggest formation of particles from dendrites, J. Neurochem., 30:1217.

Hines, J. F., and Garwood, M. M., 1977, Release of protein from axons during rapid axonal transport: an in vitro preparation, Brain Res., 125:141.

Kimura, H., McGeer, P. L., Peng, J. H., and McGeer, E. G., 1981, The central cholinergic system studied by choline acetyltransferase immunohistochemistry in the cat, J. Comp. Neurol., 200:157.

Lehmann, J., and Fibiger, H. C., 1978, Acetylcholinesterase in substantia nigra and caudate-putamen: properties and localization in dopaminergic neurons, J. Neurochem., 30:615.

Leviel, V., Chéramy, A., and Glowinski, J., 1979, Role of dendritic release of dopamine in the reciprocal control of the two nigrostriatal dopaminergic pathways, Nature, 280:236.

Levey, A. I., Wainer, B. H., Mufson, E. J., and Mesulam, M. -M., 1983, Co-localization of acetylcholinesterase and choline acetyltransferase in the rat cerebrum, Neuroscience, 9:9.

Liesli, P., Panula, P., and Rechardt, L., 1980, Ultrastructural localization of acetylcholinesterase activity in primary cultures of rat substantia nigra, Histochemistry, 70:7.

Mercer, L., Del Fiacco, M., and Cuello, A. C., 1979, The smooth endoplasmic reticulum as a possible storage site for dendritic dopamine in substantia nigra neurons, Experimentia, 25:101.

Pycock, C. J., 1980, Turning behaviour in animals, Neuroscience, 5:461.

Reubi, J. C., Iversen, L. L., and Hessell, T. M., 1977, Dopamine selectively increases GABA release from slices of rat substantia nigra in vitro, Nature, 268,652.

Shaw, S. G., 1978, D. Phil. Thesis, Oxford University.

Silver, A., 1974, "The biology of the cholinesterases," Elsevier, Amsterdam.

Singer, M., 1969, Penetration of labelled amino acids into the peripheral nerve fibre from surrounding body fluids, in: "Growth of the Nervous System," E. G. Wolstenholme and M. O'Connor, eds., Churchill, London.

Singer, M., and Steinberg, M. C., 1972, Wallerian degeneration: a reevaluation based on transected and colchicine poisoned nerves in amphibian tritures, Am. J. Anat., 133:51.

Somogyi, A. D., and Bolam, J. P., 1982, Synaptic connections of substance P-immunoreactive nerve terminals in the substantia nigra of the rat, Cell Tissue Res., 223:469.

Ungerstedt, U., 1971, Post-synaptic supersensitivity after 6-hydroxy-dopamine induced degeneration of the nigro-striatal dopamine system, Acta Phys. Scand. Suppl., 367:69.

Walker, R. J., Kemp, J. A., Yajima, H., Kitagawa, K., and Woodruff, G. N., 1976, The action of Substance P on mesenphalic reticular and substantia nigral neurons of the rat, Experientia, 32:214.

Wassef, M., Berod, A., Sotelo, C., 1981, Dopaminergic dendrites in the pars reticulata of the rat substantia nigra and their striatal input. Combined immunocytochemical localization of tyrosine hydroxylase and anterograde degeneration, Neuroscience, 6:2125.

DISEASES OF THE BASAL GANGLIA

Keith Bradley

St Andrew's Hospital and Department of Anatomy
University of Melbourne
Victoria, Australia

PARKINSON'S DISEASE

Clinical Signs

There are various conditions in which involuntary movements develop and in which pathological changes are present in the basal ganglia. Some of these conditions, and especially paralysis agitans, have been the subject of intense study during the past two decades. Most readers will be familiar with that curious physician James Parkinson, a medical practitioner who lived near London. He was a rather crusty sort of fellow, who wrote sometimes under assumed names to the newspapers, and on one occasion he questioned the Lords of the Admiralty and their imbibing of alcohol, since he thought that some of their decisions were related to that and nothing else. James Parkinson's (1817) essay, on the shaking palsy, provided an account of the symptoms to which little has been added until recently. He stated that the difference between the tremor at rest and the tremor on movement had been noted by various workers over the centuries. The clinical picture combines varying degrees of rigidity and tremor and a generalised attitude of flexion. The head is bowed on the shoulders, the trunk inclined forwards, the arms bent at the elbows and the fingers at the knuckles; which makes the average Parkinsonian recognisable on sight (Figure 1).

Parkinson's own definition included the following:-
"Involuntary tremulous motion, with lessened muscular power, in parts not in action and even when supported; with a propensity to bend the trunk forward, and to pass from a walking to a running pace; the senses and intellect being uninjured."

He did not refer to rigidity, but the posture in flexion was clearly a manifestation of the rigidity of which he was aware. The first inroads of the illness are slight and nearly imperceptible and the patient can hardly recollect the time of onset. A valuable early symptom is an infrequent blinking of the eyelids. When seated there is often noticed a general poverty of movement, maintaining a position for an unusual length of time. It was Kinnear Wilson, I believe, who reported that at a boring medical meeting another doctor in front of him did not move, whereas the rest of the audience nearby were crossing their legs, folding their arms and moving restlessly. Occasionally this gentleman just seemed to stamp his heel, but nothing else. As he walked out to morning tea, it was realised he had early manifestations of Parkinson's disease.

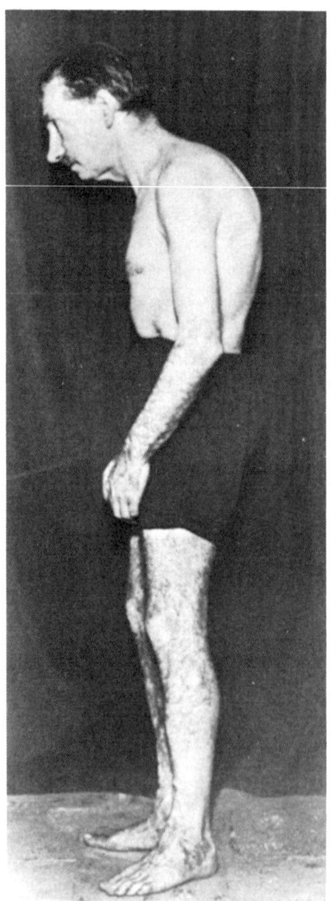

Fig. 1. Paralysis agitans.
Note the abnormal posture.

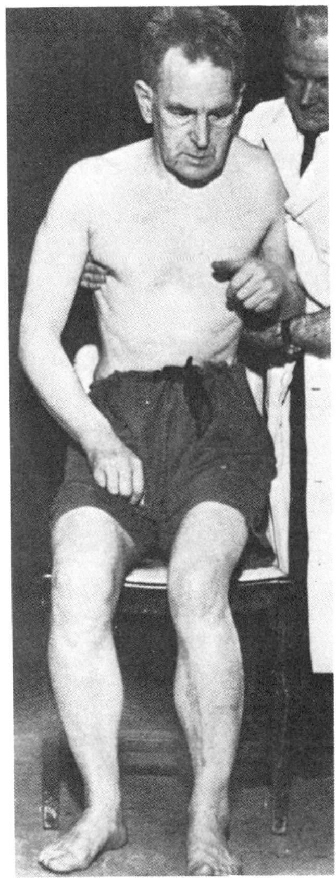
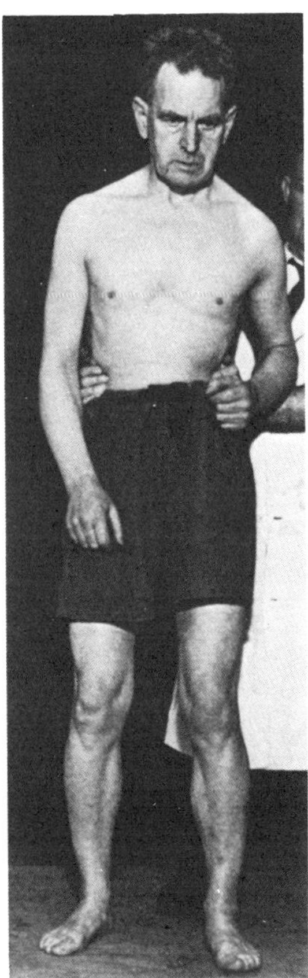

Fig. 2. Paralysis agitans.
Poverty and slowness of movement are present.

The condition changes when there is an added trembling in one of the hands or arms. The patient often believes that this was the first sign, whereas an immobility of the features is usually an earlier sign. Slowness of movement and rigidity usually precede the visible shaking (Figure 2). The early manifestations are one-sided, either in an arm or a leg, and most frequently in the arm. Stiffness and tremor, once established, gradually increase and spread. If an arm is attacked, eventually the leg and foot on the same side will be involved, and sooner or later the posture when walking is less upright. The classical triad embraces tremor, rigidity and the flexion attitude.

Tremor. The fine alternating movements of a few oscillations per second usually affect small muscle groups and, in particular, the fingers, the lips, the tongue and the larynx. The larger muscle groups are more likely to be the site of rigidity. The tremor is present during rest and disappears on purposeful movement.

Akinesia. This symptom is expressed as a general poverty of movement and is evident in all muscle actions as in writing and speaking. The patient often sits immobile and seldom crosses his legs or folds his arms. It does not depend on rigidity and is often quite noticeable when the latter is not present.

Rigidity. This increase of muscle tone is common in an advanced case. Resistance to passive movement is widespread and is especially evident in the larger joints. A patient, commonly with a kyphosis and forward position of the head, will lie for hours in the supine position without a pillow.

Bradykinesia: The slowness of movement in the Parkinsonian is quite evident and one observes that the patient is "slow off the mark". I have seen this - for example - with the unfortunate subject who is waiting at a street crossing, but cannot get off the mark quickly enough to move across before the lights change. There are often two speeds - full stop or full steam ahead. The shuffling gait, the forward flexion and the attempt to overcome the centre of gravity are all clearly shown in these cases. One found in the outpatient clinic that the best way to get them to step off was to put some object in front of their foot. They would step over it and continue walking. Others said they needed some music to keep them going. If you ask a Parkinson's patient to perform certain actions, such as touching their nose, quite often they will not do it or they will take a long time to decide to perform the movement. Alternatively, if you tickle their nose they will immediately touch it. On being asked why they did not touch their nose when asked, one or two patients have replied "There has to be a reason!".

Attitude. In erect position the trunk is bent forward giving an impression of overhanging. The arms are usually adducted and flexed at the elbow to bring the hands in front of the body. The over-hanging forward attitude causes a patient when walking to move forwards with a series of quick short steps; in an attempt to get his limbs under the centre of gravity he may break into a little run, but when asked to halt, he has difficulty in pulling up. Associated movements are abnormal. It is commonly noted that the Parkinsonian does not swing his arms when walking and if asked to do so has great difficulty in - for example - producing a swinging movement of an affected arm. He finds that it is easier not to swing his arms.

Clinical Experiences

My first experience in this area goes back quite some time, when as a neurosurgeon, I was first asked to try to control some of the manifestations of diseases of the basal ganglie. The first case I treated was a lady who had a marked tremor of the left hand and manifestations of typical Parkinson's disease.

We heard that Irving Cooper, at St. Barnabas's Hospital in New York, had observed a change in the manifestations of Parkinson's disease when he had inadvertantly ligated some of the blood supply to the globus pallidus. We decided to make a lesion of the globus pallidus, so we set out to make our own atlas of the region with the guidance of Sir Sydney Sunderland, although it was not published, since in the meantime Schaltenbrand's beautiful atlas appeared.

In this lady, we used a small device made in our workshop to set the angles of a tiny stainless steel tube which we introduced into a small trefine opening in the contralateral side of the skull. We aimed for the globus pallidus, using the third ventricle as a guide, and then introduced two drops of procaine. Although we had been told this was the right place for the lesion, we were amazed that the tremor stopped immediately. We then introduced a few drops of absolute alcohol, and this was probably the best result we ever had, despite the sophisticated sterotactic equipment we used in later cases.

A great deal has been written about the pathological changes in the basal ganglia in Parkinson's disease, and the loss of cells in the substantia nigra, with Lewy body inclusions, is characteristic of paralysis agitans.

A young man who died in 1978 and whose favourite mixture was cocaine and a derivative of Demerol was found to have lesions in his brain typical of aged Parkinson's disease. The cells of the substantia nigra were largely destroyed and there were Lewy bodies in the brain. Recently several young heroin users developed severe Parkinsonism, and it was learned that their symptoms had been caused by the same by-product of Demerol. This drug apparently specifically knocks out the nigro-striatal dopamine system.

Following the incidence of drug-induced Parkinson's disease in California, the National Institute of Mental Health scientists succeeded in producing Parkinson's disease in monkeys by giving them a compound related to Demerol. They stated that the monkeys "have all the major clinical features of Parkinson's disease in humans". This discovery will lead to a great deal of research allowing for the testing of new drugs and perhaps to an understanding of why this condition appears naturally in older people.

HUNTINGTON'S CHOREA

This condition was described by George Huntington, a physician of Long Island, U.S.A., in an address to the New York Neurological

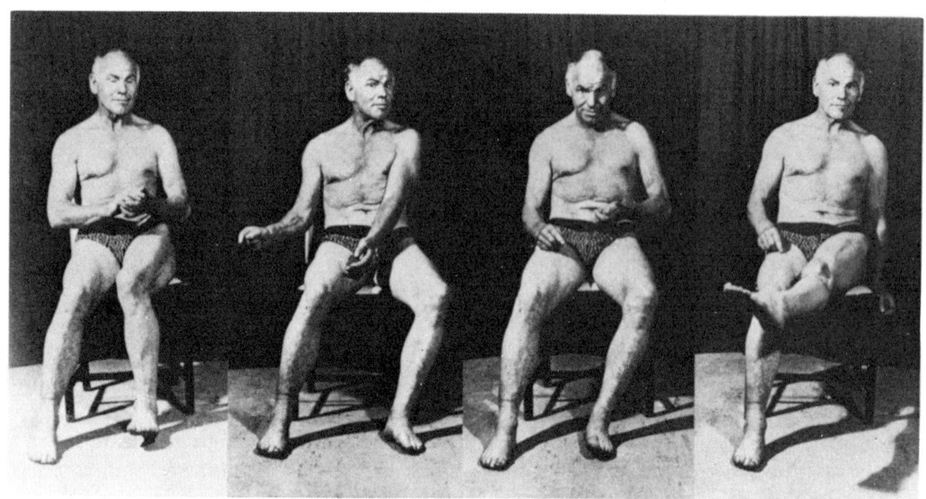

Fig. 3. Huntington's chorea.
Dementia and chorea are features of this condition.

Society in 1910. No causal factors other than heredity are known. A slowly progressive degeneration of the cells of the basal ganglia and of the cerebral cortex occurs. The condition is transmitted by both sexes and it rarely, if ever, skips a generation. The choreiform movements appear to differ in no essential way from those in rheumatic chorea. The face is rarely still, and blinking of the eyes, smacking and licking movements of the lips and tongue, and sniffs and grimaces are usually seen. Speech is slurred, and there is an incoordination of movements and gait (Figure 3). Mental changes may occur during, before, or after the onset of the chorea. There is a restlessness with continuous aggressive involuntary movement.

Our first case of Huntington's type was referred from Western Australia as one with Parkinson's disease who might benefit from a lesion to globus pallidus. As he walked through the door, it was obviously not a case of Parkinson's, although I could not make out what it was. He fidgeted, sat about, got up, sat about some more and was euphoric. When I asked about his ability to perform certain movements, he said he was fine and walked over to the bookcase, picked up a book and threw it into the air and let it crash to the deck. This event did not worry him, he just said it was a mistake. This was a patient with Huntington's Chorea. His first problems occurred when, as a schoolteacher, he had difficulty with balance and the automatic movements required for riding his bicyle. He was the only case in which I made a stereotactic lesion, because it only stopped the fidgeting. He did not have the rigidity of Parkinson's disease, and if anything he became more hypotonic.

HEMICHOREA

A well-defined syndrome is hemiballismus. Unilateral severe large-amplitude continuous involuntary movements are present and the onset is sudden. In the one subject which I treated there were wild flinging movements of the right arm, and continuous spasms of the facial and shoulder girdle muscles. The commonest responsible vascular lesion is either in or adjacent to the subthalamic nucleus.

Hemiballismus is rarely seen, and the particular case was referred to us because people at that time were making lesions in the globus pallidus and later the thalamus. The patient came from New South Wales, a man with a tremendous development of the shoulder girdle and right arm from the quite frightening movements which were exhausting him. Indeed, it is said that some die within weeks of contracting the condition. I recall our lesion stopped the hemiballismus, although he was very hypotonic. I was pleased that his thrashing around had ceased, because this farmer used to shave with a cut-throat razor held in the other hand. We suggested that there were such things as safety razors and electric shavers, but he was not convinced that this was a sensible idea.

TORSION DYSTONIA

This is a rare condition due to disease of the basal ganglia and is characterized by slow, powerful, involuntary movements which produce torsion of the limbs and the spine. The onset is gradual and appears in late childhood. The first involuntary movements are usually in the lower or the upper limbs and, at first, there may be difficulty in putting the heel to the ground when walking. Abnormalities of gait are bizarre, with flexion of the thighs and knees and twisting of the pelvis. Similar changes about the neck, trunk and shoulders produce bouts of torticollis, together with flexion of the arms and twisting and writhing movements of the trunk. Muscular tone is variable but muscle wasting is absent and tendon reflexes are normal. The curious involuntary movements of the limbs are followed by torsion movements of the head and trunk (Figure 4). Voluntary movements induce various contortions. The powerful muscle spasms produce flexion of the toes and flexion and inversion of the feet (Figure 5).

Other conditions characterised by involuntary movements, but rarely seen, are Sydenham's chorea and Wilson's disease (hepato-lenticular degeneration).

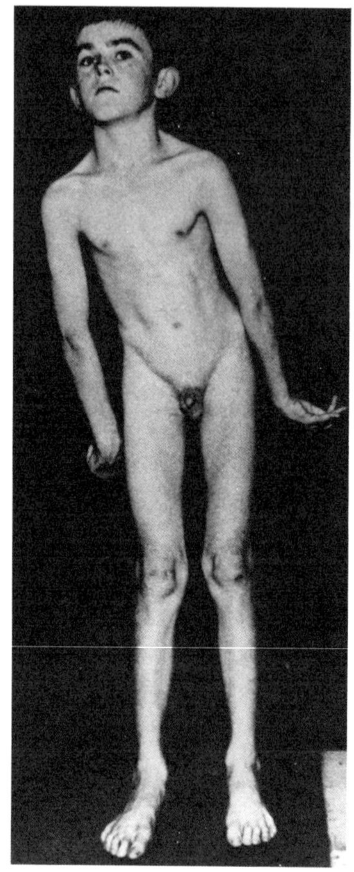

Fig. 4. Torsion Dystonia.
Involuntary movements of the limbs and torsion of the head and trunk are typical.

CONCLUDING REMARKS

This is a brief survey of some of the clinical disorders of the basal ganglia. I think there will be a time when researchers will give us a much clearer understanding of these diseases, so leading t better treatment. Neurosurgeons, using lesions to modify personalit with leucotomy or using stereotactic lesions for other purposes, realise these are mutilating operations which damage the system. I am sure that eventually we will have sophisticated methods of treatment, perhaps using normal substances that exist in the body. I am confident that the clinician will receive considerable aid in management of these conditions as a result of the active research

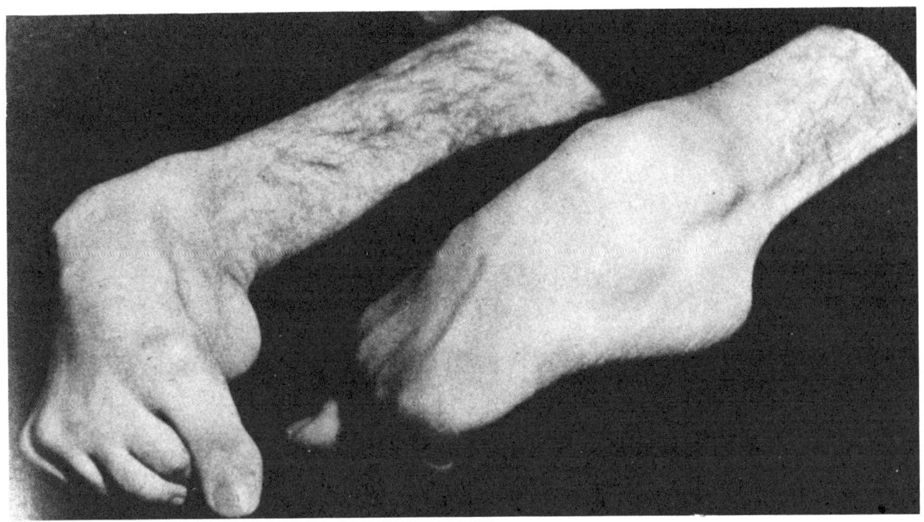

Fig. 5. Torsion dystonia.
Muscular spasms flexing the toes and feet.

into the structural characteristics, the neural mechanisms and the functions of the basal ganglia.

REFERENCES

Huntington, G., 1872, On chorea, Medical & Surgical Report, 26:317; reprinted in Adv. in Neurol., 1973, 1:33.

Huntington, G., 1910, Recollections of Huntington's chorea as I saw it in East Hampton, Long Island, during my boyhood, J. Nerv. Men. Dis. 37:255; reprinted in Adv. in Neurol., 1973, 1:37.

Parkinson, J., 1817, "An Essay on the Shaking Palsy." Sherwood, Neely and Jones, London; reprinted facsimile, 1959, Dawson, London.

LONG LATENCY REFLEXES IN PATIENTS WITH BASAL GANGLIA DISORDERS

J. Noth, H. -H. Friedemann and K. Podoll

Neurologische Klinik
Universität Düsseldorf
Moorenstr. 5
D-4000 Düsseldorf, FRG

INTRODUCTION

In basal ganglia disorders two main syndromes with contrasting clinical symptoms can be distinguished, one with hypotonia and hyperkinetic movements, the other with rigidity and bradykinesia. New insight into the pathogenesis of these syndromes has been obtained by the application of a method using imposed limb dispacements for eliciting short and long latency electromyographic (EMG) responses in the stretched muscle (Lee and Tatton, 1975; Marsden et al., 1976). In Parkinson's disease an enhanced long latency reflex response was observed by Tatton and Lee (1975) and Rothwell et al. (1983), but whether this is due to an exaggerated supraspinal or even transcortical reflex response is not yet clear (Berardelli et al., 1983).

Huntington's disease (H.D.) is a well defined basal ganglia disorder of the hypotonic-hyperkinetic type (for a recent survey see Hayden, 1981). This degenerative disorder primarily affects the neostriatum, but loss of neurones is frequently encountered in other parts of the brain including the neocortex (Lange, 1981). For several reasons growing attention has been paid to this disease during the last decade. Firstly, the symptoms of H.D. are not confined to abnormalities of motor control. Even at an early stage of the disease and not uncommonly before manifestation of the first choreatic movements, patients may exhibit a broad variety of neuropsychological alterations, and dementia is always associated with the disease. Affective changes are frequently encountered, their main manifestations including irritability and apathy. An increased prevalence of manic depressive and, more controversially, schizophrenic syndromes has been reported (Folstein and Folstein, 1983). These behavioural aspects of the symptomatology in H.D. have been taken as support for the assumption that the basal ganglia are involved in cognitive processes. Secondly, since H.D. is inherited as an autosomal dominant, fully penetrant genetic disorder, the study of the offspring of patients with Huntington's chorea provides an opportunity to investigate the early manifestations and the evolution of the symptoms leading to the fully developed clinical picture at the later stages of the disorder. Emerging neurophysiological findings in H.D. indicate that the early somatosensory evoked potentials (SEP) of these patients are pathologically reduced even at an early stage of the disease (Oepen et al., 1981, 1982; Josiassen et al., 1982; Noth et al., 1983). This also seems to hold for the visual evoked potentials (Oepen et al., 1981). This surprising observation points to a functional deficit of the cerebral cortex, which is not in line with the assumption that the cognitive impairment described in H.D. indicates the processing of these functions within the neostriatum. Neurophysiological studies including the recording of long latency reflexes have now been carried out on a sample of about 100 patients with H.D. and 100 persons at risk for H.D. (first generation), and the results will be summarized in this report.

SEP'S IN HUNTINGTON'S DISEASE

The first reports about SEPs in H.D. were controversial. All authors described a reduction in amplitude of the early cortical responses to median nerve stimulation, but a prolongation in latency was only seen by Takahashi and Okada (1972) and Josiassen et al. (1982), while Oepen et al. (1981) could not find a significant change in latency. This matter has been reinvestigated in a larger sample of 37 patients with Huntington's disease aged between 17 and 70 years (Noth et al., 1984). In addition to median nerve stimulation, the tibial nerve was also tested on both sides. For each subject the mean value of both sides was calculated. In the group of patients there was a drastic reduction in the amplitude of the first positive deflection of the cortical SEPs. The mean amplitudes of the control group (n=30) were 4.46 µV for the median nerve and 2.08 µV for the tibial nerve, and the corresponding values for the patients (n=37) were 1.30 µV and 0.59 µV, respectively. There was almost no overlap between the amplitude distributions of the patients and the control group. Figure 1 presents a typical tibial SEP in a patient with H.D. in comparison to the response of a healthy subject. The upper traces show the simultaneously recorded neck potentials. There was no difference in latency or amplitude of the neck potential between healthy subjects and patients, which proves that the afferent transmission up to the level of the dorsal column nuclei was unaltered.

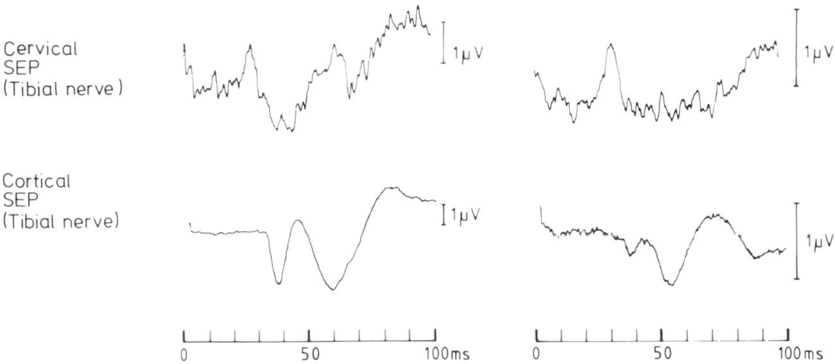

Fig. 1. Tibial-SEPs of a healthy subject of 42 years (left) and a patient of 30 years with Huntington's disease (right). Stimulation of the right tibial nerve at 5/s, cervical SEP recorded at the second cervical spinous process, cortical SEP 3 cm behind Cz (reference electrode over Fz). Number of sweeps were 512 for the normal subject and 1024 for the patient. Note the drastic diminution of the early cortical potential in the patient.

In many patients more than 500 sweeps had to be averaged before a reproducible early cortical component became manifest. This can explain why in earlier studies a significant increase in latency was described, a finding which could not be confirmed by Noth et al. (1984). The reason for this might be that Josiassen and coworkers (1982) used a fixed number of 192 averaged sweeps, which in many patients is not sufficient to produce a visible first response. If the larger second negativity is then mistaken for the first wave, an apparent increase in latency would result. The mean latency of the first cortical negativity N_{20} (median nerve stimulation) was 19.3 $\pm$ 1.32 ms for the control group (n=30) and 20.1 $\pm$ 1.8 ms for the patients (n=37). The difference can be explained by the pronounced reduction in amplitude which may lead to a blurred onset of the negativity.

SEP'S OF PERSONS AT RISK FOR HUNTINGTON'S DISEASE

A first systematic investigation of SEPs in asymptomatic first order offspring of patients with H.D. (Noth et al., 1984) confirmed the result obtained in a small sample of 9 subjects (Oepen et al., 1981). In about 42% of our subjects (n=43) a pathologically reduced amplitude of the early cortical potential was present in response to median or tibial nerve stimulation. Moreover, many subjects exhibited an abnormal between sides difference. This is demonstrated in the upper half of figure 4 for a subject at risk. Again, the cervical SEPs are normal with respect to latency and amplitude, while the cortical potentials are reduced in amplitude. The first negative/positive deflection of the left side amounts to less than 50% of the right side. The largest laterality difference found in healthy subjects was 65%.

It was surprising to find that the mean amplitude (first negative/positive response to median nerve stimulation) of the 14 subjects at risk with the smallest amplitude values was equal to the mean amplitude of 14 slightly affected patients. This indicates that the pathogenetic mechanism underlying the reduction in cortical SEPs is an early event in the disease and not necessarily linked with the onset of choreatic movements. This idea is supported by the fact that there was only a slight tendency towards a further decrease in amplitude when patients with advanced choreatic movements were compared with slightly affected patients (Noth et al., 1984).

LONG LATENCY REFLEXES IN PATIENTS WITH HUNTINGTON'S DISEASE

The controversy over the origin of the long latency EMG response to imposed limb displacements is not yet settled (Wiesendanger and Miles, 1981). It therefore seemed promising to study patients with reduced cortical SEPs but otherwise minor motor deficits. In these patients the impaired afferent pathway to the motorsensory cortex should result in a diminution of the long latency reflex as well, if the reflex is transmitted via a transcortical pathway. Similar experiments have been performed by Jenner and Stephens (1982), who showed that the second excitatory response of a distal hand muscle to cutaneous finger stimulation is indeed reduced in patients with unilaterally dimished early cortical SEP components.

Limb displacements can be imposed under isotonic conditions by randomly applied torque pulses. For distal hand muscles, isometric conditions are much more common for movements executed under natural circumstances involving the precision grip (Napier, 1956; Johansson and Westling, 1983). For studying the reflex response of the readily accessible first dorsal interosseus muscle, an isometric motor task is favourable. The principle of the motor paradigm used for the investigation of long latency reflexes is schematically depicted in figure 2. The subject was asked to exert a steady flexing force by the index finger pressing against a lever. Short triangular pulses were randomly applied to the lever extending the index finger by about 1° measured at the proximal finger joint. The surface EMG of the first dorsal interosseus muscle was rectified and averaged 64 to 128 times.

Healthy subjects exhibit two excitatory EMG responses. The first appears about 32 ms after onset of stretch and the second with

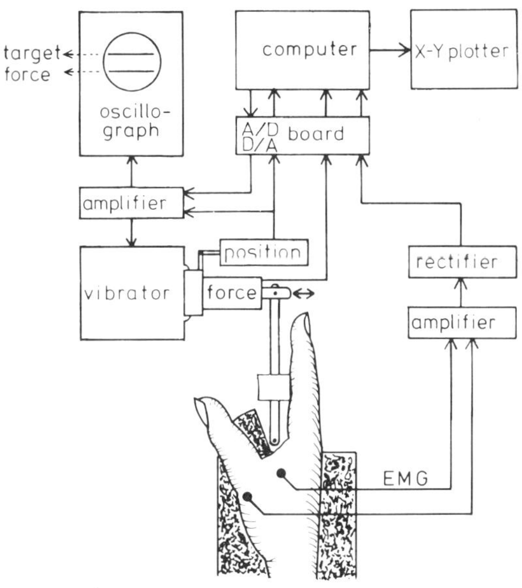

Fig. 2. Schematic diagram of the experimental apparatus. Subject exerted flexing force with the index finger against a lever (10% max. voluntary force). The subject was provided with visual feed-back on the oscillograph. For further details see text.

a latency of about 57 ms (Noth et al., 1983). A representative example is given in figure 3 for a healthy subject and a patient suffering from H.D. There was no difference in the size of the first, segmental response, while in most of the patients the second component was completely lacking. In some patients a small second response was maintained on either side or on both sides. The grand average EMG of 50 patients with H.D. did not reveal any sign of a second excitatory phase (following the short latency response) and in principle was like the response shown in figure 3 for a single patient.

Double pulses were applied in a number of patients in order to see whether the second displacement was capable of eliciting a second short latency response after a brief interval. To this end the interval between the stretches was chosen to mimic the interval between the short and long latency response in normal subjects, i.e. 25 ms. All patients tested exhibited two EMG responses (Noth et al., 1983, fig. 2). This proves that the motoneuron pool in patients with H.D. is capable of responding to a second afferent volley even when applied after a short interval. A spinal abnormality (enhanced Renshaw or short latency cutaneous inhibition, see Jenner and

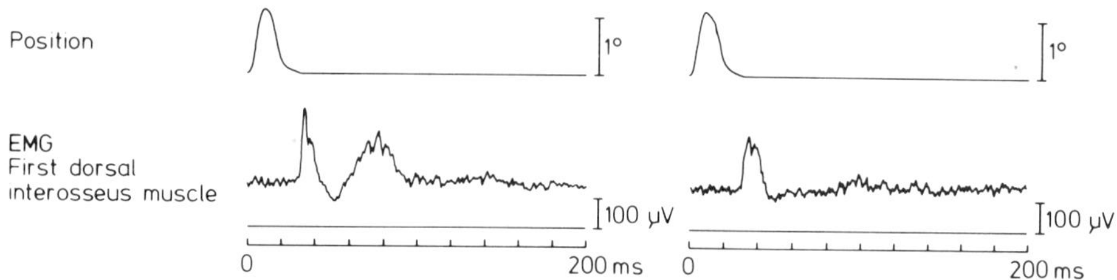

Fig. 3. Records of a healthy subject of 36 years (left) and a patient of 32 years with Huntington's disease (right). Angular displacement of the index finger (single trace) and surface EMG of the first dorsal interosseus muscle (100 times averaged) are illustrated, all records from right hand.

Stephens, 1982) can therefore be excluded. Furthermore, a marked change in the excitability of alpha-motoneurones in patients with H.D. is unlikely, since the first, monosynaptic response was of equal size in both groups.

LONG LATENCY RESPONSES IN PERSONS AT RISK FOR HUNTINGTON'S DISEASE

As mentioned above, a high percentage of subjects at risk exhibited pathologically reduced SEPs. If the long latency reflexes are transmitted via a transcortical loop, even asymptomatic subjects at risk should show abnormalities of the long latency reflexes. Sixty-three subjects with a mean age of 30.1 years (range 12-47 years) were examined. In 45 subjects (71%) a normal result was obtained on examination of both hands. In 10 subjects (16%) the long latency component was missing on one side. In these cases the size of the second response, which was maintained on the other side, was reduced in most instances (see fig. 4 as compared with fig. 3). In 8 subjects (13%) the late component was completely lacking on both sides. Thus a total of 18 subjects at risk (29%) showed a pathological finding. This percentage is still well below the 50% risk of having inherited the pathological gene. The mean age of the 45 at risk subjects exhibiting a normal result was 29.4 years (range 16-46 years). The 18 subjects with a suspicious result had a mean age of 31.7 years (range 12-47 years). This shows that the loss of the long latency reflex may occur early and mostly before the manifestation of the clinical symptoms associated with the disease. Behavioural deficits of subjects at risk lacking the long latency reflex have not so far been detected, but this may be due to the fact that adequate tests demonstrating a functional counterpart of missing long latency reflex responses have not yet been applied.

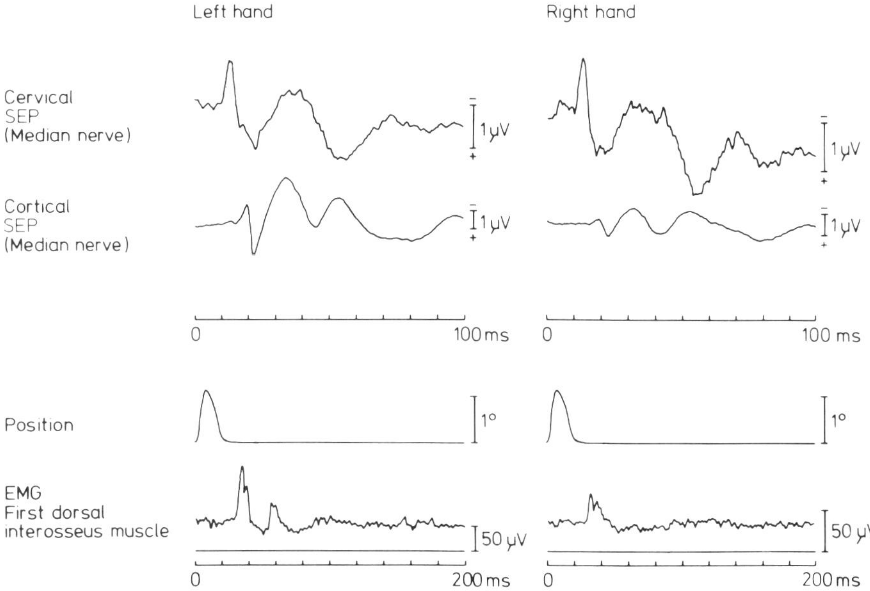

Fig. 4. Unilateral diminution of cortical somatosensory evoked potential to median nerve stimulation and corresponding unilateral absence of the long latency EMG response in a subject at risk for Huntington's disease aged 34 years. The cortical SEP is recorded over the hand area contralateral to the stimulated hand (512 times averaged). Note different time scale in the surface EMG records (100 times averaged).

OTHER BASAL GANGLIA DISORDERS ASSOCIATED WITH CHOREATIC MOVEMENTS

SEPs were studied in 2 patients with benign familial chorea (Haerer et al., 1967; Pincus and Chutorian, 1967) and 4 patients with symptomatic chorea (Wilson's disease). They exhibited normal median and tibial SEPs except for one patient with Wilson's disease, who had slightly reduced amplitudes of the tibial SEPs. Two of the patients with Wilson's disease lacked the long latency EMG reflex on both sides. A bilateral lesion of the neostriatum was verified by cranial computertomography in all patients with Wilson's disease. Normal long latency reflexes were encountered in the other two patients with Wilson's disease and in the patients with benign familial chorea.

CONCLUSIONS

With regard to the function of the basal ganglia, the question immediately arises whether our result indicates that the basal ganglia have a regulatory function on the gain of the long latency reflex. There is ample support for the assumption that at least in hand muscles the long latency reflex is of the supraspinal origin (Marsden et al., 1977; Jenner and Stephens, 1982; Noth et al., 1983; for review see Wiesendanger and Miles, 1982). The relatively short central transmission time favours a straight transcortical reflex transmission. Furthermore, it is well known that cortical neurones within area 4 including pyramidal tract neurones reveal short latency reflex responses to limb perturbations and that the activity of these neurones is related to active movements of the perturbated limb (Evarts, 1973).

If we assume the operation of such a transcortical loop, the basal ganglia could exert their influence either on thalamic relay neurones or on neurones at the cortical level. In rats, caudate nucleus stimulation can in fact suppress the spontaneous activity of cells located at the anterior motor cortex level (Cóndes-Lara et al., 1982), but the same stimulation can also exert an effect on cells located in the specific relay nuclei of the thalamus (Albe-Fessard et al., 1983). Gating of afferent pathways to the sensorimotor cortex during ballistic movements has also been described in humans (Rushton et al., 1981; Obeso et al., 1982). If it is assumed that the transcortical reflex pathway acts as a servo or servo-assistant loop during holding tasks (Phillips, 1969), the interruption of such a loop at the beginning of movements initiated by the basal ganglia would then be feasible or even necessary. The continuously occurring choreatic movements in patients with H.D. may be associated with a permanent active suppression of the transcortical reflex. In Parkinson's disease, on the other hand, rigidity may be the consequence of an inability of the basal ganglia to disrupt the transcortical pathway, resulting in a state of high transcortical reflex gain. Evidence for such a high reflex gain has been presented by several groups (Lee and Tatton, 1975; Rothwell et al., 1983).

It would be argued that the lack of the long latency EMG-response in H.D. is simply a consequence of degeneration of the lemniscal pathway to the cortex, resembling the diminution of the cortical SEPs in these patients. Although this argument cannot be refuted, it should be noted that even with subtle clinical examinations no sensory deficit could be revealed in patients with reduced SEPs and long latency reflexes. This may indicate an intact somatosensory pathway to the parietal cortex but an impaired transmission to the motor cortex (area 4), but certainly more work is needed to prove this hypothesis. For the time being a gain control of the transcortical reflex pathway exerted by the basal ganglia may serve as a working hypothesis which can be subjected to further experimental work.

Two other implications of the results presented may briefly be mentioned. One is that SEP recordings are suitable as a tool for early diagnosis of H.D. and, together with other neurological examinations, may even be employed for the investigation of subjects at risk for H.D. The other is related to the question whether the cognitive impairments observed in patients with H.D. allow one to postulate an intrinsic role of the basal ganglia in neuropsychological phenomena. As SEPs reflect the activity of cortical cells, the diminution of the SEPs in H.D. patients and even in a relatively large number of subjects at risk suggests an early involvement of the cerebral cortex, and this is also valid for the visual evoked potentials (Oepen et al., 1981). Whether this reflects a basal ganglia mediated deficiency or not, it does at least show that processing of afferent information is impaired in the cerebral cortex in H.D.

ACKNOWLEDGEMENTS

We are grateful to Dr H. W. Lange (Rheinische Landesklinik, Direktor Prof. Dr K. Heinrich) for referral of the patients in his care, to Fr. K. van Acken for typing the manuscript and to Fr. S. Brinkmann for reading through the text. The work was supported by a grant of the Deutsche Forschungsgemeinschaft (SFB 200).

The authors acknowledge the permission of Elsevier Biomedical Press to publish a figure from Noth et al., 1984, (in press).

REFERENCES

Albe-Fessard, D., Condés-Lara, M., and Sanderson, P., 1983, The focal tonic cortical control of intra-laminar thalamic neurons may involve a cortico-thalamic loop, Acta Morphol. Acad. Sci. Hung., 31:9.

Berardelli, A., Sabra, A. F., and Hallett, M., 1983, Physiological mechanisms of rigidity in Parkinson's disease, J. Neurol. Neurosurg. Psychiatry, 46:45.

Condés-Lara, M., Kesar, S., and Albe-Fessard, D., 1982, Comparison of caudate nucleus and substantia nigra control of medial thalamic cell activities in the rat, Neurosci. Lett., 31:129.

Evarts, E. V., 1973, Motor cortex reflexes associated with learned movement, Science, 179:501.

Folstein, S. E., and Folstein, M. F., 1983, Psychiatric features of Huntington's disease, Psychiatric Developments, 2:193.

Haerer, A. F., Currier, R. D., and Jackson, J. F., 1967, Hereditary nonprogressive chorea of early onset, N. Engl. J. Med., 276: 1220.

Hayden, M. R., 1981, "Huntington's Chorea," Springer, Berlin.

Jenner, J. R., and Stephens, J. A., 1982, Cutaneous reflex responses and their central nervous pathways studied in man, J. Physiol. (Lond.), 333:405.

Johansson, R. S., and Westling, G., 1984, Influences of cutaneous sensory input on the motor coordination during precise manipulation, in: "Somatosensory Mechanisms," D. Ottoson et al., eds., MacMillan Press, London.

Josiassen, R. C., Shagass, C., Mancall, E. L., and Roemer, R. A., 1982, Somatosensory evoked potentials in Huntington's disease, Electroencephalogr. Clin. Neurophysiol., 54:483.

Lange, H. W., 1981, Quantitative changes of telecephalon, diencephalon, and mesencephalon in Huntington's chorea, postencephalitic, and idiopathic parkinsonism, Verh. Anat. Ges., 75:923.

Lee, R. G., and Tatton, W. G., 1975, Motor responses to sudden limb displacements in primates with specific CNS lesions and in human patients with motor system disorders, Can. J. Neurol. Sci., 2:285.

Marsden, C. D., Merton, P. A., and Morton, H. B., 1976, Servo action in the human thumb, J. Physiol. (Lond.), 257:1.

Marsden, C. D., Merton, P. A., Morton, H. B., and Adam, J., 1977, The effect of posterior column lesions on servo responses from the human long thumb flexor, Brain, 100:185.

Napier, J. R., 1965, The prehensile movements of the human hand, J. Bone Jt. Surg., 38B:902.

Noth, J., Friedmann, H.-H., Podoll, K., and Lange, H. W., 1983, Absence of long latency reflexes to imposed finger displacements in patients with Huntington's disease, Neurosci. Lett., 35:97.

Noth, J., Engel, L., Friedemann, H.-H., and Lange, H. W., 1984, Evoked potentials in patients with Huntington's disease and their offspring. I. Somatosensory evoked potentials, Electroencephalogr. Clin. Neurophysiol., (in press).

Obeso, J. A., Rothwell, J. C., and Marsden, C. D., 1982, Gating action on somatosensory afferents during movements, Neuroscience, 7:S162.

Oepen, G., Doerr, N., and Thoden, U., 1981, Visual (VEP) and somatosensory (SSEP) evoked potentials in Huntington's chorea, Electroencephalogr. Clin. Neurophysiol., 51:666.

Oepen, G., Doerr, N., and Thoden, U., 1982, Huntington's disease: Alterations of visual and somatosensory cortical evoked potentials in patients and offspring, in: "Clinical Applications of Evoked Potentials in Neurology," J. Courjon, F. Manguiere, and M. Revol, eds., Raven Press, New York.

Phillips, C. G., 1969, Motor apparatus of the baboon's hand, Proc. R. Soc. B, 173:141.

Pincus, J. H., and Chutorian, A., 1967, Familial benign chorea with intention tremor: A clinical entity, J. Pediatr., 70:724.

Rothwell, J. C., Obeso, J. A., Traub, M. M., and Marsden, C. D., 1983, The behavior of the long-latency stretch reflex in patients with Parkinson's disease, J. Neurol. Neurosurg. Psychiatry, 46:35.

Rushton, D. N., Rothwell, J. C., and Craggs, M. D., 1981, Gating of somatosensory evoked potentials during different kinds of movement in man, Brain, 104:465.
Takahashi, K., and Okada, E., 1972, Somatosensory and visual evoked potentials in Huntington's chorea, Clin. Neurol. (Tokyo), 22:381.
Tatton, W. G., and Lee, R. G., 1975, Evidence for abnormal long-loop reflexes in rigid Parkinsonian patients, Brain Res., 100:671.
Wiesendanger, M., and Miles, T. S., 1982, Ascending pathway of low-threshold muscle afferents to the cerebral cortex and its possible role in motor control, Physiol. Rev., 62:1234.

MOTOR EFFECTS PRODUCED BY DISRUPTION OF BASAL GANGLIA OUTPUT TO THE THALAMUS

Marjorie E. Anderson and Fay B. Horak

Departments of Physiology and Biophysics
and Rehabilitation Medicine and
Regional Primate Research Center
University of Washington, Seattle
Washington, 98195, U. S. A.

INTRODUCTION

When a limb movement is made to a target, information relevant to a series of decisions or parametric values must be carried in the eventual neural output to muscles. One way to examine the role(s) of the basal ganglia in controlling movement, then, is to examine the influence of basal ganglia output on these movement-related decisions or parameters, several of which are shown diagrammatically in Figure 1.

A basic decision to be made is whether or not a movement is to occur. This "yes/no" decision, which is usually context or instruction dependent, must be translated into the production or suppression of activity in the relevant muscles. If a movement is to occur, the _trajectory_ of the movement is one of the parameters that must be determined by a combination of the neural output and the mechanical properties of muscle and associated structures. Neurally, this would be specified by the selection of the particular muscles to be activated or inhibited and the sequence in which they will be activated or inhibited. In its simplest form, often used experimentally, the trajectory decision may be reduced to determination of movement direction.

The parameters of movement _amplitude_, _acceleration_, and _velocity_ all may be scaled by the recruitment of the appropriate number of motor units and the determination of their instantaneous firing rate and/or the duration of their activity. For an unimpeded movement, the scaling of movement amplitude, acceleration (positive and

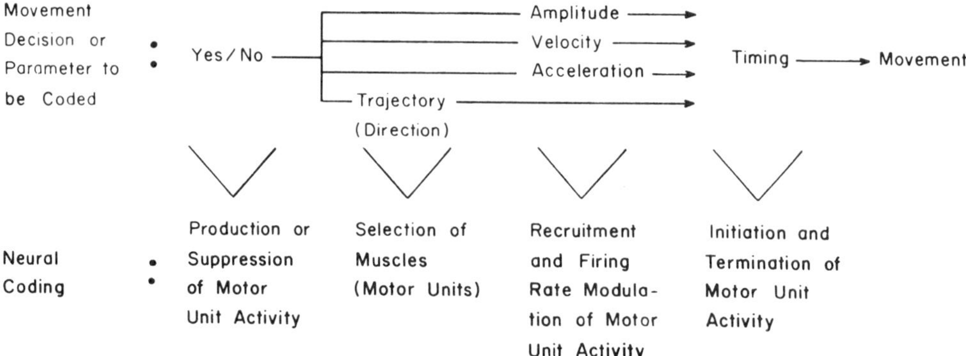

Fig. 1. Neural coding of movement decision and parameters.

negative) and velocity will be interrelated. If the amplitude of the movement is determined mechanically, however, recruitment and firing rate modulation will affect only the acceleration, velocity, and deceleration of the movement, all of which will determine the total movement time (MT).

Finally, some signal from the central nervous system must specify the time at which a movement, or a particular phase of a movement, is to be initiated or terminated. One way to evaluate the control of the timing of movement initiation is to measure the reaction time (RT).

The data to be described are relevant to deciding whether, for a stereotyped rapid movement, output from the globus pallidus carries information relevant to the selection and sequencing of muscle activity that determine movement trajectory, the scaling of motor output that determines movement time, or the timing of the initial motor activity that determines the mechanically-defined simple reaction time (Horak and Anderson, 1984a,b; Anderson and Horak, 1984).

METHODS

Monkeys (M. fascicularis) were trained to make an arm-reaching movement in a simple reaction time task. They were perched in a primate chair with a counterweighted lever at approximately waist height at their side, and they faced a panel of buttons, each of which could be back-lit with an LED. The animal's task was to depress the lever (start lever) for a variable time until one of the buttons was lit, and then to release the lever and depress the lighted button as rapidly as possible. In most experiments, a single button on the panel of six was used, and the required movement had a straight-line distance of approximately 20 cm. The mechanical reaction time (RT) was measured as the time between target-light onset and closure of a microswitch when the start lever was released, and the movement time (MT) was measured as the time between start lever release and target button depression, which also closed a microswitch. The animal's behavior was shaped in daily sessions to a minimum RT plus MT, which was less than 500 msec for each animal after training.

A stainless steel chamber was stereotaxically implanted on the skull to allow angled access to the putamen and globus pallidus contralateral to the arm used in the task. Electrodes inserted through the chamber first were used to determine the location of the globus pallidus (GP) by recognition of the high frequency tonic activity characteristic of neurons in this region (DeLong, 1971; Anderson, 1977). After the activity of individual neurons was amplified and recorded during task performance, the same electrode (insulated with a polyimide material with a high dielectric constant) could be used to stimulate within the GP or adjacent structures during task performance, using stimulus intensities of 100 microamperes or less. By varying the duration of the stimulus train, the time at which it was initiated, and the position of the electrode in and around GP, and by comparing the RTs and MTs of randomly interspersed stimulation and non-stimulation trials, any effects due to stimulus-induced changes in GP activity could be mapped as a function of both the spatial and temporal position of stimulus application (Horak and Anderson, 1984b; Anderson and Horak, 1984).

After the location of GP neurons was determined, the neurotoxic agent, kainic acid (KA) also could be introduced into GP from a microliter syringe through a needle inserted through a stainless steel cannula that prevented leakage along the needle track. MTs and RTs of trials studied before and after KA-induced lesions of GP neurons at different locations could then be compared.

The electromyographic (EMG) activity of anterior deltoid (DELT), pectoralis major (PECT), biceps brachii (BIC), triceps brachii (TRI), extensor carpi radialis (ECR) flexor carpi radialis (FCR), and deep thoracic paraspinal (THOR) muscles was recorded with bipolar stainless steel electrodes sutured into the muscles and led to a connector implanted in a vitreous carbon button (Biosnap, Bently Laboratories) that protruded through the skin of the back. Using this technique, stable EMG records could be compared pre- and post lesion, as well as during stimulation and non-stimulation trials.

All data were recorded on FM tape, and averaged data for several trials under different conditions were compared using t-tests to identify statistically significant changes in RT and MT ($p<.001$) or were examined visually to identify changes in the sequence, amplitude, or rate of rise of the underlying EMG activity.

RESULTS

Lesion-induced Changes in Movement

Kainic acid-induced lesions were made in 5 sites in 4 animals. In each case, injections of 1 ug KA in 1 ul of mock CSF were made in 0.1 ul aliquots over a 30 min period, with the animal's behavior in the reaction time task monitored continuously. Qualitatively, the results were the same in all case: the movement time was increased, but there was no consistent change in the reaction time. The increase in movement time did not become significant until approximately 0.5 ul of KA solution had been injected, and animals did not show overt changes in behavior, disturbances in the trajectory of the arm, or increased numbers of trials in which they missed the target button. Instead, the movements simply appeared slowed and there was less impact of the hand as it contacted the target buttons.

Figure 2 shows the averaged, rectified EMGs for 20 trials made by one monkey one day after KA injection, compared to data recorded just prior to KA injection (control). The striking changes are in the peak amplitude and rate of rise of EMG activity in all of the muscles examined. The time at which EMG activity began in the earliest muscles activated did not change when averaged from target onset (not shown) nor was there overt co-contraction or abnormal overlap in the activity of antagonistic muscles. Instead, a normal sequence of muscle activity persisted, even though it was spread out in time and the muscles activated late in control movements were activated even later during the slowed movements made after KA-induced lesions.

Although KA injection was followed by prolonged MTs in all animals, the persistence of the slowing varied between animals and was related to the location, and perhaps the extent, of cellular loss. The monkey with the most persistent increases in movement time (8 days, until sacrifice) was the one in which neurons were almost completely abolished in the ventrolateral half of both segments of GP, throughout approximately the posterior half of the nuclei, together with a small portion of the putamen adjacent to lateral GPe (external segment of GP). In a second animal, histological examination after a 7-day survival showed neuronal cell loss of the ventrolateral half of GPe, but not of GPi (internal segment of GP), and in this animal, the initial increase in movement time persisted through day 2 following the lesion, but by day 7, MTs had recovered considerably, although not to pre-lesion control values.

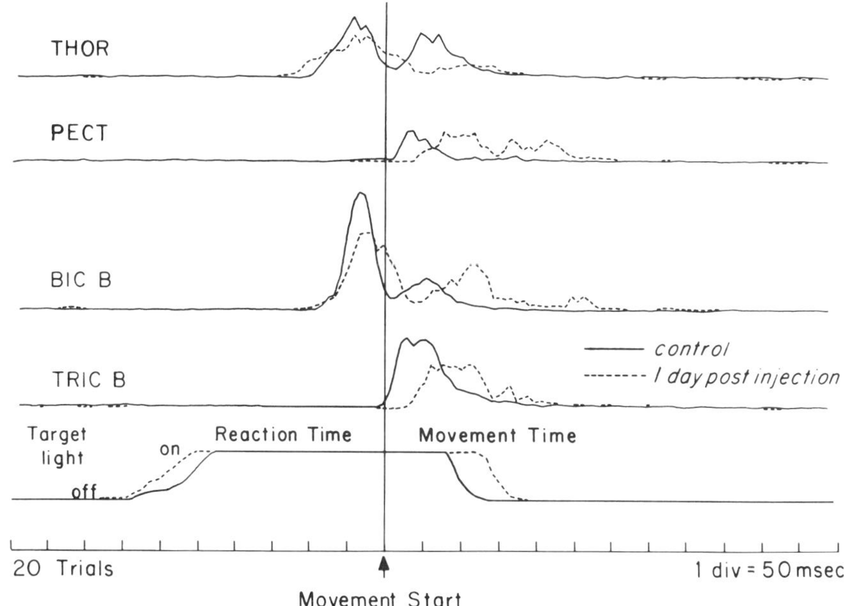

Fig. 2 Changes in electromyographic activity and movement time produced by injection of kainic acid into the globus pallidus. The averaged rectified EMG activity for thoracic paraspinal muscles (THOR), pectoralis (PECT), biceps brachii (BIC B) and triceps brachii (TRI B) recorded using the same amplifier gain for 20 trials studied just prior to injection of kainic acid (solid lines) and 20 trials studied 30 minutes after kainic acid injection (1 μg in 1 μl) was completed. Averages were centered on the mechanically-detected start lever displacement (movement start). The time between target light onset and movement start (reaction time) did not change, but the time between movement start and target contact (movement time) was prolonged after an injection of kainic acid that destroyed neurons in the ventrolateral portion of the external pallidal segment and adjacent neurons in the putamen.

The other three injections destroyed neurons in the dorsomedial half of GPe or GPi, but spared neurons in the ventrolateral portions of both nuclei. In these cases, MTs were prolonged for shorter periods of time and recovered to pre-injection control values within 1-2 days following the injections.

In summary, destruction of neurons in the globus pallidus produced increases in movement time, without changes in the time at which EMG activity or movement were initiated or in the sequence in which different muscles were activated. The slowed movements were the result of a decrease in the rate of rise and peak of EMG activity in all of the muscles examined, and the effect was most persistent when neurons in the ventrolateral half of caudal GP were destroyed by the lesion. The majority of neurons with changes in activity associated with passive or active arm movement have been reported previously to be in this portion of GPe and GPi (DeLong and Georgopoulos, 1979).

Stimulation-induced Changes in Movement

Stimulation within either segment of the globus pallidus during performance in the reaction time task also produced changes in movement time, but produced almost no change in the mechanically determined reaction time, no change in the time of earliest changes in EMG activity, and no change in the sequence in which different muscles were activated during the movement.

Long duration stimulus trains (500 msec) that encompassed both the reaction time and movement time portions of the task caused either increases or decreases in movement time, depending on the site of stimulation within GPi, GPe, or the putamen. Figure 3 shows diagramatically the MTs from trials during which stimulation was applied at various positions (circles or triangles connected by dashed line), compared with the mean movement times for all non-stimulation trials completed while the electrode was in the same track (solid line, widened for the extent over which neurons with tonic high frequency discharge had been observed). The displacement of the symbols from the solid track line indicates the magnitude of the change in MT, with filled circles depicting significantly increased MTs and filled triangles depicting significantly decreased MTs ($p<.001$). These data are from 13 tracks in 2 animals and illustrate the general pattern of effects produced by stimulation at 302 locations in and around the lenticular nuclei in 4 different animals.

The points at which stimulation produced decreased MTs (i.e. speeded movement, solid triangles), tended to be located in ventrolateral GPi, at GPe positions adjacent to ventrolateral GPi, or at locations in the ansa lenticularis, where fibers from GP pass between GPi and the optic tract, en route to the thalamus or brainstem. Stimulation at dorsomedial positions in GPi, at most locations in GPe, and at the few positions tested in the putamen either produced increases in MT (slowed movements, solid circles) or had no significant effect (open circles). Stimulation outside of the lenticular nucleus, in the internal capsule or the optic tract, either produced no detectable effect or produced stimulus-locked

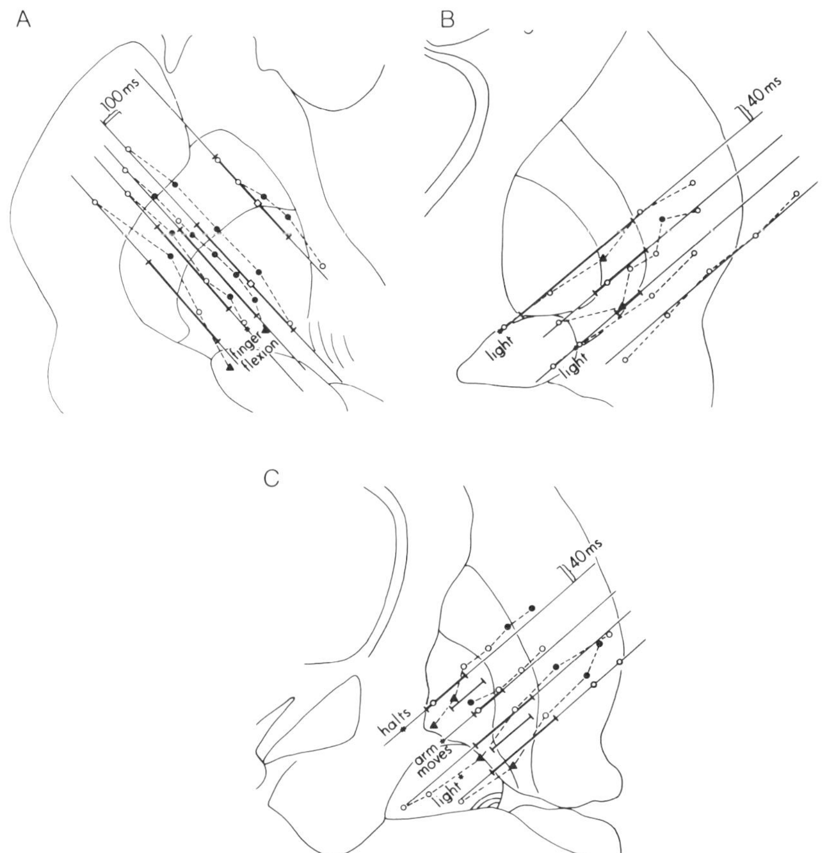

Fig. 3. Changes in movement time produced by stimulation at different locations in the globus pallidus and putamen. A, Data from stimulation in one plane in monkey N. B and C, Data from stimulation at positions in planes through mid(B) and caudal (C) GPi in monkey. Solid lines depict both the location of the track and the mean movement time for all non-stimulation trials studied while the electrode was in that track. Dashed lines connect points representing movement times measured during stimulation at corresponding depths in each track. Distance of symbols from the solid lines indicates the amplitude of the change in movement time (see calibrations). Solid circles depict significant increases in movement time; solid triangles, significant decreases in movement time; open circles, no significant change from interspersed non-stimulation control trials ($p < .001$). Asterisks depict points at which stimulation caused movement, cessation of movement (halt), or at which light-reactive fiber activity was recorded, as labeled.

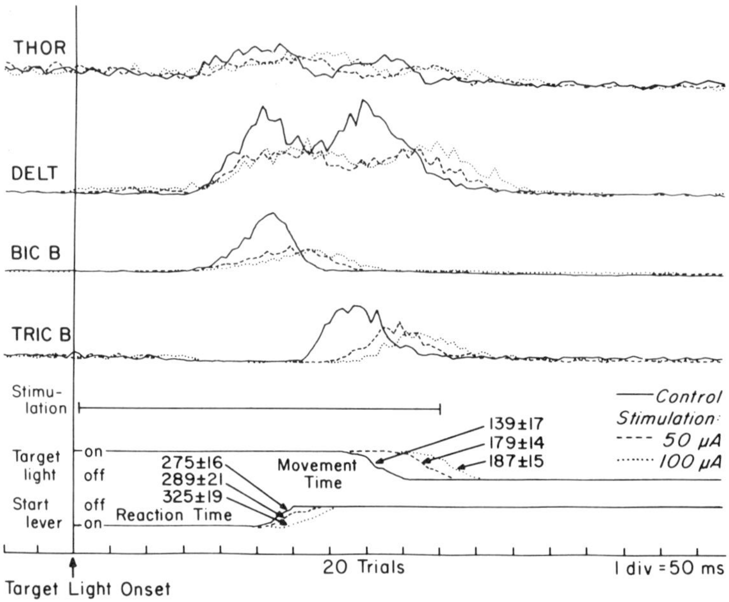

Fig. 4. Changes in electromyographic activity, movement time, and reaction time produced by stimulation in GP at two different intensities. Averaged records from 20 trials at each stimulus intensity (dotted lines, 100 μA), compared to randomly-interspersed non-stimulus control trials (solid lines). Averages were initiated when the target light was illuminated, and the mean reaction time (to start lever off) and mean movement time (start lever off to target light off) are shown for each average. Movement times were significantly increased ($p < .001$) at each stimulus intensity and were associated with depressed and delayed peak EMG activity in all muscles examined. Early changes in EMG activity were not delayed, even when reaction time was increased significantly (100 μA stimulation), although late EMG activity was delayed (DELT and TRIC B), so that there was no change in the sequence or overlap of activity in different muscles. Stimulus train duration: 500 msec. Averaged rectified EMG activity is from the thoracic paraspinal muscles (THOR), anterior deltoid (DELT), biceps brachii (BIC B), and triceps brachii (TRIC B).

movements of the head or limbs or a complete cessation of task-related behavior (stars). It was not possible, however, to convert a slowed movement to cessation of movement by increasing the stimulus intensity, even though the two effects might be produced from stimulation at points separated by only 1 mm.

Stimulation-induced changes in movement time were associated with a depression in the peak amplitude and rate of change of EMG activity in all of the muscles examined that showed phasic changes in activity during the task. This was true irrespective of the stimulus intensity or site of stimulus application. Data obtained during stimulation at two different intensities (50 µA and 100 µA) are shown in Figure 4, in which averages were initiated at the onset of the target light and the beginning of stimulation. Depression of the phasic EMG burst amplitude occurred in shoulder and elbow muscles at both stimulus intensities, and there was a delay in the onset of the late phases of activity of anterior deltoid and triceps brachii, so that co-contraction of monitored antagonistic muscles (biceps and triceps brachii) was avoided. Even in cases in which 25 µA stimulation resulted in significantly prolonged movement times (not shown), there was a depression of activity in all of the muscles examined. Furthermore, both the starting position and the limb trajectory, examined with high speed cinematography, were unchanged, compared with randomly-interspersed non-stimulation trials. When stimulation at other points produced decreased MTs (speeded movements), the EMG also changed in all muscles examined, but in these cases, the changes were _increases_ in the peak amplitude and rate of rise in EMG activity.

Thus, stimulation-induced changes in MT were not associated with a change in the selection of muscles or the pattern of movement, but with a change in the amplitude and rate of rise of EMG activity. Such a change could be described as a rescaling of EMG activity.

In most cases, stimulation was not associated with significant changes in reaction time, even though the movement time might be increased or decreased markedly. The data in Figure 4 show the most marked of the three statistically significant exceptions to that pattern, in which mean RT was increased by 14 msec (5%) during stimulation with 50 µA pulse trains and by 50 msec (18%) during stimulation with 100 µA pulse trains. Examination of the EMG data in Figure 4 shows, however, that even when mechanically-measured RTs increased significantly, the time at which the early EMG activity was initiated (biceps brachii, anterior deltoid, and perhaps thoracic paraspinal musculature) did not change. Instead, the prolonged mechanical RTs reflect the decreased rate of rise of EMG activity.

Timing of Movement-related Information

Since stimulation in GP resulted in changes in movement time, presumably because of changes in pallidal output, it was of interest to determine whether the time at which stimuli must be applied to change movement was the same as the time at which pallidal neurons showed movement-related changes in discharge. This question was examined by shortening the duration of the pulse train applied at positions from which stimulation produced increases in movement time and varying the time at which these shorter pulse trains were applied.

"Turning curves" of movement time, as a function of the time at which stimulus trains were initiated, are shown for four different stimulus train durations in Figure 5, A-D. The mean and standard deviation of movement times recorded during the randomly-interspersed non-stimulation (control) trials are shown at the left in each panel. When mean movement times for trials initiated in separate 50 msec bins were compared to control MTs, in every case, stimulus trains of 25 msec duration or longer that overlapped the period 50 to 100 msec prior to movement initiation (arrow) were associated with mean movement times that were significantly longer than controls. In some cases, MTs also were increased when stimulus trains only overlapped the bin 150 to 100 msec prior to movement initiation. Stimulus trains that were initiated <u>and</u> terminated earlier than 150 msec prior

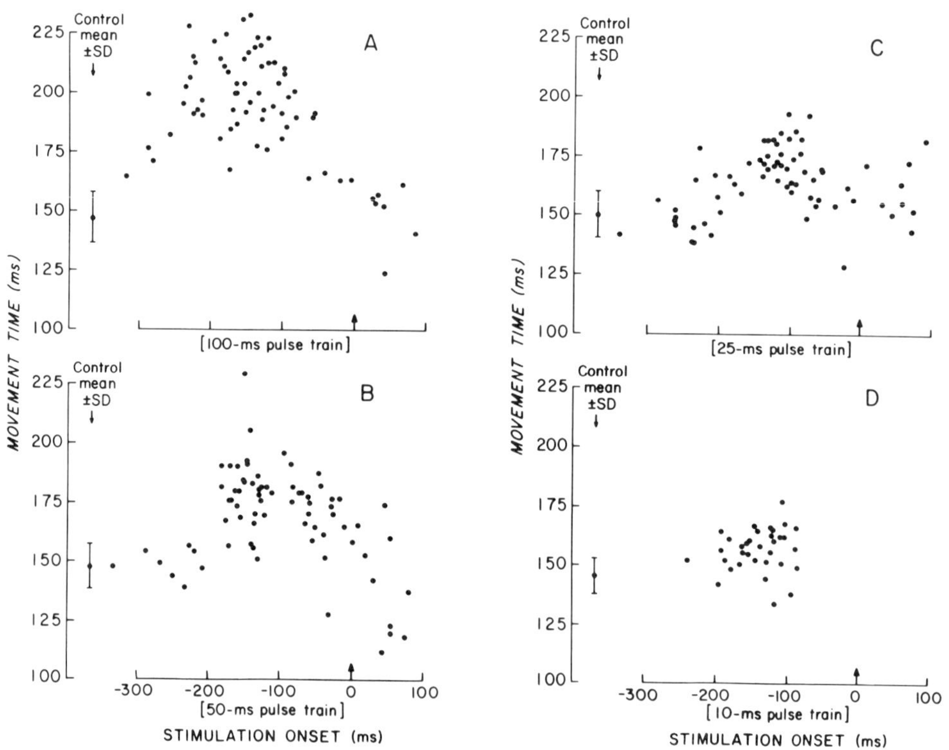

Fig. 5. Effect of pulse train duration and stimulation time on movement times. Points are plotted along the abscissa at the point at which a stimulus train of duration 100 msec (A), 50 msec (B), 25 msec (C), or 10 msec (D) was initiated, relative to the time at which movement was detected (time 0, upward arrow). The mean ± SD of movement time for randomly interspersed control trials is plotted at the left on each panel.

to movement onset did not produce significant changes in MT or RT, nor did those initiated later than 50 msec prior to mechanically-detected movement initiation. (Apparent decreases in MT with 100 msec or 50 msec pulse trains initiated after movement onset were not statistically significant, nor were they consistently observed with stimulation at various positions in GP).

The activity of 65 of the neurons studied in GPe and GPi of monkey N (from which the stimulation data of Figure 5 were obtained) showed phasic changes in firing rate that began prior to or during the movement. For 44 neurons (12 in GPi, 32 in GPe), the change was an increase in firing rate. For 21 cells (3 in GPi, 18 in GPe), the change was a decrease in firing rate. Cumulative distribution curves of the time at which these changes became significant (greater than 2 standard deviations from the mean firing rate prior to target onset, with the deviation persisting for at least 30 msec) are shown in Figure 6A and B for GPi and GPe neurons, respectively.

Only 2 neurons, 1 in GPe and 1 in GPi, showed changes in firing rate earlier than 150 msec prior to movement initiation, and between 150 and 50 msec prior to the movement, approximately 50% of the task-related neurons in each nucleus had begun to change their firing

Table I. Timing of EMG changes during arm reaching tasks

	Early Activity		Late Activity	
Muscle	Onset Time*	Peak Amplitude Time	Onset Time	Peak Amplitude Time
FCR	-110 ± 7 msec	-23 ± 23		
ECR	-99 ± 12	+1.0 ± 11		
DELT	-83 ± 17	-27 ± 11	---	+131 ± 24
THOR	-135 ± 25	-20 ± 15		
BIC	-103 ± 11	-11 ± 15		
TRI			+34 ± 13	+107 ± 17

* Times are referenced to movement initiation (time 0). Negative times precede movement initiation; positive times follow it.

rate. By the time the start lever was released (time 0), approximately 80% of the task-related neurons had significant increases or decreases in firing rate, and only a small number of neurons had firing rate changes that began during the movement.

Thus, the time at which stimulus application was most consistently effective in slowing the ensuing movement (150 to 50 msec prior to start lever release) was the same time during which approximately 50% of the task-related neurons began to change their firing rate, and most of the remainder of the neuronal firing rate changes occurred between this time and the time at which start lever displacement was detected.

Examination of the EMG activity collected during the same trials in which the neuronal activity was recorded revealed that most muscles were activated between 125 and 80 msec prior to start lever displacement and reached their peak activity just prior to the mechanically-detected movement initiation. Table I shows the mean and standard deviations for EMG onset and peak times for the trials in which the unit data of Figure 6 were obtained. It can be seen that the earliest muscles activated (thoracic paraspinals) began to discharge on the average 135 msec prior to movement initiation, and others followed at -110 msec (FCR), -103 (BIC), -99 msec (ECR), and -83 msec (DELT). Comparison of Figure 6 and Table I reveals that very few GP neurons showed initial changes in firing rate prior to initial changes in EMG activity, and it is not surprising that changing GP activity, either by KA-induced destruction of GP neurons or stimulation in GP, produced no significant changes in reaction time. Instead most GP neurons changed their activity during the buildup of EMG activity, which peaked between -30 and 1.0 msec for all of the muscles with early bursts. Changing the development of GP neuronal activity, by lesions or stimulation, did change the rate of rise and peak EMG amplitude in all of these muscles.

DISCUSSION

Examination of the activity of pallidal neurons and the effects produced by changing pallidal activity, either by destruction of pallidal neurons or by stimulation within the pallidum, has led to the following conclusions:

1. Pallidal activity does not determine the time at which initial EMG activity or arm displacement begins for arm reaching movements made in a visually-triggered simple reaction time task.
2. Pallidal activity changes markedly during the time between initial and peak EMG activity in the muscles raising the arm, and changes in pallidal activity, produced by lesions of stimulation, cause changes in the rate of rise and peak amplitude of EMG activity and a consequent change in movement time.

3. Changing pallidal activity, either by pallidal lesions or stimulation, does not change the muscles active in the movement, the sequence with which muscles are activated or the relative timing of activity in antagonistic muscles.

Destruction of GP neurons, using kainic acid, consistently produced increases in movement time, although the persistence of this effect varied, depending on the location, and perhaps the extent, of the lesion. Stimulation at many points in GP also produced increased movement times, although stimulation at some points, located primarily in the ventrolateral internal segment of GP, resulted in speeded movements (decreased movement times). The observation that

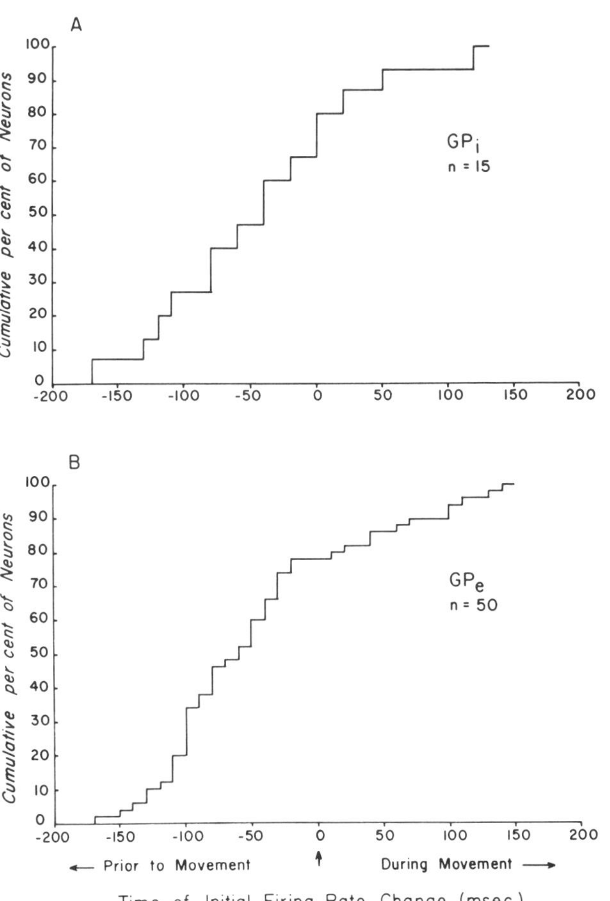

Fig. 6. Cumulative histograms of the time at which initial changes in firing frequency began for movement-related neurons in (A) internal globus pallidus and (B) external globus pallidus. Movement was mechanically-detected at time 0 (upward arrow).

both destruction of GPi neurons and stimulation in GP produced increased movement times would be inconsistent if it were assumed that the same neurons would be destroyed by the lesions and excited by the stimulation. If, however, the lesions of ventrolateral GP (the ones that produced the most persistent increases in movement time) destroyed GPi neurons critical for producing rapid arm movements, whereas stimulation produced either excitation or inhibition of the critical GPi neurons, depending on the site of stimulation, there would be no discrepancy between the changes in movement time produced by KA induced lesions and those produced by stimulation.

Kainic acid, when administered in a way that does not induce too high a concentration of KA at a particular time, is reported to cause a local destruction of cell bodies, but to have little effect on en passant axons (McGeer et al., 1978). In our experiments, the increased movement times were most persistent when KA destroyed neurons in the ventrolateral half of GPi. DeLong and Georgopoulos (1979) have reported that neurons with changes in discharge associated with active or passive movement of the contralateral arm are located primarily in the ventrolateral portions of the pallidum. Some axons from this part of GPi pass ventrally and medially through the ansa lenticularis before coursing toward the thalamus or brainstem (Kuo and Carpenter, 1973). Stimulation in the ventrolateral pallidum or adjacent ansa lenticularis, then, would be the most logical site at which to excite the arm movement related neurons or their axons directly, and the locations at which stimulation caused decreases in movement time were largely in the ventrolateral pallidum and adjacent ansa lenticularis.

Stimulation at other positions in GPi, GPe, and putamen, which, if it changed movement time, tended to prolong it, could have excited axons that exert an inhibitory action on GPi neurons critical for ensuring rapid movement. Striatal axons, that gather into dense bundles as they travel into GPe and GPi (Wilson, 1914; Szabo, 1962; Carpenter, 1976) exert a strong inhibitory action on GPi neurons (Ohye et al., 1976) that probably is mediated by GABA (Obata and Yoshida, 1973). Furthermore, there may be an inhibitory action of GPe axons on subthalamic neurons (Ohye et al., 1976) that, in turn, might excite GPi neurons (Hammond et al., 1978). Thus, stimulation in medial GPi, GPe, and portions of putamen could have reduced the activity of critical GPi neurons by an inhibitory action. If these neurons were critical for producing rapid movement in this task, then reducing their action, either by destroying them or inhibiting them, could have led to increased movement times.

Increased movement times following electrolytic lesions or cooling of GPi also were reported by Trouche et al. (1979), who studied baboons performing an arm-reaching task similar to the one used in this study. Likewise Hore and Vilis (1980) also reported that cooling the basal ganglia caused movements to be slowed in an elbow movement task, although in most of their animals the cooling probe was centered in GPe or putamen. The latter investigators also found an earlier than normal onset of EMG activity in antagonists that could have caused the slowed movements. Early antagonist activity was not a feature in our study, and Trouche et al. (1979) did not record EMG activity.

Consistent with the lack of changes in reaction time and EMG onset in the present study, Hore and Vilis (1980) found that cooling the basal ganglia produced no change in the earliest onset time of EMG activity in agonist muscles. Amato et al. (1978), however, reported that reaction times decreased when the pallidum was cooled or destroyed electrolytically and movement times were increased (Trouche et al, 1979). Their animals, however, were not forced to minimal reaction times and movement times during training and, instead, simply had to contact the target within 1 sec. In that case, if movement execution was slowed, animals could have compensated by shortening reaction times sufficiently to stay within the 1 sec criterion time. In our experiments, animals were forced to minimum reaction time plus movement time criteria that always were less than 500 msec, and in this situation, they could not compensate for slowed movements by reducing their reaction times.

As indicated in Figure 1, the acceleration and velocity of a movement, which were not measured directly in our experiments but which, for movements of constant load, amplitude and trajectory would determine movement time, would be determined by the number of motor units recruited and the rate at which they fire. If there was no concurrent change in the activity of antagonistic motor units, changing the number of motor units activated or the rate at which they fire could result in a rescaling of the movement velocity and its derivatives.

There was no clear change in the trajectory of the limb or the sequence of muscle activity during stimulation or neuronal destruction in GP, suggesting that GP activity did not play a critical role in selecting the particular muscles to be activated or inhibited. Likewise, there was no indication that GP plays a critical role in determining the timing of the initial EMG activity that determines the reaction time. The yes/no context decision was not carefully evaluated in our study.

Pallidal output could potentially influence movement by its connections to cortex through the thalamus or possibly by its connections to brainstem neurons in the pedunculopontine area. At the thalamus, pallidal axons terminate in VA and VLo (Kim et al., 1976; DeVito and Anderson, 1982; Parent and deBellefeuille, 1982). There they may exert an inhibitory action (Uno et al., 1978) on neurons in a cortico-thalamo-cortical loop connecting these portions of the thalamus with prefrontal and premotor cortical regions (Anderson and DeVito, 1983), including the supplementary motor cortex (Schell and Strick, 1984). Globus pallidus output, however, does not appear to have direct thalamo-cortical access to area 4 motor cortex, which instead, receives input from portions of the thalamus to which the anterior deep cerebellar nuclei project (Schell and Strick, 1984).

If the pallido-thalamo-cortical connections were the route by which changes in movement time were elicited by the mechanisms proposed in the present study, it would mean that lesion or inhibition-induced decreases in pallido-thalamic inhibition (i.e. a disinhibition) resulted in slowed movements, where an increase in pallido-thalamic inhibition of the relevant cortico-thalamo-cortical loop (as produced by excitation of the critical neurons in ventrolateral GPi) would result in rapid movements. The mechanisms by which these changes in movement time would be implemented are not known at present.

ACKNOWLEDGEMENTS

Supported in part by PHS grants NS15017, NS10804, GM7108, and PR00166, Department of Health and Human Services, and NIHR grant P-56818, Department of Education.

REFERENCES

Amato, G.., Trouche, E., Beaubaton, D., and Grangetto, A., 1978, The role of internal pallidal segment on the initiation of a goal directed movement, Neuroscience Letters, 9:159.

Anderson, M. E., 1977, Discharge properties of basal ganglia neurons during active maintenance of postural stability and adjustment to chair tilt, Brain Res., 143:325.

Anderson, M. E., and Horak, F. B., 1984, Influence of the globus pallidus on arm movements in monkeys, III. Timing of movement-related information, J. Neurophysiol., (submitted).

Anderson, M.E., and DeVito, J. L., 1983, Potential sources of convergent input to ventrolateral and ventral anterior nuclei of the thalamus in the cat: An anatomical study, Proc. XXIX Congr. of Congr. of Physiol Sciences, Sydney, Austr., (Abst.).

Carpenter, M. B., 1976, Anatomical organization of the corpus striatum and related nuclei, in: "The Basal Ganglia," M. Yahr, ed., Raven Press, New York.

DeLong, Mahlon, 1971, Activity of pallidal neurons during movement, J. Neurophysiol., 34:414.

DeLong, M. R., and Georgopoulos, A. P., 1979, Motor functions of the basal ganglia as revealed by studies of single cell activity in the behaving primate, Adv. Neurol., 24:131.

DeVito, J. L., and Anderson, M. E., 982, An autoradiographic study of efferent connections of the globus pallidus in Macaca mulatta, Exp. Brain Res., 46:107.

Hammond, C., Deniau, J., Rizk, A., and Feger, J., 1978, Electrophysiological demonstration of an excitatory subthalamo-nigral projection in the rat, Brain Res., 151:235.

Horak, F. B., and Anderson, M. E., 1984, Influence of the globus pallidus on arm movements in monkeys. I. Effects of kainic acid-induced lesions, J. Neurophysiol., (in press).

Horak, F. B., and Anderson, M. E., 1984, Influence of the globus pallidus on arm movements in monkeys. II. Effects of stimulation, J. Neurophysiol., (in press).

Hore, J., and Vilis, T., 1980, Arm movement performance during reversible basal ganglia lesions in monkeys, Exp. Brain Res., 39:217.

Kim, R., Katsuma, N., Jayaraman, A., and Carpenter, M., 1976, Projections of the globus pallidus and adjacent structures: An autoradiographic study in the monkey, J. Comp. Neurol., 169:263.

Kuo, J., and Carpenter, M. B., 1973, Organization of pallido thalamac projections in the rhesus monkey, J. Comp. Neurol., 151:201.

McGeer, P. L., McGeer, E. G., and Hattori, T., 1978, Kainic acid as a tool in neurobiology, in: "Kainic Acid as a Tool in Neurobiology," E. G., McGeer, J. W. Olney, and P. L. McGeer, eds., Raven Press, New York.

Obata, K., and Yoshida, M., 1973, Caudate-evoked inhibition and actions of GABA and other substances on cat pallidal neurons, Brain Res., 64:455.

Ohye, C., LeGuyader, C., and Feger, J., 1976, Responses of subthalamic and pallidal neurons to striatal stimulation: an extracellular study on awake monkeys, Brain Res., 111:241.

Parent, A., and deBellefeuille, L., 1982, Organization of efferent projections from the internal segment of globus pallidus in primate as revealed by fluorescence retrograde labeling method, Brain Res., 245:201.

Schell, G., and Strick, P., 1984, The origin of thalamic input to the premotor and supplementary motor areas, J. Neuroscience., (in press).

Szabo,, J., 1962, Topical distribution of the striatal efferents in the monkey, Exp. Neurol., 5:21.

Trouche, E., Beaubaton, D., Amato, G., Legallet, E., and Zenatti, A., 1979, The role of the internal pallidal segment on the execution of a goal directed movement, Brain Res., 175:362.

Uno, M., Ozawa, N., and Yoshida, M., 1978, The mode of pallido-thalamic transmission investigated with intracellular recording from cat thalamus, Exp. Brain Res., 33:493.

Wilson, S. A. K., 1914, An experimental research into the anatomy and physiology of the corpus striatum, Brain, 36:427.

UNILATERAL ELECTROLYTIC AND 6-OHDA LESIONS OF THE SUBSTANTIA NIGRA IN BABOONS: BEHAVIOURAL AND BIOCHEMICAL DATA

Francois Viallet, Elisabeth Trouche, André Nieoullon,
Daniel Beaubaton and Eric Legallet

Institut de Neurophysiologie et Psychophysiologie
C.N.R.S., B.P. 71, 13277 Marseille Cedex 9, France

INTRODUCTION

The role of the substantia nigra (SN) in the control of motor activity has mainly been analyzed through clinical data on Parkinson's disease in human subjects (Marsden, 1982). This disease has been known for a long time to involve progressive degeneration of the pigmented neurons in the SN (Tretiakoff, 1919; Hassler, 1938). Identification of the nigrostriatal dopaminergic projection (Anden et al., 1964) from these pigmented neurons gave rise to studies in which striatal dopamine (DA) deficit was proved to be responsible for the onset of the symptoms of Parkinsonism, particularly akinesia and rigidity (Hornyckiewicz, 1966; Bernheimer et al., 1973). Besides these clinical observations on human patients, a body of experimental work has been published on animals (see Schultz, 1982) discussing the functional role of the SN.

The SN of primates is a composite structure, the histology of which shows the dopaminergic neurons of the pars compacta (SN p.c.) to be closely intermingled with the non-dopaminergic neurons of the pars reticulata (SN p.r.). The anatomical relations established by the SN p.c. are different from those of the Sn p.r., particularly with regard to the efferents of these two structures: the neurons of the SN p.c. project onto the striatum, forming the nigro-striatal dopaminergic pathway, whereas those of the SN p.r. work on other structures such as the thalamus, the superior colliculus and the reticular formation of the midbrain (Carpenter, 1976) and were recently found probably to use GABA as a neurotransmitter (Di Chiara et al., 1979). Electrophysiological recordings have moreover shown that the functional activity of the SN p.c. neurons is different from that of the SN p.r. neurons, particularly with regard to those cellular activities having to do with movement (Schultz et al., 1982; Delong et al., 1983).

Electrolytic lesion of the SN entails destruction of both the dopaminergic neurons of the SN p.c. and the non-dopaminergic neurons of the SN p.r. (Delong and Georgopoulos, 1981). Nigral lesions of this type have been produced in monkeys both bilaterally (Carpenter and McMasters, 1964; Poirier and Sourkes, 1965; Stern, 1966; Pechadre et al., 1976) and unilaterally (Goldstein et al., 1969; Viallet et al., 1981; Gross et al., 1982). The resulting symptoms closely resemble Parkinsonian akinesia and rigidity.

Specific destruction of the dopaminergic neurons of the SN p.c. has been achieved using 6-hydroxydopamine (6-OHDA), a neurotoxin showing a specific affinity for the catecholaminergic neurons. This technique has been widely used with rats (Ungerstedt, 1968; Agid et al., 1973), and less commonly with monkeys (Crossman and Sambrook, 1978): the particular morphological characteristics of the primate SN make special injection techniques and appropriate doses of 6-OHDA necessary if this approach is to be successfully employed with these animals (Schultz, 1982).

In the present study on baboons, it was proposed to compare the behavioural and biochemical effects of unilateral electrolytic lesion of the SN with those obtained by unilateral intranigral injection of 6-OHDA.

Behavioural data were obtained by quantitative analysis of a pointing movement directed towards a spatially defined visual target (Viallet et al., 1983): investigation of this pointing movement yields precise information as to the existence of any disorders affecting the initiation and execution of a standard motor task.

Biochemical measurements were conducted on the activity of the specific synthesizing enzymes for DA, GABA and acetycholine (ACh) in dopaminergic, GABAergic and cholinergic nerve terminals respectively, at the level of the SN p.c. and SN p.r. projecting structures.

METHODS

The experimental subjects were five young *Papio papio* baboons (M1, M2, M3, M11, M12) weighing between 6 and 8 kg.

Surgical Procedure

Under Nembutal anaesthesia, the animal's head was fixed in a stereotaxic frame. Radiological and electrophysiological checks were carried out, using techniques described previously (Viallet et al., 1983).

Three animals (M1, M2, M3) were subjected to an electrolytic SN lesion, performed by juxtaposing about 40 coagulation points, each corresponding to 1 mm^3 of tissue, through which a monopolar DC current was delivered (1.5 mA for 20 s).

In the other two animals (M11, M12), a chemical SN lesion was obtained by intranigral injection of 6 OHDA. Crystalline 6 OHDA at a concentration of 1 µg/µl was extemporaneously dissolved in artificial CSF containing ascorbic acid (1.2 µg/µl). The SN lesion was then performed by juxtaposing 10 injection points, into each of which was injected 1 µl of the solution.

The Pointing Movement

Experimental arrangement (Fig. 1). The material and methods used in this experiment have been described previously (Trouche and Beaubaton, 1980; Beaubaton and Trouche, 1982).

The animal was placed in a cage designed so as to obtain a standard posture facing a vertical board. The lower medial part of this board was fitted with a handle on which the animal was trained to keep its hand during the preparatory period until the visual target appeared. On the upper part of the board was a square screen on which the visual target was presented. As soon as the luminous target appeared the animal had to release the handle and touch the target with its index finger. Each sequence was recorded on line by a microprocessor system which recorded the data and carried out the statistical processing.

Variables recorded. The temporal variables recorded were respectively the reaction time (RT) and the movement duration, called movement time (MT). The accuracy of pointing was measured by the pointing error (E). RT corresponds to the time interval between the appearance of the target and release of the handle. MT corresponds to the time interval between releasing the handle and touching the screen. E corresponds to the distance between the computed mean point of the pointing coordinates and the target.

Data acquisition. Once the performances were stabilized after training, unilateral electrolytic or 6 OHDA SN lesion was performed. Mean preoperative and postoperative values were obtained from samples of 300-400 trials for RT, MT and E respectively. Postoperative developments were observed over periods running 0-20, 20-30, 30-40, 40-60, 60-90, 90-120 and 120-200 days after SN lesion.

Biochemical assays

One year after lesions were performed, the animals were sacrificed under deep anaesthesia. The brains were then rapidly

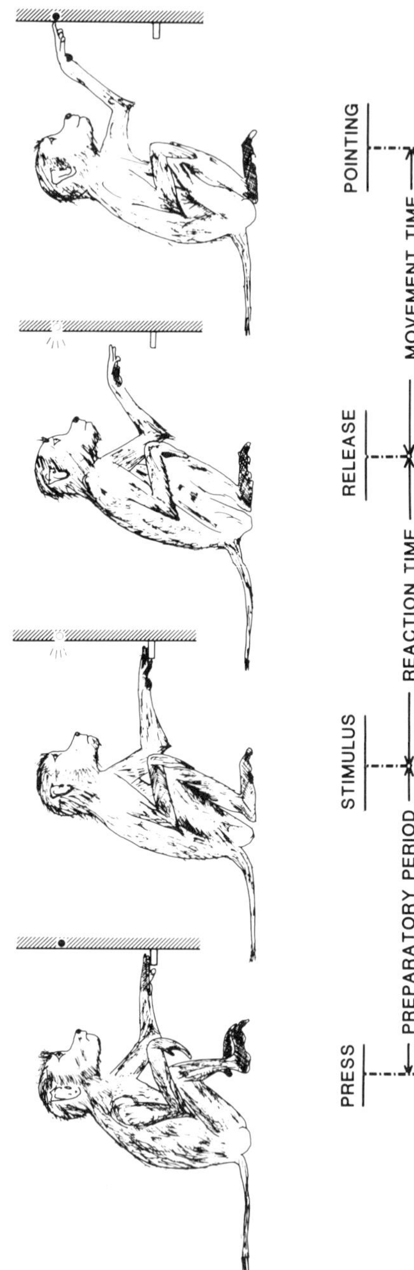

Fig. 1. Different stages in the pointing movement.

removed from the skull. A frontal transection was carried out just behind the mamillary bodies; the anterior part containing striatum, pallidum and thalamus was used for biochemical measurements, whereas the posterior part containing SN was used for the histological controls. Serial frontal sections 500 µm in thickness were made using a cryostat set at a temperature of -10°C. Microdiscs of tissue were punched out from serial slices maintained at -10°C on the frozen table of the microtome with a special stainless steel tube (diameter 0.8 mm) sharpened at the extremity, in different nuclei of the forebrain including the caudate nucleus, the putamen, the globus pallidus (external and internal parts; GPe and GPi) and the ventrolateral part of the thalamus (VL-VA area). The biochemical assays were performed on at least 90 microdiscs of tissue taken bilaterally on the successive transverse frozen sections. Activities of tyrosine-hydrolase (TH), choline acetyltransferase (CAT) and glutamic acid decarboxylase (GAD), used as specific markers for dopaminergic, cholinergic and GABAergic neurons respectively, were measured on the same tissue sample using the radioisotopic techniques described previously (Nieoullon and Dusticier, 1982; Forni and Nieoullon, 1983). The micropunches were homogenized by weak sonication in 10 volumes of an ice-cold 0.2% Triton X-1000 solution to induce a total release of enzymatic activities. TH activity was measured by the method of Waymire et al. (1971) as slightly modified by Puymirat et al. (1979). This enzymatic activity was estimated by liquid spectrometry as the rate of $^{14}CO_2$ formation from L-(1-^{14}C) tyrosine during 30 mins, at 37°C in 10 µl of tissue homogenate. GAD activity was assessed by Albers and Brady's (1959) procedure as the rate of $^{14}CO_2$ formation from L-1 (^{14}C) glutamic acid during 20 mins at 37°C in 10 µl of tissue homogenate. CAT activity was established by Fonnum's (1972) method as the rate of (^{14}C) acetylcholine (ACh) formation from ^{14}C acetyl-coenzyme A during 10 mins at 37°C in 10 ul of tissue homogenate. The protein content of each assay was determined by the method of Lowry et al. (1951). Enzymatic activities were expressed as the number of $^{14}CO_2$ nanomoles formed per hour and per mg protein for TH and GAD, and as the number of $^{14}CO_2$ nanomoles formed per hour and per mg protein for CAT. Data obtained from samples dissected from the side ipsilateral to the SN lesion were compared to those obtained from the contralateral side, which were used as controls. Statistical evaluation of the differences in enzymatic activity between both sides of the brains were achieved using Student's two-tailed t-test ($p<0.05$).

RESULTS

Behavioural data. During the post-operative period, the animals were regularly examined with a view to determining the effects of SN lesions on their spontaneous behaviour. When the animals were able to perform the pointing movement again, daily recordings were

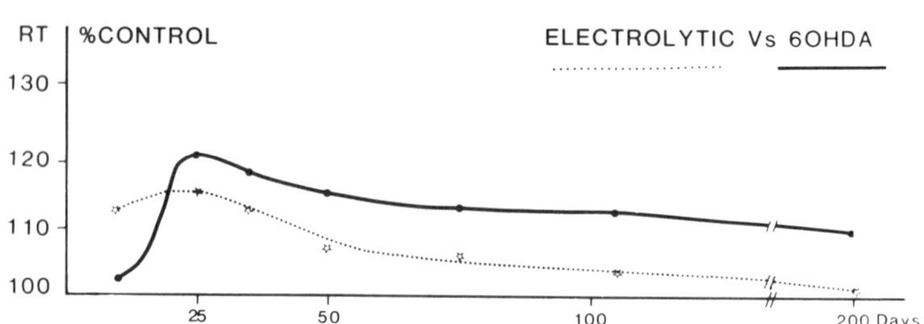

Fig. 2. Effect of electrolytic lesion and 6-OHDA lesion of the substantia nigra (SN) on reaction time (RT). On the ordinate: the percentage increase in RT with respect to the preoperative level. On the abcissa: the time interval elapsing between the moment of destruction and observation (expressed in days).

undertaken up to about six months after operation, allowing analysis of the long-term effects of SN lesion on limb movement.

Electrolytic SN lesion. (a) Spontaneous behaviour: the animals exhibited a semi-flexed posture of the contralateral body half which was accentuated in the upright position and accompanied by a dystonic inclination of the head to the side opposite the lesion. During locomotion and feeding a failure to use the contralateral body half was observed without any primary or sensory deficit. A general decrease in reactivity, as reflected by the paucity of spontaneous movements, was also observed. These akinetic and dystonic symptoms progressively improved and disappeared after one or two weeks.

(b) Pointing movement: during the initial postoperative period (0-20 days), changes in RT and MT were statistically significant on the contralateral side (Table I). RT was lengthened by 30 to 68 ms, while MT was much more markedly lengthened by 90 to 97 ms. The pointing error (E) remained unchanged. The course of these contralateral RT and MT effects up to 200 days postoperation was indicative of progressive recovery (Fig. 2). The RT values slowly regained preoperative level. The MT values began to diminish more rapidly until 60 days after operation, after which MT recovery became more progressive, reaching the preoperative level within 100 days postoperation.

6 OHDA SN lesion. (a) Spontaneous behaviour: the animals exhibited only a mild contralateral akinesia which disappeared within 48 hours. After this delay, no detectable impairment was observed during locomotion and feeding.

Table I. Unilateral electrolytic SN lesion : effects on the pointing movement (contralateral hand)

		N (trials)	Reaction time (ms)	Movement time (ms)	Pointing error (mm)
M1	Preop	355	257.84 ± 28.32	167.29 ± 22.98	4.74 ± 7.51
	Postop	331	288.34 ± 30.52	257.59 ± 45.82	4.94 ± 10.11
			(13.54)*	(32.27)*	(0.29)
M2	Preop	353	257.82 ± 47.66	138.29 ± 29.08	11.72 ± 7.47
	Postop	333	296.27 ± 50.25	235.90 ± 73.48	12.19 ± 14.18
			(10.27)*	(22.62)*	(0.54)
M3	Preop	357	219.70 ± 16.55	143.80 ± 12.25	13.08 ± 6.18
	Postop	339	288.24 ± 39.40	241.42 ± 26.98	12.70 ± 8.92
			(29.63)*	(60.90)*	(0.65)

Values are means ± SD

* $p < 0.005$ by Student's t test
Postop (0-20 days)

(b) Pointing movement: during the initial postoperative period (0-20 days), changes in MT were also statistically significant on the contralateral side; but no significant changes in RT were observed. However, 20 days after operation, a significant lengthening of RT appeared on the contralateral side. The pointing error (E) was unchanged. These results are presented in Table II. The post-operative time course of the contralateral RT and MT effects up to 200 days showed this delayed lengthening of RT (Fig. 2) to be followed by an increment in the MT values. Moreover, both RT and MT values showed a mild tendency to recover until 200 days. At this point, RT and MT were still considerably longer than the pre-operative values.

Table II. Unilateral 6 OHDA SN lesion: effects on the pointing movement (contralateral hand)

		N (trials)	Reaction time (ms)	Movement time (ms)	Pointing error (mm)
M11	Preop	372	238.75 ± 21.32	147.31 ± 18.39	3.38 ± 5.24
	Postop 1	259	240.91 ± 27.59 (1.05)	230.81 ± 29.44 (40.47)*	2.87 ± 5.71 (1.14)
	Postop 2	445	291.43 ± 43.74 (22.41)*	202.96 ± 35.29 (28.89)*	3.19 ± 5.28 (0.51)
M12	Preop	398	254.16 ± 27.17	144.25 ± 19.81	6.88 ± 6.27
	Postop 1	453	257.46 ± 26.24 (1.79)	222.49 ± 27.65 (47.84)*	6.46 ± 5.26 (1.05)
	Postop 2	433	294.96 ± 26.37 (21.93)*	190.71 ± 19.83 (33.78)*	7.35 ± 5.21 (1.17)

Values are means ± SD

*$p < 0.005$ Student's t test
Postop 1 (0-20 days); Postop 2 (21-30 days)

Histology and Biochemical data

Electrolytic SN lesion. Histological checks were conducted in order to ensure that the lesions were confined to the SN and did not damage adjacent structures. The lesion involved both the SN p.c., which showed considerable neuronal impairment, and the SN p.r. Figure 3B shows the results of the biochemical measurements. The TH activity shows a significant decrease (-47%) in the caudate nucleus, no change in the putamen and the GPi, and a slight increase in the GPe (+20%). The GAD activity shows no change in the caudate nucleus and a slight increase in the putamen (+19%), but shows a significant decrease in the thalamus VA-VL area (-34%). In all of these structures the CAT activity remains unchanged.

6 OHDA SN lesion. The histological examinations showed considerable impairment of the pigmented neurons in the SN p.c. An overall decrease in volume of the SN was also observed in comparison with the intact side. Figure 3C shows the results of the biochemical

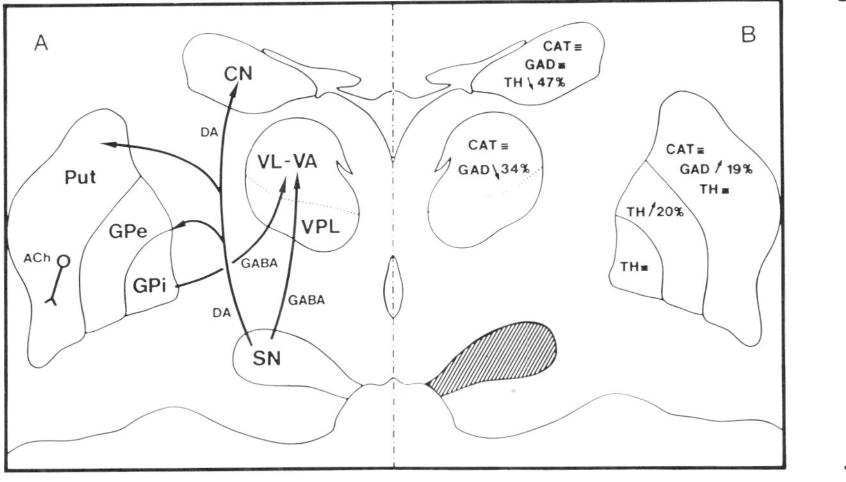

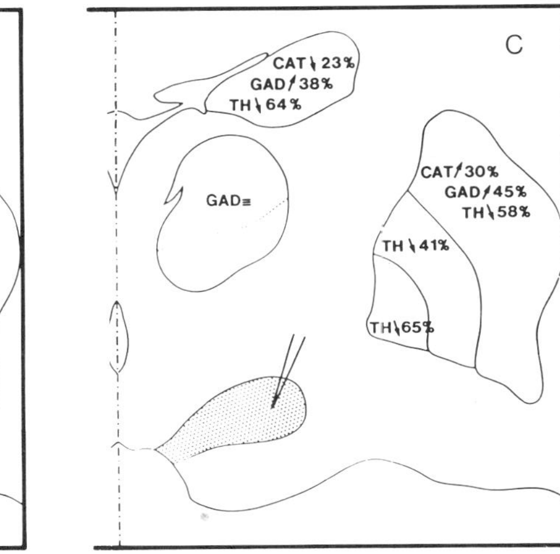

Fig. 3. Changes in tyrosine hydroxylase (TH), glutamic acid decarboxylase (GAD) and choline acetyltransferase (CAT) activities in basal ganglia and thalamic structures (A) about one year after electrolytic (B) and 6-hydroxydopamine (C) lesion of substantia nigra (SN) in baboons on one side of brain. Changes in enzymatic activities in structures ipsilateral to SN lesion expressed as percentage of activity measured in homologous structure on contralateral side. CN: caudate nucleus; Put.: putamen; GPe: globus pallidus, external part; GPi: globus pallidus, internal part; SN: substantia nigra; VL-VA: thalamus ventrolateral-ventroanterior; DA: dopamine; GABA: gamma amino-butyric acid; ACh: acetylcholine. Results are means of values collected in two animals in each group. Mean control activity (nanomoles of $^{14}C\text{-}CO_2$ or ^{14}C-ACh formed/h/mg protein):

	CN	Put	GPe	GPi	VL-VA
TH	5.2	4.5	0.5	0.6	-
GAD	117.3	127.2	-	-	107.0
CAT	81.3	108.9	-	-	22.8

measurements: a homogeneous, highly significant decrease in TH activity can be seen in the caudate nucleus, the putamen and the globus pallidus (-41% to -65%). GAD activity shows no change at the level of the thalamus, but significantly increases in the putamen (+45%) and the caudate nucleus (+38%). CAT activity also significantly increases in the putamen (+30%) but decreases in the caudate nucleus (-23%).

DISCUSSION

The results obtained after unilateral electrolytic and chemical lesions of the SN suggest that this structure plays a specific role in the control of the spontaneous motor activity of the contralateral body half and in the initiation and execution of a pointing movement performed with the contralateral hand. The biochemical data illustrate the extent to which the nigral lesion has directly affected the dopaminergic and GABAergic neurons in the substantia nigra or indirectly affected the activity of GABAergic and cholinergic neurons in the structures innervated by the nigral efferent pathways. The extent and the specificity of the lesion could therefore be correlated to the severity of the behavioural syndroms.

Lesion of the SN and Spontaneous Behaviour

Conspicuous semi-flexed posture of the contralateral body-half was shown by subjects only after electrolytic lesion of the SN. This disorder is reminiscent of the postural defects observed as a result of pallidal lesions (Denny-Brown, 1962; Delong and Coyle, 1979; Beaubaton et al., 1981). Now recent studies tend to suggest that GPi forms a functional entity with the SN p.r., from which it is separated in primates by the internal capsule (Delong and Georgopoulos, 1979; Nauta, 1979). It is thus feasible that the postural dystonia which occurs after electrolytic lesion of the SN may be partly caused by the destruction of SN p.r. non-dopaminergic neurons (Viallet et al., 1983). The fact that this symptom does not occur after 6-OHDA lesion pf the SN further supports this hypothesis. Indeed, the biochemical measurements show a significant decrease in GAD activity at thalamus level after electrolytic lesion of the SN; this result is compatible with the destruction of the nigro-thalamic neurons originating in the SN p.r. (Carpenter et al., 1976; Feger, 1981) which, it has been suggested, may use GABA as their neurotransmitter (Di Chiara et al., 1979). 6-OHDA lesion of the SN, on the contrary, does not change GAD activity at thalamus level, which suggests that the nigrothalamic neurons are not affected by the neurotoxin. Further studies are nevertheless required in order to confirm this finding.

Non-utilization syndrome of the contralateral body half was observed for 1 to 2 weeks after electrolytic lesion of the SN. The absence of any primary motor or sensory deficit prompts one to introduce the concept of sensory neglect as a means of explaining the disuse, which is so strikingly similar to the akinesia observed in human Parkinsonians. Studies performed on rats (Ljungberg and Ungersted, 1976; Marshall, 1979; Siegfried and Bures, 1979) and on cats (Feeney and Wier, 1979) have tended to focus on the concept of contralateral sensory neglect occurring after unilateral lesion of the SN. Sensory neglect may well be caused by the destruction of nigrostriate dopaminergic neurons, as has been shown by pharmacological analysis (Ljungberg and Ungerstedt, 1976; Marshall, 1979).

It is nevertheless surprising that after 6-OHDA lesion of the SN, the contralateral disuse syndrome appears only in a rather mild form. In our opinion, this syndrome is unlikely to result merely from the postural impairments. This illustrates the kind of difficulties often encountered in the observation of primates' spontaneous behaviour, and shows how quantitative analysis of the pointing movement can be a useful alternative.

The absence of spontaneous circling behaviour in our animals after either electrolytic or 6-OHDA lesions of the SN calls for some discussion. On the basis of numerous studies on rats, it has now been recognized that unilateral destruction of the SN causes circling behaviour (Ungerstedt, 1968): this destrucion can be either specific for the dopaminergic neurons if produced by 6-OHDA (Ungerstedt and Arbuthnott, 1970) or non-specific if performed by electrocoagulation (Schwartz et al., 1976). Circling can be said to result from a disturbed distribution of muscle tonus, which in conjunction with unilateral sensory neglect brings about a functional asymmetry (Glick et al., 1976). Circling behaviour of this type has been elicited in monkeys by 6-OHDA injections in the dopaminergic bundle at the level of the lateral hypothalamus, causing a widespread destruction of the nigrostriatal dopaminergic neurons (Crossman and Sambrook, 1978). It should nevertheless be specified that this type of functional asymmetry is however difficult to obtain in monkeys when the lesion is performed at the actual level of the SN. Compensatory mechanisms involving an increased activity of the neurons remaining intact have been shown to exist at the level of the nigrostriatal dopaminergic pathway (see Schultz, 1982), which make it virtually impossible for a unilateral SN lesion to cause sufficient asymmetry for circling behaviour to ensue.

After 6-OHDA lesion of the SN in monkeys, our biochemical measurements show a maximal decrease of 65% in TH activity in the ipsilateral striatum. This decrease was no doubt insufficient to entail spontaneous circling behaviour, and suggest vigorous compensatory activity on the part of the remaining dopaminergic neurons. A compensatory mechanism of this type has already been proved to exist (Agid et al., 1974; Hefti et al., 1980).

After electrolytic lesion of the SN in monkeys, the decrease in striatal TH activity was even less conspicuous (47%) and in the case of some sites sampled, entirely null. The same type of compensation may therefore be involved again.

It should thus be emphasized that in rats, functional asymmetry induced by SN lesion brings about visible, long-lasting behavioural impairments which can be easily recognized by observing spontaneous behaviour. In monkeys it is difficult to induce such a functional asymmetry. It was therefore necessary to study the pointing movement which allows better quantitative analysis of post-lesional disorders.

Lesion of the SN and the Pointing Task

Analysis of the pointing movement performed after electrolytic and 6-OHDA lesion of the SN shows a lengthening of RT and particularly of MT on the contralateral side without any change in pointing error (E). This suggests that the SN intervenes before and during execution of the movement.

SN and movement initiation. These experiments show that lesion of the SN induces a lengthening of RT. By using a system of strain gauges attached to the fixed handle to record the force applied by the animal's hand to the handle it was shown that the lengthening of RT was not due only to the reduction in the initial speed of movement. A similar lengthening of RT has also been observed in Parkinsonian patients (Heilman et al., 1976; Evarts et al., 1981). The possible interpretations of this lengthening of RT have been discussed previously (Trouche et al., 1983a; Viallet et al., 1983).

SN and movement execution. The lengthening of MT observed after lesion of the SN with no change in terminal accuracy lends itself to two complementary interpretations. On the one hand, the slowing of movement results in a decrease in the initial acceleration. This has been demonstrated previously by means of motion film recordings of the hand trajectory during the movement (Viallet et al., 1983). Furthermore, EMG recordings made during rapid movements performed by Parkinsonian patients show a decrease in the amplitude of activity bursts in the muscles involved in the movement (Hallett and Khoshbin, 1980). On the other hand, the movement is controlled to a greater extent by visual feed-back, since the final accuracy of the movement is maintained by an increase in visual feed-back (Hore et al., 1977, Viallet et al., 1983). In the same way, the accuracy of visually controlled movements is unaffected in Parkinsonian patients, wereas their errors increase considerably in the absence of visual feed-back (Flowers, 1976; Cooke et al., 1978). This body of data suggests that the control of movement execution changes over from a feed-forward to a feed-back mode of control as the result of SN lesion (Trouche et al., 1983b).

Biochemical Correlations

Comparison of the results obtained with electrolytic versus 6-OHDA lesion of the SN brings two noteworthy differences to light:
- the lengthening of RT which occurs immediately after electrolytic lesion appears only 20 days after 6-OHDA lesion;
- the recovery curves show that RT and MT disturbances last much longer after 6-OHDA lesion than after electrolytic lesion.

RT effect: electrolytic versus 6-OHDA SN lesion. If 6-OHDA lesion of the SN is considered to be specific for the dopaminergic neurons, the occurrence of a lengthening of RT after a lapse of 20 days can be assumed to result from the dopaminergic dysfunction. Now it has been shown that after 6-OHDA intranigral injection in rats, destruction of the dopaminergic neurons was achieved much more rapidly, within roughly 24 hours (Agid, 1973; McGeer et al., 1973; Forni and Nieoullon, 1983). Moreover, comparison between 6-OHDA and electrolytic lesion of the SN in the rat showed no difference with regard to the effects on TH activity at the level of the striatum (Agid et al., 1974; McGeer et al., 1973). Comparative behavioural studies on the water intake of rats have nevertheless been reported in which 6-OHDA lesion of the SN caused a deficit only in sensitization tests, whereas electrolytic lesions of the SN were accompanied by an immediate, spontaneous deficit (Fibiger et al., 1973). This type of result is comparable to our own findings. It thus seems likely that a hyperactivity of the remaining dopaminergic neurons fulfils a compensatory function immediately after lesion of the SN. This compensation may suffice to prevent the occurrence of an immediate lengthening of RT after a 6-OHDA lesion of the SN. After a lapse of 20 days the compensatory mechanism is no longer sufficient to supply the dopaminergic receptor site in the striatum, so that a dopaminergic deficit is set up, along with considerable lengthening of the RT. The intensity of the dopaminergic dysfunction might therefore explain the increases in GAD and CAT activities measured in the putamen which could correspond to the release of the dopaminergic inhibitory input on the striatal interneurons. The dopaminergic deficit thus constituted does not, however, lend itself to the induction of impairments in spontaneous behaviour, as mentioned earlier. The decrease of around 65% in striatal TH activity does not in fact correspond to a sufficiently pronounced functional asymmetry (see Schultz, 1982).

It now remains to account for the immediate lengthening of RT observed after electrolytic lesion of the SN. The postural disorders that were found to most likely result from the destruction of non-dopaminergic neurons in the SN p.r. might be implicated here. The important part played by postural disorders in the onset of oro-digestive deficits observed in rats after lesion of the GPI has already been stressed (Labuszewski et al., 1981). Hence in monkeys, a lengthening of RT after electrolytic lesion would appear to be initially caused by postural disorders more likely resulting from the lesion of the non-dopaminergic neurons, and ultimately maintained by a dopaminergic deficit, the onset of which can be delayed by the hyperactivity of the remaining dopaminergic neurons, providing a temporary functional compensation as after 6-OHDA lesion of the SN.

RT and MT recovery: electrolytic versus 6-OHDA SN lesions. The persistance of RT and MT lengthening beyond 200 days after 6-OHDA lesion of the SN would seem to indicate that the induced dopaminergic deficit was a chronic one. This assumption was moreover corroborated by the biochemical measurements, which show not only the decrease in TH activity but also an increase in the GAD and CAT activities in the putamen ipsilateral to the nigral lesion more than one year after operation. Given the existence of this chronic dopaminergic deficit, it might be feasible to consider the role possibly played by a progressively developing denervation supersensitivity at the level of the striatal dopaminergic receptor sites. Investigations on the rat have shown that this supersensitivity takes the form of an enhanced response after the administration of DA agonists; this response appears progressively and was reported to reach a maximum only 10 to 20 days after the lesion (Krueger et al., 1976; Mishra et al., 1980; Staunton et al., 1981). The dopaminergic deficit has to be at least 80 to 90% in order to trigger such a compensatory mechanism, however (see Schultz et al., 1982b). It is therefore unlikely that this mechanism came into play with our animals except in some precise and rather small areas of the brain.

After electrolytic lesion of the SN, recovery of RT and MT occurs more rapidly than in the case of 6-OHDA lesion. This result was corroborated by the biochemical data, which show that striatal TH activity was less than in the case of the neurotoxic lesion.

In conclusion, the following statements emerge from our comparison of the behavioural and biochemical data on electrolytic versus 6-OHDA lesion of the SN in monkeys.

- The postural disorders observed after SN lesion are most likely related to the destruction of the non-dopaminergic neurons in the SN p.r. They may be responsible for the immediate lengthening of RT observed after electrolytic lesion of the SN.

- Non-utilisation of the body half contralateral to the SN lesion is possibly related to the existence of a dopaminergic deficit. This deficit is probably rapidly accompanied by a functional compensation after an electrolytic lesion; whereas after a 6-OHDA lesion, although there is sufficient compensation to prevent the occurrence of spontaneous circling behaviour, it becomes inadequate after a lapse of 20 days, giving way to a lengthening of RT. The persistent dopaminergic deficit after 6-OHDA lesion is accompanied by a durable lengthening of RT and MT.

- The compensatory mechanism involved is probably a hyperactivity of the remaining dopaminergic neurons. The extent of the induced dopaminergic deficit seems to be insufficient for a denervation supersensibility to appear at the level of the striatal DA receptors.

ACKNOWLEDGEMENTS

The authors wish to thank D. Casanova, N. Dusticier, R. Massarino,, G. Poveda and A. Zenatti for their technical assistance and J. Blanc for her help with the English translation. This work was partly supported by CNRS grant (ATP neurobiologie du système nerveux central) and INSERM grant (PRC Cerveau et sante mentale).

REFERENCES

Agid, Y., Javoy, F., Glowinski, J., Bouvet, D., and Sotelo, C., 1973, Injection of 6-hydroxydopamine into the substantia nigra of the rat. II. Diffusion and specificity, Brain Res., 58:291-301.

Agid, Y., Javoy, F., and Glowinski, J., 1974, Chemical or electrolytic lesion of the substantia nigra: early effects on neostriatal dopamine metabolism, Brain Res., 74:41-49.

Albers, R. W., and Brady, R. O., 1959, The distribution of glutamic decarboxylase in the nervous system of the rhesus monkey, J. Biol. Chem., 234:926-928.

Anden, N. E., Carlsson, A., Dahlstrom, A., Fuxe, K., Hillard, N. A., and Larsson, K., 1964, Demonstration and mapping out of nigrostriatal dopamine neurons, Life Sci., 3:523-530.

Beaubaton, D., Trouche, E., Amato, G., and Legallet, E., 1981, Perturbations du declenchement et de l'éxecution d'un mouvement visuellement guidé chez le babouin au cours du refroidissement et après lésion du segment interne du globus pallidus. J. Physiol. (Paris), 77:107-118.

Beaubaton, D., and Trouche, E., 1982, Participation of the cerebellar dentate nucleus in the control of a goal-directed movement in monkeys, Exp. Brain Res., 46:127-138.

Bernheimer, H., Birkmayer, W., Hornkiewicz, O., Jellinger, K. and Seitelberger, F., 1973, Brain dopamine and the syndromes of Parkinson and Huntington. Clinical: morphological and neurochemical correlations, J. Neurol. Sci., 20:415-455.

Carpenter, M. B., and McMasters, R. E., 1964, Lesions of the substantia nigra in the rhesus monkey. Efferent fiber degenerations and behavioural observations. Amer. J. Anat., 114:293-319.

Carpenter, M. B., 1976, Anatomical organization of the corpus striatum and related nuclei, in: "The Basal Ganglia," M. D. Yahr, ed.., Raven Press, New York.

Carpenter, M. B., Nakano, K., and Kim, R., 1976, Nigrothalamic projections in the monkey demonstrated by autoradiographic technics, J. Comp. Neurol., 165:401-416.

Cooke, J. D., Brown, J. D., and Brooks, V. B., 1978, Increased dependence on visual information for movement control in patients with Parkinson's disease, Can. J. Neurol. Sci., 5:413-415.

Crossman, A. R., and Sambrook, M. A., 1978, Experimental torticollis in the monkey produced by unilateral 6-hydroxy-dopamine brain lesions, Brain Res., 149:498-502.

Delong, M. R., and Coyle, J. T., 1979, Globus pallidus lesions in the monkey produced by kainic acid: histologic and behavioural effects, Appl. Neurophysiol., 42:95-97.

Delong, M. R. and Georgopoulos, A. D., 1979, Motor functions of the basal ganglia as revealed by studies of single cell activity in the behaving primate, in: "Advances in Neurology," Vol. 24, L. J. Poirier, T. L. Sourkes and P. J. Bedard, eds., Raven Press, New York.

Delong, M. R. and Georgopoulos, A. P., 1981, Motor functions of the basal ganglia, in: "Handbook of Physiology", sect. 1: "The Nervous System", Vol. II, Part 2, V. B. Brooks, ed.

Delong, M. R., Crutcher, M. D., and Georgopoulos, A. P., 1983, Relations between movement and single cell discharge in the substantia nigra of the behaving monkey, The Journal of Neuroscience, 3 (8):1589-1606.

Denny-Brown, D., 1962, "The Basal Ganglia and Their Relation to Disorders of Movement," Oxford University Press, London.

Di Chiara, G., Porceddu, M. L., Morelli, M., Mulas, M. L., and Gessa, G. L., 1979, Evidence for a GABAergic projection from the substantia nigra to the ventromedial thalamus and to the superior colliculus of the rat, Brain Res., 176:273-284.

Evarts, E. V., Teravainen, H., and Calne, D. B., 1981, Reaction time in Parkinson's disease, Brain, 104:167-186.

Feeney, D. M., and Wier, C. S., 1979, Sensory neglect after lesions of substantia nigra or lateral hypothalamus: differential severity and recovery of function, Brain Res., 178:329-346.

Feger, J., 1981, Les ganglions de la base: aspects anatomiques et electrophysioliques, J. Physiol. (Paris), 77: 7-44.

Fibiger, H. C., Phillips, A. G. and Clouston, R. A., 1973, Regulatory deficits after unilateral electrolytic or 6-OHDA lesions of the substantia nigra, Amer. J. Physiol., 225:1282-1287.

Flowers, K. A., 1976, Visual "closed-loop" and "open-loop" characteristics of voluntary movement in patients with parkinsonism and intention tremor, Brain, 99:269-310,

Fonnum, F., 1972, Application of microchemical analysis and sub cellular fractionation technique to the study of neurotransmitters in discrete areas of mammalian brain, Adv. Biochem. Psychopharmacol., 6:75-88.

Forni, C., and Nieoullon, A., 1983, Electrochemical detection of dopamine release in the striatum of freely moving hamsters, Brain Res. (in press).

Glick, S. D., Jerussi, T. P., and Fleisher, L. N., 1976, Turning in circles: the neuropharmacology of rotation, Life Sci., 18:889-896.

Goldstein, M., Anagnoste, B., Battista, A. F., Owen, W. S., and Nakatani, S., 1969, Studies of amines in the striatum in monkeys with nigral lesions, J. Neurochem., 16:645-653.

Gross, Ch., Feger, J., Seal, J., Haramburu, Ph. and Bioulac, B., 1982, Neuronal activity in area 4 and movement parameters recorded in trained monkeys after unilateral lesion of the substantia nigra, in: "Neuronal Coding of Motor Performance," J. Massion, J. Paillard, W. Schultz, M. Wiesendanger, eds., Springer-Verlag, Berlin.

Hallett, M., and Khoshbin, S., 1980, A physiological mechanism of bradykinesia, Brain, 103:301-314.

Hassler, R., 1938, Zur pathologie der paralysis agitans und des post encephalitischen parkinsonismus, J. Psychol. Neurol., 48:387-476.

Hefti, F., Melamed, E., and Wurtman, R. J., 1980, Partial lesions of the dopaminergic nigrostriatal system in rat brain: biochemical characterization, Brain Res., 195:123-127.

Heilman, K. M., Bowers, D., Watson, R. T. and Greer, M., 1976, Reaction times in parkinson disease, Arch. Neurol. (Chic.)., 33:139-140.

Hore, J., Meyer-Lohmann, J., and Brooks, V. B., 1977, Basal ganglia cooling disables learned arm movements in monkey in the absence of visual guidance, Science, 195:584-586.

Hornyckiewicz, O., 1966, Dopamine (3-Hydroxytyramine) and brain function, Pharmac. Rev., 18:925-964.

Krueger, B. K., Forn., J., Walters, J. R., Roth, R. H., and Greengard, P., 1976, Stimulation by dopamine of adenosine cyclic 3.5-monophosphate formation in rat caudate nucleus: effect of lesions of the nigro neostriatal pathway, Molec. Pharmac., 12:639-648.

Labuszewski, T., Lockwood, R., McManus, F. E., Edestein, L. R., Lidsky, T., 1981, Role of postural deficits in oro-ingestive problems caused by globus pallidus lesions, Exp. Neurol., 74:93-110.

Ljungberg, T., and Ungerstedt, U., 1976, Sensory inattention produced by 6-Hydroxydopamine induced degeneration of ascending dopamine neurons in the brain, Exp. Neurol., 53:585-600.

Lowry, O. H., Rosebrough, N. J., Farr, A. L., Randall, R. J., 1951, Protein measurement with the folin phenol reagent, J. Biol. Chem., 193:265-275.

Marsden, C. D., 1982, The mysterious motor function of the basal ganglia: the Robert Wartenberg lecture, Neurology, 32:514-539.

Marshall, J. F., 1979, Somatosensory inattention after dopamine-depleting intracerebral 6-OHDA injections: spontaneous recovery and pharmacological control, Brain Res., 177:311-324.

McGeer, E. G., Fibiger, H. C., McGeer, P. L., Brooke, S., 1973, Temporal changes in amine synthesizing enzymes of rat extrapyramidal structures after hemitransections or 6-Hydroxydopamine administration, Brain Res., 52:289-300.

Mishra, R. K., Marshall, A. M., and Varmuza, S. L., 1980, Supersensitivity in rat caudate nucleus: effects of 6-hydroxydopamine on the time course of dopamine receptor and cyclic AMP changes, Brain Res., 200:47-57,

Nauta, H. J. W., 1979, A proposed conceptual reorganization of the basal ganglia and telencephalon, Neurosci., 4:1875-1881.

Nieoullon, A., Dusticier, N. 1982, Glutamate uptake, glutamate decarboxylase and choline acetyl-transferase in subcortical areas after sensorimotor cortical ablations in the cat, Brain Res. Bull., 10:287-293.

Pechadre, J. C., Larochelle, L:. Poirier, L. J., 1976, Parkinsonian akinesia, rigidity and tremor in the monkey. Histopathological and neuropharmacological study, J. Neurol. Sci., 28:147-157.

Poirier, L. J., and Sourkes, T. L., 1965, Influence of the substantia nigra on the catecholamine content of the striatum, Brain, 88:181-192.

Puymirat, J., Javoy-Agid, F., Gaspar, P., Ploska, A., Prochiantz, A., and Agid, Y., 1979, Post mortem stability and storage in the cold of brain enzymes, J. Neurochem., 32:449-454.

Schultz., W., 1982, Depletion of dopamine in the striatum as an experimental model of Parkinsonism: direct effects and adaptive mechanisms, Prog. in Neurobiol., 18:121-166.

Schultz, W., Aebischer, P. and Ruffieux, A., 1982, The encoding of motor acts by the substantia nigra, in: "Neural Coding of Motor Performance," J. Massion, J. Paillard, W. Schultz, M. Wiesendanger, eds., Springer-Verlag, Berlin.

Schwartz., W. J., Gunn, R. H., Sharp, F. R., and Evarts, E. V., 1976, Unilateral electrolytic lesions of the substantia nigra cause contralateral circling in rats, Brain Res., 105: 358-361.

Siegfried, B., and Bures, J., 1979, Conditioning compensates the neglect due to unilateral 6-OHDA lesions of substantia nigra in rats, Brain Res., 167:139-155.

Staunton, D. A., Wolfe, B. B., Groves, P. M., Molinoff, P. B., 1981, Dopamine receptor changes following destruction of the nigrostriatal pathway: lack of a relationship to rotational behavior, Brain Res., 211:315-328.

Stern, G., 1966, The effects of lesions in the substantia nigra, Brain, 89:449-478.

Tretiakoff, C., 1919, Contribution à l'étude de l'anatomie pathologique du locus niger de Soemmering avec quelques déductions relatives à la pathologie des troubles du tonus musculaire de la maladie de Parkinson, Thèse Médecine, Paris.

Trouche, E., and Beaubaton, D., 1980, Initiation of a goal-directed movement in the monkey, Exp. Brain Res., 40:311-321.

Trouche, E., Beaubaton, D., Amato, G., Viallet, F., Legallet, E., 1983a, Changes in reaction time after pallidal or nigral exclusion, in: "Advances in Neurology," Vol. 40, R. G. Hassler, and J. F. Christ, eds., Raven Press, New York (in press).

Trouche, E., Beaubaton, D., Viallet, F., Legallet, E., 1983b, Central coding of movements direction: participation of the neocerebellum and the basal ganglia as revealed by reaction time studies in monkeys, in: "Space Physiology." F. Lestienne and A. Barthoz, eds. (in press).

Ungerstedt, U., 1968, 6-Hydroxydopamine induced degeneration of central monoamines neurons, Eur. J. Pharmacol., 5:107-110.

Ungerstedt, U. and Arbuthnott, G. W., 1970, Quantitative recording rotational behaviour in rats after 6-Hydroxydopamine lesions of the nigrostriatal dopamine system, Brain Res., 21:485-493.

Viallet, F., Trouche, E., Beaubaton, D., Nieoullon, A., and Legallet, E., 1981, Bradykinesia following unilateral lesions restricted to the substantia nigra in the baboon, Neurosci. Lett., 24:97-102.

Viallet, F., Trouche, E., Beaubaton, D., Nieoullon, A., and Legallet, E., 1983, Motor impairment after unilateral electrolytic lesions of the substantia nigra in baboons: behavioural data with quantitative and kinematic analysis of a pointing movement, Brain Res., (in press).

Waymire, J. C., Bjur, R., Weiner, N., 1971, Assay of tyrosine hydroxylase by coupled decarboxylation of DOPA from (1 ^{14}C) L. tyrosine, Analyt. Biochem., 43:588-600.

POSTURAL AND BEHAVIORAL CHANGES RELATED TO NIGRAL CELL LOSS IN MONKEYS

C. Ohye, T. Shibazaki, T. Hirai, Y. Nagaseki, M. Hirato, H. Wada and Y. Kawashima

Department of Neurosurgery
Gunma University School of Medicine
3-39, Showa-machi, Maebashi, Gunma, Japan

INTRODUCTION

In the course of studying the neural mechanism of experimental tremor in monkeys by a mesencephalic ventromedial tegmental lesion (Ohye, 1978, 1979; Ohye et al., 1979, 1983) it was always noticed that the monkeys with tremor exhibited a characteristic "flexed posture" of the contralateral upper limb, adduction at the shoulder joint, flexion (about 90°) at the elbow joint and natural extension of the wrist and fingers (Fig. 1). The monkey usually neglected the affected limb and did not use it except in some emergency situation such as being threatened, or restriction of the normal limb. To elucidate the cause of this postural change, we have conducted several anatomical and physiological studies.

METHODS

Fourteen small monkeys (macaca irus) weighing less than 5 Kg were used. The monkey was placed in the stereotactic apparatus under light nembutal (20 mg/Kg) and Ketamine (0.2 ml) anaesthesia. A mesencephalic ventromedial tegmental (VMT) lesion was made stereotactically under radiological and physiological controls (cf. Fig. 6), using a pair of Leksell's coagulation needles and a lesion generator, or by a platinum electrode for smaller selective lesions. The monkey was kept in a cage, and its behavioral and postural changes observed periodically. Neurological examination, EMG examination and cinematographic observation were repeated at intervals. In the later series, in two monkeys, a selective small VMT lesion was made through a platinum microelectrode positioned

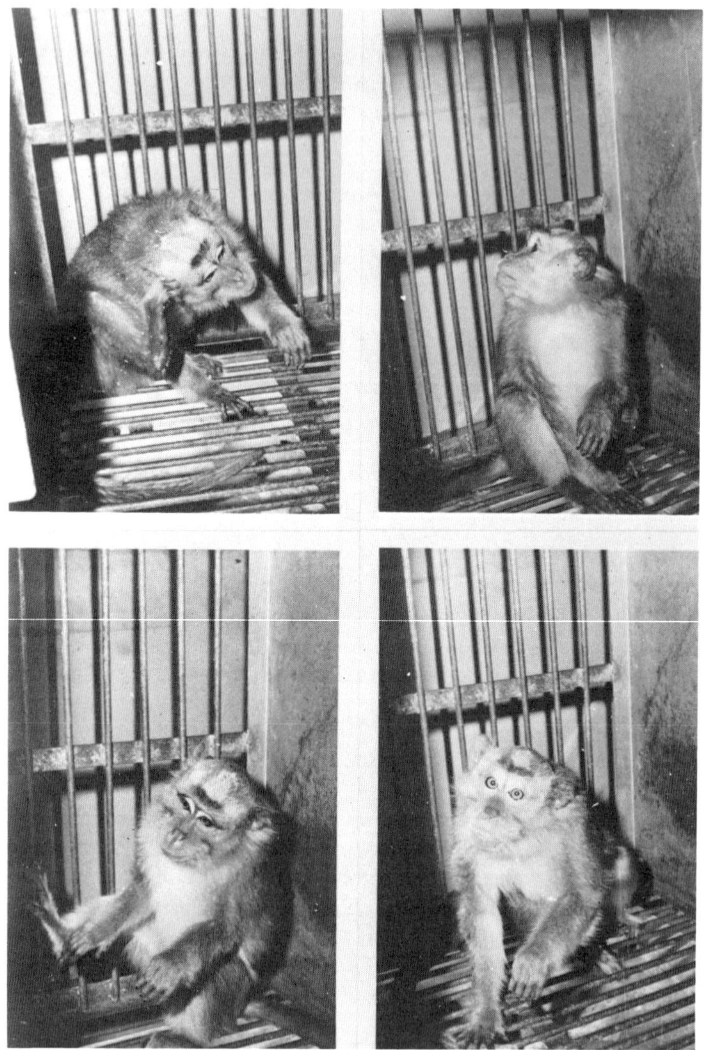

Fig. 1. Four different pictures of a lesioned monkey (Rt-VMT) showing the typical flexed posture of the left upper limb. Note that the affected limb is always in the same position regardless of his attitude.

under radiological and physiological controls. Further, in two monkeys, local injection of 6-OH-dopamine (DA) was used to destroy nigral cells as far as possible without damaging the mesencephalic reticular formation close to the substantia nigra (SN). Operated monkeys were observed for various survival times according to the

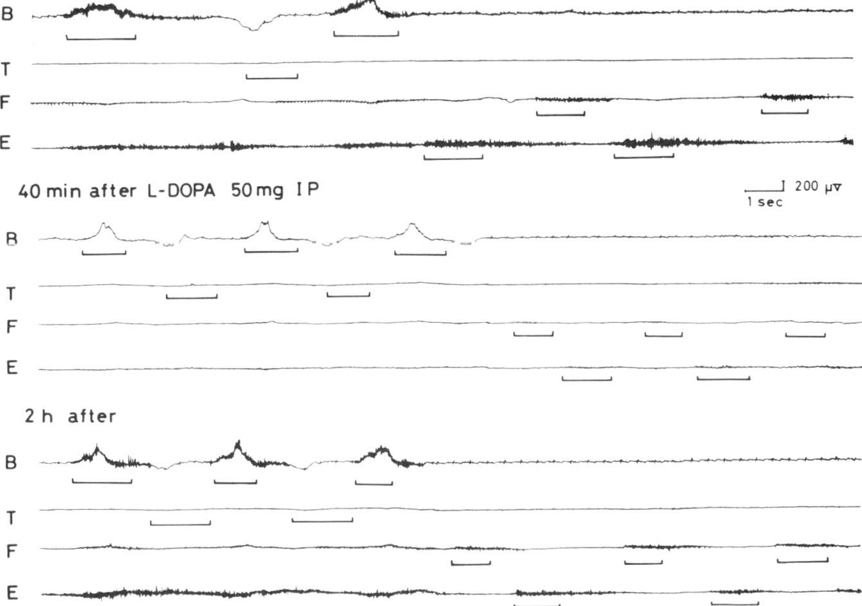

Fig. 2. EMG recordings from the right upper limb muscles of a lesioned monkey (Lt-VMT) demonstrating an exaggerated tonic stretch reflex and the effect of L-Dopa on it. Upper set of four traces: tonic stretch reflex before L-Dopa administration. Middle: Effect of L-Dopa (50 μmg, IP), stretch reflex being reduced after 40 min. Lower: Stretch reflex reappeared after 2 hr. Short horizontal bars indicate approximate moment of stretch. B: biceps brachii, T: triceps brachii, F: forearm flexor, E: forearm extensor.

experimental plan and finally sacrificed under deep nembutal anesthesia. The brain was perfused with saline and then with neutral formalin. Histological examinations were carried out on serial sections cut at 50 μm and stained by the Nissl method.

RESULTS

Operated monkeys exhibited ipsilateral midriasis, torticollis towards the operated side (immediate effects), contralateral flexed posture and tremor (late effects) in various combinations and degrees (Poirier, 1960; Ohye et al., 1979). In this paper, however, our attention is focused only on the postural changes, and other systems mentioned have been analysed elsewhere (Ohye et al., 1983).

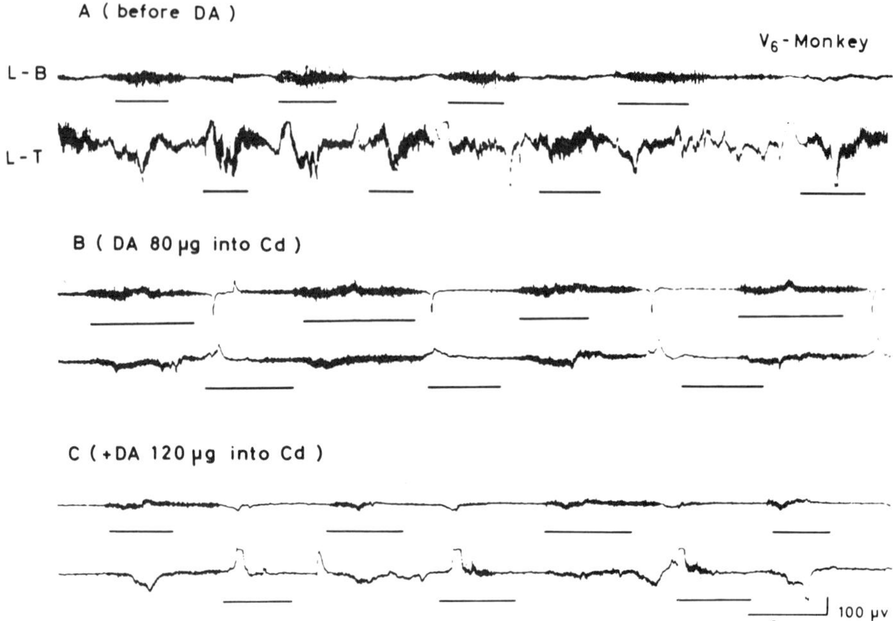

Fig. 3. EMG recordings from the left upper limb muscles of a lesioned monkey (Rt-VMT) showing the effect of DA into the CD nucleus. A: exaggerated tonic stretch reflex, before DA injection, B: slightly reduced stretch reflex after DA (80 ng) injection, C: much reduced stretch reflex after DA (120 ng) injection. B: biceps brachii, T: triceps brachii. Short horizontal bars indicate approximate moment of stretch.

The postural change usually developed gradually within one week, and it continued unchanged for several months or years (long term observations were made in another series of experiments).

EMG examination of the affected limb by surface or needle electrodes revealed exaggerated stretch reflexes of the tonic type as shown in Fig. 2A. It was generally the case that the biceps brachii and forearm extensor muscles were more affected. This kind of exaggerated tonic stretch reflex was ameliorated temporarily by intravenous administration of L-Dopa (50 mg). The effect of L-Dopa appeared around 30 min after injection and terminated after about 2 hrs (Fig. 2). In two monkeys, local injection of dopamine (DA) itself into the caudate (Cd) nucleus was tried. For this purpose, a small tube (diameter 0.7 mm) was implanted into the Cd head under radiological control, contralateral to the affected flexed limb. DA was injected with the aid of a microsyringe through a fine cannula inserted into the implanted tube. The effect of DA was checked

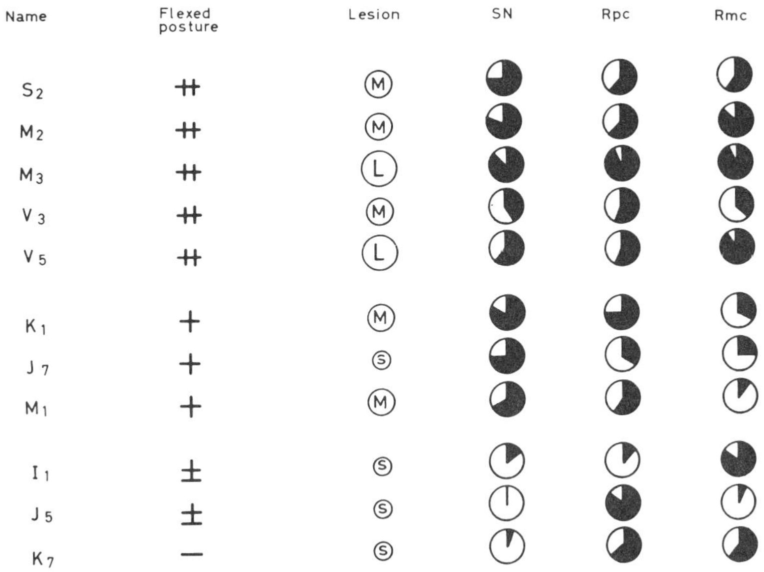

Name	Flexed posture	Lesion	SN	Rpc	Rmc
S_2	++	M			
M_2	++	M			
M_3	++	L			
V_3	++	M			
V_5	++	L			
K_1	+	M			
J_7	+	S			
M_1	+	M			
I_1	±	S			
J_5	±	S			
K_7	—	S			

Fig. 4. Correlation of the degree of flexed posture (++: very marked, +: marked, ±: not marked) and the degree of cell loss in the SN, RNpc and RNmc demonstrated by clock diagrams. Approximate volume of the lesion is also shown, L: large lesion, M: middle size lesion, S: small lesion.

repeatedly by examination of the stretch reflex by EMG through the course of DA injection. In the illustrated case in Fig 3, the tonic stretch reflex was reduced by 80 ng of DA and almost disappeared with an additional 120 ng. Therefore, an apparently dose dependent response was observed.

Histological examination revealed that the lesion was located mainly between the red nucleus (RN) and the SN, 1-3 mm lateral to the midline as expected. The lesion destroyed the RN to varying degrees and induced cell loss in the SN, mainly of the zona compacta. In serial sections, the extent of damage to the RN and cell loss in the SN were quantitatively estimated. Ten sections were selected, photographed and magnified 100 times. The number of cells counted on the lesioned side was expressed as a percentage of that on the intact side. This was correlated with the degree of clinical manifestation of flexed posture (for example, by intensity of flexion at the elbow). Fig. 4 summarizes the quantitative study in the early series of 11 cases. The manifestation of flexed posture seemed to be more related to cell loss in the SN than to destruction of the RN. For example, three cases without flexed posture, I1, J5, K7, did not show marked cell loss in SN, although destruction of either the

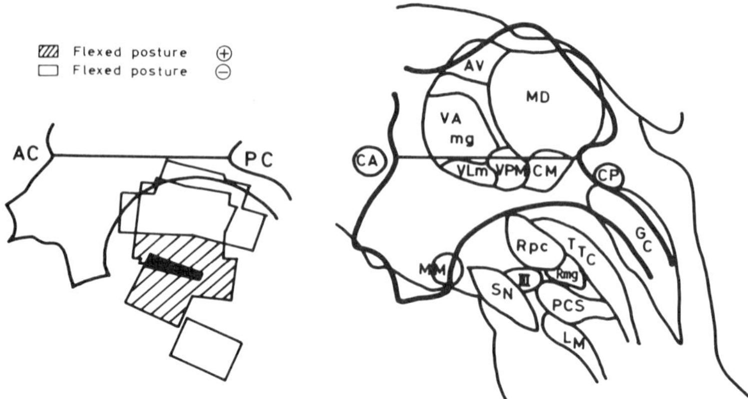

Fig. 5. Left: X-ray profiles of the coagulative lesions are superimposed referring to the third ventricle, the posterior commissure (pc) and the anterior commissure (ac). The hatched area denotes lesions producing flexed posture whereas the white area denotes lesions not producing a postural change. The small black area is an area common to all positive lesions. Right: Brain stem and thalamic structures close to the midline are shown. Directly relevant structures in this study are Rpc: parvocellular red nucleus, Rmc: magnocellular red nucleus, SN: substantia nigra, CP: posterior commissure, CA: anterior commissure.

parvocellular or magnocellular parts of the RN was obvious. This histological study led us to consider that cell loss in the SN zona compacta might cause flexed posture and DA deficiency in the striatum.

Therefore, the optimum lesion point to induce effective cell loss in the SN was looked for by superimposing the X-ray profiles of the effective and non-effective VMT lesions, with reference to the image of the third ventricle. Thus, it was concluded that the area just ventral to the parvocellular part of the RN was most likely to be responsible (Fig. 5). This area corresponds to the anatomical exit zone of the nigrostriatal fibers from the SN (Poirier and Sourkes, 1965). Next, in two cases, we attempted to make a small lesion exactly in the ventral part of the rostralmost parvocellular division of the RN. As shown in Fig. 6, using radiological control to visualize the third ventricle and the posterior commissure, several electrode tracks were made to record deep subcortical electrical activity and to find out the extent of the RN, which always exhibited very highy spontaneous activity (Fig. 6, A3,4 and B5) compared with its surrounding areas. After 2-3 penetrations, the extent of the RN, especially its rostral tip, could be approximately

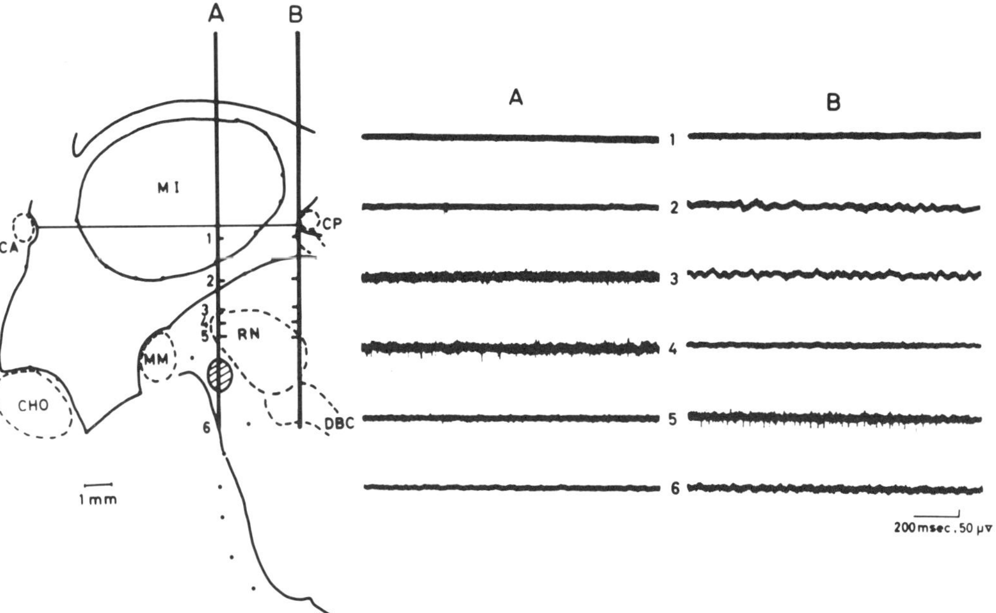

Fig. 6. Examples of depth recording along two tracks (A and B) shown on the left, referring to the third ventricle and intercommissural line (between CA and CP). Numbers (1-6) correspond to the recorded points indicated along track A. The extent of the RN thus defined is marked by a dotted line. The round hatched area is coagulated. MI: massa intermedia, CHO: optic chasma, MM: mamillary body, DBC: decussation of brachium conjunctivum.

figured out, and a small lesion was made with a fine platinum electrode without invading the RN and the SN themselves. In the monkey illustrated (Fig. 7), the lesion was placed bilaterally with 2 weeks' interval. The animal showed unequivocally a flexed posture of the left upper limb with only a slight postural change on the right. Corresponding to these postures, EMG examination revealed marked exaggeration of the stretch reflex on the left and slight exaggeration on the right as shown in Fig. 8A, B. Postmortem histological examination (Fig. 8C) supported the finding, the VMT lesion on the right (contralateral to the side of marked flexed posture) being larger than that on the left. Further, the cell loss in the right SN was more conspicuous than that on the left. Probably the VMT lesion on the left was too small, and slightly more medially located than that on the right.

Finally, to preclude the possibility of damage to the mesencephalic reticular formation inducing the postural change of the

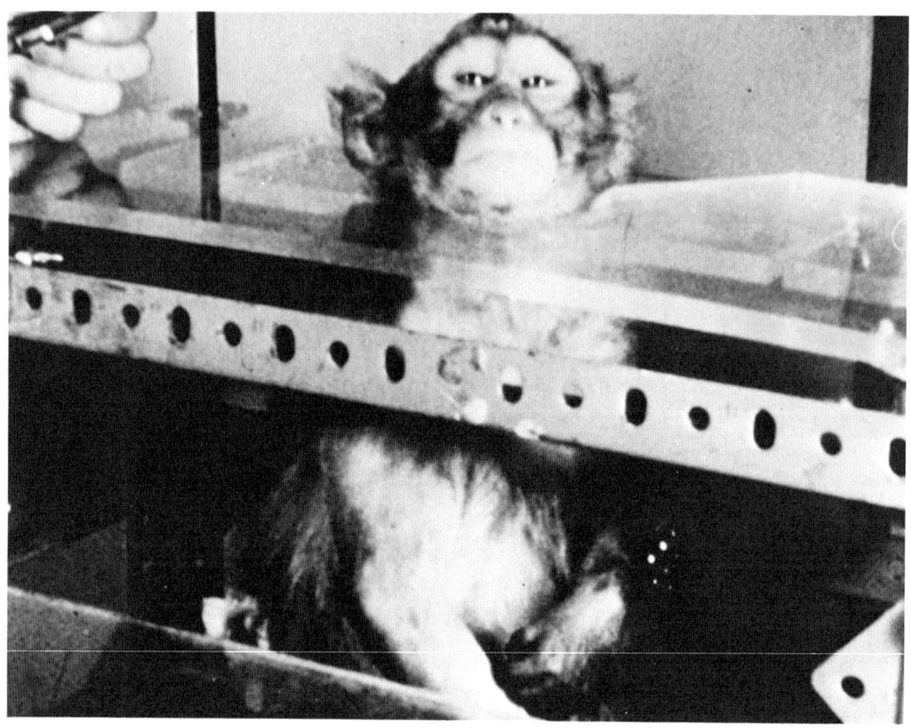

Fig. 7. Monkey (W3) with bilateral selective VMT lesions. Note that the flexed posture of the left upper limb is more marked than that of the right.

contralateral upper limb, 6-OH-DA (8 ng) was injected into the SN, with a view to destroying SN cells directly without damaging the reticular formation. For this purpose, in two monkeys, a small glass pipette was again inserted into the medial part of the SN under radiological control. Then, 6-OH-DA (8 ng) was injected slowly through a microsyringe. Within 1-2 weeks after injection, the animal exhibited a flexed posture as illustrated in Fig. 9. The morphological changes in the SN of this monkey on histological examination are shown in Fig. 10. The tip of the micropipette reached exactly into the medial part of the SN. Although the cell loss in the SN was not very marked, there were unequivocal changes in the left SN as compared with the intact side.

DISCUSSION

The present investigation demonstrated that a selective VMT lesion, which did not encroach on the RN or the SN but induced

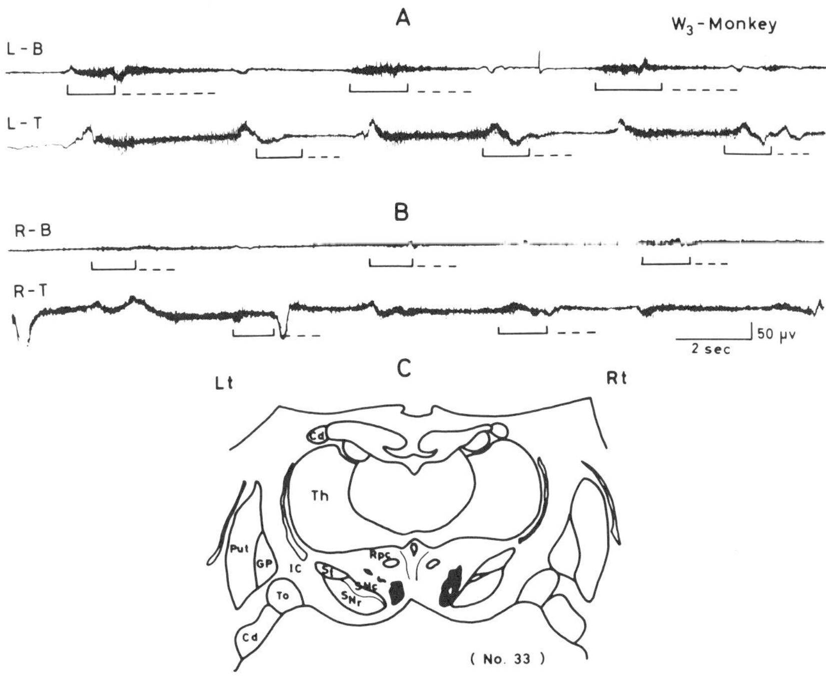

Fig. 8. EMG recordings (A,B) from the upper limbs in the lesioned monkey (bilateral VMT) shown in Fig. 7. Tonic stretch reflexes are greater in the left arm (A) than that in the right (B). C: A sketch from a histological section to show location of the bilateral VMT lesions (black area) in this case. The right lesion (on your right) is accurately located but the left one is a little too medial.

secondary cell loss in the latter structure, resulted in a characteristic postural change in the contralateral upper limb. The animal neglected the affected limb but no sensory or motor paresis was detected. Although the animal did not normally use the affected limb, for example, for taking food, he climbed up the cage by holding bars with both hands when threatened, or picked up food with the affected limb when the normal limb was restrained. This flexed posture reminded us of the characteristic postural changes with rigidity seen in Parkinson's disease. Also in monkeys, it was shown that the flexed posture was related to a DA deficiency due to SN cell loss, in the sense that the posture was temporarily ameliorated by administration of L-Dopa. The effect of L-Dopa was well demonstrated by EMG examination showing a reduction of the exaggerated tonic stretch reflexes. Moreover, in another series of experiments we observed that the experimental tremor which was always superimposed on the postural change stopped temporarily on administration of

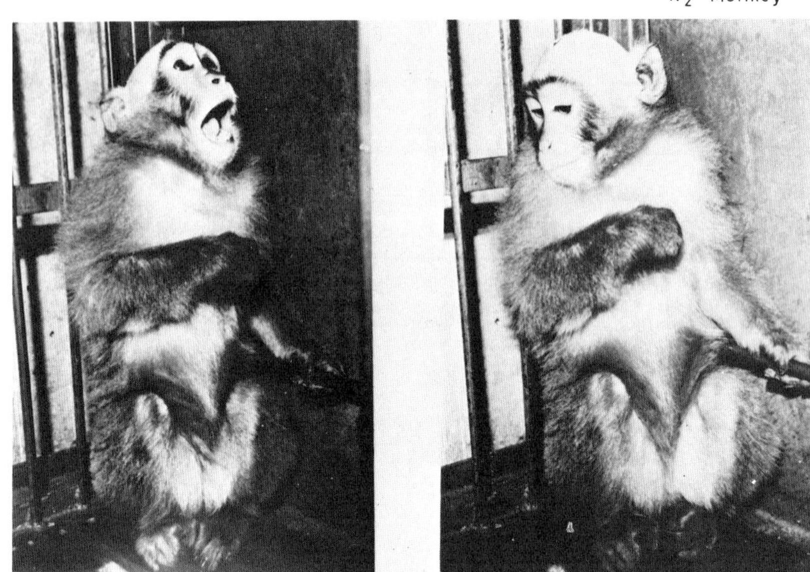

Fig. 9. Monkey (W2) with a left SN lesion by 6-OH-DA showing the typical flexed posture of the right upper limb.

L-Dopa, and that choreic movements appeared with an overdose of L-Dopa (Ohye et al., 1977). These features taken together strongly support the idea that the flexed posture with exaggerated tonic stretch reflexes in VMT lesioned monkeys might be equivalent to the postural changes seen in Parkinson's disease (Pechadre et al., 1976).

On the basis of the histological study, the pathological change responsible for the Parkinsonian-like flexed posture in this animal model was cell loss in the SN zona compacta. A restricted small lesion at the exit zone of the SN-striatal fibers, or a direct 6-OH-DA lesion in the SN, resulted in the same postural changes in the contralateral upper limb. The latter procedure was of great importance because the coagulative lesion mediodorsal to the SN inevitably destroyed a part of the mesencephalic reticular formation at the same time, which might have induced the postural change. This latter possibility was, however, almost certainly avoided by the last experiment in which 6-OH-DA was injected locally into the medial part of the SN. This kind of experiment should be repeated further to obtain a quantitative correlation between the extent of cell loss in the SN and the degree of flexed posture. Cell loss by direct destruction of the RN was not responsible for the postural change, because in this as well as in other series of experiments, we have tried RN destruction alone, which resulted in no postural change at all, confirming the previous study by Carpenter (1956).

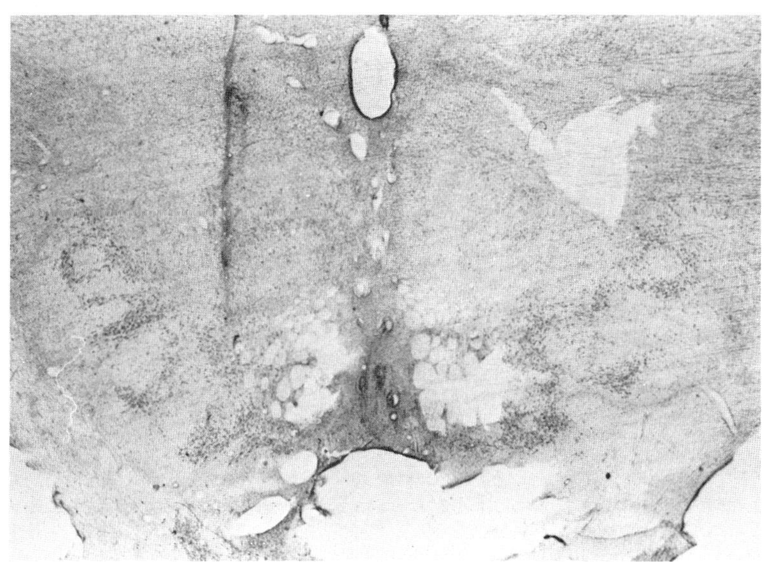

Fig. 10. Photomicrograph of the midbrain of the monkey shown in Fig. 9. The track of the injection tube is seen on the left side. Note that the medial part of the left SN is moderately damaged.

In Parkinson's disease, chronic degenerative cell loss in the SN zona compacta has been well documented (Hassler, 1938). In this regard, the experimental postural change induced by a small VMT lesion or by 6-OH-DA injection into the SN has some common features with Parkinson's disease, although the experimental procedure was an acute one. It is noteworthy, however, that the flexed posture thus induced in a monkey persists unchanged for 7-8 years, as far as we followed up in another study. Therefore, it seems likely that once the SN cells have degenerated, no compensatory recovery occurs.

The present investigation may give an answer to the question of what results from cell loss in the SN in the primate, throwing light on the possible function of the SN. In the literature, there are few studies on the functional role of the primate SN. Stern (1966) tried to make a surgical incision directly into the primate SN and concluded that hypokinesia was the most prominent symptom following destruction of the SN. Although he did not mention much about postural changes our results appear to point in the same direction as his conclusion.

REFERENCES

Carpenter, M.B., 1956, A study of the red nucleus in the rhesus monkey., J. Comp. Neurol., 105:195.

Hassler, R., 1938, Zur Pathologie der Paralysis agitans und des Postencephalitischen Parkinsonismus, J. Psychol. Neurol. (Berlin), 48:387.

Ohye, C., 1978, Pathophysiology of experimental tremor in monkeys, (in Japanese), Adv. Neurol Sci., 22:765.

Ohye, C., 1979, Primate model of Parkinsonian motor symptoms, (in Japanese), Adv. Neurol. Sci., 23:949.

Ohye, C., Imai, S., Nakajima, H., Shibazaki, T. and Hirai, T., 1979, Experimental study of spontaneous postural tremor induced by more successful tremor producing procedure in the monkey, in: "The Extrapyramidal System and Its Disorders," L. J. Poirier, T. L. Sourkes and P. J. Bedard, eds., Raven Press, New York.

Ohye, C., Isobe, I., Miyazaki, M. and Shibazaki, T., 1977, L-Dopa induced choreic movement in the tremor monkey, Excerpta med., Internat. Congr. Series No. 427, p 346.

Ohye, C., Shjibazaki, T., Hirai, T., Nagaseki, Y., Wada, H., Kawashima, Y. and Hirato, M., 1983, Possible descending pathways mediating spontaneous tremor in monkeys, Adv. Neurol., 40: (in press).

Pechadre, J. C., Larochelle, L. and Poirier, L. J., 1976, Parkinsonian akinesia, rigidity and tremor in the monkey, J. Neurol. Sci., 28:147.

Poirier, L. J., 1960, Experimental and histological study of midbrain dyskinesia, J. Neurophysiol., 23:534.

Poirier, L. J. and Sourkes, T. L., 1965, Influence of the substantia nigra on the catecholamine content of the striatum, Brain, 88:181.

Stern, G., 1966, The effects of lesions in the substantia nigra, Brain, 89:449.

NEUROBIOLOGICAL CHANGES INDUCED BY NEOSTRIATAL KAINIC ACID INJECTION: AN ELECTROPHYSIOLOGICAL AND MORPHOLOGICAL APPROACH TO THE PHYSIOPATHOLOGY OF HUNTINGTON'S CHOREA

D. Doudet*, P. Bioulac-Sage*, M. Arluison+, C. Gross* and B. Bioulac*

* Groupe Motricité, Laboratoire de Neurophysiologie
Université de Bordeaux II

+ Laboratoire de Cytologie
Université de Paris VI

INTRODUCTION

Huntington's chorea is an autosomal dominant disorder characterized clinically by progressive dementia and involuntary choreiform movements (Huntington, 1872). Pathologically, there is a loss of neurons in numerous brain areas, but the major trait is an atrophy of the corpus striatum (Bruyn, 1968).

Biochemical studies on human post-mortem choreic tissues reveal a clear alteration of cholinergic, GABAergic and substance P neurons contained in the caudate nucleus (see review by Spokes, 1980). This structure forms, with the putamen, the neostriatum which is a part of the nigro-neostriato-nigral complex (Poirier et al., 1972; Dray, 1979). This complex is an important component of the extra-pyramidal system and is mainly known for its inhibitory functions in motor control (Denny-Brown, 1962; Féger, 1981: Divac et Oberg, 1979; Steg et Johnels, 1979). Even though the pathogenesis of Huntington's chorea is completely unknown, the preceding pathological and neurochemical data added to recent pharmacological advances have suggested that hyperactivity of the dopaminergic (DA) nigro-neostriatal pathway may be involved in the appearance of hyperkinetic and choreiform movements (Klawans, 19780; Chase, 1973). Anti-dopaminergic drugs such as α-methyl-para-tyrosine or certain neuroleptics (phenothiazines and butyrophenones) ameliorate choreic movements (Constantinidis, 1972; Chase, 1973). Conversely DA agonists such as L-dopa and D-amphetamine exacerbate chorea or induce choreiform motor activities. Furthermore, the hypothesis of DA nigro-neostriatal pathway hyperactivity is easily supported by the fact that the contents of GABA and glutamic acid decarboxylase (GAD, its synthetizing enzyzme) are decreased in the neostriatum, pallidum and substantia nigra (SN) of choreic brains (Spokes, 1980). The loss of the strong inhibitory action, normally exerted upon nigral DA cells by the GABAergic neostriato-nigral pathway, may account for the hyperactivity of this neuronal pool (Constantinidis, 1972). Another method to investigate the pathophysiology of Huntington's disease is to resort to animal models. Thus, some authors, such as Coyle and Schwarcz (1976), McGeer and McGeer (1976) have proposed an experimental model of chorea in rats. For this purpose, they destroy neostriatal neurons with kainic acid (KA), a cyclic analogue of glutamic acid. KA injection into the neostriatum destroys its intrinsic neurons while sparing, to a great extent, passing fibres. Numerous morphological, biochemical, neuropharmacological and behavioural studies demonstrate that this animal model of Huntington's disease is suitable for experiments (Hattori and McGeer, 1977; Coyle et al., 1978; Fibiger, 1978; Mason and Fibiger, 1979). The aim of the present work is to test for DA nigro-neostriatal pathway hyperactivity in this experimental situation by using electrophysiological and morphological techniques.

More precisely, we have compared the electrical activity of nigral DA neurons in normal rats and in rats with KA lesion of the neostriatum. In parallel we have performed morphological studies, essentially histochemical fluorescence, in neostriatum and substantia nigra of normal and KA lesioned rats. Results obtained with these two categories of experiment support the idea that a KA lesion of the neostriatum induces a deep functional disorganization rather than a hyperactivity of the DA nigro-neostriatal pathway.

MATERIALS AND METHODS

Surgery

All surgery is performed under chloral hydrate anaesthesia (400 mg/kg i.p.). Each animal (male Wistar rat weighing 250-350 g) is positioned in a Narishige stereotaxic apparatus according to the De Groot (1959) coordinate system. Usually the KA solution (2.5 μg dissolved in 1 μl saline) is freshly prepared. Animals receive a unilateral injection of 1 μl of this solution in the left neostriatum. Injection is done stereotaxically with a 10 μl Hamilton syringe assembled with a small cannula. Infusion time is 5 min, and the cannula is left 5 min more to prevent upward movement of the solution along the cannula track. With the same parameters, the same amount of saline is injected in the caudate nucleus of a control group of rats. The injection coordinates from bregma, midline and dural surface are:- A: 1.5 mm, L: 2.8 mm, H: 4.5 mm. During the days following the striatal lesion, a striking rotational behaviour which was previously described by Coyle and Schwarcz (1976) and McGeer and McGeer (1976) is observed. Twenty-four hours after injection, rats exhibit marked rotatory behaviour away from the injected side. Two days after surgery, they reverse the direction of rotation towards this side, this circling gradually disappearing after a few days. Animals show adipsia and aphagia and it is often necessary to maintain them with daily 5 ml i.p. administration of isotonic glucose. In the ninety-six hours following surgery, death occurs in 30 to 40% of KA treated animals. Surviving KA lesioned animals and their controls are divided into two groups for electrophysiological and morphological studies.

Electrophysiological Studies

Unit recording, electrical stimulation and histological procedure. The KA and saline injected animals are prepared for unit recording 7 to 30 days after surgery. All the animals are anaesthetized with urethane (1.3 g/kg) and mounted in the stereotaxic apparatus. The skull is exposed and two small burr holes drilled at points determined by a previously calibrated electrode, one directly above the substantia nigra, and another above the neostriatum for insertion of the stimulating electrodes.

A flat bipolar stainless steel stimulating electrode (Rhodes Medical NEX 200) is implanted in the KA-injected, control, or normal (untreated) ipsi-lateral caudate nucleus. The site of stimulation is nearly the same as the site of injection of KA or saline. Stimuli used have the following parameters: single square pulses of 0.5-2.0 mA and 0.1 msec width.

Extracellular recordings from single nigral neurons are obtained using single barreled glass microelectrodes (21% pontamine sky blue dissolved in 1.5 M NaCl; tip diameter approximately 0.5-1.0 µm; impedance: 3 to 8 megohm in vitro). Unit activity is recorded, amplified and stored on magnetic tape for further analysis. Activity is monitored on a Tektronix 5113 oscilloscope. Electrode signals are passed through a Mentor N 750 spike analyser and are immediately displayed in a digitalized form on a paper chart recorder. At the end of each electrophysiological experiment, the recording site is marked by passing 15 µA of cathodal current through the recording barrel for 20 min, depositing a blue point at the electrode tip. Rats are then intracardially perfused with 10% formalin. The brain is taken out, sectioned at 50 µm thickness and stained with cresyl violet for histological verification. Precise location of the electrode tip in SN pars compacta is confirmed by histological examination.

Identification of nigral DA neurons. Recordings are obtained from DA neurons located in the pars compacta of the SN. In accord with previous reports, these neurons can be strictly identified

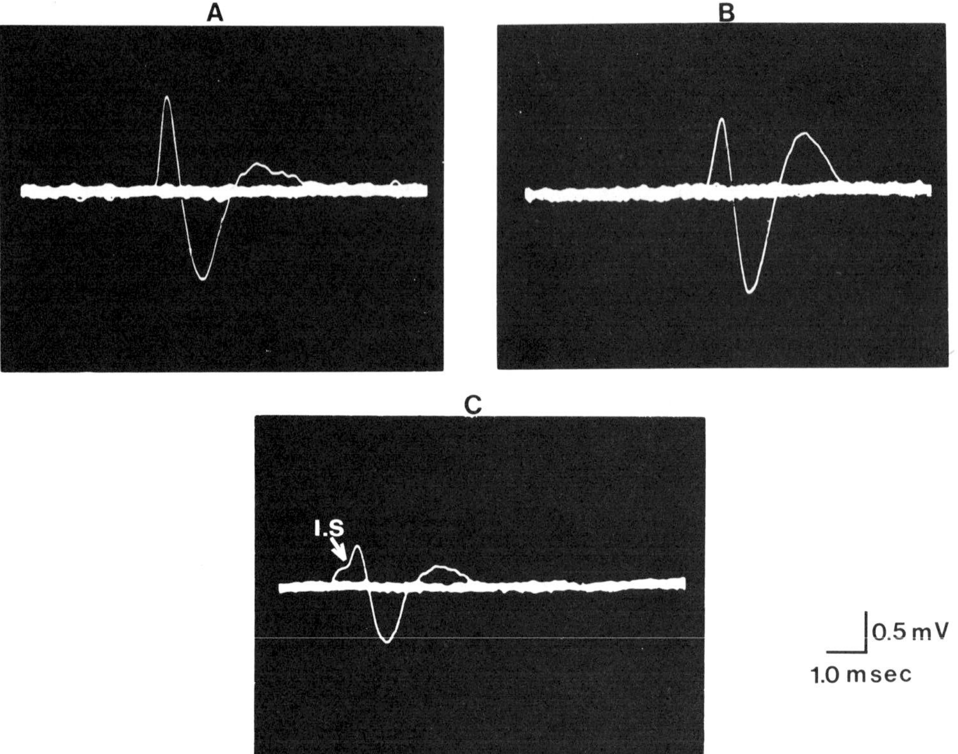

Fig. 1. The characteristic shape of three identified dopaminergic neurons located in the pars compacta of the SN. Notice the unusual width of the action potential and the late positive component. The presence of an initial segment (I.S.) is sometimes observed (C).

(Aghajanian and Bunney, 1973; Walters et al., 1975; Wilson et al., 1977; Deniau et al., 1978; Guyenet and Aghajanian, 1978) by several criteria:

a) A characteristic action potential: broad wave form (3-4 msec), often including the presence of a pronounced initial segment (I.S.) with a somatodendritic break and a late positive component (Fig. 1).
b) Slow firing rate: 1 to 7 spikes/sec (Fig. 7A.B).
c) Their antidromic invasion after stimulation in the neostriatum (Fig. 2): antidromic spikes are highly fractionated into initial segment (I.S.) and somato-dendritic component. Usually, only the initial segment is recorded. Antidromic spikes respond with fixed latency, follow stimulus frequencies of 250 Hz or more, and collide with spontaneous spikes.

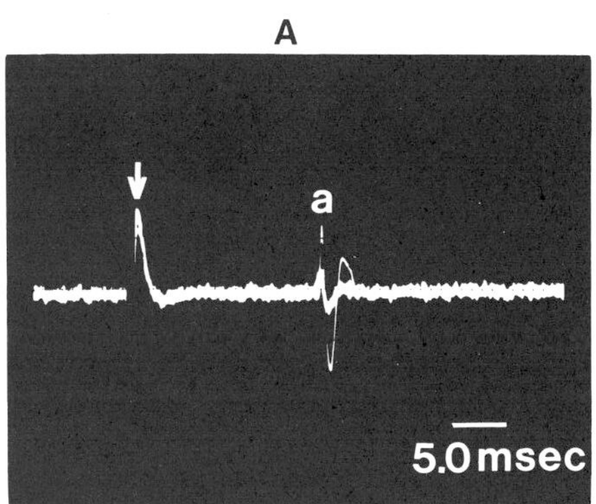

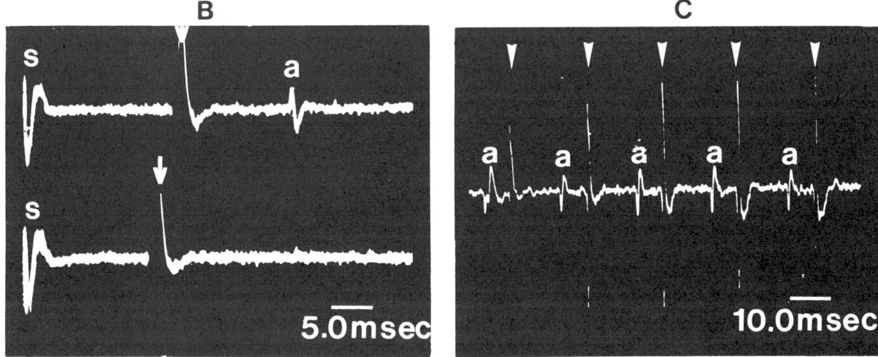

Fig. 2. Antidromic response of nigral DA neurons to neostriatal stimulation: The three criteria of antidromic invasion
A: response with fixed latency: the initial segment is generally recorded and sometimes the full spike;
B:collision with spontaneous action potential;
C:the antidromic initial segment follows high stimulation frequencies.
a: antidromic spike; s: spontaneous spike; arrow: stimulus artefact.

d) Pharmacological tests with DA agonists (Fig. 3): the responsiveness of a certain number of cells is tested with i.p. or i.v. injection of two DA agonists: apomorphine (2 mg/kg), and piribedil (2 mg/kg) which inhibits DA neurons. One cell is tested per animal and integrated activity is recorded on a paper chart recorder.

e) Location of the neurons in the SN, reconstructed after histological examination of the position of the blue point (Fig. 4).

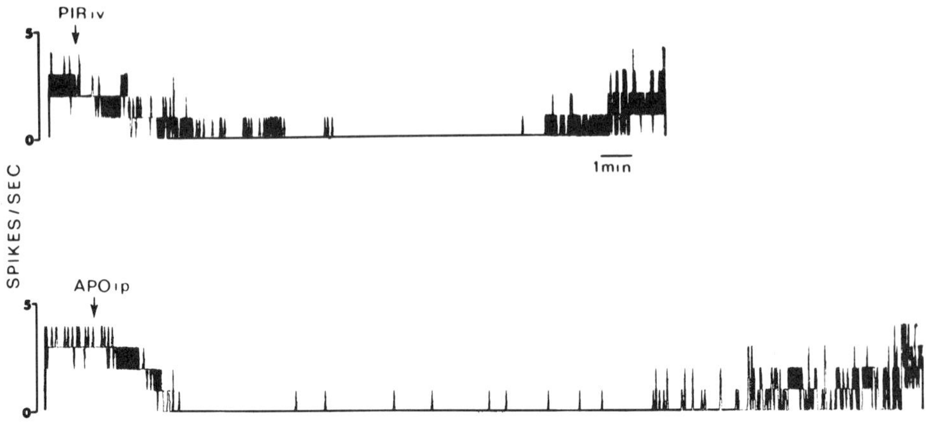

Fig. 3. Typical effect of piribedil (PIR 2 mg/kg, i.v.) and apomorphine (APO 2 mg/kg, i.p.) on the firing rate of dopaminergic nigral cells.

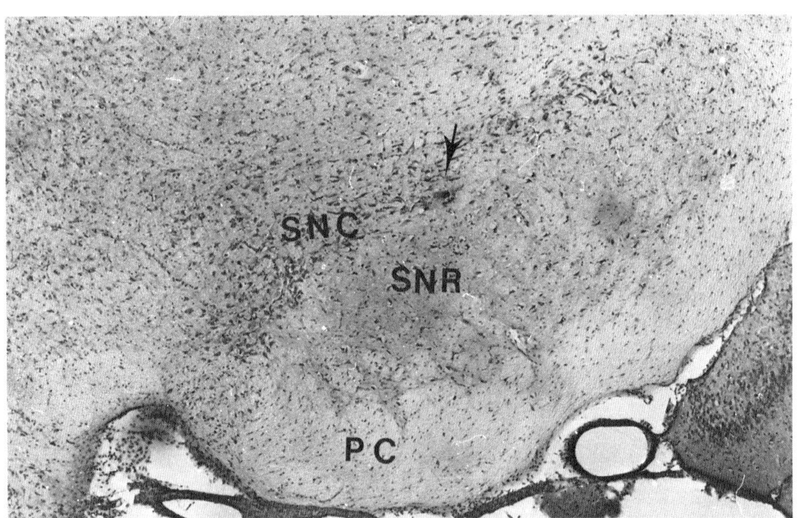

Fig. 4. Location of a recorded cell in the pars compacta of the substantia nigra. The ejection site of pontamine sky blue is indicated by the arrow. Cresyl violet (X 50).
PC: Pedunculus cerebri; SNC: substantia nigra pars compacta; SNR: substantia nigra pars reticulata.

Data analysis. The magnetic tape is replayed through the window discriminator and displayed on the oscilloscope. Standard output of the Mentor is led into a PDP 11/40 computer to generate autocorrelation and time interval histograms. Data are retained for statistical analysis only when neurons maintain a signal to noise ratio of 3/1 or better.

The autocorrelogram specifies the probability of encountering the same spike as a function of time after a given spike. A typical regular firing pattern is shown in Fig. 5, with its autocorrelogram showing an initial trough (the prolonged quiescent period) followed by periodic fluctuations with equally spaced density peaks appearing approximately every 300 msec for this example. The autocorrelogram implies that, for any action potential of the spike train, there is a high probability of finding the successive action potential at an interval of 330 msec. This autocorrelogram silhouette is characteristic of a regular firing pattern (Perkel et al., 1967; Moore et al., 1970). In fact, a property of the autocorrelation histogram is nthat it "flattens out" to a constant value slowly for regular neurons, and more rapidly for spike trains with greater variation in interval length. In the example of Fig. 6, one can see that there are no more periodic fluctuations and no initial trough. We have considered as regular or rhythmic any neuron which clearly exhibits the initial trough and first peak of the autocorrelogram. Thus, we could separate the DA neurons recorded in the normal, control, and KA lesioned rats into "organized" (regular) or "disorganized" neurons. From the time interval histogram, the frequency (F) and the standard deviation (s) of the curve, called dispersal index (D), are calculated. Means of these two parameters have been calculated for the two groups of neuronal populations recorded in normal, control and lesioned animals. Data are analyzed using the Student's t-test.

Morphological Studies

Two groups of rats are used for light and electron microscopy on one hand, and histofluorescent study on the other.

Light and electron microscopy. Ten rats are used for this study made 5 to 10 days after KA injection. Under ether anaesthesia, the rat undergoes intracardiac perfusion with a mixture of 2% paraformaldehyde-2.5% glutaraldehyde in a phosphate buffer (pH: 7.4; Karnowsky's medium). Fixed brains are rapidly removed and dissected. Caudate nuclei are removed, fixed for 2 days more in 10% formalin and processed for routine microscopy. Slices of lesioned and controlateral neostriatum are stained with the Kluver and Barrera technic (Cresyl violet and Luxol fast blue).

Ipsi- and contralateral SN are dissected under a binocular microscope and fixed for an additional two hours in Karnowsky's

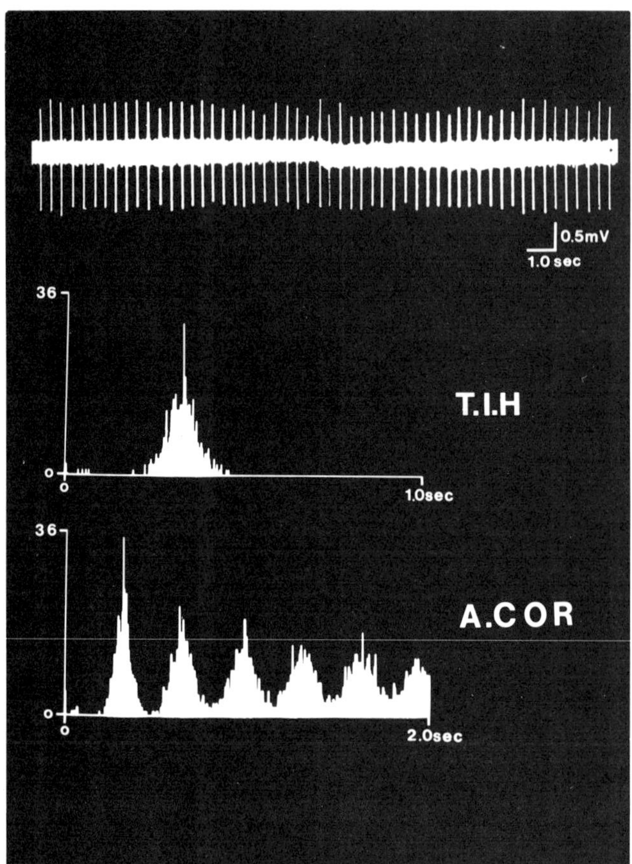

Fig. 5. Characteristics of a nigral dopaminergic neuron with a regular pattern of discharge.
Top: representative spike train showing a regular firing pattern.
Middle: time interval histogram (T.I.H) of this neuron. Bin width: 1 msec.
Bottom: autocorrelogram (A.COR) of the same neuron demonstrating the initial prolonged quiescent period (initial trough) followed by periodic fluctuations appearing approximately every 330 msec. Abscissa: analysis time. Ordinate: number of occurrences.

medium. Then tissue blocks are post-fixed in 1% osmium tetroxide, dehydrated and embedded in Epon. Ultrathin sections of SN are double contrasted with uranyl acetate and lead citrate and are examained with a Philips EM 301.

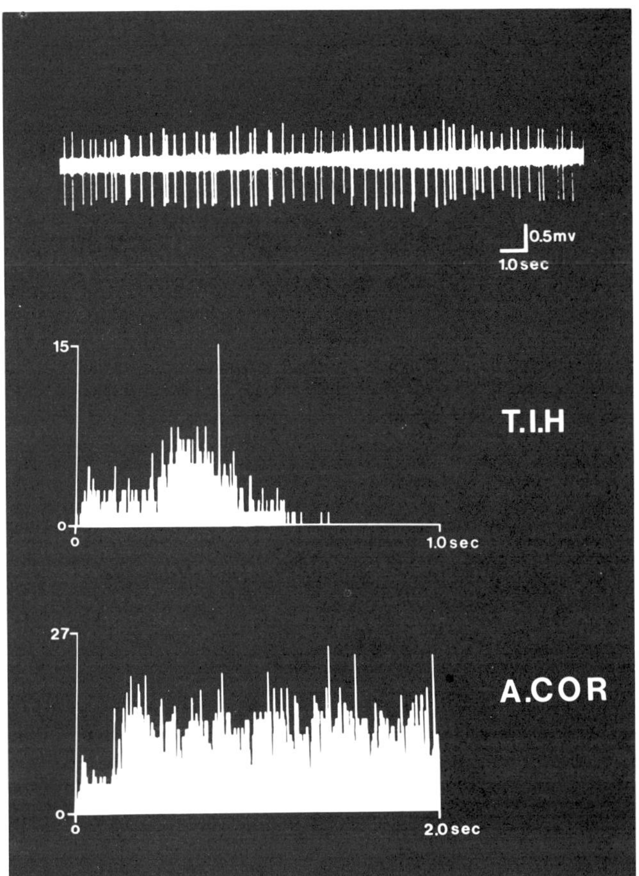

Fig. 6. Characteristics of a nigral dopaminergic neuron with an irregular pattern of discharge.
Top: representative spike train.
Middle: time interval histogram (T.I.H) (bin width: 1 msec).
Bottom: autocorrelogram (A.COR). Notice that the initial trough and the periodic fluctuations have disappeared.
Abscissa: analysis time. Ordinate: number of occurrences.

Fluorescence Microscopy. Twenty ether-anaesthetized rats are sacrificed by decapitation 1-2 weeks after the KA injection. Brains are quickly removed and coronal sections (3 to 5 mm thick) are obtained including anterior and posterior caudate nuclei, anterior and posterior SN, for the lesioned and control sides. These specimens are processed for fluorescent microscopy according to the Falck et al., protocol (1962). Blocks of brains are quickly frozen in a mixture of two-thirds propane/one-third isopentane cooled by liquid nitrogen and subsequently freeze-dried in vacuo for 5 days at

10^{-3} Torr and for one day at 10^{-6} Torr. After the condensation reaction (1 hour at 80°C with formaldehyde gas of 70% relative humidity generated from paraformaldehyde), pieces of tissues are embedded in paraplast in vacuo. Blocks are cut in serial sections 10 um thick. Every tenth section is mounted in liquid paraffin and examined with a Zeiss fluorescence microscope. Pictures are taken with 400 ASA Ektachrome film.

RESULTS

Electrophysiological Studies

Activity of DA neurons in normal and control rats. The spontaneous activity of nigral DA neurons recorded in normal and control animals was very similar. The characteristics are summarized in Table I. As previously described (Wilson et al., 1977; Deniau et al., 1978; Guyenet and Aghajanian, 1978), the mean frequency of these neurons was low (see distribution of rates; Fig. 7A and B). For a large majority, the time interval histogram had a gaussian form (see the left part of Fig. 9) and the dispersal index was generally inferior to 120 msec (see distribution of dispersal index; Fig. 8A and B).

A large percentage of identified DA neurons (96.20% in normal and 96.66% in control animals) possessed a regular and rhythmic firing pattern. This is particularly well displayed by their autocorrelograms, which exhibit the same pattern: an initial prolonged quiescent period followed by periodic fluctuations. The right part of Fig. 9 presents the autocorrelogram of three organized DA neurons of the SN. Characteristics are available for cells with slow firing rate (Fig. 9A: F = 1.60 spikes/sec) as well as for cells with higher discharge rate (Fig. 9B: F = 6.10 spikes/sec). A few DA nigral neurons (3.80% in normal and 3.34% in control animals) were considered as "disorganized" (Table I) and their autocorrelograms did not exhibit the regular firing pattern observed for the wide majority of the DA recorded cells.

Activity of DA neurons in KA lesioned rats. As shown by Table I, results obtained in KA lesioned rats were very different from those obtained in normal and control animals. The mean frequency of these neurons was drastically decreased (see distribution of rates: Fig. 7C) and the dispersal index significantly increased (see distribution of dispersal index; Fig. 8C). One can see in Table I that 51.04% of DA nigral cells were "disorganized". Fig. 10 presents the characteristics of three identified neurons with "disorganized" firing pattern. Time interval histograms of these cells did not have gaussian form and were very dispersed. Autocorrelation histograms had a flattened silhouette with neither the initial trough nor the periodic fluctuations.

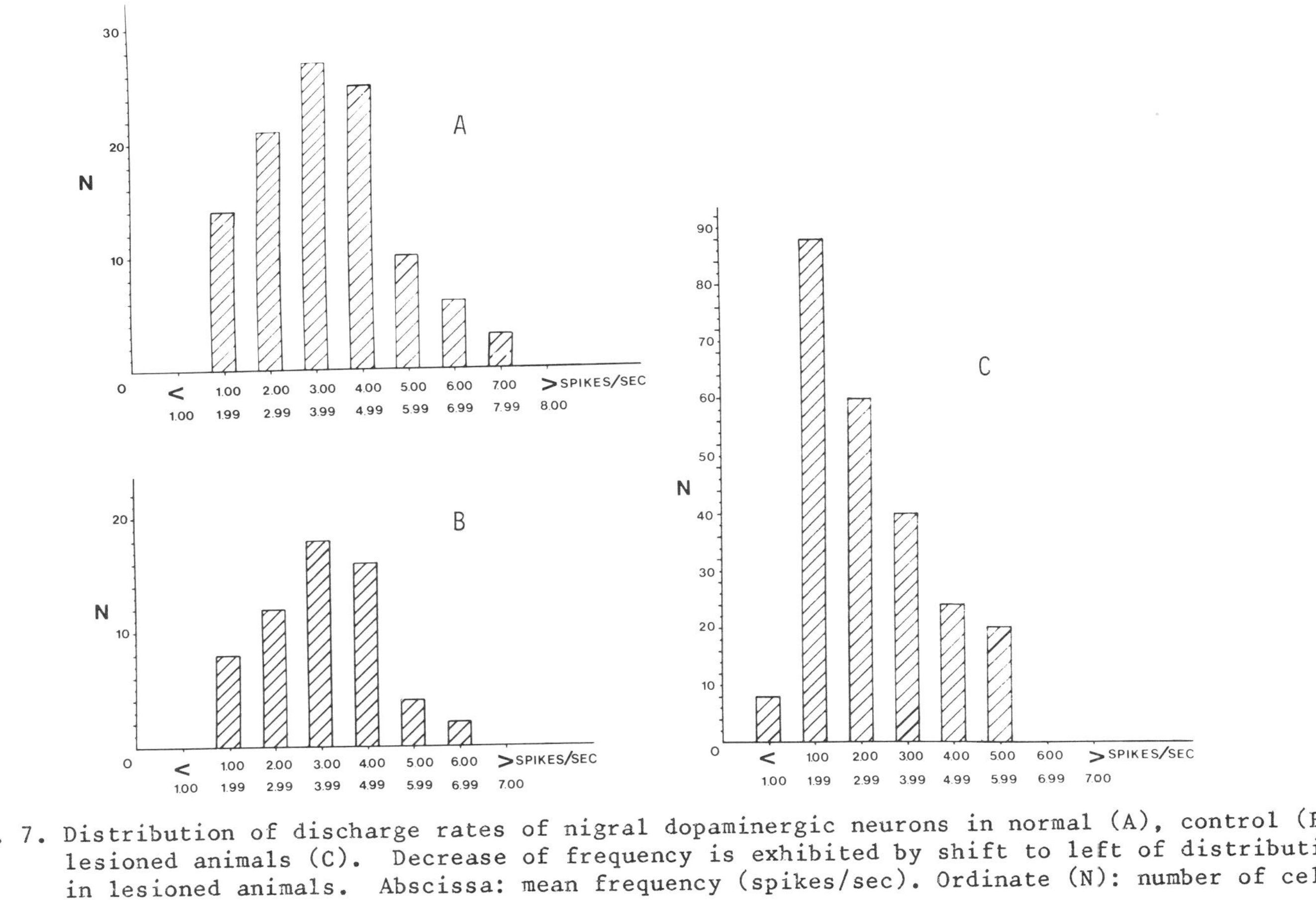

Fig. 7. Distribution of discharge rates of nigral dopaminergic neurons in normal (A), control (B), lesioned animals (C). Decrease of frequency is exhibited by shift to left of distribution in lesioned animals. Abscissa: mean frequency (spikes/sec). Ordinate (N): number of cells.

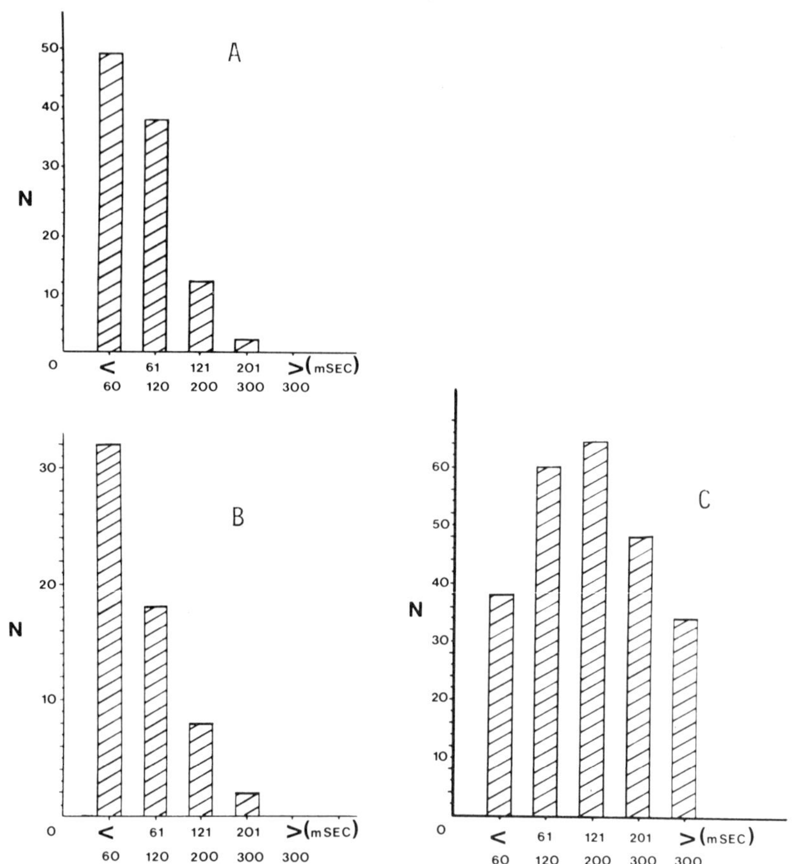

Fig. 8. Distribution of dispersal index of the nigral dopaminergic neurons in normal (A), control (B) and lesioned animals (C). The disorganization is indicated by the shift to the right of the distribution of dispersal index in lesioned animals. Abscissa: mean dispersal index (msec). Ordinate (N): number of cells.

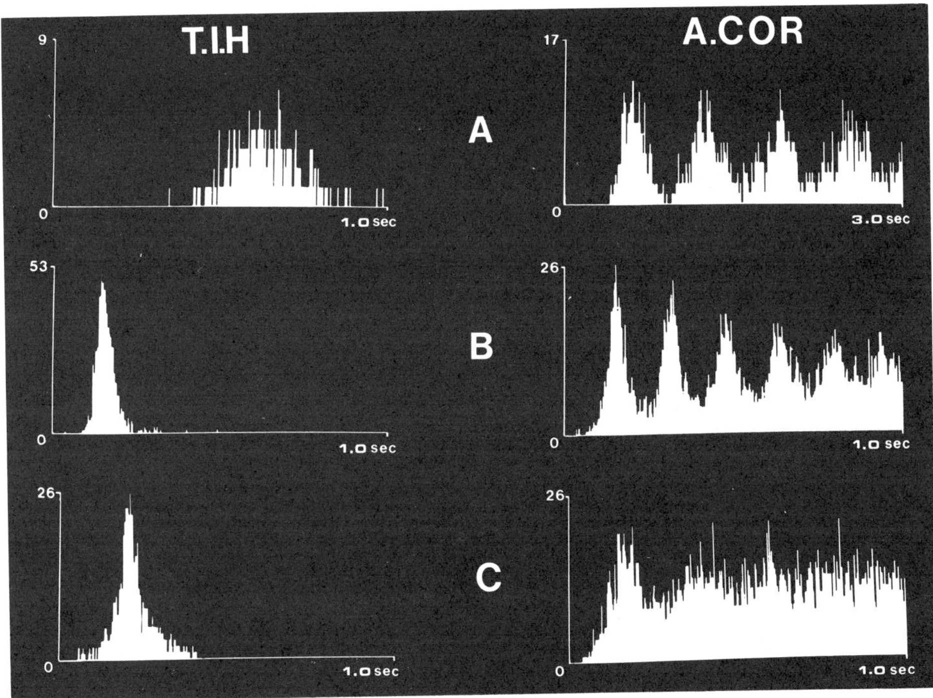

Fig. 9. Typical firing pattern of three identified "organized" dopaminergic neurons in normal (A, C) and control animals (B). Left: time interval histogram (T.I.H); note gaussian form. Bin width 1 msec. Right: autocorrelogram (A.COR). Three histograms presenting an initial trough and periodic fluctuations approximately every 625 msec for cell A, and every 165 msec for cell B. In C a nigral dopaminergic neuron exhibiting the two main criteria demanded for rhythmicity; initial trough and first peak. Although there are no periodic fluctuations this cell is considered "organized" or rhythmic. Abscissa: analysis time. Ordinate: number of occurrences.

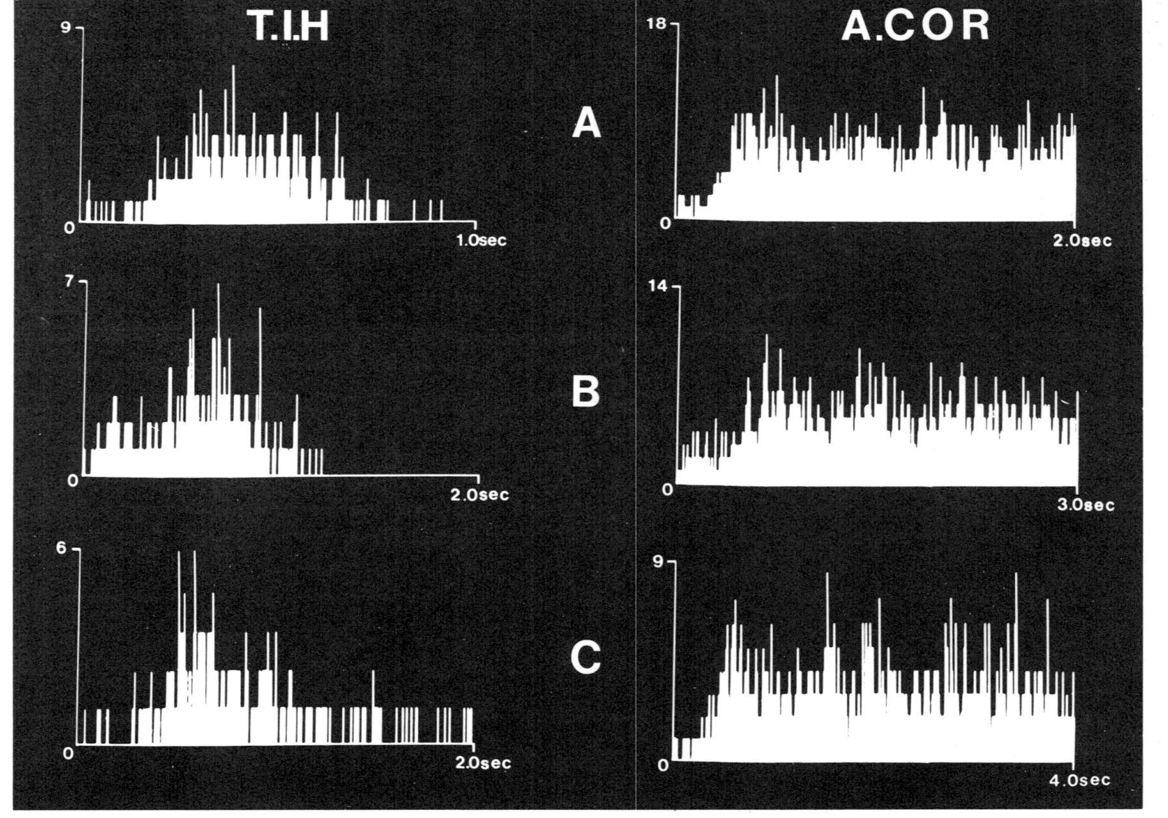

Fig. 10. Typical firing patterns of 3 "disorganized" dopaminergic neurons in lesioned animals. Left: time interval histogram (T.I.H.); gaussian form no longer apparent (bins: A, 1 msec; B-C, 2 msec). Right: autocorrelogram (A.COR); histogram does not show initial trough and periodic fluctuations. Abscissa: analysis time. Ordinate: number of occurrences.

Table I.

Animal Preparation	Number of Neurons	%	Mean frequency (F; spikes/s) Mean dispersion (D; msec)
	ORGANIZED DOPAMINERGIC NEURONS		
Normal	101	96.20	F: 3.67 ± 1.38 D: 67.49 ± 34.03
Saline Control	58	96.66	F: 3.46 ± 1.37 D: 70.06 ± 41.41
Lesioned	118	48.96	F: 3.19 ± 1.28 D: 93.08 ± 54.37
	DISORGANIZED DOPAMINERGIC NEURONS		
Normal	4	3.80	F: 3.80 ± 1.59 D: 124.40 ± 56.75
Saline Control	2	3.34	F: 4.03 ± 1.39 D: 125.36 ± 46.66
Lesioned	123	51.04	F: 2.19 ± 1.23 D: 297.31 ± 183.63

The Student's t-test indicates that there is no significant difference ($p > 0.1$) between the respective values of F and D in normal and control animals. A significant difference ($p < 0.01$) exists between the respective value of F and D in normal and control animals versus lesioned animals. These observations are particularly clear for the so-called "disorganized" dopaminergic neurons but equally true for the organized neurons. For each value of F and D, the standard deviation is given.

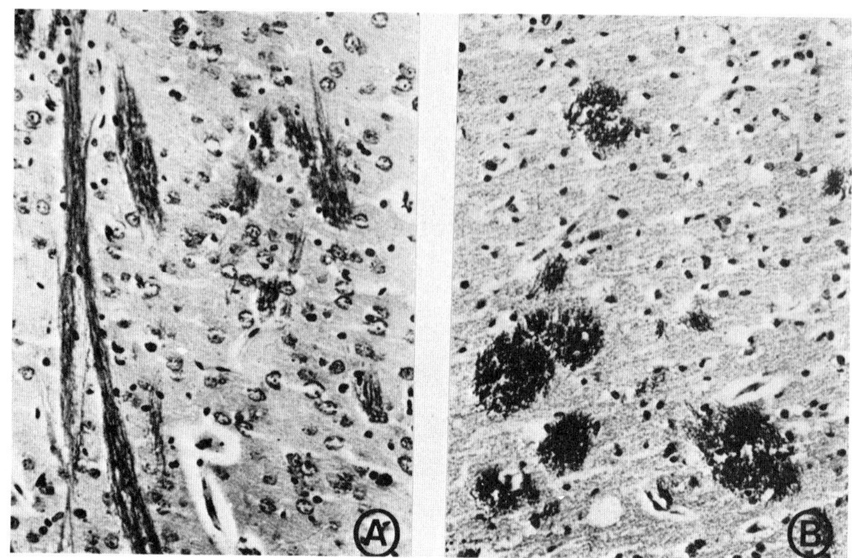

Fig. 11 Light microscopy of the neostriatum in KA injected rat. Compared to the normal histology of the controlateral caudate nucleus (11A), in the KA injected side there is a loss of neurons replaced by glial proliferation (11B). Kluver and Barrera stain (X 200).

Morphological Studies

Light Microscopy. If we compare the normal controlateral neostriatum (Fig. 11A) with the KA injected side, there was a heavy loss of intrinsic neurons in the latter, replaced by a marked glial proliferation (Fig. 11B). These histological changes involved almost the entire head of the caudate nucleus. In addition, we observed a slight loss of neurons in the justacallosal layers of the cortex above the KA injection site, and a relative enlargement of the lateral ventricle.

Areas far from the injection site, such as the tail of the caudate nucleus, pallidum, nucleus accumbens and hippocampus, were spared. The brains of the control rats had slight damage along the cannula track, otherwise they were intact. In comparison, the controlateral caudate nucleus (Fig. 11A) had normal morphological characteristics.

Electron microscopy. The ipsilateral SN revealed degenerating boutons of various types. After 5 days of KA lesion, the structure of these boutons appeared altered, with a dense and dark matrix and

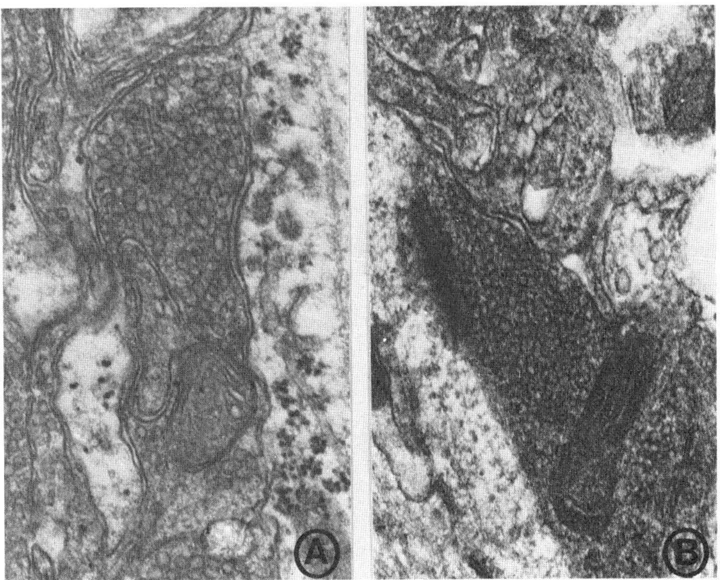

Fig.12 Electron microscopy of the substantia nigra (SN) in KA injected rat. Normal aspect of a pleiomorphic synaptic bouton in controlateral SN (12A); in ipsilateral SN, some synaptic boutons have a dense matrix with clustered synaptic vesicles (12B). Uranyl acetate and lead citrate (X 50,000).

clustered synaptic vesicles (Fig. 12B). Boutons appeared normal (Fig. 12A) in the controlateral SN or in SN of control rats where the different types of boutons are well recognized.

Fluorescence microscopy. The controlateral neostriatum exhibited its characteristic DA fluorescence which appears green, pale and homogeneously distributed (Fig. 13A). On the contrary, in the KA injected neostriatum fluorescence appeared more pronounced and heterogeneous with bright and swollen varicosities (Fig. 13B). This fluorescence contrasted with the yellow auto-fluorescence surrounding the needle track, corresponding to non-specific tissue degeneration. In both the ipsi and controlateral SN, DA fluorescence was quite clear in the nigral neurons. Within the limits of the method, no apparent differences in the intensity of fluorescence could be detected between the two sides (Fig. 14A and B).

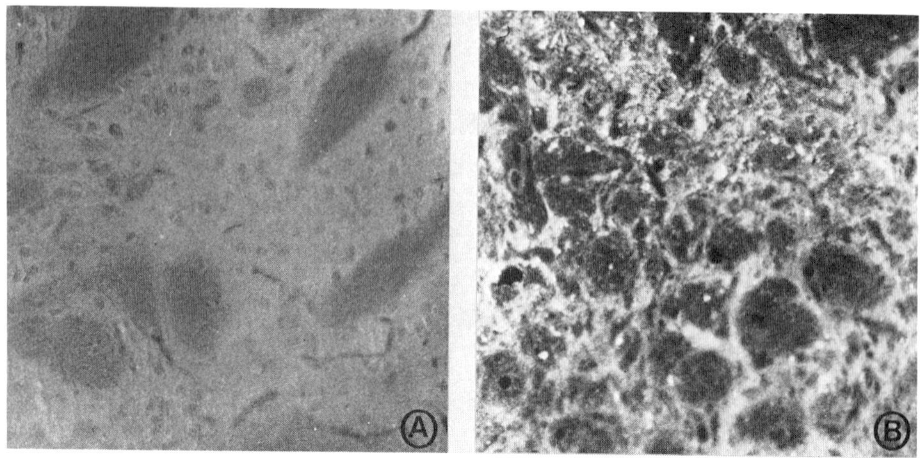

Fig.13. Histochemical fluorescence of neostriatum (NS) in KA injected rat. Compared to the normal homogeneous dopaminergic histofluorescence in contralateral NS (13A), in the injected NS, fluorescence appeared brighter, and heterogeneous with swollen varicosities (13B) (X 184).

DISCUSSION

From the results obtained with the electrophysiological and morphological experiments, we can offer three remarks:

(1) In normal animals, the electrophysiological study revealed that in their great majority the nigral DA neurons exhibit a very regular and rhythmic pattern of discharge with a low firing frequency. In addition, the histochemical method of Falk et al. (1962) shows the well known homogeneous green fluorescence of dopamine in the neostriatum.

(2) In lesioned animals, most nigral DA neurons do not present the previously described regular pattern of discharge. On the contrary, these neurons have a profound disorganization of electrical activity with a decrease of firing frequency. Concomitantly, histochemical study shows a drastic alteration of the morphological aspect of the neostriatum. Thus, the homogeneous green fluorescence is no longer observed. Terminals and varicosities appear filled with fluorescent dopamine, neurotransmitter abnormally remaining in these intraneuronal spaces. Furthermore no significant differences in fluorescence between normal and lesioned rats can be detected at the level of perikarya of nigral DA cells.

(3) The electrophysiological and morphological data obtained in

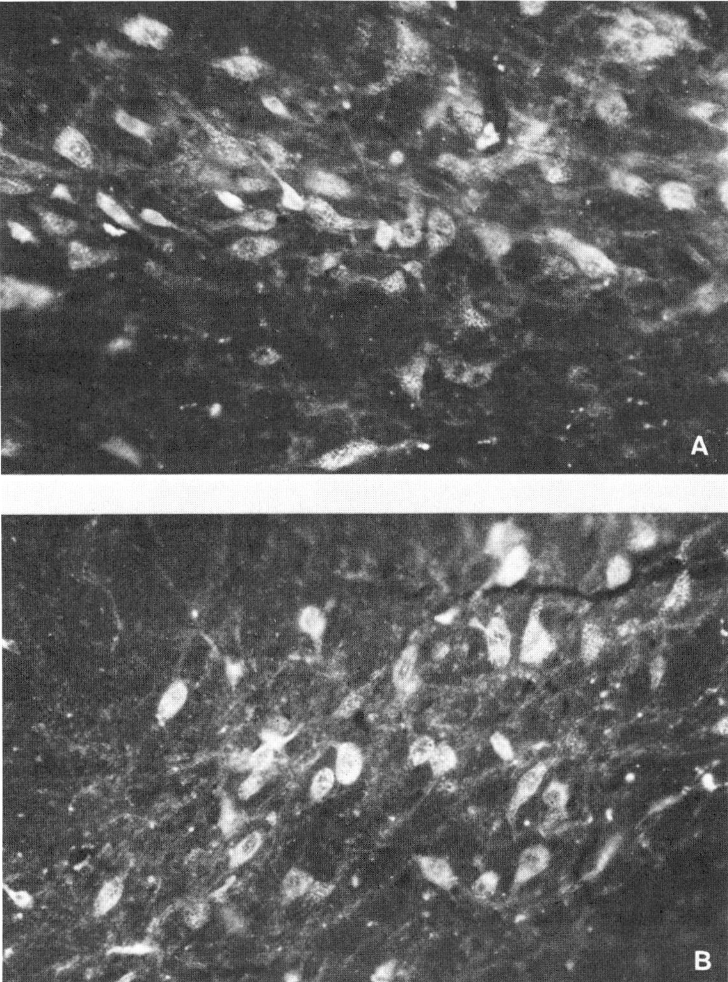

Fig.14 Histochemical fluorescence of dopaminergic neurons in substantia nigra (SN) of KA injected rat. There is no obvious difference in fluorescence intensity between controlateral (14A) and ipsilateral (14B) SN (X 230).

lesioned animals do not agree with the generally accepted hypothesis, concerning chorea pathophysiology, of a DA nigro-neostriatal pathway hyperactivity. In fact, after destruction of the neostriatum, both electrophysiological and morphological results suggest a drastic dysfunction and not hyperactivity of this DA pathway.

A first point for discussion concerns the electrophysiological and morphological results collected in normal animals. One may attempt to establish a functional relationship between the regular and rhythmic pattern of discharge of the nigral DA neurons and the homogeneous green fluorescence of dopamine in the neostriatum. Thus, the very regular and slow electrical activity of the nigral DA neurons may account for a regular and constant release of dopamine by the DA nigro-neostriatal terminals. Several authors have observed a spontaneous and regular dopamine release in the neostriatum of different species of animals (Besson et al., 1971; Chiueh and Moore, 1973; Gauchy et al., 1974). This continuous and regular dopamine release could possibly be explained by the regular arrival at synaptic boutons of trains of action potentials. Finally the regular release of dopamine in the neostriatum linked to the regular firing pattern of nigral DA cells would be responsible for the homogeneous green fluorescence of the caudate nucleus. It is noteworthy here that another member of the A.P.U.D. system presents a similar functional modality. Wang and Aghajanian (1982) have reported a rhythmicity and slow firing pattern of the serotoninergic neurons in the dorsal raphe nucleus. Similarly, Hery et al. (1979) have demonstrated spontaneous and continuous release of serotonin from the serotoninergic terminals of the neostriatum. Taken together these data suggest that the aminergic systems may have an identical functional modality and therefore may play some regulatory homeostatic role in the central nervous system.

The other main point for discussion concerning the electrophysiological and morphological data obtained in lesioned animals, could be tentatively elucidated in the light of the preceding remarks. Thus, there may exist a functional relationship between the electrical disorganization exhibited by the majority of the nigral DA cells (i.e. a real "electrophysiological syndrome") and the morphological abnormalities revealed by the histochemical fluorescence in the neostriatum.

One may, therefore, assume that the firing frequency decrease and the important irregularity of the pattern of discharge of the nigral DA neurons lead to a disturbance of dopamine release by the nigro-neostriatal DA terminals. The incorrect electrical activity may be responsible for an insufficient stimulation-release coupling (Douglas, 1968; Dreifuss et al., 1971; Baker, 1974; Nordmann, 1983) subserving the propensity of the neurotransmitter to remain in the synaptic boutons or varicosities. Previous and similar morphological work generally supports the present results. Even though Divac et al. (1978) do not find any difference in the intensity of catecholamine fluorescence, Meibach et al. (1978) observe a "loss of the homogeneous pattern of fluorescence in the neostriatum, with the presence of numerous swollen varicosities"; Gottesfeld et al. (1979) note that the neostriatal "sphere of kainic acid influence had a higher intensity of fluorescence and contained varying amounts of small intensely green fluorescent swollen spherical structures". As widely indicated above, one may assert that neostriatal destruction by KA does not induce a hyperactivity of the DA nigro-neostriatal pathway. Further strengthening this point of view is the fact that, at the level of the perikarya of nigral DA cells, there is no conspicuous difference in fluorescence intensity between normal and lesioned rats. This unchanged aspect of fluorescence intensity of DA cells would support the hypothesis of a decrease in dopamine release by the neostriatal terminals rather than an increase in dopamine synthesis by the nigral cells. Furthermore the concentration of dopamine in the lesioned neostriatum does not reveal any significant change in comparison with the values obtained from normal neostriatum (Bioulac-Sage et al., 1983).

These results agree with those of Melamed et al. (1982) who demonstrated that dopamine and homovanillic acid (the major DA metabolite) levels in the caudate nucleus are unchanged in choreic brains and in rat brains with chronic neostriatal lesions.

A last point for discussion concerns the genesis of the "electrophysiological syndrome" presented by most nigral DA cells after neostriatal destruction. Thus, one cannot interpret this profound electrophysiological disturbance only by the removal of the inhibitory GABAergic neostriato-nigral pathway. The abnormal activity exhibited by these DA neurons involves a much more complex mechanism than the simple "lifting of a brake".

Firstly, one must bear in mind that our lesion mainly affects the anterior part of the neostriatum and does not destroy only the GABAergic neostriatonigral tract. And in fact, our ultrastructural results indicate that, in the SN, numerous synaptic boutons of different types present degenerating patterns (Grofova, 1970; Hadju et al., 1973; Sotelo et al., 1973). More particularly, we may have destroyed the excitatory substance P pathway which originates from the head of caudate nucleus (Brownstein et al., 1977). Besides, recent biochemical studies performed, post-mortem, on choreic human brain tissue have shown that the level of substance P is significantly decreased in the SN and the pallidum (Kanazawa et al., 1977; Gale et al., 1978).

Secondly, numerous inputs, arising from different central structures such as pallidum, cerebellum, raphe nuclei and neocortex, converge onto the SN (Nieoullon, 1978; Gerfen et al., 1982) and still continue to exert their own effects after the destruction of the neostriato-nigral pathway. The reciprocal regulation existing between the two SN has also to be taken into consideration (Nieoullon et al., 1977). The imbalance of inputs impinging upon nigral DA cells may contribute to the genesis of the so-called electrophysiological syndrome.

Thirdly, the above-mentioned different types of inputs or influences interact upon the complex neuronal circuitry of the SN (Francois, 1979). The nigral neurons are characterized by large dendritic ramifications with important overlap of their fields. Several categories of interneurons may be interposed between the intrinsic neurons and the nigral inputs (Grace and Bunney, 1981). In particular, Cheramy et al. (1978) have postulated the existence of a glycinergic interneuron betwene the GABAergic afferents and the DA nigral cells. The disinhibition of this interneuron, occuring after the destruction of the GABAergic neostriato-nigral tract, can explain the decrease of firing frequency, one element of the electrophysiological syndrome.

One may finally conclude that lesions of the neostriatum, in the experimental model and, most likely, in Huntington's chorea and in the various choreic and hyperkinetic syndromes, induce a drastic dysfunction rather than a hyperactivity in the nigral DA cells. The main consequence of this functional impairment is the production of abnormal messages by the DA cells of the pars compacta of the SN, and very probably (due to the internal circuitry), by the nigral efferent cells of the pars reticulata. These messages, via ascending pathways (such as: the neostriato-pallido-thalamocortical tract) or descending pathways (such as: the nigro-reticulo-spinal and nigro-spinal tracts), may reach the motoneurons and participate in the genesis of the involuntary, hyperkinetic and choreic movements (Carpenter et al., 1950; Beckstead et al., 1979; Féger, 1981; Gerfen et al., 1982). The disturbance in DA release at the neostriatal terminals is directly attributable to the abnormal activity. This, lastly, brings up the question of the role exerted by the extrapyramidal DA system in motor control. The so-called regulatory homeostatic process would allow maintenance of a constant level of dopamine in certain central structures highly implicated in elaborating messages which subserve motor behaviour.

ACKNOWLEDGEMENTS

We wish to thank Mrs R. Bonhomme and Mrs J. Arsaut for the histological preparation and Mr D. Varoqueaux for preparing the illustrations. We are also indebted to Miss G. Gaurier and the late G. Labayle for their technical assistance.

This study was supported by the C.N.R.S. (E.R.A. 493 and A.T.P. No. 3262) and the Unite I.N.S.E.R.M., U 176.

A preliminary report was presented at the XXIXth Congress of Physiological Sciences, Sydney, September 1983.

REFERENCES

Aghajanian, G. K., and Bunney, B. S., 1973, Central dopaminergic neurons: morphological identification and responses to drugs, in: "Frontiers in Catecholamine Research," E. Usdin and S. Snyder, eds., Pergamon Press, Oxford.
Baker, P. F., 1974, Excitation-secretion coupling, in: "Recent Advances in Physiology," Churchill-Livingstone, London.
Beckstead, R. H., Domesick, V. B., and Nauta, W. J. H., 1979, Efferent connections of the substantia nigra and ventral tegmental area in the rat, Brain Res., 175:191.
Besson, M. J., Chéramy, A., Feltz, P., and Glowinski, J., 1971, Dopamine spontaneous and drug-induced release from the caudate nucleus in the cat, Brain Res., 32:407.
Bioulac-Sage, P., Arluison, M., Simonnet, G., Doudet, D., Gross, C., and Bioulac, B., 1983, Effect of intrastriatal injection of kainic acid (KA) in the rat. A histofluorescent study, Proceedings of the International Union of Physiological Sciences, XV:272.
Brownstein, M. J.., Mroz, E. A., Tappaz, M. L., and Leeman, F., 1977, On the origin of substance P and glutamic acid decarboxylase (GAD) in the substantia nigra, Brain Res., 135:315.
Bruyn, C. W., 1968, Huntington's chorea: historical, clinical and laboratory synopsis, in: "Handbook of Clinical Neurology," P. J. Vinken, and G. W. Bruyn, eds., North-Holland Publishing Company, Amsterdam, 6:298.
Carpenter, M. B., Whittier, J. R., and Mettler, F. A., 1950, Analysis of choreoid hyperkinesia in the rhesus monkey, J. Comp. Neurol., 92:293.
Chase, T. N., 1973, Biochemical and pharmacological studies of monoamines in Huntington's chorea, in: "Advances in Neurology," A. Barbeau, T. N. Chase, and G. W. Paulson, eds., Raven Press, New York.
Chéramy, A., Nieoullon, A., and Glowinski, J., 1978, Inhibition of dopamine release in the cat caudate nucleus by nigral application of glycine, Europ. J. Pharmacol., 47:141.
Chiueh, C. C., and Moore, K. E., 1973, Release of endogenously synthetized catecholamines from the caudate nucleus by stimulation of the nigro-striatal pathway and by administration of d-amphetamine, Brain Res., 50:221.
Constantinidis, J., 1972, Monoamines et syndromes choréiques, in: "Les Médiateurs Chimiques. Leurs Rôles dans la Physiopathologie de la Motricité de la Vigilance et du Comportement." Rapports de la XXIX réunion neurologique internationale, Masson et Cie, Paris.
Coyle, J. T., and Schwarcz, R., 1976, Lesion of striatal neurones with kainic acid provides a model for Huntington's chorea, Nature, 263:244.

Coyle, J. T., Mc Geer, E. G., Mc Geer, P. L., and Schwarcz, R., 1978, Neostriatal injections: a model for Huntington's chorea, in: "Kainic Acid as a Tool in Neurobiology," E. G. Mc Geer, J. Olney, and P. L. Mc Geer, eds., Raven Press, New York.
De Groot, J., 1959, "The Rat Forebrain in Stereotaxic Coordinates", North Holland Publishing Company, Amsterdam.
Deniau, J. M., Hammond, C., Riszk, A., and Féger, J., 1978, Electrophysiological properties of identified output neurones of the rat substantia nigra (pars compacta and pars reticulata): evidences for the existence of branched neurons, Exp. Brain Res., 32:409.
Denny-Brown, D., 1962, "The Basal Ganglia and Their Relation to Disorders of Movement," Oxford University Press, London.
Divac, I., Markowitsch, H. J., and Pritzel, M., 1978, Behavioural and anatomical consequences of small intrastriatal injections of kainic acid in the rat, Brain Res., 151:523.
Divac, I., and Oberg, R., 1979, Current conceptions of neostriatal functions: History and an evaluation, in: "The Neostriatum," I. Divac and R. G. E. Oberg, eds., Pergamon Press, Oxford.
Douglas, W. W., 1968, Stimulus-secretion coupling: the concept and clues from chromaffin and other cells, Brit. J. Pharmacol., 34:451.
Dray, A., 1979, The striatum and substantia nigra: a commentary on their relationships, Neuroscience, 4:1407.
Dreifuss, J. J., Kalnins, I., Kelly, J. S., and Ruf, K. B., 1971, Action potentials and release of neurohypophysial hormones in vitro, J. Physiol. (Lond.), 215:805.
Falck, B., Hillarp, N. A., Thieme, G., and Ford, A., 1962, Fluorescences of catecholamines and related compounds condensated with formaldehyde, J. Histochem. Cytochem., 10:348.
Féger, J., 1981, Les ganglions de la base: aspects anatomiques et électrophyisologiques, J. Physiol. (Paris), 77:7.
Fibiger, H. C., 1978, Kainic acid lesions of the striatum: a pharmacological and behavioural model of Huntington's disease, in: "Kainic Acid as a Tool in Neurobiology," E. G. McGeer, J. W. Olney, P. L. McGeer, eds., Raven Press, New York.
François, C., 1979," Anatomie Topographique, Subdivision Cytoarchitectonique Types Neuronaux de la Substantia Nigra et Localisation des Neurones Nigro-striés chez le Primate", Thèse de 3ème cycle, présentée à l'Université Pierre et Marie Curie, Paris.
Gale, J. S., Bird, E. D., Spokes, E. G., Iversen, L. L., and Jessel, T., 1978, Human brain substance P: distribution in controls and Huntington's chorea, J. Neurochem., 30:633.
Gauchy, E., Bioulac, B., Chéramy, A., Besson, M. J., Glowinski, J., and Vincent, J. D., 1974, Estimation of chronic dopamine release from the caudate nucleus of the Macaca mulatta, Brain Res., 77:257.

Gerfen, C. R., Staines, W. A., Arbuthnott, G. W., and Fibiger, H. C., 1982, Crossed connections of the substantia nigra in the rat, *J. Comp. Neurol.*, 207:283.

Gottesfeld, Z., and Jacobowitz, D. M., 1979, Kainic acid induced neurotoxicity in the striatum: a histofluorescent study, *Brain Res.*, 169:513.

Grace, A. A., and Bunney, B. S., 1981, Paradoxical GABA excitation of nigral dopaminergic cells: indirect mediation through reticulata inhibitory neurons, *Europ. J. Pharmacol.*, 59:211.

Grofova, I., and Rinvik, E., 1970, An experimental electron microscopy study on the striatonigral projection in the cat, *Exp. Brain Res.*, 11:249.

Guyenet, P. G., and Aghajanian, G. K., 1978, Antidromic identification of dopaminergic and other output neurons of the rat substantia nigra, *Brain Res.*, 150:69.

Hadju, F., Hassler, R., and Bax, I. J., 1973, Electron microscopy study of the substantia nigra and the strio-nigral projection in the rat, *Z. Zellforsch.*, 146:207.

Hattori, T., and Mc Geer, E. G., 1977, Fine structural changes in the rat striatum after local injections of kainic acid, *Brain Res.*, 129:174.

Hery, F., Simonnet, G., Bourgoin, P., Soubrie, P., Artaud, F., Hamon, M., and Glowinski, J., 1979, Effect of nerve activity on the in vivo release of (3H) serotonin continuously formed from L - (3H) tryptophan in caudate nucleus of the cat, *Brain Res.*, 169:317.

Huntington, G., 1872, The medical and surgical reporter, XXVI, 320.

Kanazawa, I., Bird, E., O'Connel, R., and Powell, D., 1977, Evidence for a decrease in substance P content of substantia nigra in Huntington's chorea, *Brain Res.*, 120:387.

Klawans, Jr. H. L., 1970, A pharmacological analysis of Huntington's chorea, *Europ. Neurol.*, 4:148.

Mason, S. T., and Fibiger, H. C., 1979, Kainic acid lesions of the striatum in rats mimics the spontaneous motor abnormalities of Huntington's disease, *Neuropharmacol.*, 18:403.

McGeer, E. G., and McGeer, P. L., 1976, Duplication of biochemical changes of Huntington's chorea by intrastriatal injections of glutamic and kainic acid, *Nature*, 263:517.

Meibach, R. C., Brown, L., and Brooks, F. H., 1978, Histofluorescence of kainic acid-induced striatal lesions, *Brain Res.*, 148:219.

Melamed, E., Hefti, F., and Bird, E. D., 1982, Huntington chorea is not associated with hyperactivity of nigrostriatal dopaminergic neurons: studies in post-mortem tissues and in rats with kainic acid lesions, *Neurology*, 32:640.

Moore, G. P., Segundo, J. P., Perkel, D. H., and Levitan, H., 1970, Statistical sign of synaptic interaction in neurons, *Biophys. J.*, 10:876.

Nieoullon, A., Chéramy, A., and Glowinski, J., 1977, Interdependence of the nigrostriatal dopaminergic systems on the two sides of the brain in the cat, *Science*, 198:416.

Nieoullon, A., 1978, "Etude de la Régulation de l'Activité de la Voie Dopaminergique Nigro-striée", Thèse presentée à l'Universite d'Aix Marseille III.

Nordmann, J. J., 1983, Stimulus-secretion coupling, in: "The Neurohypophysis Structure, Function and Control Progress in Brain Research," B. A. Cross, and G. Leng, eds., Elsevier Science Publishers, 60:281.

Perkel, D. H., Gerstein, G. L., and Moore, G. P., 1967, Neuronal spike trains and stochastic point process. I. The single spike train, Biophysical J., 7:391.

Poirier, L. J., Bedard, P., Langelier, P., Larochelle, L., Parent, A., and Roberge, A., 1972,, Les circuits neuronaux impliqués dans la physiopathologie des syndromes parkinsoniens, in: "Les Médiateurs Chimiques. Leurs Roles dans la Physiopathologie de la Motricité, de la Vigilance et du Comportement", Rapports de la XXIX réunion neurologique internationale," Masson et Cie, eds., Paris.

Sotelo, C., Javoy, F., Agid, Y., and Glowinski, J., 1973, "Injection of 6-hydroxy-dopamine in the substantia nigra of the rat. Morphological study, Brain Res., 58:269.

Spokes, E. G. S., 1980, Neurochemical alterations in Huntington's chorea. A study of post-mortem brain tissue, Brain, 103:179.

Steg, G., and Johnels, B., 1979, Motor functions of the striatum, in: The Neostriatum, I. Divac and R. G. E. Oberg, eds., Pergamon Press, Oxford.

Walters, J. R., Bunney, B. S., and Roth, R. H., 1975, Piribedil and apomorphine: pre- and postsynaptic effects on dopamine synthesis and neuronal activity, in: "Dopaminergic Mechanisms, Advances in Neurology," D. Calne, T. N. Chase, and A. B. Barbeau, eds., Raven Press, New York.

Wang, R. Y., and Aghajanian, G. K., 1982, Correlative firing patterns of serotoninergic neurons in rats dorsal raphe nucleus, J. Neuroscience, 2:11.

Wilson, C. J., Young, S. J., and Groves, P. H., 1977, Statistical properties of neuronal spike trains in the substantia nigra: cell types and their interactions, Brain Res., 136:243.

DOPAMINE NEURONE DEGENERATION-LIKE DEFICITS

PRODUCED BY INTRAHYPOTHALAMIC DOPAMINE INJECTIONS

Gregory L. Willis and Graeme C. Smith

Monash University
Department of Psychological Medicine
Prince Henry's Hospital
Melbourne, 3004

SUMMARY

When nigro-striatal and mesolimbic dopamine neurones degenerate, a loss of dopamine in various forebrain terminal fields occurs which is accompanied by increased amines in the degenerating axons that traverse the hypothalamus. While the behavioural deficits that occur after nigro-striatal degeneration have been attributed to the loss of dopamine neurotransmission in the striatum, recent evidence supports the contention that the amine accumulation in the degenerating axons represents functional neurotransmitter which may be released from these axons and may thereby participate in the production of behavioural deficits. To test this hypothesis further, albino rats were injected bilaterally with 200 nmol of dopamine in a location just rostral to the diencephalon/mesencephalon border, where amine accumulation is commonly observed following lateral hypothalmic lesions that reduce forebrain dopamine content. The effect of these injections upon motor performance and thermoregulation was determined 40 minutes after dopamine injection. In a second study, pargyline (50 mg/kg i.p.) was administered 30 minutes before intracerebral dopamine to determine whether this treatment would increase the severity of motor and thermoregulatory deficits which occurred after dopamine injections alone. Deficits in locomotion, rearing, and the ability to regulate body temperature were seen after the dopamine injections. The behavioural deficits displayed by pargyline-pretreated, dopamine-injected animals were slightly but not significantly more severe than those displayed by animals receiving dopamine injections alone. These results add further support to the hypothesis that amines which accumulate in degenerating neurones may be neuroactive and may thereby participate in the production of behavioural deficits attributed previously to the loss of dopamine neurotransmission.

INTRODUCTION

When the ascending nigro-striatal and mesolimbic dopamine (DA) systems degenerates a loss of neurotransmitter is seen in the axon terminals of the nucleus caudatus putamen (NCP) and other forebrain terminal fields, while the axons which traverse the lateral hypothalamus (LH) become enlarged with accumulated amines (Ungerstedt, 1971b; Willis and Smith, 1982a, 1982b; Zigmond and Stricker, 1973). It is now well accepted that the reduced neurotransmitter content is the degenerative phenomenon which underlies the occurrence of consummatory, thermoregulatory and motor deficits that are seen in experimental animals bearing DA depleting lesions (Oltmans and Harvey, 1972; Zigmond and Stricker, 1973).

It has been suggested more recently that the hypothalamic accumulations may be areas of functional neurotransmitter produced during degeneration which may also participate in the production of behavioural deficits previously attributed only to the loss of functional neurotransmitter in the terminal fields (Singer and Willis, 1977; Willis and Smith, 1982a, 1982b, 1983; Willis et al, 1983). That this accumulated neurotransmitter may be neuroactive and possibly affecting adjacent catecholamine (CA) receptors is supported by the finding that the injection of CA and CA agonists into the LH produce satiety or reductions in food intake which are similar to those produced by degenerative lesions (Leibowitz and Rossakis, 1979a, 1979b, 1979c). When we discovered that the accumulation and the severity of behavioural deficits were directly related we hypothesised that the accumulation represented an area of brain tissue functioning like an "artificial synapse" to release neurotransmitter, thereby affecting adjacent receptor systems. It has been hypothesized that a similar phenomenon could be functioning in axotomized peripheral nerves (Alvarez, 1968).

There are numerous examples of how degenerating neurons in the brain and peripheral nerves release neuroactive substances (Brown et al., 1978; Diamond, 1959; Evetts et al., 1972; Ungerstedt, 1971a). In particular, information from studies on peripheral adrenergic nerves indicates that during the acute stages of degeneration, axons which innervate the iris can release neurotransmitter and thereby cause the nictitating membrane to contract (Langler, 1966; Trendelenberg and Wagner, 1971). This is a transient phenomenon lasting only a few hours after which time changes in the sensitivity of post-synaptic noradrenaline (NA) receptors occur. Release of NA from ligated hypoglossal nerve has also been shown to occur in response to veratridine, changes in K^+ concentration, or electrical stimulation (Esquerro et al., 1980a, 1980b). While the build-up of

NA which has been observed histochemically in ligated peripheral nerves is intraneuronal, the accumulated neurotransmitter need not necessarily remain inside the neuron. In fact it has been demonstrated that upon stimulation of the nerve trunks, functional neurotransmitter is released from the nerve above the ligation just as it is released from an intact nerve ending (Esquerro et al., 1980a, 1980b).

It has also been shown that the release of CA produced by central 6-OHDA injection can result in eating within a few hours after such injections (Evetts et al., 1972). Increased gnawing, motor activity, circling behaviour as well as transient hypothermia all occur within a few hours after 6-OHDA treatment and can also be observed with other drugs, such as colchicine, which do not produce loss of neurotransmitter within that time (Avrith et al., 1970).

In consideration of the temporal characteristics of CA accumulation, the amount of neurotransmitter at the site of accumulation would be gradually increased during the first few hours of degeneration, as would be revealed by the gradually increasing amount of accumulation seen at this time (Andén et al., 1966; Hokfelt and Ungerstedt, 1973). As the quantity of available neurotransmitter gradually increased at the site of accumulation, the stimulatory effects on behaviour followed by behavioural inhibition might be seen as a dose-response relation with increasing levels of endogenously released CA. This interpretation fits in well with the information available as to the behavioural effects of intrahypothalamic injections of CA (Booth, 1982a, Grossman, 1962). Dose-response curves generated by these studies indicate that increases in eating behaviour are observed with increasing doses of CA. However, when the dose becomes very large, decreased food intake and motor abnormalities are seen. There is additional evidence that hypothalamic CA systems are involved in satiety when they are stimulated by microinjection of exogeneous CA (Leibowitz and Rossakis, 1979a, 1979b, 1979c). If CA neurotransmitter release occurred at the site of amine accumulation, then it might activate these hypothalamic DA systems, and thus result in hypophagia and hypodypsia characteristic of DA-depleting lesions. It is also possible that amines released from areas of accumulation could affect hypothalamic receptor mechanisms which have been proposed to function in amphetamine anorexia (Booth, 1968b; Leibwitz, 1975). While the reduced food intake that occurs after amphetamine administration has been attributed to the release of DA, hypothalamic lesions or chronic amphetamine treatment reduce forebrain DA levels and attenuate amphetamine anorexia (Bittner et al., 1981; Blundell and Lesham, 1975; Ellison et al., 1978).

If amine accumulation participates in the production of behavioural deficits, then intrahypothalamic injections of CA into areas where accumulation is seen might also be expected to produce thermoregulatory and motor impairment similar to that displayed by animals with DA degeneration. It was the object of this study to examine the effects of intrahypothalamic DA injections on thermoregulation and motor performance.

METHOD

Twenty male Sprague Dawley rats ranging in weight from 250 to 300 g were housed individually in wire mesh cages and allowed ad lib access to rat pellets (Clarke King, Melbourne) and tap water. A 12 hour light dark cycle was maintained throughout the experiment with lights on at 0700h. Room temperature was maintained at 22°C (±2°C).

Surgery

All animals were anaesthetised with 84 mg/kg of alphaxalon i.p. and then placed in a stereotaxic instrument. Cannulae were aimed at the posterior aspect of the lateral hypothalamus (PLH), bilaterally (coordinates A=-0.8 mm, L=±1.9 mm, D+-6.1 mm) where amine accumulation is typically observed during DA degeneration (Hökfelt and Ungerstedt, 1973; Singer and Willis, 1977; Ungerstedt, 1971b). The cannulae were in the plane of Pellegrino et al (1979) and the injection needle extended 2 mm beyond the cannula tip. Animals were allowed at least 2 weeks to recover from surgery before commencing the experiment and were allowed ad lib access to food and water throughout the study.

Injection

A 200 mM solution of 3,4 dihydroxyphenylethylamine hydrochloride (Dopamine: Sigma) was mixed in distilled water containing 0.2 mg/ml of ascorbic acid to retard oxidation, and injected in a volume of 1 µl. This dose was chosen on the basis of previous work (Leibowitz and Rossakis, 1979a, 1979b). Control injections were made with isotonic saline solutions containing 0.2 mg/ml of ascorbic acid. Each animal received 1 µl of the control solution, bilaterally.

Intraperitoneal injections were made with 50 mg/ml solutions of pargyline hydrochloride (Sigma) or with isotonic saline, and all solutions were administered in the volume of 1 ml/kg. Previous work has demonstrated that this dose of pargyline potentiates the behavioural deficits produced by DA injection (Leibowitz and Rossakis, 1979a). All i.p. injections were made 30 min before intracerebral injections.

Procedure

In the first study 5 animals received bilateral intracranial injections of DA while another 5 animals received a bilateral

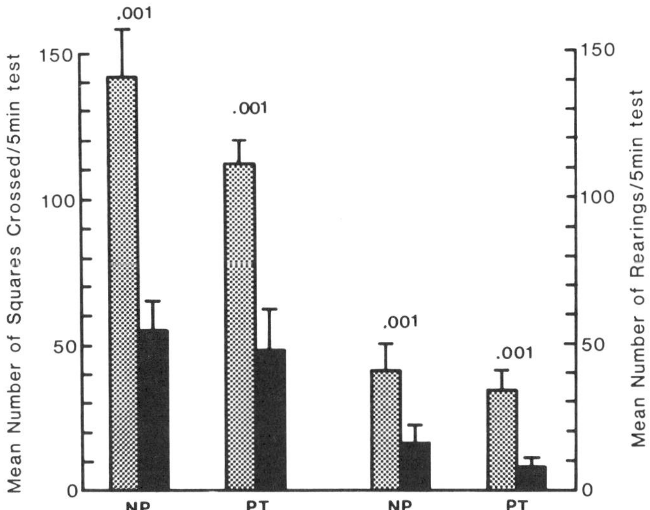

Fig. 1. The open field behaviour of animals receiving bilateral injection of dopamine (filled), or vehicle (stippled) into the posterior lateral hypothalamus. while some of the animals were pretreated with 15 mg/kg of pargyline (PT), the others were not (NP). The T-bars represent the standard error of the mean. Drug treated groups were significantly different from their respective controls at the "P" levels indicated.

injection of isotonic saline. 45 minutes after the intracerebral injections the performance of all animals was assessed in open field and motor reflex tests as described previously (Willis and Smith, 1982a; Willis et al., 1983). The number of squares crossed and the number of rearings during a 5 min test period were recorded. After this the latency to retract an elevated limb, to step up or down from a raised platform and to ambulate from within a prescribed area were measured. Rectal temperature was taken 1.5h after intracranial injections. Two days after this the drug and vehicle treatment groups were reversed and the performance of all animals on the motor tests was assessed and rectal temperature was measured. The paradigm employed in the second study utilizing 10 more animals was the same as that described in experiment 1 except that all animals were pretreated with either 15 mg/kg of pargyline or vehicle (1 mg/ml), 30 minutes before receiving the intracerebral injections.

At the completion of the experiments, all animals were sacrificed and their brains removed, stored in 10% formalin and then sectioned and stained using cresyl violet and luxol fast blue. The areas of necrosis resulting from injection were plotted on plates from the atlas of Pellegrino et al (1979). Statistical comparisons

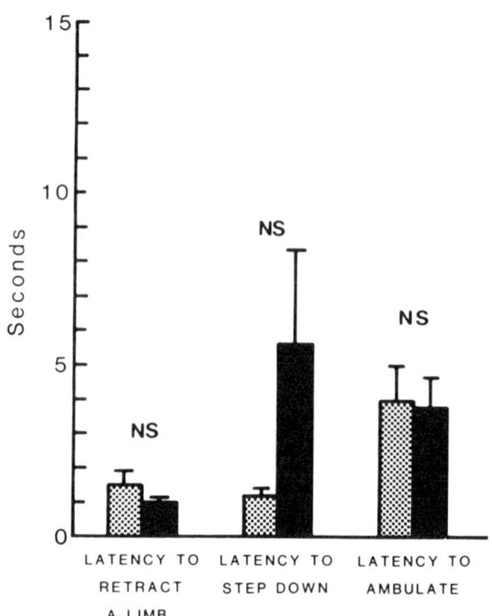

Fig. 2. The latency to retract a limb, to step down and to ambulate in animals injected bilaterally with dopamine (filled) or vehicle (stippled). The t-bars represent the standard error of the mean, N.S. indicates that there was no significant difference between drug and vehicle treatment.

were made between vehicle injection days and DA injection days for all measures using dependent t-tests.

RESULTS

As shown in figure 1, the open field behaviour of DA injected rats was significantly depressed when compared with vehicle-injected controls. The number of squares crossed during the 5 min test period was significantly less in the unpretreated ($p<.005$) and pargyline pretreated animals ($p<.005$) when compared to their respective controls. The number of rearings was greatly reduced in those animals receiving DA injections with ($p<.005$) and without ($p<.005$) pargyline pretreatment. Pargyline pretreatment tended to increase the severity of locomotor deficits when compared to those animals receiving no pretreatment but this difference was not significant.

As shown in figures 2 and 3, the performance of all DA injected groups, in latency to retract a limb, latency to step down, and latency to ambulate tests was not significantly impaired by the DA

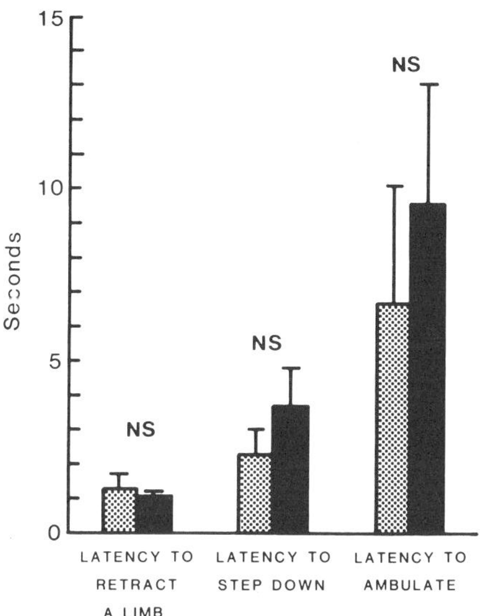

Fig. 3. The latency to retract a limb, to step down and to ambulate in animals injected bilaterally with dopamine (filled), or vehicle (stippled) 30 min after pretreatment with 15 mg/kg of pargyline. The T-bars represent the standard error of the mean, N.S. indicates that there was no significant difference between drug and vehicle treatments.

injection. The performance of animals pretreated with pargyline was in some cases slightly impaired when compared with control animals but the differences were not significant.

Intracerebral injections of DA produced a significant drop in rectal temperature 1.5h after injection as compared to the vehicle-injected controls (fig. 4, $p<.005$). Pargyline pretreated, DA injected animals also displayed a significant reduction in rectal temperature when compared to their vehicle injected counterparts ($p<.005$). Pargyline pretreatment alone did not reduce significantly the parameter of body temperature.

Histological verification of the injection sites of the animals used in the behavioural studies confirmed that the sites of DA injection were in the proximity of the medial forebrain bundle in the PLH.

DISCUSSION

These results demonstrate that hypothermia and deficits in motor performance can be produced by injections of DA into the PLH. Given that these deficits are similar to those produced by DA-degenerating lesions which produce accumulations proximal to the site of injection in the PLH, we suggest that the unregulated release of CA from the sites of accumulation may aid in producing the behavioural deficits usually attributed only to the loss of functional neurotransitter in the striatum. Decreases in food intake can be seen when similar doses of DA are injected into the perifornical hypothalamus (Leibowitz and Rossakis, 1979a, 1979b, 1979c). This anatomical location of injection is very close to the site where degeneration-associated amine accumulations are found in animals bearing DA-depleting lesions and which suffer severe deficits in food and water intake (Singer and Willis, 1977; Willis and Smith, 1982a, 1982b; Willis et al., 1983). The results of this work taken together with that from other studies supports the hypothesis that DA released from sites of accumulation may be affecting hypothalamic CA systems that regulate satiety and which may be activated by the peripheral administration of CA agonists and enzyme inhibitors that produce anorexia (Booth, 1968b; Leibowitz and Rossakis, 1979a; Willis and Smith, 1982b). While we have not yet verified that the amine accumulated in degenerating neurons is DA, the results from the present study and from previous work (Willis et al., 1983) suggest that DA may be the CA which is accumulated in impaired animals.

It has been reported previously that the severity of hypothermia which occurs 1.5h after 6-OHDA treatment correlates well with the volume of accumulation which is seen in the hypothalamus (Singer and Willis, 1977). The larger the volume of accumulation the more severe the hypothermia. On the basis of this observation it was concluded that the neurotransmitter may be released at these sites rather than from DA terminals in the striatum or preotic area of the hypothalamus which is thereby producing the hypothermic reponse (Cox et al., 1978; Nakamura and Thoenen, 1971). While the results from the present study lend support to our previous hypothesis, it is not clear whether lesion-induced hypothermia results from activation of DA systems involved in the homeostatic control of thermoregulation (Cox et al., 1978).

It is difficult to explain why severe deficits in locomotor behaviour were seen in the DA injected animals while their performance on the 3 motor reflex tests was normal. It could be that the mechanisms governing more finely tuned aspects of motor control, as detected by these tests, are controlled by the nigrostriatal system and were not affected by the intracerebral DA injection. It could be that only locomotory and consummatory behaviours are altered by the accumulation of amines in the hypothalamus. It is also possible that the quantity of neuroactive DA after these intrahypothalamic injections was not as great as that occurring after DA depleting lesions. Fluorescent histochemical observations of amine accumulation after dopamine injections indicated that the

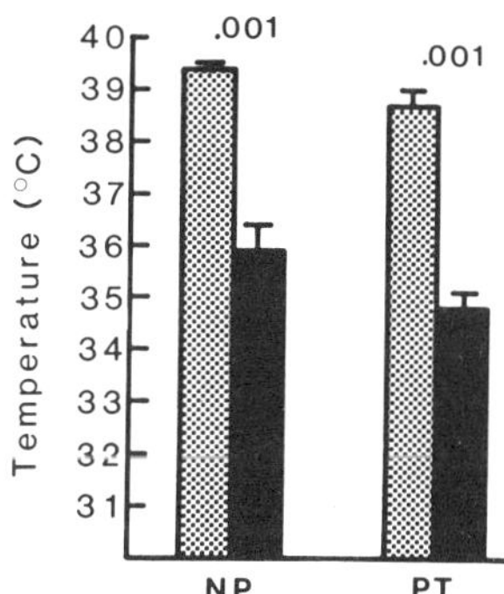

Fig. 4. The mean rectal temperature of animals receiving bilateral intracerebral injections of dopamine (filled) or vehicle (stippled) with (PT) and without (NP) pargyline pretreatment 30 min prior to intracerebral injection. The T-bars represent the standard error of the mean. Drug-treated groups were significantly different from their respective controls at the "P" levels indicated.

increase in fluorescence is much less than that observed in impaired animals with DA-depleting lesions of the hypothalamus (Willis and Smith, unpublished observations) and therefore may not be sufficient to affect motor reflex control.

It is entirely possible that some process other than transmitter release from areas of accumulation is responsible for producing the behavioural deficits which are associated with the build-up of CA after CA-depleting lesions. One possible contender is ephaptic transmission or "cross talk" between adjacent axotomized neurons. This process has been shown to occur in sensory and motor fibres in the peripheral nervous system (Blumberg and Jänig, 1982; Rasminsky, 1978, 1980). These are only a few of the numerous studies demonstrating that when fibre bundles are cut in experimental animals electrical potentials can arise spontaneously from the cut end of afferent nerves and are presumably caused by potentials arising in adjacent efferent fibres. Furthermore, ephaptic transmission in the CNS has been hypothesised to be responsible for producing many of the symptons of progressive demyelination disease (Patterson et al., 1980). As we have suggested previously (Willis and Smith 1982a), an alternative explanation could be that the local swelling of axons at the site of accumulation may be having an effect on various axons of several adjacent systems. In that the medial forebrain bundle contains a wide variety of neurochemically discreet systems, this axon swelling may be affecting many of the axons contained in it. When the CA accumulation decreases over time (Willis and Smith, 1982a), then normal function of adjacent neurons is gradually restored. However, no information is available as to what effects such swelling might have on adjacent neurons. Finally, it is also possible that accumulation is a marker of some as yet unidentified change in neuronal function elsewhere in the brain.

In conclusion, the results from this work and from other studies (Singer and Willis, 1977; Willis and Smith, 1982a, 1982b, 1983; Willis et al., 1983) indicate that many of the deficits in thermoregulation, food intake and locomotor behaviour which occur after DA-depleting treatments can also be produced by the intracerebral injection of DA. Furthermore, CA precursors and agonists which are administered peripherally can also produce deficits in locomotor behaviour, consummatory behaviour and thermoregulation similar to those seen after DA degeneration or intracerebral DA injections, supporting the hypothesis that the hypothalamus may be the site of action of these compounds which cause a release of CA.

REFERENCES

Alvarez, J., 1968, A possible artificial axo-axonic chemical synapse, Acta Physiol. Lat. Amer., 18:366.

Andén, N. -E., Dahlström, A., Fuxe, K., Larson, K., Olson, L., and Ungerstedt, U., 1971, Ascending monoamine neurons to the telencephalon and diencephalon, Acta Physiol. Scand., 65:313.

Avrith, D., Haas, H. L., and Mogenson, C. J., 1979, Behavioural and electro-physiological changes following microinjections of colchicine into the substantia nigra, Neurosci., 4:227.

Bittner, S. E., Wagner, G. C., Aigner, T. G., and Seiden, L. S., 1981, Effects of a high-dose treatment of methamphetamine on caudate dopamine and anorexia in rats, Pharmac. Biochem. Behav., 14:481.

Blumberg, H., and Jänig, W., 1982, Activation of fibres via experimentally produced stump neuromas of skin nerves: Ephaptic transmission or retrograde sprouting? Exp[. Neurol, 76:468.

Blundell, J. E., and Lesham, M. B., 1975, Hypothalamic lesions and drug-induced anorexia, Postgrad. Med. J., 51:45.

Booth, D. A., 1968a, Mechanisms of action of norepinephrine on eliciting an eating response on injection into the rat hypothalamus, J. Pharmacol. Exp. Ther., 160:336.

Booth, D. A., 1968b, Amphetamine anorexia by direct action on the adrenergic feeding system of rat hypothalamus, Nature, 217:869.

Brown, M. C., Holland, R. L., and Ironton, R., 1978, Degenerating nerve products affect innervated muscle fibres, Nature, 275:65.

Cox, B., Kerwin, R., and Lee, T. F., 1978, Dopamine receptors in the central thermoregulatory pathways of the rat, J. Physiol., 282:471.

Diamond, J., 1959, The effects of injecting acetylcholine into normal and regenerating nerves, J. Physiol., 145:611.

Ellison, G., Eison, M. S., Humberman, H. S., and Daniel, F., 1978, Long-term changes in dopaminergic innervation of caudate nucleus after continuous amphetamine administration, Science, 201:276.

Esquerro, E., Cena, V., Sanchez-Garcia, P., Kirpakar, S. M., and Garcia, A. G., 1980a, Release of noradrenaline from the ligated hypogastric nerve, Eur. J. Pharmacol., 61:183.

Esquerro, E., Schiavone, M., Garcia, A. G., Kirpekar, S. M., and Prat, J. C., 1980b, Release of endogenous noradrenaline from ligated cat hypogastric nerve by veratridine, Europ. J. Pharmacol., 66:367.

Evetts, K. D., Fitzsimmons, J. T., and Setler, P. E., 1972, Eating caused by 6-hydroxyodopamine-induced release of noradrenaline in the diencephalon of the rat, J. Physiol., 223:35.

Grossman, S. P., 1962, Direct adrenergic and cholinergic stimulation of hypothalamic mechanisms, Amer. J. Physiol., 202:872.

Hökfelt, T. and Ungerstedt, U., 1973, Specificity of 6-hydroxydopamine-induced degeneration of central monoamine neurones: an electron and fluorescence microscopic study with special reference to intracerebral injection on the nigrostriatal dopamine system, Brain Res., 60:269.

Langler, S. Z., 1966, The degeneration contraction of the nictitating membrane in the unaesthetized cat, J. Pharmacol. Exp. Ther., 151:66.

Leibowitz, S. F., 1975, Catecholaminergic mechanisms of the lateral hypothalamus: their role in the mediation of amphetamine anorexia, Brain Res., 98:529.

Leibowitz, S. F., and Rossakis, C., 1979a, L-Dopa feeding suppression: effect on catecholamine neurons of the perifornical lateral hypothalamus, Psychopharmacol., 61:273.

Leibowitz, S. F., and Rossakis, C., 1979b, Mapping of brain dopamine- and epinephrine-sensitive sites which cause feeding suppression in the rat, Brain Res., 172:101.

Leibowitz, S. F., and Rossakis, C., 1979c, Pharmacological characterisation of perifornical hypothalamic dopamine receptors mediating feeding inhibition in the rat, Brain Res., 115.

Nakamura, K., and Thoenen, H., 1971, Hypothermia induced by intraventricular administration of 6-hydroxydopamine in rats, Europ. J. Pharmacol., 16:46.

Oltmans, C. A., and Harvey, J. A., 1972, LH syndrome and brain catecholamine levels after lesions of the nigrostriatal bundle, Physiol. Behav., 8:69.

Patterson, V. H., Foster, D. H., and Heron, J. R., 1980, Variability of visual threshold in multiple sclerosis: Effect of background luminance on frequency of seeing, Brain, 103:139.

Pellegrino, L., Pellegrino, A., and Cushman, A., 1979, "A Stereotaxic Atlas of the Rat Brain," Plenum Press, New York.

Rasminsky, M., 1978, Ectopic generation of impulses and cross-talk in spinal nerve roots of "dystrophic" mice, Ann. Neurol., 3:351.

Rasminsky, M., 1980, Ephaptic transmission between single nerve fibres in the spinal nerve roots of dystrophic mice, J. Physiol., 305:151.

Singer, G., and Willis, G., 1977, Biochemical and pharmacological basis for the lateral hypothalamic syndrome, Brain Res. Bull., 2:485.

Trendelenberg, U. and Wagner, K., 1971, The effect of 6-hydroxydopamine on the cat's nictitating membrane: degeneration contraction and sensitivity to (-)-noradrenaline, in: "6-Hydroxydopamine and Catecholamine Neurons," T. Malmfors and H. Thoenen, North Holland Publ., Amsterdam.

Ungerstedt, U., 1971a, Striatal dopamine release after amphetamine or nerve degeneration revealed by rotational behaviour, Acta Physiol. Scand. Suppl., 367:49.

Ungerstedt, U., 1971b, Adipsia and aphagia after 6-hydroxydopamine induced degeneration of the nigro-striatal dopamine system, Acta Physiol. Scand., 367:95.

Willis, G. L., and Smith, G. C., 1982a, The behavioural effects of intra-hypothalamic multistage versus single injection of 6-hydroxydopamine, Brain Res., 245:345.

Willis, G. L., and Smith, G. C., 1982b, Anorexic properties of three monoamine oxidase inhibitors, Pharmac. Biochem, Behav., 17:1135.

Willis, G. L., and Smith, G. C., 1983, Short-term effects of intra-hypothalamic colchicine, Brain Res., 276:119.

Willis, G. L., Smith, G. C., and McGrath, B. P., 1983, Deficits in locomotor behaviour and motor performance after central 6-hydroxy-dopamine or peripheral L-Dopa, Brain Res., 266:279.

Zigmond, M. J., and Stricker, E. M., 1973, Recovery of feeding and drinking by rats after intraventricular 6-hydroxydopamine or lateral hypothalamic lesions, Science, 182:717.

OUTPUT PATHWAYS MEDIATING BASAL GANGLIA FUNCTION

Gaetano Di Chiara and Micaela Morelli

Institute of Experimental Pharmacology and Toxicology
University of Cagliari
Viale A. Diaz 182, 09100 Cagliari, Italy

In the past seven years work from our laboratory (Di Chiara et al., 1979b and 1981) and from others (Kilpatrick et al., 1982, Leigh et al., 1983) has provided considerable information on the transmitter-specific pathways which mediate the output of motor and behavioural information arising in the basal ganglia.

BASIC ASSUMPTIONS

The basic assumption of this work is that certain components of the motor and behavioural syndromes induced in the rat by systemic administration of dopamine (DA)-receptor stimulants (particularly apomorphine) or DA-receptor blockers (neuroleptics) are a specific expression of basal ganglia activity. The syndromes in question are the sterotyped behaviour (sniffing, licking, gnawing syndrome), evoked in the rat by systemic apomorphine (Di Chiara and Gessa, 1979) and the catalepsy induced by systemic haloperidol (Fog et al., 1970). To these syndromes one should add the turning behaviour, which can be elicited by administration of apomorphine to a rat unilaterally denervated of DA projections to basal ganglia by injection of 6OHDA in the medial forebrain bundle (Ungerstedt, 1971). The above assumption derives its strength from the evidence that these three motor-behavioural syndromes result from stimulation or blockade of DA-receptors specifically located in the basal ganglia (Di Chiara and Gessa, 1979; Ungerstedt, 1971; Kelly et al., 1975; Fog et al., 1979). On this basis we have assumed that any alteration in the pattern and intensity of these syndromes produced by experimental manipulation of discrete brain regions would be due to interference with the specific pathways mediating their expression. Therefore, studies of this kind would provide insight on the role of specific brain areas and neurotransmitter systems in the mediation of basal ganglia function. A further assumption which theoretically derives from the previous one and is confirmed by experiment is that the eliciting of a behavioural syndrome phenomenologically similar to the pharmacologically-induced one, or to some of its components, by experimental manipulation of discrete brain regions, is evidence that the brain area in question is involved in the expression of that specific syndrome and that, by the same mechanism, it might play a role in the expression of basal ganglia function. Therefore our approach consists in studying the interaction of localized manipulation of brain function with drug-induced syndromes, or their ability to evoke them in non-treated animals.

COMMENTS AND CAVEATS

The advantage of using pharmacologically-induced syndromes rather than spontaneous ones derives from the fact that while the pharmacological syndromes can be ascribed with certainty to basal ganglia, it is not known which spontaneous motor or behavioural function should be taken as an index of basal ganglia activity. The reason for this is that the normal functions of the basal ganglia are not known with certainty and are therefore the subject of much debate.

We would stress here that by no means can our results be taken to establish the role of specific neurotransmitter circuits in specific functions of the basal ganglia in normal motor behaviour. Such information could be derived from the application of our experimental approach to spontaneous or conditioned behaviour, but again depends on advances in the knowledge of the role of basal ganglia in normal motor behaviour and on the possibility of devising motor and behavioural paradigms appropriate for its study.

A basic feature of our approach is the idea that by lesioning or depressing brain areas one can obtain important information on the way the brain functions. This approach is not dissimilar from that utilized more than a century ago by Hughlings Jackson (1875). Indeed some of his principles can be useful for analysing our results. Hughlings Jackson stated that following localized brain lesions only negative symptoms (deficits) are related to the inherent function of the lesioned area, while positive symptoms simply indicate that the lesioned area is normally able to influence another area which is the real "center" of the positive symptom and of its related function. According to this reasoning, the eliciting of apomorphine-like stereotypy or turning or neuroleptic-like catalepsy, by a localized brain lesion or by reversible inactivation of neural activity as after local injections of muscimol, is not evidence that the lesioned or depressed area is intrinsically related to the syndrome evoked.

Thus, the fact that lesions of the substantia nigra pars reticulata produce spontaneous stereotyped behaviour (if bilateral) or contraversive turning (if unilateral) indicates, (Di Chiara et al., 1979a), according to the Jacksonian principle, that the substantia nigra is not the "center" for stereotypy and for contraversive turning, but rather exerts an inhibitory influence on another area which is in turn the actual "center" for these behaviours. However, the brain does not exert its functional role simply by virtue of the activity of the "centers" for certain motor patterns; a major part of brain function consists in modulating, through phasic activity of inhibitory areas, the function of the "centers". Such integrative function is just as essential as that of the "centers", and might be the essence of the function of certain brain areas. Basal ganglia appear to fall in the category of such modulatory systems of the brain. Thus, after nigral kainate lesions, when the acute effects of the lesion have disappeared one can show a deficit both in apomorphine stereotypy (Morelli et al., 1980) and in turning behaviour (Di Chiara et al., 1979a) as well as in haloperidol-catalepsy (Morelli et al., 1981b). Indeed, the lesions can abolish the expression of those syndromes arising from the basal ganglia. This means to us that an important way in which basal ganglia exert their effects on motor behaviour is through the modulation of centers located outside them. In terms of neural mechanisms, stereotypy, turning and catalepsy result from inhibition or disinhibition of nigral inhibitory units with consequent disinhibition or inhibition of the areas of nigral projection. In view of this, the Jacksonian principle of "center" appears rather reductionist since even drug-induced behaviour cannot be simply described as the result of the disinhibition of a certain number of "centers", but rather as the result of the ability of the basal ganglia to "organize" the disinhibition of such "centers". Therefore the basal ganglia appear to provide, even in drug-induced behaviour, a level of integration which transcends the individual function of "centers" otherwise unrelated one to the other. This ability is, by itself, a function of the basal ganglia which is not inherent in any of its effector "centers", and might justify their appraisal as an integrative "center" on their own. On the other hand, the function of the basal ganglia can hardly be only that of modulating the "centers" involved in drug-induced sniffing, licking and turning, although we cannot dismiss the possibility that basal ganglia can control normal gnawing, sniffing, or licking and head-turning possibly through the same pathways utilized in drug-induced syndromes. However, the problem of the relevance of the drug-induced syndromes to normal basal ganglia function is beyond the limits of our approach and is not essential for it, since we assume that, no matter which "centers" the basal ganglia modulate during their normal function, such modulation takes place through the same basic neurochemical and neuroanatomical paradigms, and such basic paradigms we can disclose with our experimental analysis.

BASIC METHODOLOGIES

The approach utilized in our studies of basal ganglia consists of two basic techniques: localized brain lesions and local intracerebral injection of potent receptor-specific drugs. The brain lesion technique, although rather old and classic, has in the last few years undergone an important development with the introduction of selective neurotoxins as lesioning tools. Such neurotoxins can be injected through thin (0.2, 0.3 mm, o.d.) injection cannulas, dissolved in small volumes (0.5-0.25 µl), into selected brain regions using stereotaxic coordinates. In this manner, one can selectively lesion dopaminergic or noradrenergic neurons by 6OHDA (Hökfelt and Ungerstedt, 1973; Ungerstedt, 1971), serotonergic neurons by 5-7 dihydroxytryptamine (Bjorklund et al., 1973; Baumgarten and Lachenmayer, 1974), cholinergic neurons by aziridinium mustard (Fisher et al., 1982) or destroy neuronal perikarya without interrupting fibers of passage with the excitotoxins, particularly kainic acid, the most potent, cheap and widely used one (Coyle et al., 1978). All these lesioning techniques have their pitfalls; however, one can devise selected conditions and appropriate controls in order to optimize their effects. In the case of kainic acid, one difficulty in its use, viz. distant brain damage (Wuerthele et al., 1978) related to spread of excitation to specific convulsion-evoking pathways, can be overcome by the use of low amounts of kainate (0.15-0.25 µg) injected at each site and by concurrent peripheral administration of diazepam as anticonvulsant (Ben-Ari et al., 1979). The second difficulty is that of performing really discrete lesions. Also this can be overcome by the use of low amounts of kainate (0.15-0.25 µg), but the effectiveness depends on the sensitivity to kainate of the area to be lesioned compared to the neighbouring ones. Use of appropriately small injection volumes (0.25-0.5 µl) and slow infusion speed (e.g. 1 µl/6 min.) has proven to be essential for effectiveness and reproducibility of the lesions. Some workers prefer to use ibotenic acid instead of kainate because it is expected to produce more discrete lesions and to avoid distant brain damage (Schwarcz et al., 1979). In general we find that discreteness of ibotenate versus kainate lesions mainly reflects the much lower potency of ibotenate (10-50 times) compared to kainate; however, the ratio of their potency varies greatly depending on the area, probably because the receptors mediating their neurotoxicity are different. On the other hand, in certain areas such as the substantia nigra, the nucleus fastigii, the inferior olive, the superior colliculus and the reticular formation, kainate is more effective and reliable than ibotenate (Di Chiara et al., unpublished observations). Therefore we regard kainate as the "first choice" excitotoxin to be used as a lesioning agent.

It is essential if the lesioning technique is to provide meaningful information to apply sensitive and reliable methods for evaluating the specificity, degree and topography of the lesions and, for each animal, to correlate such information with the behavioural results. These methods might be behavioural, neurochemical and histological, and can be applied individually or collectively depending on the neurotoxin and on the topography of the lesion performed.

For example, in order to evaluate kainate-induced unilateral nigral lesions one can utilize a behavioural index such as the intensity of spontaneous postural asymmetry (contraversive turning), a neurochemical index such as the loss of glutamic acid decarboxylase (GAD) in the superior colliculus and a histological index such as the loss of neuronal perikaria in the substantia nigra and in extranigral areas (Di Chiara et al., 1979a,b,c). Another technique used in these studies is the intracerebral drug-injection technique through stereotaxically placed cannulas. The success of this technique has been made possible by the availability of potent and specific drugs capable of interfering with transmitter-specific receptors (e.g. muscimol, bicuculline) or with the coupling mechanisms of these receptors with ion channels (e.g. picrotoxin). The method used to perform such technique varies among investigators but two basic procedures have been adopted: acute intracerebral injection in ether or halothane-anaesthetized animals placed on the stereotaxic apparatus, or injection in awake animals by injectors inserted into chronic guide cannulas secured to the skull by dental cement. The first procedure is fairly popular because of its simplicity but one should be aware of its pitfalls and liability to produce artifacts. The main problem of the acute injection technique is the interference of the anaesthesia with the behavioural effects of the drugs injected intracerebrally. Recovery from anaesthesia after completion of the acute intracerebral injection involves at least three phases, an initial phase during which the animal is still anaesthetized and which cannot be eliminated unless the anaesthetic is discontinued some minutes before dismounting the animal from the stereotaxic frame. This procedure is not advisable both for ethical reasons and for being extremely stressful to the animal. Since the animals are still under the depressant effect of the anaesthesia during the first 5-10 min. after its discontinuation, which coincides with the end of the injection, it is impossible by this technique to evaluate the latency of the behavioural effects of the drugs injected. This is of fundamental importance because estimation of the latency and its correlation with the topography of the injection is essential to establish the area of origin of the effects observed. In fact, one of the main disadvantages of the intracerebral drug-injection technique is that the effects observed are not necessarily due to an action of the drug at the point of injection but rather might result from diffusion to an adjacent area. Indeed, certain effects can be evoked by injection into sites distant from the active one provided sufficient time is given for diffusion. This applies, for example, to the catalepsy evoked by intrathalamic injection of muscimol (Di Chiara et al., 1979d) or to the contraversive turning evoked after intramesencephalic muscimol (Scheel-Kruger et al., 1977); an accurate study of the latency of the effects will show that the area providing the lowest latency is the VM thalamus in the first case and the substantial nigra pars reticulata in the second. Another criterion for establishing topography is to find the area which responds to minimally effective doses of the drug, i.e. the most sensitive area for a certain effect. The application of this criterion is made difficult by the use of the acute intracerebral injection procedure, since the depressant influence of anaesthesia can reduce the sensitivity for a particular effect, thus making impossible the use of minimally effective doses of the drugs. The second post-anaesthetic phase is characterized by hyperactivity and ataxia while the third phase, rather long-lasting, is marked by behavioural depression. For this reason, anaesthesia can influence in a complex, often biphasic manner, the effect of drugs injected intracerebrally. Thus, we have found (Morelli and Di Chiara, in preparation) that opiates injected acutely in the nigra of one side result in a higher contraversive turning intensity (number of turns) in the first 10 min and in a lower turning intensity thereafter as compared to injection into awake rats through chronic guide cannulas. We interpret this as due to a potentiating effect exerted by the anaesthetic-induced hypermotility and by an inhibitory effect exerted by the late post-anaesthetic depressant phase. In some cases, anaesthesia strongly potentiates certain depressant effects as is the case with the catalepsy evoked by intrathalamic muscimol (Di Chiara et al., 1979d), or inhibits the expression of certain stimulatory effects such as the stereotyped gnawing obtained after intracollicular picrotoxin (Imperato and Di Chiara, 1981) or after injection of muscimol in the ventral ponto-mesencephalic reticular formation (Imperato and Di Chiara, in preparation).

The use of the acute injection procedure in anaesthetized animals is probably the reason for some of the disagreement existing in the literature. For example, according to Kilpatrick et al. (1982), picrotoxin injected in the superior colliculus of anaesthetized rats produces active contralateral circling, while we have obtained in awake rats tight posturing and only very low intensity circling (Imperato and Di Chiara, 1981). In the case of Kilpatrick et al. (1982), anaesthesia might have provided that hypermotility which, in the presence of a strong postural asymmetry, could result in active circling. Conversely, we (Imperato et al., in preparation) have failed to observe the inhibitory effect of intrareticular muscimol on apomorphine stereotypy reported by Childs and Gale (1983) after acute injection in anaesthetized rats. Rather, we have found that intrareticular muscimol injected at the same site as by Childs and Gale (1983), but in awake rats bearing chronic guide cannulas, results in spontaneous gnawing and in potentiation of apomorphine stereotypy.

Apart from avoiding the interference of anaesthesia, use of

chronically implanted guide cannulas makes it possible to perform, in the same animal, multiple injections of the same compound at different depths in order to compare the effect of different compounds at the same site. For all these reasons we suggest that the use of acute intracerebral injections be limited to those cases in which the results obtained with this technique have been found to be superimposable on those obtained with chronic guide cannulas.

EVALUATION OF BEHAVIOUR

It is obvious that the conclusions one can draw from the study of the behavioural effects of experimental manipulations of brain function is highly dependent on the methods used to evaluate the behaviour itself. In general, automated and quantitative estimation of behaviour in strictly controlled and reproducible conditions is advisable. Stereotyped behaviour, turning behaviour and catalepsy should be in principle amenable to estimation in such manner, and indeed various automated apparatus for such assessment have been devised and are available. It is important however to stress that, even when an automatic apparatus is used, behaviour should be assessed in parallel by an observational method. In this way one often gains precious information for explaining the mechanism of changes recorded by the automatic apparatus. For example, direct and indirect DA-stimulants produce a biphasic effect on motility counts, with an increase at lower doses and a decrease at higher ones. Such decrease, however, is associated with heightened behavioural stimulation, and the replacement of exploratory sniffing by confined, higher intensity sniffing and licking or gnawing (Di Chiara and Gessa, 1978). Therefore, potentiation or inhibition of the stimulant action of DA-ergic drugs might result in the same effect, i.e. reduction, when only motility counts are taken.

In view of this, we believe that the best way to assess behaviour is to integrate when possible an automated with an observational procedure. For this purpose, video-taping of animal behaviour and its retrieval is a very sensitive development of the observational method for estimating behaviour, eventually coupled to a multiple channel event recorder (Morelli et al., 1980). This set-up is more time-consuming than automatic procedures and requires experienced observers, but is the one that provides the most complete and accurate information on behaviour.

ESTIMATION OF TURNING BEHAVIOUR

In the case of turning behaviour the use of automated rotometers and the simple estimation of the number of turns/minute (Ungerstedt and Arbuthnott, 1970) leads to overlooking what we consider to be an important component of turning, i.e. the degree of postural asymmetry (posturing). In fact there is evidence that turning behaviour is the result of two components, the postural component and the locomotor component, which are reflected respectively by the circling intensity in turns/minute and by the score for posturing (Kelly et al., 1975; Costall et al., 1976; Pycock and Marsden, 1978). Recent evidence indicates that these two components can be affected differentially by brain lesions or pharmacological manipulations, and therefore it is advisable to estimate both.

ESTIMATION OF STEREOTYPY

Stereotyped behaviour in response to DA-receptor stimulants is currently estimated by a system which assigns a progressive score to each item of the stereotyped syndrome from locomotion to sniffing, licking and gnawing (Di Chiara and Gessa, 1978). This method is based on the assumption that the full syndrome of stereotypy is a continuum of responses, of increasing intensity from locomotion to sniffing, licking and gnawing. In normal animals it is conceivable that the increase of stereotypy score reflects an increasing DA-receptor occupancy and that an item of higher score is likely to occlude the expression of those of lower score, therefore in principle one could accept the use of the scoring system to measure the intensity of central DA-receptor stimulation in otherwise intact animals. The method does not, however, seem appropriate when studying the changes induced by brain lesions or intracerebral drug injections. In fact there is evidence that the different items of stereotypy originate from different brain areas; thus locomotion appears to originate from stimulation of DA-receptors in mesolimbic areas, while confined sniffing and gnawing are evoked from the caudate-putamen (see below). Therefore it is quite possible that brain manipulations will influence independently the different items of stereotypy. The possibility of an independent segregation of stereotypy items makes it necessary to evaluate the intensity of each item separately. This approach has been followed by Ljundberg and Ungerstedt (1977) using an automated apparatus, and by us using an observational method which records the percent of time spent by the animal in performing each stereotyped item (Morelli et al., 1980). In spite of their separate evaluation and of the possibility that they segregate independently, the various stereotypy items compete with one another for expression. Therefore it is expected that inhibition or facilitation of one item will respectively favour or inhibit the expression of the others. In this case it is necessary to recognize which item is primarily affected and which is modified only secondarily. For example, lesions of substantia nigra pars reticulata reduce gnawing but increase sniffing and locomotion (Morelli et al., 1980). This change is due to a primary inhibition of the expression of gnawing and its replacement by sniffing and locomotion, and not to a primary stimulation of sniffing and locomotion; in fact the lesions do not potentiate the ability of apomorphine to elicit sniffing and locomotion at doses that do not elicit gnawing (Morelli et al., 1980).

SUBSTANTIA NIGRA AS AN OUTPUT STATION FOR STRIATAL IMPULSES

A large body of experimental evidence now indicates that turning behaviour, stereotyped behaviour of the apomorphine-type, and catalepsy of the neuroleptic-type, are mediated by the striato-nigral pathway (Marshall and Ungerstedt, 1977; Garcia-Munoz et al., 1977) via modulation of nigral pars reticulata non-dopaminergic neurons (Di Chiara et al., 1977, 1978, 1979a,b, 1981). Results obtained by the intranigral injection of GABA-receptor blocking drugs suggest that contraversive turning and sterotypy in response to stimulation of striatal DA-receptors is mediated by an activation of striato-nigral GABA-ergic neurons with secondary inhibition of nigra pars reticulata neurons (Di Chiara et al., 1978). Catalepsy, on the other hand, would be due to reduction of the activity of GABA-ergic striato-nigral neurons with consequent disinhibition of nigra pars reticulata neurons. De Montis et al. (1980) have suggested that striato-nigral GABA-ergic projections expressing behaviour do not impinge directly upon nigral output neurons of pars reticulata but rather upon cholinergic interneurons in turn synapsing with pars reticulata projection neurons. This however is unlikely because choline acetyltransferase-containing perikarya cannot be shown in the nigra (Sofroniew et al., 1982) and the only cholinergic structures of nigra are synapses innervating mainly the pars compacta (Kimura et al., 1981). Indeed, all the available evidence points to a direct connection of striato-nigral neurons with nigral output neurons (Deniau et al., 1976; Somogyi et al., 1979).

ON THE OUTPUT ROLE OF THE ENTOPEDUNCULAR NUCLEUS

Recently it has been reported that the entopeduncular nucleus, the homologue of the medial pallidal segment, has a role similar to that already described for the substantia nigra and it has been suggested that striatal impulses can be expressed also by activation or inhibition of striato-entopeduncular GABA-ergic pathways with secondary inhibition or disinhibition of pallidal output neurons (Scheel-Kruger et al., 1981). While this scheme is in agreement with neurochemical, immunohistochemical and electrophysiological data on the existence of a GABA-ergic striato-entopeduncular pathway (Scheel-Kruger et al., 1981), the behavioural data supporting it should be considered with caution, in view of the small size of the entopeduncular nucleus in the rat and of its central position within the forebrain and critical location within the fibers of the cerebral peduncle, just along the course of the striato-nigral pathway. Thus, kainate-lesions or local injections in the entopeduncular nucleus are expected to damage or to influence other critical areas such as the

VM-thalamus, globus pallidus external segment and nucleus subthalamicus, which have a role of their own in basal ganglia functions. On the other hand, electrolytic lesions or electrical stimulation of the entopenduncular nucleus are likely to impinge on or spread to the striato-nigral fibers.

Scheel-Kruger et al. (1981) failed to obtain turning after injection of muscimol in the entopeduncular nucleus of the rat. Bilateral injections of muscimol in this nucleus resulted in a syndrome of behavioural activation similar to that evoked by systemic administration of amphetamine (Scheel-Kruger et al., 1981). On the other hand electrolytic lesions of entopeduncular nucleus in cats results in ipsilateral circling in response to dopaminergic agonists (Pazo et al., 1982). Moreover, electrical stimulation of the entopeduncular nucleus results in contraversive head turning and circling (Filion and Hebert, 1983). These data would suggest that the entopeduncular nucleus plays a role in turning behaviour; however, it is notable that lesions restricted to the known targets of the entopeduncular projections, i.e. the ventral thalamus and the pedunculo-pontine nucleus, alone or in combination, failed to influence the turning elicited from the nucleus (Filion and Hebert, 1983). In order to be effective, the lesions had to impinge upon the superior colliculus and the adjacent mesencephalic reticular formation. Since this area is not a target of entopeduncular but rather of nigral projections, these data, coupled to the possibility that the turning elicited from stimulation of the entopeduncular nucleus could be due to current spread to the adjacent striato-nigral fibers and to involvement of nigral output pathways, cast serious doubts on the role of entopeduncular nucleus in turning behaviour. Some additional information indirectly points to the same negative conclusions. Thus, decortication, both ipsilateral and bilateral, in rats lesioned unilaterally with 6OHDA in the MFB, fails to influence the contraversive turning evoked by systemic apomorphine (Crossman et al., 1977). Moreover, the other efferent target of entopeduncular nucleus, the pedunculo-pontine nucleus, does not appear suitable for providing an output since it mainly projects back to the thalamus and to the basal ganglia themselves (Jackson and Crossman, 1983; Saper and Loewy, 1982).

OUTPUT OF STRIATAL IMPULSES BEYOND THE SUBSTANTIA NIGRA

Beyond the substantia nigra, striatal impulses can be channelled through the thalamus (ventro-medial nucleus of the rat), superior colliculus and reticular formation Becksteadt et al., 1979; Bentivoglio et al., 1979). The terminations of nigral efferents in the superior colliculus are in the deep layers (intermedialis and profundum) (Graybiel, 1978; Jayaraman et al., 1977). In the reticular formation, various sites of nigral projections have been reported. Retrograde and anterograde tracer studies have shown that the substantia nigra projects to the dorsal mesencephalic reticular formation, the deep tegmental nucleus of Castaldi (1923) and of Veazy and Severin (1980a and b) or the cuneiform nucleus of Edwards (1980). Another site of nigral projection is the nucleus tegmenti peduculopontinus pars compacta (PPN), a very discrete group of neurons just adjacent to the brachium conjunction which also receives projections from the medial pallidal segment (Jackson and Crossman, 1983; Kim et al., 1976). Some authors identify the whole of the nigroreticular projections with the nigro-PPN projection (Beckstead, 1983); this apparently restrictive view derives in part from the tendency to include in PPN, which should be identified strictly speaking with the site of termination of entopeduncular projections (Kim et al., 1976; Saper and Loewy, 1982), a much larger area, embracing also the deep tegmental nucleus.

Nigral efferents have been reported also to the lateral part of the periaqueductal grey (PAG) (Hopkins and Niessen, 1976; Rinvik et al., 1976) and to the gigantocellular tegmental field (Gonzalez-Vegas, 1981). However, more recent studies have not confirmed these two last claims (Duggal and Barasi, 1983). Particularly relevant is the fact that two studies, performed by intra-PAG microiontophoretic administration of HRP, have failed to show a direct nigro-PAG projection (Marchand and Hagino, 1983; Mantyh, 1980). Previous findings therefore are probably due to spread of HRP to the deep layers of the superior colliculus and to the dorsal MRF. Biochemical and electrophysiological studies indicate that nigro-collicular and nigro-thalamic projections are GABA-ergic (Vincent et al., 1978; Di Chiara et al., 1979c).

ROLE OF THE SUPERIOR COLLICULUS AND DORSAL MESENCEPHALIC RETICULAR FORMATION

In contrast to previous reports and views expressed by Reavill et al. (1979) and by Crossmann and Sambrook (1978), studies from our laboratory and from others have indicated that the deep layers of the superior colliculus are an important output station for turning and stereotyped behaviour (Di Chiara et al., 1982; Kilpatrick et al., 1982; Pope et al., 1980; Redgrave et al., 1980; Taha et al., 1982). In addition, our studies suggest that dorsal mesencephalic reticular formation (MRF) plays a role even more critical than that of the above areas in the expression of striatal activity (Mulas et al., 1981; Morelli et al., 1981a, Imperato et al., 1981). Jenner et al. (1980) and Reavill et al. (1981) have instead indicated the area composed by the lateral periaqueductal grey and the cuneiform nucleus as the critical site for the expression of turning and have created for it the new term "angular complex". We have already expressed some reservations on the use of the term "angular complex", mainly on the basis of our observations that lesions of PAG produce motor asymmetries only when they impinge upon the dorsal MRF, indicating that this area and not the PAG is the area critical for postural asymmetry (Di Chiara et al., 1982). These reservations now appear even more justified in view of the recent anatomical studies showing that while there is a definite projection from the nigra to the dorsal MRF (Shammah-Lagnado et al., 1983; Veazey and Severin, 1982), no such projection has been demonstrated to the PAG (Mantyh, 1980; Marchand and Hagino, 1983). In view of the anatomical studies showing that the dorsal MRF (cuneiform nucleus or deep tegmental nucleus) receives profuse projections from the adjacent deep collicular layers (Edwards, 1980), it is not unlikely that the role of the dorsal MRF in the expression of striatal functions derives in part from direct nigro-reticular projections and in part from an indirect nigro-colliculo-reticular pathway. In view of this and of the anatomical and physiological similarities of the deep collicular layers and of the dorsal MRF, recently summarized by Edwards (1980), these areas can be viewed as a complex involved in sensory-motor integration and, in particular, the expression of striatal function.

ROLE OF THE THALAMUS

In spite of the presence of large nigro-thalamic and pallido-thalamic projections, the role of the thalamus in drug-induced behaviours such as stereotypy, turning and catalepsy is still unclear, and the studies appearing since our previous review on the subject (Di Chiara et al., 1981), rather than illuminating the field have projected onto it an even darker shade than previously.

Here the only clear point is the fact that the sniffing, licking and gnawing syndrome characteristic of DA-receptor stimulation of the striatum cannot be evoked from the thalamic nuclei receiving nigral and pallidal projections, nor can it be influenced by lesions or local drug injections in these thalamic nuclei (Di Chiara et al., 1979d; Starr and Summerhayes, 1983a). Therefore the thalamus is not likely to have a role in stereotypy, as in the case of the superior colliculus and adjacent dorsal MRF which appear heavily implicated in the expression of this behaviour. On the other hand, much confusion exists as to the role of the thalamus in turning behaviour and catalepsy. This arises in part from discrepancies in the results obtained by different authors and even by the same author in different studies.

For example, Kilpatrick et al. (1980) reported that 1 week old unilateral electrolesions of VM drastically reduce the contralateral circling evoked after ipsilateral intranigral injection of muscimol. More recently Starr and Summerhayes (1983b) report that the same lesions fail to affect either circling intensity or posture in response to intranigral muscimol. Similar discrepancy exists for the effect of VM-electrolesions ipsilateral to a unilateral intrastriatal injection of apomorphine, which were reported to completely prevent contraversive turning by Kilpatrick et al. (1980), and to be ineffective by Starr and Summerhayes (1983b). The differences in experimental conditions of the two studies (use of 400 ng(!) versus 40 ng of muscimol for intranigral injection, or the use of normally innervated versus 6OHDA-denervated rats) do not justify the clear-cut difference in results obtained.

Careful reading shows for example that in the paper by Garcia-Munoz et al. (1983) no indication is given of the time elapsed between lesions of VM-thalamus and behavioural testing, a critical issue in view of the apparent short-lasting effect of thalamic electrolesions on apomorphine-induced turning reported by Starr and Summerhayes (1983b).

In contrast with the previous inconsistencies, our seminal observation that bilateral intra-VM injection of muscimol induces strong and long-lasting catalepsy (Di Chiara et al., 1979d) has been confirmed by Scheel-Kruger et al. (1981) and by Starr and Summerhayes (1983a). Our next finding, that unilateral intra-VM muscimol results in tight and high-intensity ipsiversive circling upon peripheral apomorphine administration (Di Chiara et al., 1979d) has been also confirmed by Starr and Summerhayes (1983a) and by Reavill et al (1981). These apparently robust and easily reproducible findings brought us to suggest am involvement of VM-thalamus in posture and motor behaviour (Di Chiara et al., 1979d). However, the relationship of the VM-thalamus with these functions does not appear a simple one.

The effect of lesions of the VM-thalamus on turning behaviour elicited from the striatum is much debated. Reavill et al. (1981a) and Kilpatrick et al. (1980) apparently agree that chronic (from 1 week to 20 days old) electrolesions of VM ipsilateral to a striatum made hyperactive by 6OHDA denervation plus peripheral apomorphine injection greatly reduce turning intensity. In none of these studies has posture been assessed independently and therefore we do not know if the lesions also influence the postural component of turning. More recently, Starr and Summerhayes (1983b) report that electrolesions of VM ipsilateral to hyperactive striatum reduce turning intensity only if performed acutely during the turning test or 3 hours before it. From the day after the VM-lesion up to at least a week later, rats had resumed their pre-lesion turning frequency in response to apomorphine. Garcia-Munoz et al. (1983) report a marked reduction of tuning intensity in response to apomorphine after electrolytic or kainate-lesions of VM ipsilateral to unilateral 6OHDA denervation of the striatum, but since no mention is made of the interval between VM-lesion and behavioural testing we cannot utilize these data for comparative purposes.

When considering the effect of thalamic lesions on the turning elicited from the nigra, further controversy is found. Thus, while Kilpatrick et al. (1980) report that chronic electrolesions of ipsilateral VM drastically impair contraversive circling in response to intranigral muscimol, Starr and Summerhayes (1983b) report the same lesions to be ineffective unless performed acutely after the intranigral injection of muscimol. According to the same authors, chronic kainate-lesions of VM ipsilateral to intranigral muscimol influence turning intensity (turns/min) but not posture. It is difficult to reconcile these discrepancies unless one suggests differences in the topography and extent of the lesions. Although the charted topography and extent of the lesions seem strikingly similar and almost superimposable in these different studies, one might suppose that, in view of the difficulties in assessing the exact extent of a lesion, the "effective" lesion is, in some studies, larger than represented (see also Starr and Summerhayes, 1983b). This possibility is almost a certainty in the case of intra-VM kainate which, in our experience, is likely to lesion an area much larger than VM, sometimes involving extra-thalamic areas as well. For example, intra-VM kainate even in low amounts (0.15 μg/0.2 μl) destroys the ipsilateral reticular nucleus and part of the whole ventro-basal complex of the thalamus and of Forel's field (Di Chiara et al., unpublished observations). For this reason we would be prone to reject the results obtained with kainate-lesions of VM if these lesions turn out to be more effective than electrolesions of VM strictly confined to this nucleus.

If we now combine the results of the chronic VM-lesions obtained by Kilpatrick et al. (1980) and Reavill et al. (1981a) with the results of acute VM-lesions obtained by Starr and Summerhayes (1983b) as well as those of Garcia-Munoz et al. (1983), it appears that unilateral VM lesions do reduce turning behaviour evoked from the ipsilateral side, although not permanently. According to Starr and Summerhayes (1983b) this influence is on turning intensity rather than on posture, which would suggest that if VM has a role in turning behaviour such a role is confined to the motility component of turning. This lack of involvement of VM in the postural component evoked from the ipsilateral nigra or striatum is deduced also from the results obtained by Starr and Summerhayes (1983a) after unilateral intra-VM injections of muscimol, which were found to reduce the intensity of turning but not the degree of posture. These rats "circled more slowly even though their body posture and stereotypy were as strong as before".

The concept that VM-thalamus conveys the motility component but not the postural component of turning behaviour does not readily explain why, after unilateral intra-VM injection of muscimol or unilateral VM electrolesions, peripheral apomorphine induces tight and fast ipsilateral circling (Di Chiara et al., 1979d), or why intra-VM muscimol or VM-electrolesions contralateral to the dominant side increase the degree of postural asymmetry (Starr and Summerhayes, 1983a and b). In fact, these results could be explained more convincingly if one postulates that the lesions reduce the postural influence of the non-dominant striatum, thus increasing the degree of asymmetry between the two sides.

In conclusion, as in our previous review (Di Chiara et al., 1981), we are again left with the necessity of postulating that the thalamus mediates a motility component of turning which, like the postural one, is lateralized and contains a certain degree of postural competence in the sense that it can amplify the degree of postural asymmetry. In order to understand this concept we would like the reader to imagine a rat as provided, like a ship, with a rudder as well as two propellers, one on each side; the postural component would be analogous to an action on the rudder and the locomotor component to an action on the propellers. For any given steering angle the tightness of turning will be higher the larger the difference in the activity of one propeller in respect to the contralateral one. In this manner we might explain why interference with the lateralized motility component of turning would amplify the degree of motor asymmetry, and even produce motor asymmetry, during active locomotion.

SEGREGATION OF SPECIFIC BEHAVIOURAL COMPONENTS BEYOND THE SUBSTANTIA NIGRA

The results of the above studies suggest that the different components of turning and stereotyped behaviour as well as catalepsy, while funnelled together from the striatum to the substantia nigra through the massive striato-nigral pathway, tend to segregate beyond the nigra into the different targets of nigral output. Thus muscimol induces catalepsy when injected into the VM thalamus (Di Chiara et al., 1979d) but not into the deep collicular layers or into dorsal MRF (Imperato and Di Chiara, 1981). Conversely, picrotoxin or bicuculline evoke postural asymmetry and stereotyped gnawing from the deep collicular layers and dorsal MRF (Imperato and Di Chiara, 1981a; Dean et al., 1982), but not from the VM thalamus from which, instead, treadmill-like forward locomotion is evoked (Di Chiara et al., 1979b; Starr and Summerhayes, 1983a). Moreover, while intracollicular muscimol specifically blocks apomorphine-induced gnawing and converts the apomorphine effect into a syndrome similar to that evoked by intra-thalamic picrotoxin (Imperato and Di Chiara, 1981; Dean et al., 1982), intrathalamic muscimol fails to inhibit apomorphine-gnawing, in spite of the induction of catalepsy (Di Chiara et al., 1979d; Starr and Summerhayes, 1983a). On the basis of these results we proposed (Di Chiara et al., 1981) that the nigro-collicular and nigro-reticular projections mediate the expression of the postural component of turning and of the high-intensity component of stereotypy (licking and gnawing), while the nigro-thalamic and pallido-thalamic (including ventral pallidum) mediate the motility component of turning and the locomotor response to apomorphine. Such segregation of the information arising from the caudate through separate nigral projection channels might reflect a segregation of these components in the striatum itself. This concept is consistent with the different origin, in the striatal complex, of the various components of stereotypy and turning. Thus, while drug-induced exploratory motility appears to arise from the nucleus accumbens-tuberculum olfactorium complex, confined sniffing and gnawing are likely to originate from the caudate-putamen (Kelly et al., 1975; Asher and Aghajanian, 1974; Pijnenburg and Van Rossum, 1973). Similar origins have been suggested for the locomotor and the postural components of turning, respectively (Kelly, 1975; Pycock and Marsden, 1978). The view that separate channels exist for certain specific components of striatal origin is also in agreement with recent histochemical data showing that the striatum has a non-homogeneous structure, being composed of a mosaic-like arrangement of chemically distinct domains (Graybiel and Ragsdale, 1978; Graybiel et al., 1981). Moreover, anterograde and retrograde tracer studies show the existence of a topographical arrangement which remains fairly well-preserved from the areas of input, like the cortex, through the striatum itself, and finally to the areas of output, like the substantia nigra and the globus pallidus (Graybiel et al., 1979; Parent, this volume; Faull, this volume).

The above concept of the existence of separate behavioural channels from specific striatal domains to different extranigral outputs is apparently in contrast with the reported branching of nigral projections (Anderson and Yoshida, 1977). However, all the existing evidence indicates that only a very restricted percent (5-17%) of nigral neurons undergo branching to different targets (Beckstead, 1983); moreover, from studies made in the cat but with two very different approaches, electrophysiological (Niijma and Yoshida, 1982) versus anatomical (Beckstead, 1983), it has been shown that nigro-tectal neurons very rarely branch also to the thalamus or to PPN, while nigro-thalamic neurons branch more commonly to PPN and vice versa. This characteristic, coupled to morphological and topographical differences, should lead, according to Beckstead (1983), to viewing the nigro-tectal projection as quite distinct from nigro-thalamic and nigro-PPN projections (Beckstead et al., 1981; Beckstead and Frankfurter, 1982). This conclusion is remarkably in agreement with our views expressed above on the role of nigro-thalamic and nigro-collicular projections in striatal motor syndromes. The view of an analogy of nigro-PPN and nigro-thalamic projections is also in agreement with the notion that PPN is a relay which feeds basal ganglia information back to thalamus and cortex rather than to downstream centers. This stresses the differences between PPN and dorsal MRF for their role in basal ganglia functions.

To summarize, we suggest that the locomotor component of turning as well as the hypermotility induced by DA-agonists is mediated by stimulation of DA-receptors located in the nucleus accumbens-tuberculum olfactorium complex which, by local circuit mechanisms, results in activation of an inhibitory output terminating in the ventral pallidum and in the substantia nigra. This results in inhibition of a further inhibitory pool of nigro-thalamic and pallido-thalamic neurons with secondary disinhibition of their target areas in the thalamus.

The postural component of turning and the high-intensity components of stereotypy in response to DA-agonists are postulated to be due to stimulation of DA-receptors in the caudate-putamen, with activation of striato-nigral GABA-ergic projections and inhibition of nigro-collicular and nigro-reticular projections.

Catalepsy in response to DA-receptor blockage might originate mainly from the caudate-putamen but, strangely enough, does not seem to follow the nigro-collicular or nigro-reticular route, but rather the pallido-thalamic and nigro-thalamic route.

REFERENCES

Anderson, M., and Yoshida, M., 1977, Electrophysiological evidence for branching nigral projections to the thalamus and superior colliculus, Brain Res., 137:361.

Asher, I. M., and Aghajanian, G. K., 1974, 6-hydroxydopamine lesions of olfactory tubercles and caudate nuclei: effect on amphetamine-induced stereotyped behaviour in rats, Brain Res., 82:1.

Baumgarten, H. G., and Lachemayer, L., 1974, 5,7-dihydroxy-tryptamine: improvement in chemical lesioning of indoleamine neurons in the mammalian brain, Z. Zellforsch., 135:399.

Beckstead, R. M., Domesick, V. B., and Nauta, W. J. H., 1979, Efferent connections of the substantia nigra and ventral tegmental area in the rat, Brain Res., 175:191.

Beckstead, R. M., Edwards, S. B., and Frankfurter, A., 1981, A comparison of the intranigral distribution of nigrotectal neurons labeled with horseradish peroxidase in the monkey, cat and rat, J. Neurosci., 1:121.

Beckstead, R. M., and Frankfurter, A., 1982, The distribution and some morphological features of substantia nigra neurons that project to the thalamus, superior colliculus and pedunculopontine nucleus in the monkey, Neuroscience, 7:2377.

Beckstead, R. M., 1983, Long collateral branches of substantia nigra pars reticulata axons to thalamus, superior colliculus and reticular formation in monkey and cat. Multiple retrograde neuronal labeling with fluorescent dyes, Neuroscience, 10, 3:767.

Ben-Ari, Y., Tremblay, E., Ottersen, O.P., and Naquet, R., 1979, Evidence suggesting secondary epileptogenic lesions after kainic acid: pretreatment with diazepam reduced distant but not local brain damage, Brain Res., 165:362.

Bentivoglio, M., Van Der Kooy, D., and Kuypers, H. G. J. M., 1979, The organization of the efferent projections of the substantia nigra in the rat. A retrograde fluorescent double labelling study, Brain Res., 174:1.

Bjorklund, A., Nobin, A., and Steveni, U., 1973, The use of neurotoxic dihydroxytryptamines as tools for morphological studies and localized lesioning of central indoleamine neurons, Z. Zellforsch., 145:479.

Castaldi, L., 1923, Studi sulla struttura e sullo sviluppo del mesencefalo, Ricerche in Cavia Cobaya, Parte I, Arch. Ital. Anat. Embriol., 20:23.

Childs, J. A., and Gale, K., 1983, Evidence that the nigrotegmental gabaergic projection mediates stereotypy induced by apomorphine and intranigral muscimol, Life Sci., 33:1007.

Costall, B., Marsden, C. D., Naylor, R. J., Pycock, C. J., 1976, The relationship between striatal and mesolimbic dopamine dysfunction and the nature of circling responses following 6-hydroxydopamine and electrolytic lesions of the ascending dopamine system of rat brain, Brain Res., 118:87.

Coyle, J. T., Molliver, M. E., and Kuhar, M. J., 1978, In situ injection of kainic acid: a new method for selectively lesioning neuronal cell bodies while sparing axons of passage, J. Comp. Neurol., 180:301.

Crossman, A. R., Sambrook, M. A., Gergies, S. W., and Slater, P., 1977, The neurological basis of motor asymmetry following unilateral 6-hydroxydopamine brain lesions in the rat: the effect of motor decortication, J. Neurol. Sci., 34:407.

Crossman, A. R., and Sambrook, M. A., 1978, The neurological basis of motor asymmetry following unilateral nigrostriatal lesions in the rat: the effect of secondary superior colliculus lesions, Brain Res., 159:211.

Dean, P., Redgrave, P., and Eastwood, L., 1982, Suppression of apomorphine-induced oral stereotypy in rats by microinjection of muscimol into midbrain, Life Sci., 30:2171.

De Montis, G., Olianas, M. C., Serra, G., Tagliamonte, A., and Scheel-Kruger, J., 1979, Evidence that a nigral GABAergic-cholinergic balance controls posture, Eur. J. Pharmacol., 53:181.

Deniau, J. M., Feger, J., and Le Guyader, C., 1976, Striatal evoked inhibition of identified nigro-thalamic neurons, Brain Res., 104: 152.

Di Chiara, G., and Gessa, G. L., 1978, Pharmacology and neurochemistry of apomorphine, in: "Advances in Pharmacology and Chemotherapy", Vol. 15, Academic Press, New York.

Di Chiara, G., Olianas, M. C., Del Fiacco, M., Spano, P. F., and Tagliamonte, A., 1977, Intranigral kainic acid is evidence that nigral non-dopaminergic neurons contol posture, Nature (Lond.), 268:743.

Di Chiara, G., Morelli, M., Porcedu, M. L., and Gessa, G. L., 1978, Evidence that nigral GABA mediates behavioural responses elicited by striatal dopamine receptor stimulation, Life Sci., 23:2045.

Di Chiara, G., Porceddu, M. L., Morelli, M., Mulas, M. L., and Gessa, G. L., 1979a, Substantia nigra as an output station for striatal dopaminergic responses: role of a GABA-mediated inhibition of pars reticulata neurons, Naunyn Schmiedebergs Arch. Pharmacol., 306:153.

Di Chiara, G., Porceddu, M. C., Morelli, M., Mulas, M. L., and Gessa, G. L., 1979b, Strio-nigral and nigro-thalamic GABA-ergic neurons as output pathways for striatal responses, in: "GABA-neurotransmitters, Pharmacochemical, Biochemical and Pharmacological Aspects," P. Krogsgaard-Larsen, J. Scheel-Kruger and H. Kofod, eds., Alfred Munksgaard, Copenhagen.

Di Chiara, G., Porceddu, M. L., Morelli, M., Mulas, M. L., and Gessa, G. L., 1979c, Evidence for a GABAergic projection from the substantia nigra to the ventromedial thalamus and to the superior colliculus of the rat, Brain Res., 176:273.

Di Chiara, G., Morelli, M., Porceddu, M. L., and Gessa, G. L., 1979d, Role of thalamic γ-aminobutyrate in motor functions: catalepsy and ipsiversive turning after intrathalamic muscimol, Neuroscience., 4:1453.

Di Chiara, G., Porceddu. M. L., Imperato, A., and Morelli, M., 1981, Role of GABA neurons in the expression of striatal motor functions, in: "GABA and the Basal Ganglia," G. Di Chiara and G. L. Gessa, eds., Raven Press, New York.

Di Chiara, G., Morelli, M., Imperato, A., and Porceddu. M. L., 1982, A re-evaluation of the role of the superior colliculus in turning behaviour, Brain Res., 237:61.

Duggal, K. N., and Barasi, S., 1983, Investigation of the connection between the substantia nigra and the medullary reticular formation in the rat, Neuroscience Lett., 36:237.

Edwards, S. B., 1980, The deep layers of the superior colliculus: their reticular characteristics and structural organization, in: "The Reticular Formation Revisited: Specifying Function for a Nonspecific System," J. A. Hobson and M. A. B. Brazier, eds., Raven Press, New York.

Filion, M., and Hebert, R., 1983, Redundancy in ascending and descending pathways mediating head turning elicited by entopeduncular stimulation in the cat, Neuroscience, 10:169.

Fisher, A., Mantione, C. R., Abraham, D. J., and Hanin, I., 1982, Long-term central cholinergic hypofunction induced in mice by ethylcholine aziridinium ion (AF64A) in vivo, J. Pharmacol. Exp. Therap., 222:140.

Fog, R., Randrup, A., and Pakkenberg, H., 1970, Lesions in corpus striatum and cortex of rat brains and the effect on pharmacologically induced stereotyped aggressive and cataleptic behaviour, Psychopharmacologia (Berl.), 18:346.

Garcia-Munoz, M., Nicolaou, N. M., Tulloch, I. F., Wright, A. K., and Arbuthnott, G. W., 1977, Feedback loop or output pathway in striato-nigral fibres? Nature (Lond.), 265:363.

Garcia-Munoz, M., Patino, P., Wright, A. K., and Arbuthnott, G. W., 1983, The anatomical substrate of the turning behaviour seen after lesions in the nigrostriatal dopamine system, Neuroscience, 8:87.

Gonzalez-Vegas, J. A., 1981, Nigro-reticular pathway in the rat - an intracellular study, Brain Res., 207:170.

Graybiel, A. M., 1978, Organization of the nigrotectal connection: An experimental tracer study in the cat, Brain Res., 143:339.

Graybiel, A. M., and Ragsdale, C. W., Jr., 1978, Histochemically distinct compartments in the striatum of human, monkey and cat demonstrated by acetylthiocholinesterase staining, Proc. Nat. Acad. Sci. (USA), 75:5723.

Graybiel, A. M., Ragsdale, C. W., Jr., and Moon Edley, S., 1979, Compartments in the striatum of the cat observed by retrograde cell-labelling, Exp. Brain Res., 34:189.

Graybiel, A. M., Ragsdale, C. W., Jr., Yoneoka, E. S., and Elde, R. P., 1981, An immunohistochemical study of enkephalins and other neuropeptides in the striatum of the cat with evidence that the opiate peptides are arranged to form mosaic patterns in register with the striosomal compartments visible by acetylcholinesterase staining, Neuroscience, 6:377.

Hökfelt, T., and Ungerstedt, U., 1973, Specificity of 6-hydroxydopamine induced degeneration of central monoamine neurons: an electron and fluorescence microscopic study with special reference to intracerebral injection on the nigrostriatal dopamine system, Brain Res., 60:269.

Hopkins, D. A., and Niessen, L. W., 1976, Substantia nigra projections to the reticular formation, superior coliculus and central gray in the rat, cat and monkey, Neurosci. Lett., 2:253.

Imperato, A., Porceddu, M. L. Morelli, M., Faa, M., and Di Chiara, G., 1981, Role of dorsal mesencephalic reticular formation and deep layers of superior colliculus as output stations for turning behaviour elicited from the substantia nigra pars reticulata, Brain Res., 216:437.

Imperato, A., and Di Chiara, G., 1981, Behavioural effects of GABA-agonists and antagonists infused in the mesencephalic reticular formation - deep layers of superior colliculus, Brain Res., 224:185.

Jackson, A., and Crossman, A. R., 1983, Nucleus tegmenti pedunculopontinus: Efferent connections with special reference to the basal ganglia, studied in the rat by anterograde and retrograde transport of horseradish peroxidase, Neuroscience., 10:725.

Jackson, J. H., 1875, Selected writings, 1:37-76, Churchill, London.

Jayaraman, A., Batton, R. R., and Carpenter, M. B., 1977, Nigrotectal projections in the monkey, Brain Res., 135:147.

Jenner, P. Leigh, N., Marsden, C. D., and Reavill, C., 1980, Involvement of the periqueductal grey in dopamine-mediated circling behaviour, Proc. B.P.S., 492.

Kelly, P. H., Seviour, P. W., and Iversen, S. D., 1975, Amphetamine and apomorphine responses in the rat following 6-OHDA lesions of the nucleus accumbens septi and corpus striatum, Brain Res., 94:507.
Kelly, P. H., and Moore, K. E., 1976, Mesolimbic dopaminergic neurones in the rotational model of nigrostriatal function, Nature, 263:695.
Kilpatrick, I. C., Starr, M. S., Fletcher, A., James, T. A., and MacLeod, N. K., 1980, Evidence for a GABAergic nigrothalamic pathway in the rat - behavioural and biochemical studies, Expl. Brain Res., 40:45.
Kilpatrick, I. C., Collingridge, G. L., and Starr, M. S., 1982, Evidence for the participation of nigrotectal γ-aminobutyrate-containing neurones in striatal and nigral-derived circling in the rat, Neuroscience, 7:207.
Kim, R., Nakano, K., Jayaraman, A., and Carpenter, M. B., 1976, Projections of the globus pallidus and adjacent structures: an autoradiographic study in the monkey, J. Comp. Neurol., 169:263.
Kimura, H., McGeer, P. L., Peng, F., and McGeer, E. G., 1981, The central cholinergic system studied by choline acetyltransferase immunohistochemistry in the cat, J. Comp. Neurol., 200:151.
Leigh, P. N., Reavill, C., Jenner, P., and Marsden, C. D., 1983, Basal ganglia outflow pathways and circling behaviour in the rat, J. Neural Transmission, 58:1.
Ljundberg, T., and Ungerstedt, U., 1977, A method for simultaneous recording of eight behavioural parameters related to monoamine neurotransmission, Pharmacol. Biochem. Behav., 8:483.
Mantyh, P. W., 1980, Afferent and efferent connections of periaqueductal grey in the monkey, Soc. Neurosci. Abs., 10th annual meeting.
Marchand, J. E., and Hagino, N., 1983, Afferents to the periaqueductal gray in the rat. A horseradish peroxidase study, Neuroscience, 9(1):95.
Marshall, J. F., and Ungerstedt, U., 1977, Striatal efferent fibres play a role in maintaining rotational behaviour in he rat, Science, (New York), 198:62.
Moon Edley, S., and Graybiel, A. M., 1980, Connections of the nucleus tegmenti pedunculopontinis, pars compacta (TPc) in cat, Anat. Rec., 196:129A.
Morelli, M., Porceddu, M. L., and Di Chiara, G., 1980, Lesions of substantia nigra by kainic acid: effects on apomorphine-induced stereotyped behaviour, Brain Res., 191:67.
Morelli, M., Imnperato, A., Porceddu, M. L., and Di Chiara, G., 1981a, Role of dorsal mesencephalic reticular formation and deep layers of superior colliculus in turning behaviour elicited from the striatum, Brain Res., 215:337.
Morelli, M., Porceddu, M. L., and Di Chiara, G., 1981b, Role of substantia nigra pars reticulate neurons in the expression of neuroleptic-induced catalepsy, Brain Res., 217:375.

Mulas, A., Longoni, R., Spina, L., Del Fiacco, M., and Di Chiara, G., 1981, Ipsiversive turning behaviour after discrete unilateral lesions of the dorsal mesencephalic reticular formation by kainic acid, Brain Res., 208:468.
Niijima, K., and Yoshida, M., 1982, Electrophysiological evidence for branching nigral projections to pontine reticular formation, superior colliculus and thalamus, Brain Res., 239:279.
Pazo, J. H., Medina, J. H., O'Donnel, P., and Dvorkin, M. A., 1982, Study of the neural basis of circling behavior induced by L-Dopa in lesioned entopeduncular cats, Brain Res., 233:337.
Pijnenburg, A. J. J., Honig, W. M. M., and Van Rossum, J. M.,. 1975, Inhibition of D-amphetamine-induced locomotor activity by injection of haloperidol into the nucleus accumbens of the rat, Psychopharmacologia (Berl.), 41:87.
Pijnenburg, A. J. J., and Van Rossum, J. M., 1973, Stimulation of locomotor activity following injection of dopamine into the nucleus accumbens, J. Pharm. Pharmacol., 25:1003.
Pope, S. G., Dean, P., anmd Redgrave, P., 1980, Dissociation of d-amphetamine induced locomotor activity and stereotyped behaviour by lesions of the superior colliculus, Psychopharmacol, 70:297.
Pycock, C. J., and Marsden, C. D., 1978, The rotating rodent, a two component system? Europ. J. Pharmacol., 47:167.
Reavill, C., Leigh, N., Jenner, P., and Marsden, C. D., 1979, Dopamine-mediated circling behaviour does not involve the nigro-tectal pathway, Exp. Brain Res., 37:309.
Reavill, C., Jenner, P., Leigh, N., and Marsden, C. D., 1981a, The role of nigral projections to the thalamus in drug-induced circling behaviour in the rat, Life Sci., 28:1457.
Reavill, C., Leigh, P. N., Jenner, P., and Marsden, C. D., 1981b, Drug-induced circling after unilateral 6-hydroxydopamine lesions of the nigro-striatal pathway is mediated via the midbrain periaqueductal gray and adjacent reticular formation (angular complex), Life Sci., 29:2357.
Redgrave, P., Dean, P., Donohoe, T. P., and Pope, S. P.., 1980, Superior colliculus lesions selectively attenuate apomorphine-induced oral stereotypy: a possible role for the nigrotectal pathway, Brain Res., 196:541.
Rinvik, E., Grofova, I., and Ottersen, O. P., 1976, Demonstration of nigrotectal and nigroreticular projections in the cat by axonal transport of proteins, Brain Res., 112:388.
Saper, C. B., and Loewy, A. D., 1982, Projections of the pedunculopontine tegmental nucleus in the rat: evidence for additional extrapyramidal circuitry, Brain Res., 252:367.
Scheel-Kruger, J., Arnt, J., Magelund, G., 1977, Behavioural stimulation induced by muscimol and other GABA agonists injected into the substantia nigra, Neuroosci. Letters, 4: 351-356.

Scheel-Kruger, J., Magelund, G., and Olianas, M. C., 1981, Role of GABA in the striatal output system: globus pallidus, nucleus entopeduncularis, substantia nigra and nucleus subthalamicus in: "GABA and the Basal Ganglia," G. Di Chiara and G. L. Gessa, eds., Raven Press, New York.

Schwarcz, R., Hökfelt, T., Fuxe, K., Jonsson, G., Goldstein, M., and Terenius, L., 1979, Ibotenic acid-induced neuronal degeneration: a morphological and neurochemical study, Exp. Brain Res., 37:199.

Shammah-Lagnado, S. J., Ricardo, J. A., Sakamoto, N. T. M. N., and Negrao, N., 1983, Afferent connections of the mesencephalic reticular formation: a horseradish peroxidase study in the rat, Neuroscience, 9:391.

Sofroniew, M. V., Eckenstein, F., Thoenen, H., and Cuello, A. C., 1982, Topography of choline acetyltransferase containing neurons in the forebrain of the rat, Neurosci. Lett., 33:7.

Somogyi, P., Hodgson, A. J., and Smith, A. D., 1979, An approach to tracing neuron networks in the cerebral cortex and basal ganglia, combination of Golgi staining, retrograde transport of horseradish peroxidase and anterograde degeneration of synaptic boutons in the same material, Neuroscience, 4:1805.

Starr, M. S., and Kilpatrick, E. C., 1981, Distribution of γ-aminobutyrate in the rat thalamus: specific decreases in thalamic γ-aminobutyrate following lesion or electrical stimulation of the substantia nigra, Neuroscience, 6:1095.

Starr, M. S., and Summerhayer, M., 1983a, Role of the ventromedial thalamus in motor behaviour. I. Effects of focal injections of drugs, Neuroscience, 10:1157.

Starr, M. S., Summerhayer, M., 1983b, Role of the ventromedial nucleus of the thalamus in motor behaviour. II. Effects of lesions, Neuroscience, 10:1171.

Taha, E. B., Dean, P., and Redgrave, P., 1982, Oral behaviour induced by intranigral muscimol is unaffected by haloperidol but abolished by large lesions of superior colliculus, Psychopharmacology, 77:272.

Ungerstedt, U., and Arbuthnott, G. W., 1970, Quantitative recording of rotational behaviour in rats after 6-hydroxydopamine lesions of the nigro-striatal dopamine system, Brain Res., 24:485.

Ungerstedt, U., 1971, Postsynaptic supersensitivity after 6-hydroxydopamine induced degeneration of the nigro-striatal dopamine system in the rat brain, Acta Physiol. Scand., 82: Suppl. 367:69.

Veazey, R. B., and Severin, C. M., 1980, Efferent projections of the deep mesencephalic nucleus (pars lateralis) in the rat, J. Comp. Neurol., 190:231.

Veazey, R. B., and Severin, C. M., 1980, Efferent projections of the deep mesencephalic nucleus (pars lateralis) in the rat, J. Comp. Neurol., 190:245.

Veazey, R. B., and Severin, C. M., 1982, Afferent projections to the deep mesencephalic nucleus in the rat, J. Comp. Neurol., 204:134.
Vincent, S. R.., Hattori, T., and McGeer, E. G., 1978, The nigrotectal projection: A biochemical and ultrastructural characterization, Brain Res., 151:159.
Wuerthele, W. M., Lowell, K. L., Jones, M. Z., and Moore, K. E., 1978, A histological study of kainic acid-induced lesions in the rat brain, Brain Res., 149:489.

RESPONSES OF NEURONS IN DIFFERENT REGIONS OF THE STRIATUM OF THE BEHAVING MONKEY

Edmund T. Rolls

University of Oxford
Department of Experimental Psychology
South Parks Road
Oxford OX1 3UD, England

Despite rapid recent advances in knowledge of the anatomy of the striatum, and the disruption of behavior produced by damage to it, its functions are still poorly understood, with some investigators proposing motor (Marsden, 1980), and others cognitive functions (Divac and Oberg, 1979; Oberg and Divac, 1979). To investigate its function, recordings are being made from single neurons in different regions of the striatum during the types of behavior known to be disrupted by striatal damage, as reviewed in this paper.

Anatomically, the striatum can be divided into the neostriatum, consisting of the caudate nucleus and putamen, and the ventral striatum, consisting of the nucleus accumbens, olfactory tubercle and islands of Calleja (Heimer and Wilson, 1975). The neostriatum receives major inputs from almost all areas of the neocortex, and has major efferent connections with the globus pallidus, which in turn is connected to the ventral group of thalamic nuclei and thus to the premotor cortex, and also has connections to brainstem motor regions (Kemp and Powell, 1970, 1971; Carpenter, 1976; Niewenhuys, 1977; Nauta and Domesick, 1979). This pattern of connections suggests that the striatum provides one important route through which the cortex can influence motor structures (Divac and Oberg, 1979; Oberg and Divac, 1979). Studies in the monkey show that there is some topography in the cortico-striate projection, in that for example the prefrontal cortex projects to the caudate nucleus, the motor cortex (area 4) and somatosensory cortex (areas 3, 1 and 2) project somatotopically to the putamen, and the temporal cortex projects to regions in the tail and head of the caudate nucleus (Kemp and Powell, 1970, 1971; Kunzle, 1975, 1977, 1978; Kunzle and Akert, 1977; Jones, Coulter, Burton and Porter, 1977; Goldman and Nauta, 1977). Although studies with autoradiography indicate that each region of the cortex projects to a wider region of the striatum (Goldman and Nauta, 1977; Yeterian and Van Hoesen, 1978) than originally shown by the fibre degeneration method (Kemp and Powell, 1970), nevertheless there is considerable topograhy retained; and the fact that in the primate the sensorimotor cortex projects almost exclusively to the putamen, and the prefrontal cortex almost exclusively to the caudate nucleus, suggests that the information processed by these different parts of the striatum may be different. It was therefore one of the aims of the present series of neurophysiological investigations (Rolls, Thorpe, Maddison, Roper-Hall, Puerto, and Perrett, 1979; Rolls, Thorpe and Maddison, 1983; Caan, Perrett and Rolls, 1984; Rolls, Thorpe, Boytim, Szabo and Perrett, 1984; Rolls, Ashton, Williams, Thorpe, Mogenson, Colpaert, and Phillips, 1982) to compare the activity of single neurons in different regions of the striatum under the same testing conditions. The striatum also receives projections from the substantia nigra, pars compacta, via the nigrostriatal bundle which is largely dopaminergic, and from the intralaminar thalamic nuclei, and has a projection to the substantia nigra, pars reticulata. Because of the importance of the dopamine input to the striatum for its normal function, the effects of dopamine on the responses of striatal neurons were also considered in the present series of neurophysiological investigations (Rolls, Thorpe, Boytim, Szabo and Perrett, 1984).

Damage to the striatum produced effects which suggest that it is involved in orientation to stimuli and in the initiation and control of movement. Lesions of the dopamine pathways which deplete the striatum of dopamine lead to a failure to orient to stimuli, a failure to initiate movements (which is associated with catalepsy), and a failure to eat and drink (Marshall, Richardson and Teitelbaum, 1974). In man, depletion of striatal dopamine is found in Parkinson's disease, in which there is tremor and akinesia, that is a lack of voluntary movement (Hornykiewicz, 1973). Extensive lesions of the caudate nucleus in the cat lead to a syndrome in which the signs appear to be opposite to those produced by nigrostriatal bundle damage, and include an increased tendency to orient to and 'lock on' to environmental stimuli, with little habituation to them, so that there is "compulsory approach" towards auditory and especially visual stimuli, and hyperactivity (Villablanca and Marcus, 1975; Villablanca, Marcus and Olmstead, 1976a,b,c). Similar effects have been reported in monkeys with smaller ablations (Denny-Brown and Yanagisawa, 1976. There is some evidence that partial caudate lesions can produce selective effects related to the function of the cortex that projects to the region of the caudate affected. For example, in the monkey lesions of the anterodorsal part of the caudate nucleus head disrupted delayed alternation performance, which is also impaired by damage to the corresponding cortical region, the dorsolateral prefrontal cortex. Lesions of the ventrolateral part o the head of the caudate nucleus, as of the orbitofrontal cortex which projects to it, impaired object reversal performance. Lastly, lesions of the tail of the caudate nucleus, as of the inferior temporal visual cortex which projects to this part of the caudate, produced a visual pattern discrimination deficit (Divac, Rosvold and Szwarcbart, 1967; Iversen, 1979).

One aim of the present investigations is to investigate the functions of the striatum by analysing the activity of striatal neurons during the types of behavior known to be impaired by striatal lesions, such as during the initiation of feeding, during visual discrimination performance and its reversal, and during the presentation of visual and other stimuli. A second aim is to obtain evidence on the possible localization of function within the striatum by comparing the activity of neurons in different regions of it during the performance of the same set of tasks. A third aim is to

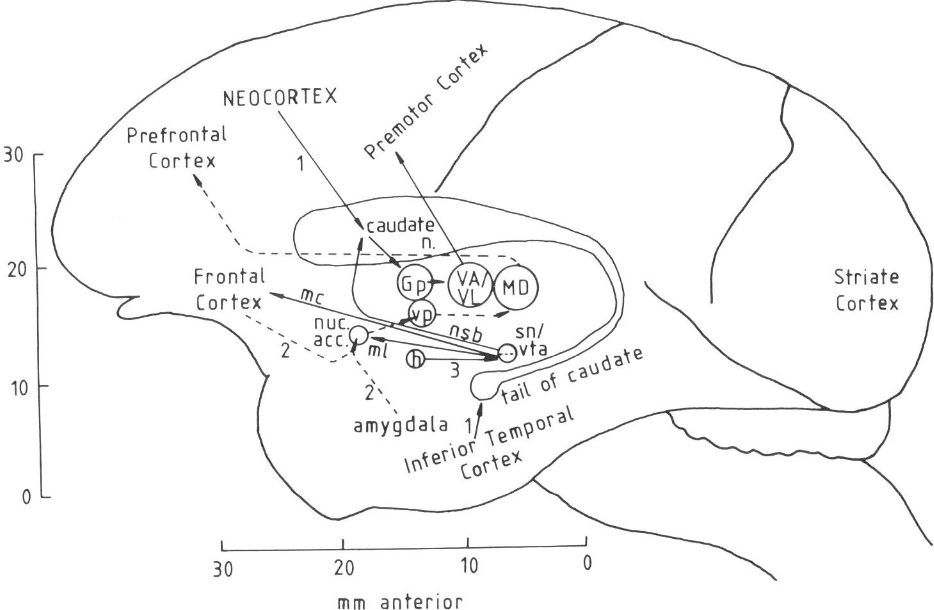

Fig. 1. Some of the striatal and connected regions in which the activity of single neurons is described shown on a lateral view of the brain of the macaque monkey brain. Gp = globus pallidus; h = hypothalamus; Sn = substantia nigra, pars compacta (A9 cell group), which gives rise to the nigrostriatal dopaminergic pathway, or nigrostriatal bundle (nsb); vta = ventral tegmental area, containing the A10 cell group, which gives rise to the mesocortical dopamine pathway (mc) projecting to the frontal and cingulate cortices, and to the mesolimbic dopamine pathway (ml), which projects to the nucleus accumbens (nuc acc).

investigate the transfer function of the striatum by comparing the activity of striatal neurons with the activity of neurons in the regions of cortex which project into the corresponding region of the striatum, and with the activity of neurons in the globus pallidus and substantia nigra, to which the striatum projects. A fourth aim is to investigate how dopamine influences the transmission of information through the striatum. A fifth aim is to investigate the role of the striatum in feeding by comparing the activity of striatal neurons with that of neurons in the hypothalamus and related structures where neuronal repsonses are found which precede and predict feeding behavior initiated by the sight of food (Rolls, Burton and Mora, 1976; Rolls Burton and Mora, 1980; Rolls, Sanghera and Roper-Hall, 1979; Rolls, 1981a,b). This aim follows from the evidence that alteration of striatal function impairs feeding, that striatal neurons may respond to food (see above), and that there is a projection from the hypothalamus to the substantia nigra, pars compacta, through which hypothalamic feeding-related neurons might influence striatal function and thus feeding (Nauta and Domesick, 1978).

In what follows, the activity of single neurons in different regions of the striatum of the macaque monkey, and the effects of dopamine on the normal responses of these neurons, are described. Then the implications of the findings for our understanding of the functions of the striatum are discussed.

HEAD OF THE CAUDATE NUCLEUS

The activity of 394 neurons, in the head of the caudate nucleus and most anterior part of the putamen, was analysed in three behaving rhesus monkeys (Rolls, Thorpe and Maddison, 1983). Of these neurons, 64.2% had responses related to environmental stimuli, movements, the performance of a visual discrimination task, or eating. However, only relatively small proportions of them had responses unconditionally related to visual (9.6%), auditory (3.5%), or gustatory (0.5%) stimuli, or to movements (4.1%). Instead, the majority responded conditionally in relation to stimuli or movements, in that the responses occurred in only some test situations, and were often dependent on the performance of a task by the monkeys. Thus it was found that in a visual discrimination task 14.5% of the neurons responded during a 0.5s tone/light cue which signalled the start of each trial; 31.1% responded in the period in which the discriminative visual stimuli were shown, with 24.3% of these responding more to either the visual stimulus which signified food reward or to that which signified punishment; and 6.2% responded in relation to lick responses. Yet these neurons typically did not respond in relation to the cue stimuli, to the visual stimuli, or to movements, when these occurred independently of the task or when performance of the task was prevented. Similarly, although 25.8% of the neurons tested during feeding responded when the food was seen by the monkey, 6.2% when he tasted it, and 22.4% during a cue given by the experimenter that a food or non-food object was about to be presented, only few of these neurons had responses to the same stimuli presented in different situations. Further evidence on the nature of these neuronal responses was that many of the neurons with cue-related responses only responded to the tone-light cue stimuli when they were cues for the performance of the task or the presentation of food, and some responded to the different cues used in these two situations.

These findings indicate that the head of the caudate nucleus and most anterior part of the putamen contain populations of neurons which respond to cues that enable preparation for the performance of tasks such as feeding and tasks in which movements must be initiated, or which respond during the performance of such tasks in relation to the stimuli used and the responses made, yet that the majority of these neurons do not have unconditional sensory or motor responses. It has therefore been suggested (Rolls, Thorpe and Maddison, 1983) that the anterior neostriatum contains neurons which are important for the utilization of environmental cues for the preparation for behavioural responses, and for particular behavioral responses made in particular situations to particular environmental stimuli, that is in stimulus-motor response habit formation. Different neurons in the cue-related group often respond to different subsets of environmentally significant events, and thus convey some information which would be useful in switching behavior and in preparing to make different responses.

It may be suggested that deficits in the initiation of movements following damage to striatal pathways may arise in part because of interference with these functions of the anterior neostriatum. Thus the akinesia or lack of voluntary movement produced by damage to the dopaminergic nigrostriatal bundle in animals, and present in Parkinson's disease in man (Hornykiewicz, 1973), may arise at least in part because of dysfunction of a system which normally is involved in utilizing environmental stimuli that are used as cues in preparation for the initiation of movements. Such preparation may include postural adjustments. The movement disorder may also be due in part to dysfunction of the system of neurons in the head of the caudate nucleus that appears to be involved in the generation of particular responses to particular environmental events (see also Rolls, Thorpe, Maddison, Roper-Hall, Puerto and Perrett, 1979).

It is of interest that the responses of neurons in the head of the caudate (Rolls, Thorpe and Maddison, 1983) are different in some respects from those in the orbitofrontal cortex (Thorpe, Rolls and Maddison, 1983), from which it receives projections. For instance, 32.4% of orbitofrontal units were classified as being visually reponsive compared with only 9.6% in the anterior striatum. Also, responses in the striatum were less likely to be selective, in that only 31.6% of the small number of neurons with visual responses were selective, whereas in the orbitofrontal cortex, 66.9% were selective. Further, orbitofrontal neurons recorded in the same three animals were often selective even for particular foods or aversive stimuli (Thorpe, Rolls and Maddison, 1983), and this degree of selectivity was not found in the anterior striatal neurons. This indicates that whereas the orbitofrontal cortex could perform computations specific to particular stimuli, the anterior neostriatum is unlikely to be involved in the fine differentiation of stimuli. Further, information about whether reinforcement had been received was represented in the responses of 36 orbitofrontal neurons (7.9%) with gustatory responses (Thorpe, Rolls and Maddison, 1983), whereas only two (0.5%) were found in the anterior neostriatum. Also, unlike the orbitofrontal cortex, no units were found in the neostriatum with clear bimodal sensory responses to, for example, the sight and taste of the same food (Thorpe, Rolls and Maddison, 1983). Although neurons in the orbitofrontal cortex (Thorpe, Rolls and Maddison, 1983) and in the dorsolateral prefrontal cortex (investigations of G. Baylis and E. T. Rolls) did respond in the cue period, so that the cue related responses of anterior striatal neurons could be derived partly from these, some classes of orbitofrontal neurons with activity related to reversal behavior were not found in the anterior striatum. Such orbitofrontal neurons included those that responded when an error was made and punishment was received in reversal or when non-reward was obtained in extinction, neurons that did not reverse their responses (i.e. coded for the stimulus), as well as those that did, and other neurons which responded to particular stimuli only if they had been associated with reward on the previous trials. Thus neuronal responses that appear to be used for the alteration of behavioral responses, to stimuli no longer associated with reinforcement, were present in the orbitofrontal cortex (Thorpe, Rolls and Maddison, 1983), but were not found in the striatum. This difference supports the view that complex neuronal mechanisms involved in the performance of tasks such as visual discrimination reversal are situated in cortical regions such as the orbitofrontal cortex, and that subcortical areas such as the head of the caudate nucleus are not directly involved, but may relay the results of such cortical decision-making to the motor system. Thus, although lesions in both areas might disrupt performance on a visual discrimination reversal task (Divac, Rosvold and Szwarchart, 1967), the reason for the disruption of performance is not the same. It could be that after orbitofrontal lesions, the neuronal mechanism for the switching of central sets is damaged, whereas after neostriatal lesions the pathways by which this mechanism actually influences behavior and the selection of behavioral responses are disrupted.

TAIL OF THE CAUDATE NUCLEUS

The projections from the inferior temporal cortex and the prestriate cortex to the striatum arrive mainly, although not exclusively, in the tail of the caudate nucleus and in the posteroventral portions of the putamen (Kemp and Powell, 1970; Yeterian and Van Hoesen, 1978; Whitlock and Nauta, 1956; Reitz and Pribram, 1969; Van Hoesen, Yeterian and Lavizzo-Mourey, 1981). Since these regions of the caudate nucleus and putamen are adjacent and have a common anatomical input they are referred to together as the caudal neostriatum. There have been few studies of the functions of these visual projections directly into the caudal neostriatum and their importance in visually controlled behavior. Divac, Rosvold and Szwarcbart (1967) reported that stereotaxic lesions placed in the tail of caudate nucleus, in the region that receives input from the inferior temporal visual cortex, produced a deficit in visual discrimination learning. The lesion did not produce impairment in an object reversal task, though in two out of four monkeys it did disturb delayed alternation peformance. Buerger, Gross and Rocha-Miranda (1974) found that lesions of the ventral putamen in the monkey produced a deficit in the retention of a pre-operatively learnt visual discrimination problem but did not disturb retention of auditory discrimination or delayed alternation tasks. The deficit produced by both neostriatal lesions seems therefore to reflect predominantly a loss of visual functions rather than a general loss of cognitive functions.

Since so little is known about the nature of visual processing within the caudal neostriatum and the fate of the visual input from inferior temporal cortex to this area, the activity of single neurons was recorded in the tail of the caudate nucleus and adjoining part of the ventral putamen (Caan, Perrett and Rolls, 1984). Of 195 neurons analyzed in two macaque monkeys, 109 (56%) responded to visual stimuli, with latencies of 90-150 ms for the majority of the neurons. The neurons responded to a limited range of complex visual stimuli, and in some cases responded to simpler stimuli such as bars and edges. Typically (in 75% of cases) the neurons habituated rapidly, within 1-8 exposures, to each visual stimulus, but remained responsive to other visual stimuli with a different pattern. This habituation was orientation specific, in that the neurons responded to the same pattern shown in an orthogonal orientation. The habituation was also relatively short-term, in that at least partial dishabituation to one stimulus could be produced by a single intervening presentation of a different visual stimulus. These neurons were relatively unresponsive in a visual discrimination task, having habituated to the stimuli which had been presented in the task on many previous trials.

The main characteristics of the responses of these neurons in the tail of the caudate nucleus and adjoining part of the putamen thus were rapid habituation to specific visual patterns, sensitivity to changes in visual pattern, and the relatively short-term nature of habituation to a particular pattern, dishabituation occurring with as little as one intervening trial using another stimulus. Hence it may be suggested that these neurons are involved in short term pattern-specific habituation to visual stimuli. This system would be distinguishable from other habituation systems (involved for example in habituation to spots of light) in that it is specialized for patterned visual stimuli which have been highly processed through visual cortical analysis mechanisms, as shown not only by the nature of the neuronal responses, but also by the fact that the system receives inputs from the inferior temporal visual cortex. It may also be suggested that the sensitivity to visual pattern change may have a role in alerting the monkey's attention to new stimuli. This suggestion is consistent with the changes in attention and orientation to stimuli produced by damage to the striatum. Thus, damage to the dopaminergic nigrostriatal bundle produces an inability to orient to visual and other stimuli in the rat (Marshall, Richardson and Teitelbaum, 1974), and damage to the neostriatum itself can lead to compulsive attention to stimuli with a failure to habituate normally in the cat (Villablanca, Marcus and Olmstead, 1976a) and monkey (Denny-Brown, 1962).

Since the recordings described here were mainly from the more anterior part of the tail of the caudate nucleus and adjoining part of the putamen, on anatomical grounds it is likely that the cells studied receive inputs directly from the inferior temporal cortex but not directly from the prestriate cortex (Kemp and Powell, 1970; Van Hoesen, Yeterian and Lavizzo-Mourey, 1981). The properties of the neurons recorded in this study also suggest that they received inputs from the inferior temporal visual cortex. Thus these neostriatal neurons responded with latencies which were mainly between 100 and 150 ms, latencies similar to those for responses of neurons in the inferior temporal visual cortex recorded in the same testing conditions (Rolls, Judge and Sanghera, 1977). A difference between the response properties of caudal neostriatal neurons and of inferior temporal neurons, which could be related to the function of the caudal neostriatal neurons suggested above, was that whereas rapid habituation was characteristic of caudal neostriatal neurons, it appeared to be less marked for inferior temporal neurons recorded in the same testing conditions. For example, Rolls, Judge and Sanghera (1977) found that neurons in the inferior temporal cortex that responded to stimuli used in a visual discrimination continued to maintain their responsiveness for thousands of trials. Also, in the present study a number of tracks were made into the anterior inferior temporal cortex of the monkeys from which neurons were recorded in the caudal neostriatum, in order to compare directly the neuronal responses in these two connected regions; the inferior temporal cortex cells that were studied did not habituate markedly compared to the neostriatal cells, and would often respond for more than 120 consecutive trials. Furthermore, cells in brain structures that receive from the inferior temporal cortex, as indicated by anatomical connections and the types of cell response found (Rolls, 1981a), in many cases do not habituate to repeated visual stimuli in the testing conditions used here. This makes it very unlikely that neurons in the inferior temporal visual cortex all show marked habituation of the type described here for caudal neostriatal neurons (see Caan, Perrett and Rolls, 1984).

Although lesions to caudal neostriatal structures can produce a mild but significant impairment of visual discrimination performance, no clear relation was found between cellular activity, in either of the caudal neostriatal structures, and reward contingencies or behavioral performance in a visual discrimination task. Many cells in the tail of the caudate nucleus and ventral putamen simply did not respond at all to the visual stimuli used in the discrimination task. The neurophysiological findings described here suggest that in so far as the caudal neostriatum is involved in visual discrimination performance, its role is not in enabling associations of the stimuli with reinforcement to be determined, but rather in enabling detection of changes in patterned stimuli in the environment, which may be important when learning and performing a visual discrimination task, as well as in many other types of response to visual stimuli.

The data we present suggest that the caudal neostriatum is not directly involved in the development and maintenance of reward associations to stimuli, but may aid discrimination peformance by its sensitivity to change in visual stimuli. As a population, these neurons could indicate whenever the stimulus on a given trial is different from that of the previous trial. It is clear that a brain system such as the caudal neostriatum which can detect change in visual stimuli over a short time scale can provide a mechanism for one form of visual short term memory, and may participate in a variety of cognitive operations in different behavioral situations.

PUTAMEN

The putamen receives inputs from the sensorimotor cortex, areas 3, 2, 1, 4 and 6 (Kunzle, 1975, 1977, 1978; Kunzle and Akert, 1977; Jones, Coulter, Burton and Porter, 1977; DeLong et al, 1983). In previous investigations, it has been shown that the activity of some neurons in the putamen is related to movements (DeLong, 1972, 1973; DeLong and Strick, 1974; Anderson, 1978; Liles, 1979). Part of the aim of the present series of recordings in the putamen (Rolls, Thorpe, Boytim, Szabo and Perrett, 1984) was to enable direct comparison with the activity of neurons recorded in other parts of the striatum in the same testing conditions, including the initiation of movements in a visual discrimination task, in order to obtain further evidence on specialization of function within the striatum (Rolls, Thorpe, Maddison, Roper-Hall, Puerto, and Perrett, 1979; Rolls, Thorpe and Maddison, 1983; Caan, Perrett and Rolls, 1984; Rolls, Thorpe, Boytim, Szabo and Perrett, 1984; Rolls, Ashton, Williams, Thorpe, Mogenson,Colpaert, and Phillips, 1982; Rolls, 1983).

Of 234 neurons recorded in the putamen of two macaque monkeys during the performance of a visual discrimination task and the other tests in which other striatal neurons have been shown to respond (Rolls et al, 1983; Caan et al, 1984), 68 (29%) had activity which was phasically related to movements (Rolls, Thorpe, Boytim, Szabo and Perrett, 1984). Many of these responded in relation to mouth movements such as licking. The neurons did not have activity related to taste, in that they responded during tongue protrusion made to a food or non-food object. Some of these neurons responded in relation to the licking mouth movements made in the visual discrimination task, and always also responded when mouth movements were made during clinical testing when a food or non-food object was brought close to the mouth. Their responses were thus unconditionally related to movements, in that they responded in whichever testing situation was used, and were therefore different from the responses of neurons in the head of the caudate nucleus (Rolls, Thorpe and Maddison, 1983).

Of the 68 neurons in the putamen with movement-related activity in these tests, 61 had activity related to mouth movements, and 7 had activity related to movements of the body. Of the remaining neurons, 24 (10%) had activity which was task related in that some change of firing rate associated with the presentation of the tone cue or the opening of the shutter occurred on each trial (see Rolls, Thorpe, Boytim, Szabo and Perrett, 1984), 4 had auditory responses, 1 responded to environmental stimuli (see Rolls, Thorpe and Maddison, 1983), and 137 were not responsive in these test situation.

This investigation provides further evidence that differences between neuronal activity in different regions of the striatum are found even in the same testing situations, and that the inputs which activate these neurons are derived functionally from the cortex that projects into a particular region of the striatum. Thus it was found that the majority of responsive neurons in the main part of the putamen (see Rolls et al, 1984) had responses related to movements made by the monkey. This is consistent with the inputs to these regions from sensori-motor cortex, areas 3, 1, 2, 4 and 6. Also, neuronal activity related to the preparation for and initiation of behavioural responses to environmental cues was found in the head of the caudate nucleus (Rolls, Thorpe and Maddison, 1983; Rolls, Thorpe, Maddison, Roper-Hall, Puerto and Perrett, 1979), whereas such neurons were relatively rare (10%) in the putamen, and instead neurons with activity unconditionally associated with movements made in the same test situations were common (29%) in the putamen. This difference in the type of neuronal response found in the putamen and the head of the caudate nucleus in the same testing situations provides further evidence that the responses of neurons in the head of the caudate nucleus are not unconditionally movement-related responses, but are related to cues used in the preparation for and initiation of movements (Rolls, Thorpe and Maddison, 1983).

VENTRAL STRIATUM

Great interest has focussed on the anatomy of the ventral striatum, which includes the nucleus accumbens, the olfactory tubercle (or anterior perforated substance of primates), and islands of Calleja, since the paper of Heimer and Wilson (1975) showing that the ventral striatum receives inputs from limbic structures such as the amygdala and hippocampus and projects to the ventral pallidum. The ventral pallidum may then influence output regions by the subthalamic nucleus/globus pallidus/ventral thalamus/premotor cortex route, or via the mediodorsal nucleus of the thalamus/prefrontal cortex route (Heimer et al, 1982). The ventral striatum may thus be for limbic structures what the neostriatum is for neocortical structures, that is a route for limbic structures to influence output regions. The dopamine pathways have a critical position in these systems, for the nigro-striatal pathway projects to the neostriatum, and the mesolimbic dopamine pathway projects to the ventral striatum (Ungerstedt, 1971). Indeed, Nauta and Domesick (1978) have provided evidence that the ventral striatum also provides a route for limbic information to influence the neostriatum, via the projections of the ventral striatum to the substantia nigra (A9), and thus via the dopamine pathways to the neostriatum. This pattern of connections of the ventral striatum appears to occur not only in the rat (Heinmer and Wilson, 1975; Newman and Winans, 1980a,b), but also in the primate (Hemphill et al., 1981. In addition, it is now clear that the olfactory tubercle is in the anterior perforated substance in the primate (Heimer et al., 1977), and that while a small part of it related to the olfactory tract does receive olfactory projections, a much larger part of it receives a heavy projection from the inferior temporal visual cortex (Van Hoesen, Mesulam and Haaxma, 1976), and could thus provide a link from temporal lobe association cortex to output regions.

There is some pharmacological evidence that links the ventral striatum to schizophrenia, and suggests that the neuroleptic drugs used to treat schizophrenia act on the ventral striatum. The ventral striatum receives one of the main dopaminergic pathways of the brain, the mesolimbic dopamine pathway (Ungerstedt, 1971). It is known that neuroleptic drugs, which in general block dopamine receptors, do not show tolerance in their effects on the nucleus accumbens (on GABA turnover) or on schizophrenia (Costa et al, 1978). (Their effects on GABA turnover in the neostriatum, and on the Parkinsonian symptoms produced as side effects, do show tolerance with repeated administration.) It is also being found that in the brains of schizophrenics there can be elevated dopamine levels (Bird et al., 1977) or increased dopamine receptor binding (Lee et al, 1978; Crow, 1979; Matthyse, 1981). Although this, and other (Matthyse, 1981; Stevens, 1979) evidence linking the ventral striatum, neuroleptic drugs, dopamine receptor blockade, and schizophrenia can of course not be regarded as final, it does point to the need for investigation of the functions of the ventral striatum and of the role of dopamine in the ventral striatum. Although it is known that locomotor activity can be influenced by manipulation of the nucleus accumbens (Kelly, Seviour and Iversen, 1975), particularly when this is in response to novel stimuli (Fink and Smith, 1975), relatively little is yet known of the functions of the ventral striatum, particularly in the primate. The activity of neurons has therefore been recorded in the ventral striatum of the behaving monkey (Rolls et al, 1982).

To analyse the functions of the ventral striatum, the activity of more than 400 single neurons has been recorded in a region that includes the nucleus accumbens and olfactory tubercle, in 4 macaque monkeys in test situations in which lesions of the amygdala, hippocampus and ITC produce deficits (Rolls, Ashton, Williams, Thorpe, Mogenson, Colpaert and Phillips, 1982). The following types of neuronal response have so far been identified. First, there are neurons which respond to novel visual stimuli. Different neurons in this category show pattern-specific habituation over 1-10 trials, and show retention of this habituation over 1-14 intervening trials. A small number of neurons respond to familiar rather than to novel visual stimuli. Second, others respond to visual stimuli of emotional significance, such as faces and some other aversive stimuli for some neurons, and to certain rewarding stimuli for example some foods, for other neurons. These neurons do not respond simply in relation to arousal produced by inputs from different modalities. Third, some neurons respond differentially in a visual discrimination task, either to the stimulus associated with punishment or to that associated with reward. Fourth, some respond in relation to somatosensory stimuli or movement. Fifth, some neurons respond, as in the head of the caudate nucleus (Rolls et al., 1983), to cues such as a 0.5 s tone which the monkey uses to prepare for the performance of the visual discrimination task. Sixth, some respond in relation to arousal, however produced. A number of neurons responded in more than one category, responding maximally for example to relatively novel aversive stimuli.

These findings are consistent with the hypothesis that the ventral striatum is involved in certain behavioral responses to novel and emotion-provoking visual stimuli, which influence it through its inputs from limbic structures.

IONTOPHORESIS

The caudate nucleus and putamen receive a dopaminergic input from the substantia nigra, pars compacta. Its degeneration in man occurs in Parkinson's disease in which there is akinesia (an inability to initiate voluntary movements) (Hornykiewicz, 1973), and its destruction in animals leads to catalepsy and an inability to orient to environmental stimuli (Ungerstedt, 1974; Marshall, Richardson and Teitelbaum, 1974; Iversen, 1979), but as yet it is unclear how this dopaminergic input influences the normal functioning of the striatum.. To investigate this, the effects of dopamine, administered microiontophoretically, on the normal responses shown by neurons in the striatum of the behaving animal have been measured (Rolls, Thorpe, Boytim, Szabo and Perrett, 1984). Microiontophoretic application was used in order to ensure that the effects observed were not due to altered behavior of the animal, which would be produced by systemic administration of dopamine agonists or antagonists, and would mean that any alteration of neuronal activity could be a result of the altered behavior rather than of alteration of the dopamine in the region of the single neuron from which recordings were being made. The recordings were made in the behaving animal because only in these conditions could the effect of the dopamine be determined not only on the spontaneous activity of the neurons, but also on the normal responses of the neurons to their inputs. Only by determining how dopamine influences the responsiveness of striatal neurons to their normal inputs, as well as its effects on spontaneous activity, can its functions in the striatum be determined. Although dopamine has been applied to striatal neurons in the anaesthetized animal previously, usually with a decrease of spontaneous activity (Dray, 1980) and of responses evoked by electrical stimulation, some investigators have claimed that its action is excitatory (Kitai, Sugimori and Kocsis, 1976; Norcross and Spehlman, 1978).

First, it was found in the behaving monkey that iontophoretically applied dopamine decreased the spontaneous firing rates in 178 of 267 neurons (67%) tested in the putamen, caudate nucleus, and the adjacent prefrontal cortex which also receives a dopaminergic projection (Rolls et al, 1984). Second, it was found that dopamine also decreased the magnitude of the responses of these neurons. Third, it was found that this decrease in the responses was of approximately the same magnitude in spikes per second as the decrease in the spontaneous firing rate of the neurons produced by the same current of dopamine. Some of the evidence for this is shown in Fig. 2, in which each point shows the reduction in the neuronal response compared to the reduction in the spontaneous activity for each neuron. In fact, the mean decrease in the response of the neurons produced by dopamine was 5.97 ± 0.99 (mean$\pm$sem) spikes/sec, and the mean decrease produced in the spontaneous activity by the same current of dopamine was 6.52 ± 1.12 spikes/sec. In that the reductions of the spontaneous and response-associated firing rates produced by dopamine were so similar, the most parsimonious explanation of the effect of dopamine is that it reduced the spontaneous firing rate and the response of the neurons equally.

The effects of dopamine on the responses relative to the spontaneous activity of the neurons can also be considered

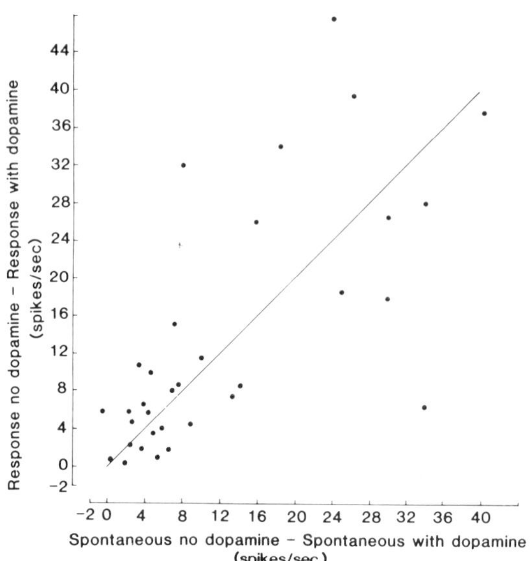

Fig. 2. The reduction in the responses of each neuron (spikes per second) produced by dopamine (ordinate) is plotted against the reduction in the spontaneous activity of that neuron produced by the same current of dopamine (abscissa). Each point provides data from one neuron. The line is drawn at 45 degrees.

quantitatively in terms of its effects on the ratio of these values. This ratio (i.e. the response/the spontaneous firing rate), with and without dopamine, is shown for each neuron in Fig. 3, from which it is clear that dopamine tended to increase the value of this ratio. The mean ratio without dopamine was 1.92 ± 0.14 and with dopamine was 4.03 ± 0.67 ($t=3.5$, $df=33$, $p<0.002$; $T=65$, $n=34$, $z=3.97$, $p<0.001$). Thus, if this ratio is considered analogous to the signal to noise ratio, this was significantly increased by the dopamine.

The effect of this action of dopamine on how information is processed within the striatum will be considered in two ways. First, the effect of dopamine on signal to noise ratio will be considered. The signal to noise ratio can be represented by the ratio of signal plus noise to noise (Green and Swets, 1966; Egan, 1975). If the signal to noise ratio of the neuron is represented by the ratio of the response of the neuron to the spontaneous firing rate, then dopamine can be considered to increase the signal to noise ratio of the neurons (see Fig. 3). (Note that the response of the neuron is therefore represented by the firing rate of the neuron while the neuron is responding, rather than this firing rate minus the spontaneous rate, in order to be analogous to signal plus noise.)

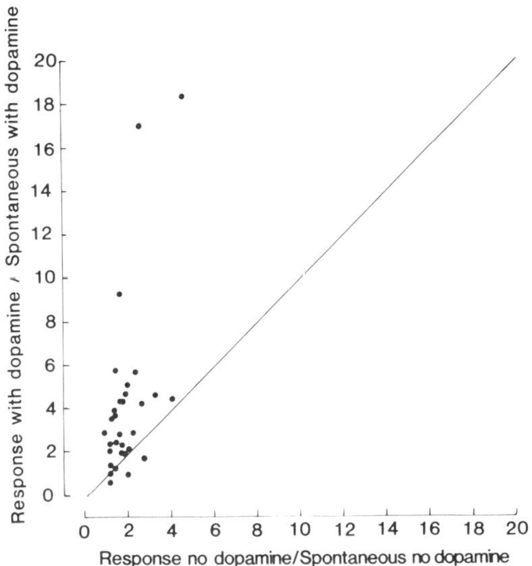

Fig. 3. The response of each neuron divided by its spontaneous activity is plotted without dopamine (abscissa) and with dopamine (ordinate). Each point provides data from one neuron. The line is drawn at 45 degrees.

This result is what is obtained if the effect of the dopamine is to decrease the spontaneous activity and the response of the neurons in a subtractive way, for as the spontaneous activity tends to zero, so this signal to noise ratio tends to a high value. Indeed, if this analysis is followed, it may be proposed that one of the functions of dopamine in the striatum is to actively set the spontaneous firing rate of striatal neurons to its normal low level, so that effectively the signal to noise ratio of the information transmission system is maximal. This maintenance of the firing rate at a low level may require the ongoing release of dopamine under normal conditions in the behaving animal, in that it was possible to show in at least some cases for neurons in the prefrontal cortex (which also receives a dopaminergic input), that the application iontophoretically of the dopamine receptor blocking agent trifluoperazine increased the firing rate of the neurons (see Rolls et al., 1984). Second, the effect of dopamine on signal detectability can be considered. In signal detection theory, the discriminability of the signal from noise, d', can be represented by the mean value of the signal minus the mean value of the noise all divided by a measure of the standard deviation of the noise and signal distributions (Green and Swets, 1966; Egan, 1975; Lindsay and Norman, 1977). For a neuron, this could be represented by the mean firing rate when the neuron is responding minus the mean spontaneous firing rate all divided by the standard deviation of these firing rates (and making appropriate adjustments if the standard deviations are different, see Green and Swets, 1966). If this analysis is used, then subtraction of a constant from the spontaneous and the response firing rates, which is the effect of dopamine implied by the data shown in Fig. 2, would not normally affect the discriminability of the signal from the noise. However, if a relatively large amount of dopamine decreased the spontaneous activity of the neuron to zero, and the response firing rate of the neuron to a low value, then d' would be reduced to a low value. Thus, if analysed in this way, dopamine could influence transmission though the striatum by setting the threshold. Regardless of which type of analysis is used, the above discussion indicates that two ways in which dopamine could act to influence transmission through the striatum are by influencing the magnitude of the responses relative to the spontaneous activity, or by setting the threshold.

These findings of regional differences in the types of neuronal resonse found in the striatum, and of how dopamine influences the responses of striatal neurons, leads to the suggestion that dopamine may not only be involved in setting the sensitivity of transmission of information received from the cortex through the striatum to output structures, but may also be involved in shaping the activity of different parts of the striatum and thus in the type of behavioral response selected. For example, if activity were high in the tail of the caudate nucleus as a result of a sudden change in the pattern of a visual stimulus (Caan, Perrett and Rolls, 1984), then this could, through differential effects on dopaminergic neurons projecting to this as compared to other parts of the striatum, set transmission in this part of the stiatum to a high level, and at the same time reduce the sensitivity of other parts of the striatum to a low level. This would result in the selection of behavioral orientation to the changed patterned visual stimulus, with the suppression of other competing but weaker responses being processed in other parts of the striatum. It may be noted (in line with the discussion on discriminability and signal to noise ratio above) that this setting to a low level could be performed by a large release of dopamine which would reduce the activity of the striatal neurons so much that they would not only show a zero spontaneous firing rate, but would also have a small change of firing rate to their normal input; or it could be achieved by decreasing the release of dopamine, so that the spontaneous firing rate of the neurons increased and thus the signal to noise ratio detected by the next stage of processing decreased. It is possible that some of the changes of firing rate observed in neurons in the striatum during the performance of behavioral tasks could be related to this resetting of sensitivity of different parts of the striatum by dopamine, which may be part of the mechanism of response selection. In this way, only one response would be emitted by the animal. It may also be noted that this selective setting of the sensitivity of transmission through different parts of the striatum could be achieved by differential influences on the firing rates of dopaminergic neurons from other systems, such as the basal forebrain.

Some of the disorders of striatal function may be related to alterations in the ability of dopamine to select and regulate transmission through the striatum. Thus the akinesia of Parkinson's disease may be considered to be due to reduced processing and transmission through the striatum of cortical information to output regions, resulting from a decreased level of dopamine in the striatum, and thus a decreased signal to noise ratio of the transmission. In contrast, increased dopamine function, produced by for example too much release of dopamine or enhanced receptor sensitivity to dopamine, would increase the signal to noise ratio of all the different signals reaching the putamen. This would mean that the normal response selection function of the striatum, involving perhaps competition between the inputs and thresholding, could not be performed adequately, with consequent oversensitivity to environmental stimuli.

DISCUSSION

Segregation of function within the striatum

The present investigation provides further evidence on neuronal activity in the striatum, on differences between neuronal activity in different regions of the striatum, and that the inputs which activate these neurons are derived functionally (as well as anatomically, Kemp and Powell, 1970, 1971) from the cortex which overlies a particular region of the striatum. Thus in this series of investigations it was found that the majority of neurons in the main part of the putamen (see Rolls et al., 1984) had responses related to movements made by the monkey (see also DeLong et al., 1983). This is consistent with the inputs to these regions from sensori-motor cortex, areas 3, 1, 2, 4 and 6 (Kemp and Powell, 1970, 1971; DeLong et al, 1983). In contrast, though the same testing methods were used, neuronal activity related to visual stimuli and which showed rapid habituation was found in the tail of the caudate nucleus and adjoining part of the ventral putamen, which receive from the inferior temporal visual cortex (Caan, Perrett and Rolls, 1984). Also, neuronal activity related to the preparation for and initiation of behavioral responses in response to environmental cues was found in the head of the caudate nucleus (Rolls, Thorpe and Maddison, 1983; Rolls et al., 1979), whereas such neurons were relatively rare (10%) in the putamen, and instead neurons with activity unconditionally associated with movements made in the same test situations were common (29%) in the putamen. This difference in the type of neuronal response found in the putamen and the head of the caudate nucleus in the same testing situations provides further evidence that the responses of neurons in the head of the caudate nucleus are not unconditionally movement-related responses, but are related to cues used in the preparation for and initiation of movements (Rolls, Thorpe and Maddison, 1983). Again, a similarity in the responses of neurons in one part of the striatum to those in the cortical regions from which it receives is found, in that the head of the caudate nucleus receives projections from the prefrontal cortex (see Rolls, Thorpe and Maddison, 1983), in which neurons which respond to the same environmental cues are found (experiments of G. Baylis and E. T. Rolls, in progress). Further, the activity of some neurons in the ventral striatum, which receives inputs from limbic structures such as the amygdala and hippocampus, occurs to stimuli related to emotional and novel environmental events (Rolls, Ashton, Williams, Thorpe, Mogenson, Colpaert and Phillips, 1982) which are probably processed through limbic structures (Rolls, 1981, 1984).

The present studies thus provide further evidence that neurons in different regions of the striatum have different types of response in the behaving primate. Moreover, the results also provide further evidence that the processing in a given region of the striatum is related to the particular inputs it receives from the regions of cortex which project into it (Kemp and Powell, 1970, 1971), and that it is these cortical inputs which are particularly important in producing the responses of striatal neurons normally shown when the system is operating physiologically in the behaving animal.

The importance of this at least partial segregation of neuronal response types in different major parts of the striatum for our understanding of striatal function is that it shows that there is at least partial segregation of function within different regions of the striatum, and that it is not functionally homogeneous. A consequence of this is that damage to different regions of the striatum (including for example regional depletion of dopamine) would not be expected necessarily to lead to the same type of disorder. For example, it might be expected that movement disorders would be produced by depletion of dopamine in the putamen, but that other more complex disorders would result if there were depletion of dopamine in other regions of the striatum, such as the head of the caudate nucleus, the tail of the caudate nucleus, or the ventral striatum.

Further issues important for our understanding of striatal function raised by the present results are whether within the striatum there is the possibility for different regions to interact, and whether the partial functional segregation is maintained in processing beyond the striatum. For example, is the segregation maintained throughout the globus pallidus and thalamus with projections to different premotor and even prefrontal regions reached by different regions of the striatum, or is there convergence at some stage during this post-striatal processing? This will be an important theme for future understanding of the functions of the basal ganglia.

The nature of processing within the striatum

Given this evidence for at least partial segregation of types of neuronal response in different regions of the striatum, with neuronal activity in some regions related to movements, but in other regions to more complex events, it may be asked whether some of these latter regions might not have a cognitive function. Indeed, the experiment of Divac, Rosvold and Szwarcbart (1967) in which lesions to different parts of the caudate nucleus produced different cognitive deficits related to the function of the connected cortical region (see Introduction) provides some support for this possibility. However, it is possible to ask whether these parts of the striatum are actually performing a cognitive computation (for the present purposes one in which neither the inputs nor the outputs are directly related to sensory or motor function - see also Oberg and Divac, 1979), or whether these parts of the striatum provide an essential output route for a cortical area with a cognitive function, but do not themselves have cognitive functions.

One way to obtain evidence on this is to analyse neurophysiologically the computation being performed by a part of the striatum, and relate this to the computation being performed in its input and output regions. One part of the striatum for which such evidence is available is the head of the caudate nucleus, in which neuronal activity can be compared with neuronal activity in the overlying prefrontal cortex. As described above, information necessary for the computation that a visual stimulus is no longer associated with taste reward reaches the orbitofrontal cortex, and the putative output of such a computation, namely neurons which respond in this non-reward situation, are found in the orbitofrontal cortex. However, such neurons which represent the necessary sensory information for this computation, and neurons which respond to the non-reward, were not found in the head of the caudate nucleus. Instead, in the head of the caudate nucleus, neurons in the same test situation responded in relation to whether the monkey had to make a response on a particular trial, that is many of them responded more on Go than on NoGo trials. This could represent the output of a cognitive computation performed by the orbitofrontal cortex, indicating whether on the basis of the available sensory information, the current trial should be a Go trial, or a NoGo trial because a visual stimulus previously associated with punishment had been shown.

A similar comparison can be made for the tail of the caudate nucleus. Here the visual responses shown by neurons typically habituated to zero within a few trials, whereas such marked habituation was less common in neurons in the inferior temporal visual cortex, which projects to the tail of the caudate nucleus. In this case, the signal being processed by the striatum thus occurred when a patterned visual stimulus changed, and this could be of use in switching attention or orienting to the changed stimulus.

In both these parts of the striatum in which a comparison can be made of processing in them with that in the cortical area which projects to that part of the striatum, it appears that the full information represented in the cortex does not reach the striatum, but that rather the striatum receives the output of the computation being performed by a cortical area, and could use this to switch or alter behavior. Thus, if the orbitofrontal cortex computed that a visual signal was no longer associated with reward, it might send the output of this computation to the striatum to indicate that behavior should be switched from the non-rewarded stimulus. Similarly, a signal from the inferior temporal cortex might indicate that behavior should be switched to a new patterned stimulus. Comparably, the cue-related neurons in the head of the caudate nucleus which respond to significant environmental events such as a tone cue which precedes the onset of a trial (see above) may reflect an output from the dorsolateral prefrontal cortex which may be involved in this computation and in which similar neurons are found (see above). The head of the caudate nucleus would thus again be in receipt of a signal decoded by the cortex to which it might be appropriate to switch or initiate behavior. Although the evidence is thus only starting to become available, the possibility arises from these findings that the striatum receives the output of cognitive computations, but does not itself perform them, and instead is involved in switching behavior as appropriate as determined by the different, sometimes conflicting, information it receives. Thus on this view, the striatum would be particularly involved in the selection of behavioral responses, and in producing one coherent stream of behavioral output, with the possibility to switch if a higher priority input was received. Dopamine could play an important role in setting the sensitivity of this response selection function as discussed above.

On this hypothesis, different regions of the striatum, or at least the outputs of such regions, would need to interact. Given the anatomy of the striatum, this could happen in a number of different ways. One would be for each part of the striatum to receive at least some input of different cortical regions. There is some evidence that although the major input from a cortical region is to a particular part of the striatum, there are nevertheless some more widespread connections (Van Hoesen, Yeterian and Lavizzo-Mourey, 1981). A second way would be for the short intrastriatal connections to produce a widespread influence, perhaps by spreading like a lateral inhibition signal, so that the strongest signal would lead to coherent behavior directed towards that. A third mechanism might be via the dopaminergic pathway, through which a signal which has descended from for example the ventral striatum might influence other parts of the striatum (Nauta and Domesick, 1973). Because of the slow conduction speed of the dopaminergic neurons, this system would probably not be suitable for rapid switching of behavior, but only for more tonic, longterm, adjustments of sensitivity.

An alternative view of striatal function is that the striatum might be organized as a set of segregated and independent transmission routes, each one of which would receive from a given region of the cortex, and project finally to separate premotor or prefrontal regions. On this view, an area of the striatum such as part of the head of the caudate nucleus which received from the prefrontal cortex and projected back to the prefrontal cortex might be expected more than on the previous view to have cognitive functions. More investigations are needed to lead to further understanding of these conceptual ideas on the function of the striatum.

Disturbance of striatal function

The findings described here on neuronal activity in different regions of the striatum have implications for our understanding of striatal dysfunction. If the transmission of information from the cortex through the striatum to its outputs was decreased, by for example depletion of dopamine, then the disturbance of behavior which resulted would not represent just a movement disorder. For example, the akinesia of Parkinson's disease could be partly due to an inability to respond to environmental cues normally used in the preparation for and initiation of movement. Similarly, the sensorimotor deficit produced by damage to the nigrostriatal bundle could occur partly because neurons which normally respond to changing patterned visual stimuli are insensitive to these stimuli.

According to this analysis, the opposite type of functional disorder, in which transmission through the striatum was abnormally elevated, might lead to increased responsiveness to temporal lobe signals, with failure of these to habituate normally so leading perhaps to hallucinations; to increased reactivity to environmental signals, even perhaps when these were irrelevant, so that distractability, inability to maintain attention and concentrate on one behavior, and an increased tendency to switch between behaviors might result. Moreover, because there is some segregation of function within the striatum, and some topography of the dopamine projection, these symptoms might be dissociable. Thus in some cases, cognitive dysfunctions as compared to motor dysfunctions might predominate. It is thus possible that some of the symptoms of schizophrenia could be related to overresponsiveness of these striatal systems. In line with this hypothesis, increased dopamine receptor binding, implying increased sensitivity to dopamine, has been reported in the striatum of some schizophrenic patients (even when they have not been treated with neuroleptic drugs which might lead to supersensitivity) (Lee et al., 1978; Matthyse et al., 1981). Further, the neuroleptic drugs used to treat schizophrenia do block dopamine receptors (see Matthyse, 1981), and might thus be expected to normalize some of the overresponsiveness described above. Although this parallel between neuronal activity in the striatum and some of the symptoms of schizophrenia is of interest, there are also dopamine projections to the prefrontal cortex and amygdala, and it is unlikely that altered neuronal responsiveness in the striatum plays an exclusive role in the symptons observed.

Thus this analysis of neuronal activity in the striatum leads to new suggestions about the symptoms which follow altered striatal function, and suggests a number of concepts of striatal function for future investigation.

REFERENCES

Anderson, M. E., 1978, Discharge patterns of basal ganglia neurons during active maintenance of postural stability and adjustment to chair tilt, Brain Res., 143:325-338.

Bird, E. D., Spokes, E. G., Barnes, J., Mackay, A. V., Iversen, L., and Shephard, M., 1977, Increased brain dopamine and reduced glutamic acid decarboxylase and choline acetylase activity in schizophrenia and related psychoses, Lancet, 2:1157-9.

Buerger, A. A., Gross, C. G., and Rocha-Miranda, C. E., 1974, Effects of ventral putamen lesions on discrimination learning by monkeys, J. Comp. Physiol. Psych., 86:440-446.

Caan, W., Perrett, D. I., and Rolls, E. T., 1984, Responses of striatal neurons in the behaving monkey. II. Tail of the caudate nucleus, Brain Res., in press.

Carpenter, M. B., 1976, Anatomical organization of the corpus striatum and related nuclei, in: "The Basal Ganglia," M. D. Yahr, ed., Res. Publ. Assoc. Nerv. Ment. Dis., 55:1-35.

Costa, E., Cheney, D. L., Mao, C. C., and Moroni, F., 1978, Action of antischizophrenic drugs on the metabolism of gammaminobutyric acid and acetylcholine in globus pallidus, striatum and n. accumbens, Fed. Prod., 2408-2414.

Cowey, A.., and Gross, C. G., 1970, Effects of foveal prestriate and inferotemporal lesions on visual discrimination by rhesus monkeys, Exp. Brain Res., 11:128-144.

Crow, T. J., 1979, What is wrong with dopaminergic transmission in schizoophrenia? Trends Neurosci., 2:52-55.

DeLong, M. R., 1972, Activity of basal ganglia neurons during movement, Brain Res., 40:127-135.

DeLong, M. R., 1973, Putamen: activity of units during slow and rapid arm movements, Science, 179:1240-1242.

DeLong, M. R., 1974, Motor functions of the basal ganglia: single unit activity during movement, in: "The Neurosciences. Third Study Program," F. O. Schmitt and F. G. Worden, eds., MIT Press, Cambridge, Mass.

DeLong, M. R., and Strick, P. L., 1974, Relation of basal ganglia, cerebellum and motor cortex units to ramp and ballistic movements, Brain Res., 71:327-335.
DeLong, M. R., Georgopoulos, A. P., and Crutcher, M. D., 1983, Cortical-basal ganglia loops and coding of motor performance, Exp. Brain Res., Suppl. 7:30-40.
Denny-Brown, D., 1962, "The Basal Ganglia", Oxford University Press, Oxford.
Denny-Brown, D., and Yanagisawa, N., 1976, The role of the basal ganglia in the initiation of movement, in: "The Basal Ganglia," M. D. Yahr, ed., Res. Publ. Assoc. Nerv. Ment. Dis., 55:115-148.
Divac, I., Rosvold, H. E., and Szwarcbart, M. K., 1967, Behavioral effects of selective ablation of the caudate nucleus, J. Comp. Physiol. Psych., 63:184-190.
Divac, I., and Oberg, R. G. E., Current conceptions of neostriatal functions, in: "The Neostriatum," I. Divac, and R. G. E. Oberg, eds., Pergamon, New York.
Dray, A., 1980, The physiology and pharmacology of the mammalian basal ganglia, Progr. Neurobiol., 14:221-335.
Egan, J. P., 1975, "Signal Detection Theory and ROC Analysis," Academic Press, New York.
Fink, J. S., and Smith, G. P., 1980, Mesolimbic and mesocortical dopaminergic neurons are necessary for normal exploratory behavior in rats, Neurosci. Lett., 17:61-65.
Goldman, P. S., and Nauta, W. J. H., 1977, An intricately patterned prefrontocaudate projection in the rhesus monkey, J. Comp. Neurol., 171:369-386.
Graybiel, A. M., and Ragsdale, C. W., 1979, Fiber connections of the basal ganglia, Prog. Brain Res., 51:239-283.
Green, D. M., and Swets, J. A., 1966, "Signal Detection Theory and Psychophysics," Wiley, New York.
Hemphill, M., Holm, G., Crutcher, M., Delong, M., and Hedreen, J., 1981, Afferent connections of the nucleus accumbens in the monkey, in: "The Neurobiology of the Nucleus Accumbens," R. B. Chronister, and J. F. DeFrance, eds., Haer, Brunswick, N. J.
Heimer, L., and Wilson, R. D., 1975, The subcortical projections of the allocortex: similarities in the neuronal associations of the hippocampus, the pyriform cortex and the neocortex, in: "Golgi Centennial Symposium: Perspectives in Neurobiology," M. Santini, ed., Raven Press, New York.
Heimer, L., Van Hoesen, G. W., and Rosene, D. L., 1977, The olfactory pathways and the anterior perforated substance in the primate brain, Int. J. Neurol., 12:42-52.
Heimer, L., Switzer, R. D., and Van Hoesen, G. W., 1982, Ventral striatum and ventral pallidum. Additional components of the motor system? Trends in Neurosciences, 5:83-87.
Hornykiewicz, O., 1973, Dopamine in the basal ganglia: Its role and therapeutic implications, Brit. Med. Bull., 29:172-178.

Iversen, S. D., 1979, Behaviour after neostriatal lesions in animals, in: "The Neostriatum," I. Divac and R. G. E. Oberg, eds., Pergamon, Oxford.

Jones, E. G., Coulter, J. D., Burton, H., and Porter, R., 1977, Cells of origin and terminal distribution of corticostriatal fibres arising in sensory motor cortex of monkeys, J. Comp. Neurol., 181:53-80.

Kelly, P. K., Seviour, P. W., and Iversen, S. D., 1975, Amphetamine and apomorphine responses in the rat following 6-OHDA lesions of the nucleus accumbens septi and corpus striatum, Brain Res., 94:507-522.

Kemp, J. M., and Powell, T. P. S., 1970, The cortico-striate projections in the monkey, Brain, 93:525-546.

Kemp, J. M., and Powell, T. P. S., 1971, The connections of the striatum and globus pallidus: synthesis and speculation, Phil. Trans. Roy. Soc., B, 262:441-457.

Kitai, S. T., Sugimori, M., and Kocsis, J. D., 1976, Excitatory nature of dopamine in the nigro-caudate pathway, Exp. Brain Res., 24:351-363.

Kunzle, H., 1975, Bilateral projections from precentral motor cortex to the putamen and other parts of the basal ganglia, Brain Res., 88:195-209.

Kunzle, H., 1977, Projections from primary somatosensory cortex to basal ganglia and thalamus in the monkey, Exp. Brain Res., 30:481-482.

Kunzle, H., 1978, An autoradiographic analysis of the efferent connections from premotor and adjacent prefrontal regions (areas 6 and 9) in Macaca fascicularis, Brain Behav. Evoln., 15:185-234.

Kunzle, H., and Akert, K., 1977, Efferent connections of area 8 (frontal eye field) in Macaca fascicularis, J. Comp. Neurol., 173:147-164.

Lee, T., Seeman, P., Tourtelotte, W. W., Farley, I. J., and Hornykiewicz, O., 1978, Binding of 3H-neuroleptics and 3H-apomorphine in schizophrenic brains, Nature, 274:897-900.

Lindsay, P. H., and Norman, D. A., 1977, "Human Information Processing," 2nd Edition, Academic Press, New York.

Liles, S. L., 1979, Topographic organization of neurons related to arm-movement in the putamen, Adv. Neurol., 23:155-162.

Marsden, C. D., 1980, The enigma of the basal ganglia and movement, Trends in Neurosciences, 3:284-287.

Marshall, J. P., Richardson, J. S., and Teitelbaum, P., 1974, Nigrostriatal bundle damage and the lateral hypothalamic syndrome, J. Comp. Physiol. Psychol., 87:808-830.

Matthyse, S., 1981, Nucleus accumbens and schizophrenia, in: "The Neurobiology of the Nucleus Accumbens," R. B. Chronister and J. F. DeFrance, eds., Haer, Brunswick, N. J.

Nauta, W. J. H., and Domesick, V. B., 1978, Crossroads of limbic and striatal circuitry: hypothalamonigral connections, in: "Limbic Mechanisms," K. E. Livingstone and O. Hornykiewicz, eds., Plenum, New York.

Nauta, W. J. H., and Domesick, V. B., 1979, The anatomy of the extrapyramidal system, in: "Dopaminergic Ergot Derivatives and Motor Function," K. Fuxe and D. B. Calne, eds., Pergamon, Oxford.
Newman, R., and Winans, S. S., 1980, An experimental study of the ventral striatum of the golden hamster. I. Neuronal connections of the nucleus accumbens, J. Comp. Neurol, 191:167-192.
Newman, R., and Winans, S. S., 1980, An experimental study of the ventral striatum of the golden hamster. II. Neuronal connections of the olfactory tubercle, J. Comp. Neurol, 191:193-212.
Niewenhuys, R., 1977, Aspects of the morphology of the striatum, in: "Psychobiology of the Striatum," A. R. Cools, A. H. M. Lohman and J. H. L. Van Den Bercken, eds., Elsevier/North Holland, Amsterdam.
Norcross, K., and Spehlman, R., 1978, A quantitative analysis of the excitatory and depressant effects of dopamine on the firing of caudatal neurons: electrophysiological evidence for the existence of two distinct dopamine-sensitive receptors, Brain Res., 156:168-174.
Oberg, R. G. E., and Divac, I., 1979, `Cognitive' functions of the neostriatum, in: "The Neostriatum," I. Divac and R. G. E. Oberg, eds., Pergamon, New York.
Reitz, S. L., and Pribram, K. H., 1969, Some subcortical connections of the inferotemporal gyrus of the monkey, Exp. Neurol., 23:632-645.
Rolls, E. T., 1981a, Processing beyond the inferior temporal visual cortex related to feeding, learning, and striatal function, in: "Brain Mechanisms of Sensation," Y. Katsuki, R. Norgren and M. Sato, eds., Wiley, New York.
Rolls, E. T., 1981b, Central nervous mechanisms related to feeding and appetite, Brit. Med. Bull., 37:131-134.
Rolls, E. T., 1983, The initiation of movements, Exp. Brain Res., Suppl., 7:97-113.
Rolls, E. T., 1984, Connections, functions and dysfunctions of limbic structures, the prefrontal cortex, and hypothalamus, in: "The Scientific Basis of Clinical Neurology," M. Swash and C. Kennard, eds., Churchill Livingstone, London.
Rolls, E. T., Burton, M. J., and Mora, F., 1976, Neuronal responses associated with the sight of food, Brain Research, 111:53-66.
Rolls, E. T., Judge, S. J., and Sanghera, M. K., 1977, Activity of neurones in the inferotemporal cortex of the alert monkey, Brain Res., 130:229-238.
Rolls, E. T., Sanghera, M. K., and Roper-Hall, A., 1979, The latency of activation of neurons in the lateral hypothalamus and substantia innominata during feeding in the monkey, Brain Research, 164:121-135.

Rolls, E. T., Thorpe, S. J., Maddison, S., Roper-Hall, A., Puerto, A., and Perrett, D., 1979, Activity of neurones in the neostriatum and related structures in the alert animal, in: "The Neostriatum," I. Divac and R. G. E. Oberg, eds., Pergamon Press, Oxford.
Rolls, E. T., Perrett, D. I., Thorpe, S. J., Maddison, and Caan, W., 1979, Activity of striatal neurones in the behaving monkey: implications for the neural basis of schizophrenia and Parkinsonism, Soc. Neurosci. Abstr., 5:1167.
Rolls, E. T., Burton, M. J., and Mora, F., 1980, Neurophysiological analysis of brain-stimulation reward in the monkey, Brain Research, 194:339-357.
Rolls, E. T., Thorpe, S. J., Maddison, S., Caan, W., Wilson, F., and Ryan, S., 1981, Neuronal responses in the striatum of the behaving monkey: implications for understanding striatal function and dysfunction, in: "Advances in Physiological Sciences, Vol. 2 Regulatory Functions of the CNS," J. Hamori and M. Palkovits, eds., Pergamon Press, Oxford.
Rolls, E. T., Ashton, J., Williams, G., Thorpe, S. J., Mogenson, G. J., Colpaert, F., and Phillips, A. G., 1982, Neuronal activity in the ventral striatum of the behaving monkey, Soc. Neurosci. Abstr., 8:169.
Rolls, E. T., Thorpe, S. J., and Maddison, J., 1983, Responses of striatal neurons in the behaving monkey. 1. Head of the caudate nucleus, Behav. Brain Res., 7:179-210.
Rolls, E. T., Thorpe, S. J., Boytim, M., Szabo, I., and Perrett, D. I., 1984, Responses of striatal neurons in the behaving monkey. 3. Effects of iontophoretically applied dopamine on normal responsiveness, Neuroscience, in press.
Stevens, J. R., 1979, Schizophrenia and dopamine regulation in the mesolimbic system, Trends in Neurosciences, 2:103-105.
Thorpe, S. J., Rolls, E. T., and Maddison, S. P., 1983, Neuronal responses in the orbitofrontal cortex of the behaving monkey, Exp. Brain Res., 49:93-115.
Ungerstedt, U., 1971, Stereotaxic mapping of the monoamine pathways in the rat brain, Acta Physiol. Scand (82 Suppl.), 367:1-48.
Ungerstedt, U., 1974, Brain dopamine neurones and behaviour, in: "The Neurosciences. Third Study program", F. O. Schmitt and F. G. Worden, eds., MIT Press, Cambridge, Mass.
Van Hoesen, G. W., Mesulam, M.-M., and Haaxma, R., 1976, Temporal cortical projections to the olfactory tubercle in the rhesus monkey, Brain Res., 109:375-381.
Van-Hoesen, G. W., Yeterian, E. H., and Lavizzo-Mourey, R., 1981, Widespread corticostriate projections from temporal cortex of the rhesus monkey, J. Comp. Neurol., 199:205-219.
Villablanca, J. R., and Marcus, R. J., 1975, Effects of caudate nuclei removal in cats. Comparison with effects of frontal cortex ablation, in: "Brain Mechanisms and Mental Retardation," N. A. Buchwald and M. A. B. Brazier, eds., Academic Press, New York.

Villablanca, J. R., Marcus, R. J., and Olmstead, C. E., 1976a, Effects of caudate nuclei or frontal cortical ablations in cats. I. Neurology and gross behavior, Exp. Neurol., 52:389-420.

Villablanca, J. R., Marcus, R. J., and Olmstead, C. E., 1976b, Effects of caudate nuclei or frontal cortical ablations in cats. II. Sleep-wakefulness, EEG, and motor activity, Exp. Neurol., 53:31-50.

Villablanca, J. R., Marcus, R. J., and Olmstead, C. E., 1976c, Effects of caudate nuclei or frontal cortical ablations in cats. III. Recovery of limb placing reactions, including observations in hemispherectomized animals, Exp. Neurol., 53:289-303.

Whitlock, D. G., and Nauta, W. J. H., 1956, Subcortical projections from the temporal neocortex in Macaca mulatta, J. Comp. Neurol., 106:183-212.

Yeterian, E. H., and Van Hoesen, G. W., 1978, Cortico-striate projections in the rhesus monkey: the organization of certain cortico-caudate connections, Brain Res., 139:43-63.

CONSEQUENCES OF DISTURBED GABA-ERGIC TRANSMISSION IN SUBSTANTIA NIGRA PARS RETICULATA IN FREELY MOVING CATS ON THEIR MOTOR BEHAVIOUR, AND IN ANAESTHETIZED CATS ON THEIR SPINAL MOTOR ELEMENTS

K.-H. Sontag, C. Heim, M. Schwarz, R. Jaspers*,
A. R. Cools* and P. Wand

Max-Planck-Institute for Experimental Medicine
Goettingen (F.R.G.)

*Institute of Pharmacology, University of Nijmegen
(The Netherlands)

INTRODUCTION

Early hypotheses about the role of basal ganglia in motor behaviour have been strongly influenced by clinicopathological studies that revealed a correlation between lesions of the nuclei of basal ganglia and a variety of disorders of movement such as akinesia, tremor, involuntary movements and disturbances of posture, equilibrium and muscle tone (for ref. Jung and Hassler, 1960; DeLong and Georgopoulos, 1981). Clinical and experimental studies over the last twenty years have confirmed the importance of basal ganglia in motor behaviour and suggested a participation of these structures in control of aspects of social or cognitive behaviour (Divac, 1977; Iversen, 1977; Oberg and Divac, 1977; Cools et al., 1981; Cools et al., 1984b).

A progressive deterioration of the basal ganglia, especially a deficit in the dopaminergic transmission within the striatum, is assumed to underly the appearance of symptoms of Parkinson's disease (Jung and Hassler, 1960; Hornykiewicz, 1963; Marsden, 1980). These symptoms include motor deficits described as tremor, bradykinesia and rigidity. Careful clinical studies revealed that Parkinsonians have, besides these motor deficits, programme disorders in motor functioning, i.e. difficulties in sequencing motor strategies used to carry out two motor tasks simultaneously and/or successively (Schwab et al., 1954; Horne, 1973; Cools et al., 1981; Cools et al., 1984b). The difficulty in switching the motor programme was manifest in motor behaviour that was not directed by external stimuli (Cools et al.,

1984b). This conclusion fits in with Stark's assumption that Parkinsonians are defective in performing open-loop movements (i.e. without sensory feedback) and operate instead in a closed-loop mode (Stark, 1968). The patients are not able to perform initial ballistic movements in a special task, but rather crept slowly towards the goal position (Flowers, 1976). Taken together, the data might offer an explanation for the finding of Flowers, that Parkinsonians have difficulties in planning and anticipating motor behaviour (Flowers, 1978).

Although the results of these studies may well apply to the deficits of Parkinson's disease, they cannot be generalized to the motor function of the striatum, since the pathology of patients with Parkinson's disease is not restricted to this nucleus (Marsden, 1980). In order to resolve this problem, experimental data of animal studies were necessary to formulate testable hypotheses about the physiological role of other nuclei of the basal ganglia.

Numerous studies have shown that the striatal dopaminergic neurotransmission plays a critical role in the control of motor behaviour (for ref. see Schultz, 1982). Depletion of the striatal dopamine system in animals induces bradykinesia, dysfunction of postural regulation, and sensory neglect. Recently, experiments on rats provided evidence that an increase of dopaminergic activity of the striatum facilitated the animal's ability to select the best strategy by accelerating the process of switching different motor patterns in order to escape a stressful situation (Cools, 1980). While stimulation of striatal DA-receptors accelerated the process of switching strategies under pressure of factors intrinsic to the organism, inhibition of striatal DA-receptors inhibited this process but did not inhibit the process of switching as long as factors extrinsic to the organism could be used. The doses of striatally applied apomorphine or haloperidol which affected the animal's ability to switch from one motor programme to another were lower than those to elicit deficits in the motor performance. This observation points to the hypothesis that the presence of motor deficits reflects a greater degree of pathology of striatal dopaminergic function than disturbances in motor programmes (Cools, 1980).

Changes in neuronal activity within the striatum are at least partly funneled out of this nucleus via two GABAergic projections to the main output stations of basal ganglia, the pallidum and the substantia nigra (DeLong and Georgopoulos, 1981). It has been previously considered that the striatonigral GABAergic neurons participate in motor control by serving as feedback inhibition of the dopaminergic nigrostriatal projection which arises in the pars compacta of the substantia nigra (SNC) (Anden and Stock, 1973; Dray, 1979). Recently, however, data have been presented indicating that the striatonigral GABAergic pathway rather modulates nigral influences on postural and motor mechanisms mediated through

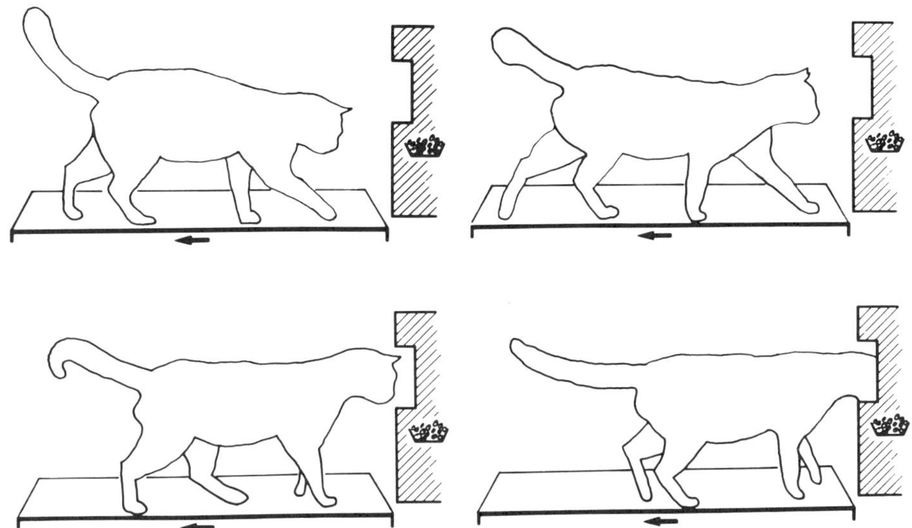

Fig. 1. Cat walking on the treadmill in front of a food dispenser. In order to get the food pellets, the cat changes the motor pattern from normal walking to accelerated walking, to walking in front of the dispenser with fast short steps of the forelimbs and slow, long steps of the hindlimbs and to bending the head through a hole in the dispenser.

non-dopaminergic nigral efferents which originate from the reticular part of SN (SNR) (DiChiara et al., 1977; Garcia-Munoz et al., 1977; Olianas et al., 1978a,b; Cools et al., 1983) and connect the striatum with thalamus, reticular formation, periaqueductal gray and the segmental pedunculo-pontine nuclei (Hopkins and Niessen, 1976; Rinvik et al., 1976; Kultas-Ilinsky et al., 1978; Ilinsky et al., 1982). It has been demonstrated that injections of GABA agonists (or antagonists) into the SNR of rats elicited motor symptoms largely resembling those induced by intrastriatal injection of dopamine agonists (or antagonists) (DiChiara et al., 1981; Kilpatrick et al., 1981; Reavill et al., 1981; Scheel-Kruger et al., 1981). These results led to the tempting assumption that GABA in the striatonigral pathway functions as mediator of behavioural effects induced by dopamine receptor stimulation (DiChiara et al., 1981; Kilpatrick et al., 1981; Reavill et al., 1981; Scheel-Kruger et al., 1981). Biochemical experiments fit in with the latter assumption: a decrease in dopaminergic activity of striatum is at least in part expressed via a decrease in GABAergic activity within the SNR (Gale and Casu, 1981).

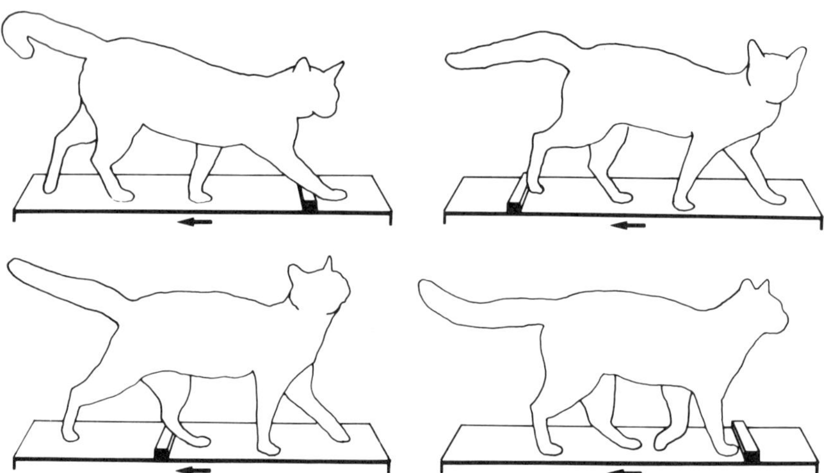

Fig. 2. Cat walking on the treadmill. The animal has to correct the step length to avoid touches of small obstacles which are fixed on the surface of the belt of the treadmill in a pseudo-random manner.

Given the insight that programme disorders of behaviour and motor deficits differ with respect to the degree of pathology of dopaminergic transmission within the striatum (Cools, 1980), the basic question arises, whether or not both disorders are mediated via the same neuroanatomical pathway, i.e. the GABAergic striatonigral projection. In order to test this hypothesis, we investigated the influence on motor programming and motor deficits of local block of GABAergic transmission with picrotoxin within the SNR. For this purpose, two test situations have been developed which allow the investigation of (1) the flexibility of sequencing and ordering different motor programmes and (2) the sensorimotor capacity necessary for a correct motor performance. In the first series of experiments, cats, that were trained to walk with a constant speed on a roofed treadmill, were given access to food pellets from a food dispenser which was attached to one end of the treadmill. In order to collect the food pellets the animals had to switch their motor pattern in a typical way (Figure 1): (1) Acceleration of walking to reach the front of the apparatus; (2) walking in front of the food-dispenser with fast, short steps of the forelimbs and slow, long steps of the hindlimbs; (3) bending the head through a hole in the opaque front of the treadmill-enclosure. In order to exclude that in this experimental set-up changes in the flexibility of sequencing the different patterns after intranigral injection of picrotoxin are due to an impairment in those sensorimotor capacities required for a correct motor performance, the motor behaviour of the animals was 'controlled' in a second series of experiments without a food dispenser. In this experimental set-up the animals had to avoid touching small obstacles, which were fixed on the surface of the belt of the treadmill in a pseudo-random manner (Figure 2).

METHODS AND RESULTS

Figure 3 illustrates the time scale of an observation session recorded by a remote controlled video system (above: treadmill with food dispenser; below: treadmill with obstacles). In each experiment the cats were allowed to get accustomed to the surroundings and the roofed treadmill for 10 minutes. Subsequently, the first 10-minute recording session consisted of 5 minutes during which the spontaneous motor behaviour of the animals was observed and another period of 5 minutes during which the animals' motor behaviour was recorded on the moving belt of the treadmill. Within these periods either the number of touched obstacles ('mistakes' in the obstacle test) or the number of switched motor programmes in order to reach the food dispenser (food-test) were counted and served as controls. After intranigral

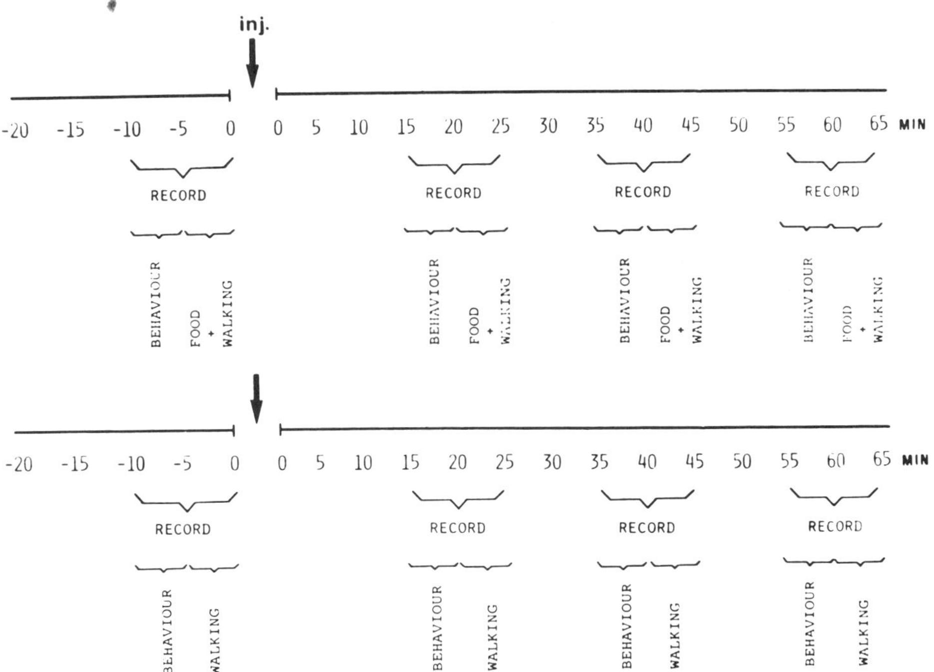

Fig. 3. Time scale of observation sessions recorded by a remote controlled video system.
Above: treadmill with food dispenser;
below: treadmill with obstacles.

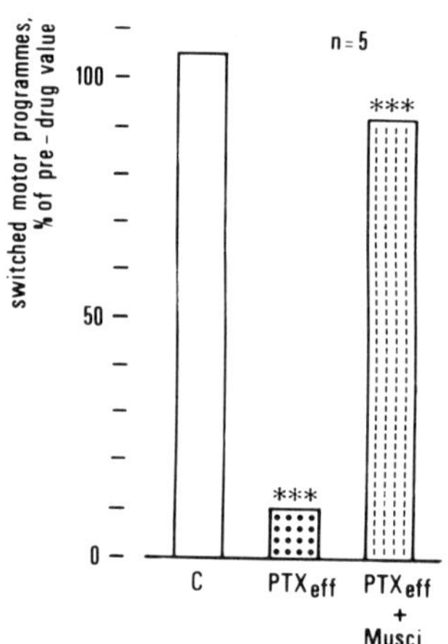

Fig. 4. Influence of intranigral injection of solvent (C), picrotoxin (PTX_{eff}) or picrotoxin + muscimol (PTX_{eff} + Musci) on the relative rate of switched motor programmes in comparison with the pre-injection control value (100%). PTX_{eff} is the smallest effective dose of picrotoxin (250-500 ng/0.5 µl). PTX_{eff} vs. C ($p < 0.001$) and PTX_{eff} + Musci vs. PTX_{eff} ($p < 0.001$); p-values from Student's t-test.

injection of solvent or drugs, recording sessions of 10 minutes were repeated 15, 35 and 55 mins after injection.

Following the unilateral injection of a 'small' dose of picrotoxin (500 ng dissolved in 0.5 µl distilled water; injection time 3.5 mins) the cats showed a characteristic spontaneous behaviour pattern: akinetic periods with body freezing, contralateral static head turning, and eye fixation. Usually cats sat on the treadmill in a totally frozen posture. At the initiation of the movement of the treadmill belt cats immediately jumped from the frozen sitting position and walked with constant speed in an undisturbed manner. However, the number of switched motor programmes after intranigral injection of picrotoxin was significantly lower in comparison with the pre-injection control values (9%; $p < 0.001$ Student's t-test) (Figure 4). This experiment showed that a low dose of picrotoxin within the SNR did prevent the arbitrary switching from one motor programme to another. The decreased amount of arbitrary switches may indicate that the cats have difficulties in sequencing motor programmes used to carry out two or more motor tasks simultaneously and/or successively. Careful observation revealed that after intranigral injection of picrotoxin, the number especially of those switches was reduced, that were not directed by observable external stimuli. The animals were, however, able to use external stimuli for switching. In that case they fixed their eyes on the food dispenser and changed the motor programmes.

Cats affected by picrotoxin walked to the food dispenser following a short pause for orientation after the treadmill stopped. This result may indicate that the cats 'remember' the pellets within the food dispenser and were able to combine and to transform this into motor programme and performance.

The picrotoxin effects in SNR, i.e. preventing switches of motor programmes, were antagonized by 200 ng muscimol dissolved in 1 µl water injected at the same site within the SNR through the same implanted cannula (Figure 4). This result makes it most likely that the effect of picrotoxin was caused by a specific block of GABAergic transmission within the SNR.

The question arises, whether only the switching from one motor programme to another is disturbed or if this is accompanied by deficits in motor performance. Therefore, we tested the capacity of walking animals (see Figure 3) to avoid touches of randomly occurring obstacles on the treadmill. Surprisingly, cats which received the same amount of picrotoxin injected into the SNR (500 ng/0.5 µl) had no difficulties in avoiding the obstacles. On the contrary, the number of touches (mistakes) was significantly lower than during the control period (57%; $p < 0.001$, Student's t-test). Additional analysis revealed that this is due to a closer attention of the cats to the obstacles (Figure 5).

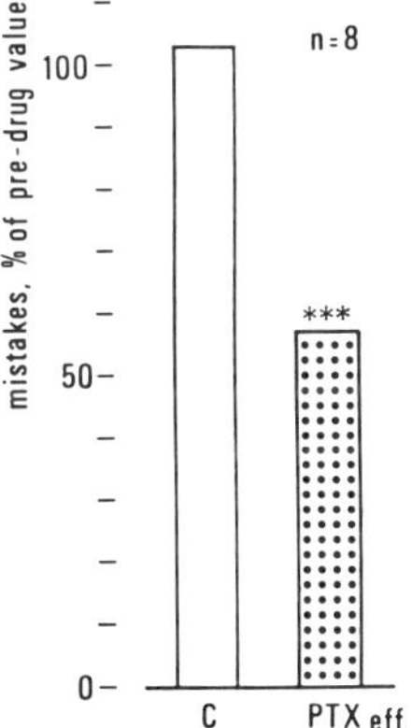

Fig. 5. Influence of intranigral injection of solvent (C) or picrotoxin (PTX_{eff}) on the relative number of touched obstacles (mistakes) in comparison with the pre-injection control value (100%). PTX_{eff} see Fig. 4. PTX_{eff} vs. C ($p < 0.001$); p-value from Student's t-test.

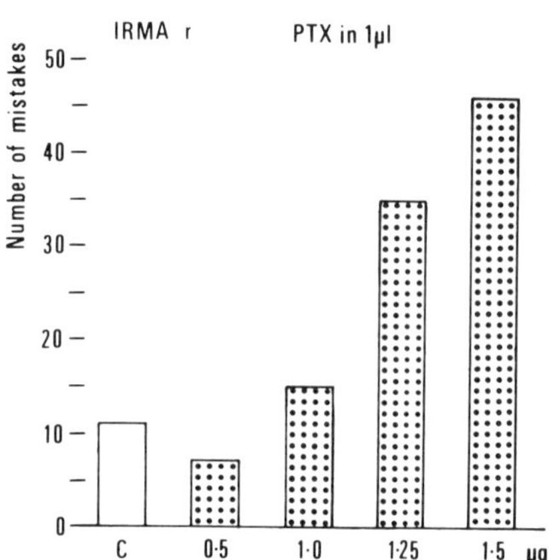

Fig. 6. Influence of intranigral injection of solvent (C) or different doses of picrotoxin on the number of touched obstacles (mistakes).

The influence of increasing picrotoxin doses on motor performance in the random-obstacle test is illustrated in Figure 6 for a typical experiment. In this cat picrotoxin elicited dose-dependently difficulties in avoiding the obstacles. The normally light and relaxed movements during walking, which were also present in the vast majority of cats treated with 500 ng picrotoxin, became stiffer with increasing doses: the flexion phase in the hindlegs was disturbed and shortened. Flexion movements of both hind- and forelegs evoked the impression of choppy walking. To obtain precise information about the performance of these movements, gait analysis with a high speed camera or e.m.g. studies are required. Cats which had received higher doses of picrotoxin froze the ongoing movement instantaneously when the treadmill stopped and persisted in that posture for up to 30 secs. However, if the treadmill was started up again, animals were still able to perform all movements necessary for walking or trotting, even if the speed of the treadmill had been changed. The cats did not lose their position in the middle of the roofed treadmill. These results show that animals which received a 'higher' dose of intranigrally injected picrotoxin were able to maintain their posture and position on the running treadmill when the external locomotion stimulus was present. On the other hand, the animals were immobile without such a stimulus.

It is known that after injection of higher doses of picrotoxin (2 μg) into the SNR cats have difficulties bending their hindlegs to climb onto a wooden bar placed 2 meters above the floor, after the experimenter hangs them there by the forepaws (Wolfarth et al., 1981; Cools et al., 1983). The inability of animals to perform correct

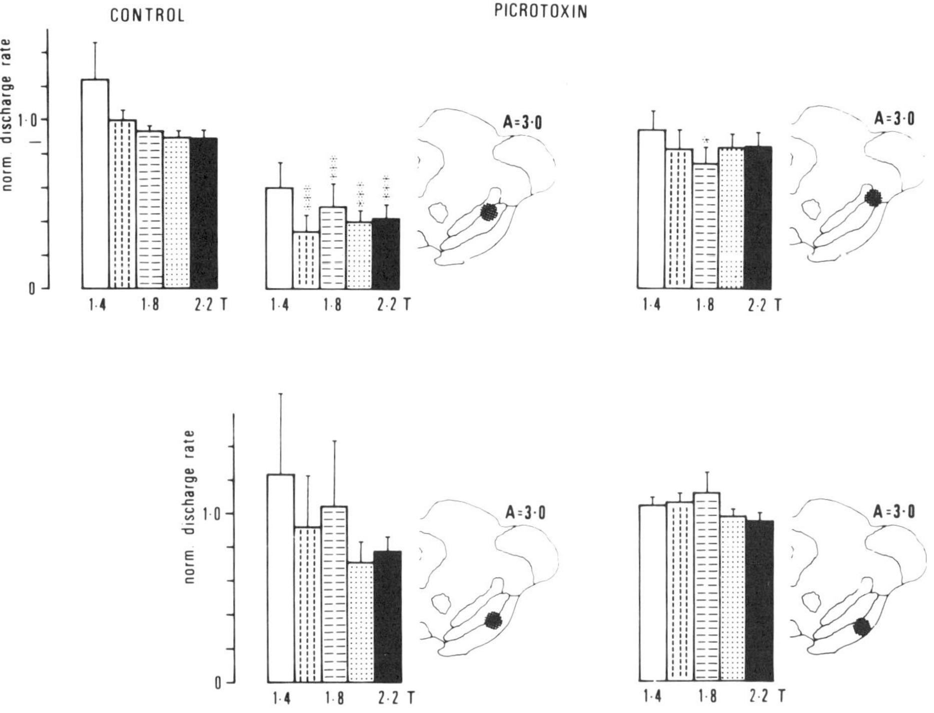

Fig. 7. Influence of intracerebral injections of solvent (control) or picrotoxin (2 µg/1 µl) on the reflex discharge rate of single pretibial flexor alpha-motoneurones in anaesthetized and laminectomized cats.
Abscissa: stimulation strength (T); T = threshold of afferent fibres of homonymous and synergistic nerves; ordinate: normalized discharge rate in comparison with pre-injection value; * $p < 0.05$, *** $p < 0.001$ vs. control; p-values from Student's t-test; insets: dotted areas represent injection sites.

flexion movements of the hindlegs suggested a decreased excitability of skeletomotor (alpha) and/or fusimotor (gamma) motoneurones of pretibial flexor muscles. The influence of intranigral injections of 2 µg/1 µl picrotoxin on the excitability of alpha-motoneurones was investigated in ketamine anaesthetized cats. The depth of anaesthesia was carefully kept constant by controlling blood pressure, rate of breathing, pinna reflex and threshold to painful stimuli. After laminectomy, single axons of motoneurones of M. tibialis anterior (TA) and M. extensor digitorum longus (EDL) were isolated and identified in ventral roots. The discharge rate of alpha-motoneurones was recorded during electrical stimulation of the proximal stump of the cut homonymous and synergistic nerve (frequency 100 mp/s; pulse width 0.2 ms, stimulation period 2s, strength 1.4 - 2.2 threshold of the afferent fibers).

Injections of picrotoxin confined to the posterolateral SNR (A = 2.5-3.5; L = 5.5; D = -4.0) significantly reduced the reflex-induced discharge rate of pretibial alpha-motoneurones (Figure 7). The decrease in discharge was observed with stimulus strengths either selectively exciting group I afferents or exciting both group I and II afferents. Injections lateral or ventral to the SNR within the cerebral peduncle, however, failed to induce significant decreases of discharge rate of flexor alpha-motoneurones. On the other hand small differences in the anterior-posterior position did not prevent the picrotoxin-induced effect.

In another set of experiments single spindle primary endings of pretibial flexor muscles TA and EDL were isolated in dorsal root filaments. Discharge of primary afferents and muscle length were monitored during sinusoidal stretching of the receptor-bearing muscles at 1 Hz (peak-to-peak amplitude 100 μm and 2 mm).

A digital computer averaged the response of the afferents to a number of cycles of sinusoidal stretching and displayed a 'cycle histogram' (Goodwin et al., 1975). The histogram was fitted by the method of least squares with a sine curve and its parameters wre printed out (center frequency (CF), modulation depth (MD), phase). Changes in fusimotor action after intranigral injection of picrotoxin were deduced from changes in spindle sensitivity (imps^{-1} mm^{-1}, defined as modulation depth (MD) per stretch amplitude), when compared with the de-efferented preparation, i.e. after having cut the ventral roots.

In the ketamine anaesthetized cat there is a tonic fusimotor action on pretibial flexor muscle spindle primary endings as indicated by the high center frequency of the fitted sinusoid and the low dynamic sensitivity (MD/stretch amplitude), when compared with the de-efferented preparation (Fig. 8). Provided that stimulus parameters were carefully kept constant, spindle sensitivity to sinusoidal stretching remained largely unaffected (changes $\pm$ 10%) throughout an adequate time frame after intranigral injection of solvent. Block of GABAergic transmission by injection of picrotoxin (2 μg/1 μl) into the SNR, however, increased the modulation depth to sinusoidal stretching of the primary sensory endings of flexor muscle spindles, so as largely to resemble the sensitivity of the spindle in the de-efferented preparation. The increase of spindle sensitivity after picrotoxin was observed in twenty out of twenty-six primary endings and could occur bilaterally. As an increase in modulation depth was found during both small- and large-amplitude stretching, it may be deduced that intranigral injection of picrotoxin increased the spindle sensitivity by <u>reducing</u> a previously present static tonic

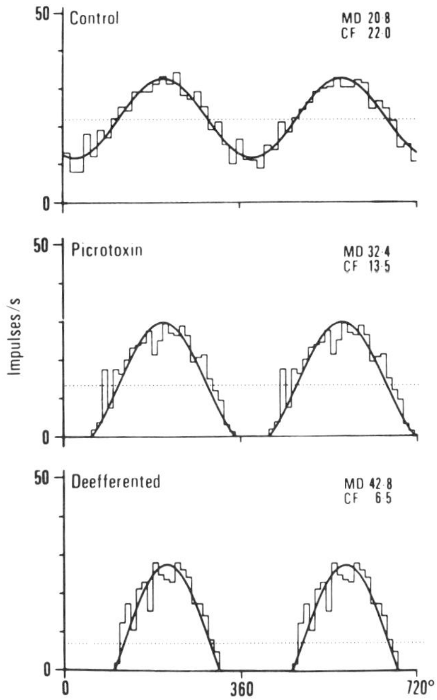

Fig. 8. Histogram showing averaged responses of a TA muscle spindle primary ending to small amplitude (100 μm), low frequency ($1\ s^{-1}$) sinusoidal stretching. Firing probability before (control) and after intranigral injection of picrotoxin (2 μg/1 μl) and after de-efferentation. 270° on the abscissa corresponds to the point of maximum extension of the muscle. Inset values give center frequency (CF; indicated by dotted lines) and modulation depth (MD) of the fitted sinusoid.

fusimotor action on the primary endings. A combination of decreased static and increased dynamic fusimotor action on the sensory endings due to the manipulation of neurotransmission within the SNR can be largely excluded by the fact that the spindle sensitivity either remained unchanged or further increased after de-efferentation, i.e. after removal of all fusimotor drive from the spindles. Histological examination revealed that all effective intranigral injection sites were located at the target area (A 2.0-3.5; L 4.5-6.0; D 4.0-5.5). This region was identical with the effective area in the other tests reported earlier.

DISCUSSION

Small impairments of GABAergic transmission within the SNR, by injections of low doses of picrotoxin, elicited difficulties in switching motor programmes required to get food during walking on the treadmill. These difficulties in switching were not due to disturbances of sensorimotoric capacities necessary for a correct motor performance, as shown in separate experiments, since the animals had no difficulty in changing the walking programme in order to avoid touching obstacles on the belt of the treadmill. Therefore, the experimental set-up presented here might be appropriate to distinguish between disturbances in motor programming and motor deficits.

Higher doses of picrotoxin, however, caused difficulties in the obstacle test and led to impairments of flexion movements of the hindlimbs. These experiments indicated that the presence of motor deficits reflects a greater dysfunction of GABAergic transmission within the SNR than the presence of abnormalities in programming motor behaviour. Perhaps this result taken together with comparable results concerning the striatum (Cools, 1980; Cools et al., 1984a) will open new perspectives for the development of motor tests appropriate to diagnose patients with disorders of basal ganglia in an early stage of Parkinson's disease.

The similarity of effects on motor programming elicited by impaired intranigral GABAergic transmission to those elicited by impaired intrastriatal dopaminergic transmission, supports the hypothesis that GABA in the striatonigral pathway functions as mediator of behavioural effects induced by dopamine receptor stimulation in the striatum (DiChiara et al., 1981; Kilpatrick et al., 1981; Reavill et al., 1981; Scheel-Krüger et al., 1981). These results, together with the observations of Scheel-Krüger, lead to the suggestion that the dopamine hypothesis about disturbances in the basal ganglia system for Parkinson's disease should be extended to include a dysfunction in the efferent GABAergic pathways beyond the striatum (Scheel-Krüger, 1984).

However, the dysfunctions of motor programming after impairment of GABAergic nigral transmission are not identical to those elicited by an impairment of intrastriatal dopaminergic neurotransmission. Inhibition of striatal dopamine receptors by local injections of haloperidol inhibited the self-generated switching of motor programmes, but did not inhibit the process of switching as long as external stimuli could be used (Cools, 1980; Cools et al., 1984a). Inhibition of nigral GABA receptors by local injections of picrotoxin likewise reduced the number of self-generated switches, but the animals' inability to switch spontaneously was only in part compensated by the use of external stimuli. These results might support observations made on cats in open field tests, suggesting that the substantia nigra is an output station of the striatum, which not only transmits, but also transforms incoming signals for its behavioural expression (Cools et al., 1983; Cools et al., 1984a). The results of the obstacle test demonstrated that intranigral application of low doses of picrotoxin did not prevent the use of external stimuli. In this test situation the animals were forced to follow external stimuli.

Single cell recordings of neurons of the SNR in the awake cat revealed that these neurons process sensorimotor information (Schwarz et al., 1984). SNR cells responded with short latency to innnocuous stimulation of the skin, passive manipulation of limbs and to moving visual stimuli. Furthermore, these neurons showed modulations in discharge during treadmill walking, phase-related to the step cycle. Disturbances of such movements by touching the obstacles on the treadmill, including the resulting compensation, were also transmitted with short latency to SNR neurones. A fraction of these cells disharged almost exclusively prior to and during self-generated movements. The most powerful modulation in firing rate was observed whenever an animal tried to overcome an external impediment. These results led to the suggestion that SNR neurones process multimodal sensory information and participate in setting appropriate gains on spinal motor systems to adequately deal with changing motor requirements (Schwarz et al., 1984).

The results on ketamine anaesthetized cats presented here demonstrate that manipulation of GABAergic transmission within the SNR with high doses of picrotoxin does affect spinal motor mechanisms. The picrotoxin-induced decrease of excitability of alpha-motoneurones, which is tested by stimilation of their homonymous or synergistic excitatory afferents from the periphery, indicates that the peripheral reflex loop is affected. This is particularly accentuated by the depressant effect of intranigral injections of picrotoxin on static gamma-motoneurones which innervate the muscle spindle. Static gamma-motoneurones are more potent in activation of homonymous and synergistic alpha-motoneurones via the muscle spindle loop than are dynamic gamma-motoneurones (Hulliger, 1981). Therefore, it can be concluded that the additive effect of both the decreased reflex-evoked discharge rate of pretibial flexor alpha-motoneurones and the diminished excitatory potency of muscle spindle afferents, due to a decreased activity of static gamma-motoneurones, lead to a deficit in active and reflex induced innervation of these flexor muscles. Thus, the SNR may be a variable 'gain regulator' at motoneuronal level in order to obtain an optimal resolution in force generation. Deficits in function of the SN have indeed been shown to be associated with deficits in grading of voluntary force. It was demonstrated in e.m.g. studies that the difficulty experienced by Parkinsonian patients with fast movements is due to an inability to adjust the amplitude of the first burst of alpha-motoneuron activation initiating a fast movement (Hallet and Koshbin, 1980). In this context the decreased excitability of pretibial flexor alpha-motoneurones and static motoneurones presented here is of particular interest, since walking movements are generally initiated by activation of flexor muscles in the legs, which is an essential prerequisite for locomotion.

REFERENCES

Anden, N. E. and Stock, G., 1973, Inhibitory effect of gamma hydroxybutyric acid and gamma aminobutyric acid on the dopamine cells in the substantia nigra, Naunyn-Schmiedeberg's Arch. Pharmacol., 279:89.

Cools, A. R., 1980, Role of the neostriatal dopaminergic activity in sequencing and selecting behavioural strategies: facilitation of processes involved in selecting the best strategy in a stressful situation, Beh. Brain Res., 1:361.

Cools, A. R., van den Bercken, J., Horstink, M., van Spaendonck, K., and Berger, H., 1981, The basal ganglia and the programming of behaviour, TINS, 124.

Cools, A. R., Jaspers, R., Kolasiewicz, W., Sontag, K. -H., and Wolfarth, S., 1983, Substantia nigra as a station that not only transmits, but also transforms, incoming signals for its behavioural expression: striatal dopamine and GABAmediated responses of pars reticulata neurons, Beh. Brain Res., 7:39.

Cools, A. R., Jaspers, R., Schwarz, M., Sontag, K.-H., Vrijmoed-de Vries, M. and van Hoof, J., 1984a, Caudate nucleus and disorders in switching motor programmes, Adv. Beh. Biol., (in press).

Cools, A. R., van den Bercken, J., Horstink, M., van Spaendonck, K., and Berger, H., 1984b, Cognitive and motor shifting aptitude disorders in Parkinson's disease, J. Neurol. Neurosurg. Psychiatry, (in press).

DeLong, M. R., and Georgopoulos, A. P., 1981, Motor functions of the basal ganglia, in: "Handbook of Physiology," Section 1, Vol. II, V. B. Brooks, (ed)., Bethesda, Maryland.

DiChiara, G., Porceddu, M. L., Imperato, A., and Morelli, M., 1981, Role of GABA neurones in the expression of striatal motor functions, in: "GABA and the Basal Ganglia". Advances in Biochem. Psychopharmacol., Vol. 30, G. DiChiara, and G.I. Gessa, eds., Raven Press, New York.

DiChiara, G.., Olianas, M., Del Fiacco, M., Spano P. F., and Tagliamonte, A., 1977, Intranigral kainic acid is evidence that nigral non-dopaminergic neurons control posture, Nature (London), 268:743.

Divac, I., 1977, Does the neostriatum operate as a functional entity? in: "Psychobiology of the Striatum," A. R. Cools, A. H. M. Lohmann, and J. H. L. van den Bercken, eds., Elsevier, Amsterdam.

Dray, A., 1979, The striatum and substantia nigra: a commentary on their relationship, Neurosci., 4:1407.
Flowers, K. A., 1976, Visual 'closed' loop and 'open' loop characteristics of voluntary movement in patients with parkinsonism and intention tremor, Brain, 99:269.
Flowers, K. A., 1978, Lack of prediction in the motor behaviour of parkinsonism, Brain, 101:35.
Gale, K., and Casu, M., 1981, Dynamic utilization of GABA in substantia nigra: regulation by dopamine and GABA in the striatum, and its clinical and behavioural implications, Mol. Cell. Biochem., 39:369.
Garcia-Munoz, M. Nicolaou, N. M., Tulloch, I. F., Wright, A. K., and Arbuthnott, G. W., 1977, Feedback loop or output pathway in striato-nigral fibers, Nature (London), 265:363.
Goodwin, G. M., Hulliger, M., and Matthews, P. B. C., 1975, the effects of fusimotor stimulation during small amplitude stretching on the frequency-response of the primary ending of the mammalian muscle spindle, J. Physiol., 253:175.
Hallett, M., and Koshbin, S., 1980, A physiological mechanism of bradykinesia, Brain, 103:301.
Hopkins, D. A., and Niessen, L. W., 1976, Substantia nigra projections to the reticular formation, superior colliculus and central gray in the rat, cat and monkey, Neurosci. Lett., 2:253.
Horne, D. J., 1973, Sensorimotor control in parkinsonism, J. Neurol. Neurosurg. Psychiatry, 36: 742.
Hornykiewicz, O., 1963, Die topische Lokalisation und das Verhalten von Noradrenalin und Dopamin (3-Hydroxytyramin) in der Substantia nigra des normalen und parkinsonkranken Menschen, Wien Klin. Wschr., 75:309.
Hulliger, M., 1981, Muscle spindle afferent units. Functional properties with possible significance in spasticity, in: "Therapie der Spastik," H. J. Bauer, W. P. Koella, and A. Struppler, eds., Verlag für angewandte Wissenschaften, Munchen.
Ilinsky, J. A. Kultas-Ilinsky, K., and Smith, K. R., 1982, Organization of basal ganglia inputs to the thalamus. A light and electron microscopic study in the cat, Appl. Neurophysiol., 45:230.
Iversen, S. D., 1977, Brain dopamine systems and behaviour, in: "Handbook of Psychopharmacology," Vol. 8, L. L. Iversen, S. D. Iversen, and S. H. Snyder, eds., Plenum, New York.
Jung, R.,, and Hassler, R., 1960,, The extrapyramidal motor system, in: "Handbook of Physiology," Sect. 1, Vol. II, J. Field, H. W. Magoun and V. E. Hall, eds., Am. Physiol. Soc., Washington, D.C.
Kilpatrick, J. C., Starr, M. S., James, T. A., and MacLeod, N. K., 1981, Evidence for the involvement of nigrothalamic GABAneurones in circling behaviour in the rat, in: "GABA and

the Basal Ganglia." Advances in Biochem. Psychopharmacol., Vol. 30, G. DiChiara, and G. L. Gessa, eds., Raven Press, New York.

Kultas-Ilinsky, K. Ilinsky, J. A., Massopust, L. C., Young, P. A., and Smith, K. B., 1978, Nigrothalamic pathway in the cat demonstrated by autoradiography and electron microscopy, Expl. Brain Res., 33:481.

Marsden, C. D., 1980, The enigma of the basal ganglia and movement, TINS, 3:284.

Oberg, R. G. E., and Divac, I., 1979, 'Cognitive' functions of the neostriatum, in: "The Neostriatum," I. Divac and R. G. E. Oberg, eds., Pergamon, Oxford.

Olianas, M. C., DeMontis, G. M., Concu, A., Tagliamonte, A., and DiChiara, G., 1978a, Intranigral kainic acid: Evidence for nigral non-dopaminergic neurons controlling posture and behaviour in a manner opposite to the dopaminergic ones, Europ. J. Pharmacol., 49:223.

Olianas, M. C., DeMontis, G. M., Mulas, G., and Tagliamonte, A., 1978b, The striatal dopaminergic function is mediated by the inhibition of a nigral, non-dopaminergic neuronal system via strionigral GABAergic pathway, Europ. J. Pharmacol., 49:233.

Reavill, C., Leigh,, N., Jenner, P., and Marsden, C. P., 1981, Critical role of midbrain reticular formation in the expression of dopamine-mediated circling behaviour, in: "GABA and the Basal Ganglia". Advances in Biochem. Psychopharmacol., G. DiChiara, and G. L. Gessa, eds., Raven Press, New York.

Rinvik, E., Grofova, J., and Ottersen, O. P., 1976, Demonstration of nigrotectal and nigroreticular projections in the cat by axonal transport of proteins, Brain Res., 112:388.

Scheel-Krüger, J., 1984, The GABA receptor and animal behaviour, in: "GABA Receptors," S. J. Enna, ed., Humana Press, Clifton, New Jersey, (in press).

Scheel-Krüger, J., Magelund, G., and Olianas, M. C., 1981, Role of GABA in the striatal output system globus pallidus, nucleus entopeduncularis, substantia nigra and nucleus subthalamicus, in: "GABA and the Basal Ganglia." Advances in Biochem. Psychopharmacol, Vol. 30, G. DiChiara, and G. L. Gessa, eds., Raven Press, New York.

Schultz, W., 1982, Depletion of dopamine in the striatum as an experimental model of Parkinsonism: direct effects and adaptive mechanisms, Progr. Neurobiol., 18: 121.

Schwab, R. S., Chafetz, M. E., and Walker, S., 1954, Control of two simultaneous voluntary acts in normal and in parkinsonism, Arch. Neurol. Psychiatry, 72:591.

Schwarz, M., Sontag, K.-H. and Wand, P., 1984, Sensory-motor processing in substantia nigra pars reticulata in conscious cats, J. Physiol. (Lond)., 347, (in press).

Stark, L., 1968, Neurological control systems - studies in bioengineering, Plenum, New York.

Wolfarth, S., Kolasiewicz, W., and Sontag, K.-H., 1981, The effects of muscimol and picrotoxin injections into the cat substantia nigra, Naunyn-Schiemedebergs, Arch. Pharmacol., 317.

BASAL GANGLIA AND SWITCHING MOTOR PROGRAMS

A. R. Cools, R. Jaspers, M. Schwarz*, K. H. Sontag*,
M. Vrijmoed-de Vries and J. van den Bercken.

Department of Pharmacology
University of Nijmegen
P. O. Box 9101, 6500 HB Nijmegen
The Netherlands

and

Max Planck Institut für Experimentelle Medizin
Göttingen, F.R.G.*

1. INTRODUCTION

Recently we have introduced a new frame of reference in the search for "rules of order" in the cerebral organization of behavior (Cools and van den Bercken, 1977; Cools, 1981). The cornerstone of this frame is Powers' definition of behavior: behavior is the control of the input of the organism. By this definition, behavior is conceived of as a process by which the organization inside the organism controls its input; the brain is thereby seen as a hierarchy of negative feedback systems controlling this input (Powers, 1973). It is this concept that allows us to describe how information sent downstream in the hierarchy is successively transformed at each level by adding more and more details about the behavior to be executed.

<u>The purpose of this article is to discuss how information that is sent to the caudate nucleus is successively transformed at the level of the caudate nucleus, substantia nigra (pars reticulata), and colliculus superior (deeper layers) by adding more and more details about the motor behavior to be executed.</u>

First, the basic terminology intrinsic to Powers' concept is recalled in section 2. In the next section some salient aspects of describing motor behavior in terms of movement coordinate systems are summarized. In section 4 the function of the caudate nucleus is discussed. We attempt to show that this brain structure reduces the degree of freedom in programming motor behavior, by adding information about the serial ordering of motor behavior. In section 5 the function of the substantia nigra, pars reticulata is discussed. We attempt to show that this structure, which receives information from the caudate nucleus via striato-nigral, GABA-ergic fibers, reduces the degree of freedom in programming motor behavior by adding information derived from proprioceptive stimuli. In section 6 the function of the superior colliculus, especially its deeper layers, is discussed. We attempt to show that this structure, which receives information from the substantia nigra via the nigro-collicular GABA-ergic fibers, once more reduces the degree of freedom in programming motor behavior by adding information derived from exteroceptive stimuli. The overall impact of the transformation which goes downstream in the hierarchy is discussed in the final section.

2. KEY POINTS OF POWERS' CONCEPT

By definition, the organization inside the organism (brain) is an integration of feedback systems allowing us to operate according to principles and definitions of servomechanism theory. Thus the organism receives information on its current state, i.e. "input signals"; compares it with information on the "desired" state, i.e. "reference signals"; produces "error signals" representing the difference between current and desired states; and as a result processes "output signals" directing motor behavior. In principle, each level in the hierarchy deals with input, reference, error and output signals (for details: Powers, 1973).

By definition, only inputs of the lowest-order systems within the hierarchy are signals emanating from the physical environment (exteroceptive stimuli), interior of the body (interoceptive stimuli), and muscles, tendon organs and joints (proprioceptive stimuli). In contrast, all input signals of higher-order systems are analogues of quantities derived from input signals of lower-order systems. Accordingly the degree of abstraction from the observable physical effects increases at each higher-order level in the hierarchy. This lays the foundation for getting from "distal" to "proximal" stimuli. Since input signals reaching the lowest-order systems are also transformed into input signals of higher-order systems, the organism has ultimately at its disposal the weighted sum of all input signals, i.e. signals received by the highest-order systems. As the latter signals are derived from "distal" stimuli, the resulting "proximal" stimuli are still analogues of the organism's "world" and, accordingly, represent the integrated whole of all aspects of this world at the highest-order level. Since such signals are abstract, invariant functions constructed by the lower-order systems themselves, they cannot be observed in any one of the incoming stimuli.

By definition, the reference signals in a feedback system are the controlled quantities of the system. When conceiving of behavior as a process by which an integrated whole of hierarchically ordered feedback systems (brain) controls its input, it follows that the reference signals for the lowest-order systems are determined by the output signals of higher-order systems. By the same token it follows that output signals of hierarchically higher-order systems are reference signals for hierarchically lower-order systems. Only the lowest-order output signals are, by some output function, directly transformed into motor behavior. In this way one gets from "programs" to motor behavior, with the restriction that "program" is defined as a nested set of rules reducing the degree of freedom in programming motor behavior. As only the lowest-order output signals direct motor behavior, in consequence of a particular interaction between input signals and internal organization of the organism, it is proposed to label these signals "motor commands", in contrast to output signals of higher-order systems reaching the lowest order systems, which we propose to label "motor programming signals".

When one is dealing with a hierarchy of feedback control systems it is clear that information available for directing motor commands is minimal at the highest-order level, which simply contains reference signals for lower-order systems. In such a hierarchy it is evident that the information going downstream carries more and more details about the motor behavior to be executed. In other words, the information available for directing motor commands increases at each lower level in the hierarchy, and reaches its maximum value at the lowest-order level in the hierarchy. The reverse holds true for the degree of freedom in programming motor behavior. This degree of freedom is maximal at the level of the highest-order systems, reduces at each lower-order level, and ultimately becomes zero at the lowest-order level. This has great impact for the programming of ongoing motor behavior: information going downstream can be continuously updated according to changes occurring in the input signals of levels that have not yet been set by their incoming reference signals.

3. MOVEMENT COORDINATE SYSTEMS

Since classical descriptions of motor behavior ignore many relevant dimensions in the organization of what is actually very complex behavior, it is appropriate to recall that there are in principle three possible sources for the patterning of complex movements: exteroceptive stimuli for "exteroceptively directed" movements, proprioceptive stimuli for "proprioceptively directed movements", and coordinating mechanisms within the brain. Although the latter mechanisms receive information derived from exteroceptive and proprioceptive stimuli, they produce movements that are not

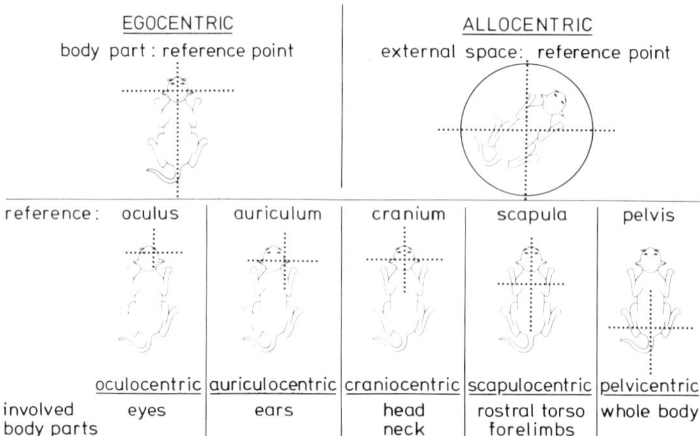

Fig. 1. Movement coordinate systems: prerequisites for describing movements. In the upper left the axes of the egocentric coordinate system are given: in principle the body (or a part) serves as point of reference. In the lower part several egocentric coordinate subsystems are depicted: the subsystems are classified according to the particular body part serving as point of reference.
In the upper right the allocentric coordinate system is illustrated: in principle the frame of reference is prescribed by the immediate surroundings of the animal. The dotted lines are allocentric coordinates which deviate from the axes of the egocentric system.
Note: in contrast to proprioceptively directed movements which can be only described by reference to the egocentric coordinate system, exteroceptively directed movements can be described by reference to either the egocentric system or the allocentric system (see section 3).

stimulus-directed. Since the absence only of exteroceptive, not of proprioceptive or interoceptive stimuli, can be registered, it is proposed to label a class of "non-exteroceptively directed movements". In the case of conditioning we must include here conditioned stimuli directing a whole chain of movements: in that case we are dealing with "triggered movements".

In principle there are two coordinate systems for describing movements: the "egocentric coordinate system" and the "allocentric coordinate" system (Fig. 1).

The egocentric coordinate system uses one or another part of the body as point of reference. To operate with an egocentric coordinate

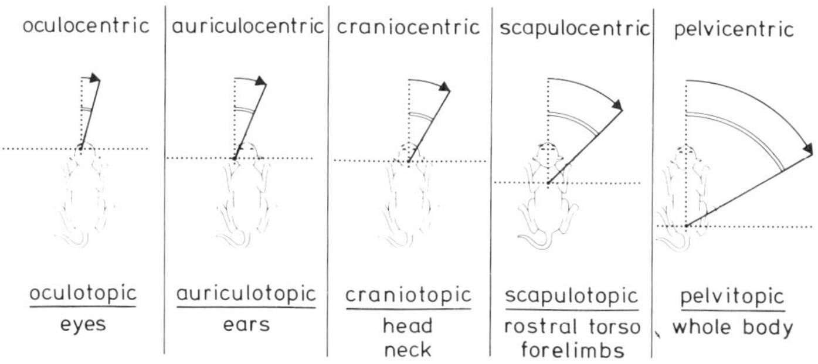

Fig. 2. Propriotopic movements are movements directed towards a point of which the egocentric coordinates are prescribed by a propriotopic code, i.e. a code derived from proprioceptive stimuli. In principle this code varies according to the egocentric coordinate subsystem involved from oculotopic to pelvitopic. Characteristic features are indicated by an asterisk.

system implies that the point of departure for describing movements is defined by reference to the axes of the egocentric coordinate system. Thus, the fully symmetric position serves as point of departure for spontaneously occurring movements. Proprioceptively directed movements can be described solely in terms of this system: they are directed towards a point for which the egocentric coordinates are fixed (Fig. 2). Apparently the organism can produce a code enabling it to switch to movements directed by proprioceptive input signals. For the time being it is proposed to label this code "propriotopic", i.e. a code prescribing egocentric coordinates of movements in terms of abstract, invariant functions which are normally constructed from proprioceptive stimuli. By definition, "propriotopic movements" are thus prescribed by such a propriotopic code. In principle this code may vary, according to the egocentric subsystem involved, from oculotopic to pelvitopic (Fig. 2).

The allocentric coordinate system is a system in which the frame of reference is prescribed by the immediate surroundings of the organism (Fig. 1). Use of an allocentric coordinate system implies that the axes are instantaneously determined by the position taken by the organism in space: the coordinates of its axes are identical to those of the axes of the egocentric system. As soon as the organism starts to approach a spatial point with coordinates deviating from the egocentric axes, the allocentric coordinates of the spatial point have to be transformed into egocentric coordinates directing the

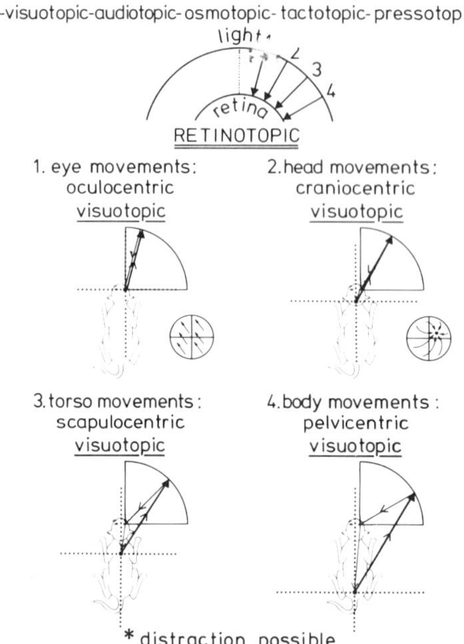

Fig. 3. Exterotopic movements are movements directed towards a spatial point whose egocentric coordinates are prescribed by an exterotopic code, i.e. a code prescribing how allocentric coordinates of the spatial point have to be transformed to egocentric coordinates, to be executed in order to reach that point. The code itself is derived from exteroceptive stimuli and varies according to the involved egocentric coordinate subsystem from visuotopic to pressotopic. This figure shows four different codes, each of them derived from light illuminating the retina in four different ways (1, 2, 3 and 4). The main attribute of spontaneously occurring exterotopic movements is indicated by an asterisk.

movements required. For the time being it is proposed to label the information controlling this process of transformation the so-called "exterotopic code", i.e. a code prescribing how allocentric coordinates of a spatial point have to be transformed into egocentric coordinates of movements to be executed in order to reach that point (Fig. 3). It will be evident that the resulting egocentric coordinates of the spatial point will vary according to ongoing changes in the distance between the point of departure and the point to be reached. In this respect exteroceptively directed movements differ from propriotopic movements which are directed at points with fixed, egocentric coordinates. The exteroceptively directed movements under discussion are thus "exterotopic movements", i.e.

movements directed towards a spatial point whose egocentric coordinates are prescribed by an exterotopic code. The code itself is derived from exteroceptive stimuli and may vary according to the allocentric coordinate subsystem involved, from visuotopic to pressotopic (Fig. 3).

4. CAUDATE NUCLEUS AND SWITCHING MOTOR PROGRAMS

Recently evidence has been found that the basal ganglia, especially the caudate nucleus, enable the organism to rearrange arbitrarily the serial ordering of non-exteroceptively directed behavior (van den Bercken and Cools, 1982; Cools, 1980, 1981a,b). Increased striatal activity dissociates behavior programs, making possible a rearrangement of the resulting elements: it improves the organism's ability to alter non-exteroceptively directed behavior. Decreased striatal activity prevents such dissociation and, consequently, impedes rearrangement; still, the organism remains able to shift exteroceptively and proprioceptively directed behavior.

Furthermore, it has been found that behavioral consequences of changes in striatal activity can be manifest at different levels in the hierarchy of behavioral organization. In man with hypofunctioning basal ganglia, for instance, such consequences manifest themselves at the cognitive and motor levels (Cools et al., 1984).

Finally, pathological changes in the composition of behavior programs at hierarchically lower levels have been found to require greater change in striatal activity than analogous effects at hierarchically higher levels. For instance, in Java monkeys, a slight dysfunction of the caudate nucleus results solely in a changed capacity to switch non-exteroceptively directed social interactions, whereas a more severe caudate dysfunction results in pure motor deficits such as tremor (Vrijmoed-de Vries and Cools, 1983). Analogously, a slight dysfunction of the neostriatum in rats results solely in changed capacity to switch non-exteroceptively directed behavior strategies, whereas a more severe hypofunctioning produces motor disturbances such as abduction of the hindlimbs (Cools, 1980).

Against this background we have advanced three questions:

1. Does the above-mentioned function of the caudate nucleus in programming the serial order of non-exteroceptively directed behavior also manifest itself at the level of motor behavior?
2. Does dysfunction of the caudate nucleus have consequences for brain structures such as the substantia nigra pars reticulata and the superior colliculus, brain structures intercalated between the caudate nucleus and the spinal cord?
3. Does knowledge about the function of the caudate nucleus offer perspectives for increasing our insight into Parkinson's disease?

Only the first question will be discussed in this section, because the second and third can be dealt with only after having considered the function of the substantia nigra pars reticulata (section 5) and superior colliculus (section 6).

To answer the first question it became necessary to develop a design allowing the animal to arbitrarily switch the serial order of non-exteroceptively directed movements. As illustrated below the so-called treadmill test for cats provides such a situation. To produce a specific and selective dysfunction of the caudate nucleus in cats we lowered the dopaminergic activity within the rostromedial part of the head of the caudate nucleus, by means of bilateral, intracaudate injections of the dopaminergic antagonist haloperidol (12.5 ug/5.0 ul). Drug concentration, volume and injection sites were selected on the basis of the fact that such injections selectively inhibit the transmission of information from nigro-striatal, dopaminergic fibers by blocking their postsynaptic receptors; the effects are specific for the intracaudate region under discussion, are dopamine-specific, and are elicited at the postsynaptic level because they are mimicked by procaine, i.e. an anesthetic that suppresses neuronal activity at the pre- and postsynaptic level (Cools, 1971; Cools et al., 1976). As the behavioural consequences of a selectively disturbed brain structure are ultimtely the result of changing the information leaving the affected structure, the drug-induced changes could be used to establish the role of the caudate nucleus in programming motor behavior. For details about the procedure used for injecting drugs into the caudate nucleus of freely moving cats the reader is referred to previously published articles (Cools et al., 1976). Only experiments in which the injection sites were restricted to the head of the caudate nucleus, rostromedial part, are discussed below (see Fig. 1 in Cools et al., 1976).

Treadmill Test

Cats - walking on a treadmill - are continuously free to alter their ongoing motor behavior by approaching the front panel of the treadmill, pushing the head through a small opening in this panel, and collecting food pellets by bending the head towards a hidden food dispenser; no auditory, visual or olfactory stimuli signalling the presence of food behind the panel are present. Normally, cats trained to walk on the treadmill and fasted for a minimum period of 24 hr are able to collect food by switching from walking to movements allowing them to get the pellets.

Before considering the drug-induced effects it is useful to point out that this design allows us to describe behavioral consequences in relation to input signals derived from available stimuli (cf. section 3). First, the cats are permanently forced to adjust their position and posture during walking; thus alterations due to a reduced capacity to use proprieceptive stimuli can be easily detected. Second, the cats can switch to exteroceptively directed movements, thereby continuously matching tactile and visual stimuli inherent in the treadmill apparatus; thus alterations due to a reduced capacity to use exteroceptive stimuli can be detected. Third, the cats can switch to non-exteroceptively directed movements allowing them to collect food: they switch without using any observable exteroceptive stimulus for directing their motor behavior; thus, alterations due to a reduced capacity to switch arbitrarily to non-exteroceptively directed movements can be detected. Fourth, the cats can switch to exteroceptively triggered movements allowing them to collect food: they are using available exteroceptive stimuli for directing the whole chain of required movements, a phenomenon that occurs only in cats accustomed to the experimental set-up. In other words, the motor behavior is directed by exteroceptive stimuli that have become conditined. Hence alterations due to a reduced capacity to use conditioned stimuli can also be detected. And, finally, only fasted cats show the food collecting behavior; hence alterations due to a reduced capacity to use stimuli dependent on deprivation can be detected.

Haloperidol-induced Effects

Haloperidol produced the following effects. First, all cats tested (n=16) showed normal postural adjustments indicating that proprioceptive stimuli were still used for adopting postures. Second, the cats were able to switch to exteroceptively directed and exteroceptively triggered movements, indicating that exteroceptive and conditioned stimuli were still used for directing movements. Third, the cats collected at least as many food pellets as solvent-treated cats (n=15; Table I); in fact, they switched significantly more to exteroceptively directed movements indicating that stimuli inherent in deprivation were still used for directing movements. Finally, the only deficit seen in haloperidol-treated cats was a reduced capacity for switching to non-exteroceptively directed movements allowing them to collect food (Fig. 4). In fact, some cats did not make any attempt to switch to non-exteroceptively directed movements, whereas other cats showed a significantly reduced capacity to do so (Jaspers et al., 1983).

Function of the Caudate Nucleus

As a whole, the above findings imply that haloperidol, blocking nigrostriatal dopaminergic transmission, reduces the capacity to switch to non-exteroceptively directed movements (Jaspers et al., 1983). The additional finding that such animals are still able to switch to exteroceptively directed movements allowing them to collect food excludes the possibility that pure motor disturbances or a reduced appetite underlie the observed effects.

Table I. Relative frequency of food collecting attempts of haloperidol-treated and solvent-treated cats (bilateral injections into the caudate nucleus): only data collected in cats showing a drug-induced decrease in non-exteroceptively directed movements are shown. Mann-Whitney U-test: no significant differences.

	solvent	haloperidol (12.5 μg)
number of cats tested	10	15
ratio $\frac{\text{post-injection period}}{\text{pre-injection period}}$	0.5±0.3	0.4±0.1

It was already known that experimentally induced disturbances in programming the serial ordering of non-exteroceptively directed behavior can be manifest at the level of social communication in monkeys as well as at the level of behavior strategies in rats (Van den Bercken and Cools, 1982; Cools, 1980). The latter findings together with the present data allow the conclusion that the caudate nucleus enables the organism to programme the serial ordering of non-exteroceptively directed behavior in its broadest sense: any behavioral consequence should be regarded as a manifestation of a deficient capacity to "program the serial ordering of non-exteroceptively directed behavior".

Patients with dysfunction of the basal ganglia, e.g. patients suffering from Parkinson's disease, also show a reduced capacity to program the serial ordering of non-exteroceptively directed behavior (Cools et al., 1984). Since the latter deficit manifests itself at the level of movements and cognition, such motor and cognitive phenomena should no longer be regarded as separate entities: they also should be regarded as manifestations of a single dysfunctioning process that influences different levels of cerebrally organized behavior. On the basis of this insight it is no longer justifiable to regard cognition and movement as two independent variables, as far as their programming is concerned!

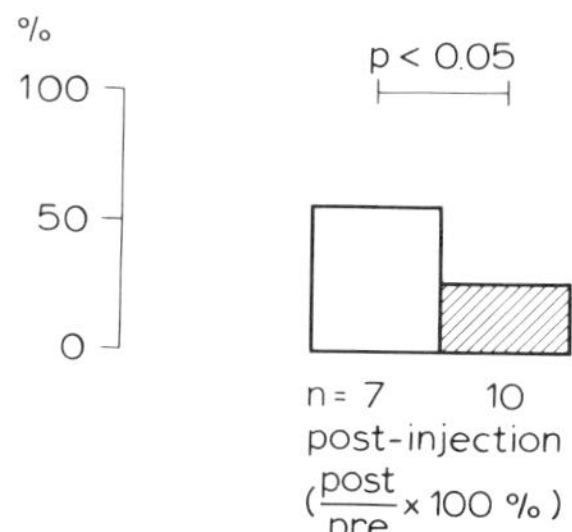

Fig. 4. Effects of bilateral, intracaudate injections of haloperidol (12.5 µg/0.5 µl per side) in freely moving cats trained to walk on a treadmill and having the freedom to switch to movements allowing them to collect food.
The number of non-exteroceptively directed movements allowing them to collect food, which are displayed during the post-injection interval of five minutes, is expressed as a percentage of those displayed during a pre-injection interval of five minutes; the pre-injection and post-injection intervals begin respectively 15 min before and 5 min after the injection.
Haloperidol-treated cats (hatched columns) show significantly fewer non-exteroceptively directed movements than solvent-treated cats (open column; solvent = distilled water; Mann-Whitney U-test).

Since behavioral consequences of experimentally induced dysfunction of the caudate nucleus are ultimately consequences of changing the information leaving the caudate nucleus, it becomes evident that the observed deficit in programming motor behavior, i.e. the reduced capacity to program the serial ordering of non-exteroceptively directed movements, somehow results from changes in the output signals of the caudate nucleus. As mentioned, haloperidol-treated cats are still able to collect food by switching to movements directed by exteroceptive stimuli intrinsic to the treadmill apparatus. In other words, such cats are still able to program their motor behavior by switching to exteroceptively directed movements in order to get their food. From this point of view it becomes evident that the caudate nucleus reduces the degree of freedom in programming behavior by adding information about the serial ordering of non-exteroceptively directed motor behavior. In this way the caudate nucleus contributes to the necessary transformation of motor programming signals into motor commands.

The fact that haloperidol-treated cats are still able to program their motor behavior by switching to exteroceptively directed movements implies that systems hierarchically superior to the caudate nucleus send their output signals not only to the caudate nucleus, but also to other structures, bypassing the latter structure. Given the notion that the main afferents of the caudate nucleus have their origin in the cortex, it is reasonable to assume that the output signals of systems hierarchically superior to the caudate nucleus, i.e. the reference signals for the caudate nucleus, are derived from the cortex. From this point of view it appears to be the cortex that contains the code for arbitrary or "voluntary" programming of behavior (cf. Eccles, 1982; Roland et al., 1980). Indeed, studies on man with lesions in the prefrontal lobes have shown such patients to have lost precisely this capacity. Since there are not only cortico-striatal but also cortico-cortical, cortico-nigral, cortico-collicular, cortico-reticular, and cortico-spinal fibers (Kuypers, 1978), it is reasonable to assume that the latter are also able to transmit the cortical code for programming behavior to other parts of the cortex, the substantia nigra pars reticulata, the colliculus superior (especially its deeper layers), the reticular formation, and even the spinal cord, bypassing the caudate nucleus. In view of the function of the cortex, the nigral system and the collicular system, it may now become understandable why animals with a hypofunctioning caudate nucleus are able to switch arbitrarily to exteroceptively triggered movements (cortical function), proprioceptively directed movements (nigral function; section 5), and exteroceptively directed movements (collicular function; section 6).

Once acquainted with the code carried by the output signals of a particular hierarchical sub-system, it becomes possible to investigate how this information is carried downstream in the hierarchy, and how it is ultimately transformed into motor commands. Tracing the efferents of the hierarchical sub-system under study is the tool of choice for delineating sub-systems inferior to it and as nearly as possible next in order. Since the output signals of higher-order systems are reference signals for lower-order systems (section 3), knowledge about these output signals provides an excellent starting point for describing the function of lower-order systems. Although it is not yet possible to map the whole process of transformation of motor program signals into motor commands, we will consider two steps downstream in the hierarchy by following information that leaves the caudate nucleus.

5. SUBSTANTIA NIGRA PARS RETICULATA AND SWITCHING MOTOR PROGRAMS

Over recent years it has become evident that the substantia nigra pars reticulata (SNR) is an important output station of the caudate nucleus, from which it receives a monosynaptic, GABA-ergic input (Cools et al., 1983; for review: Scheel-Kruger, 1983). With rather selective chemical tools available, to enhance GABA-ergic activity with muscimol (a direct GABA-ergic receptor agonist) and to attenuate it with picrotoxin (a drug closing the chloride channels

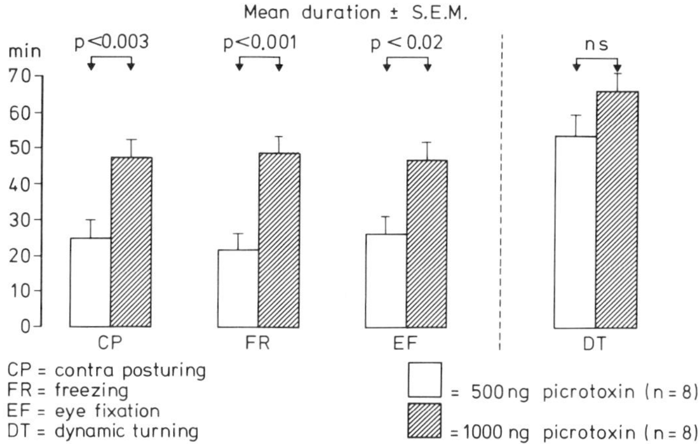

Fig. 5. Dose-dependent effects elicited by unilateral picrotoxin injections into the substantia nigra, pars reticulata (SNR) of freely moving cats in an open-field test (for description of observation cage: Cools et al., 1976; for details of the induced behaviors: Cools et al., 1983; Wolfarth et al., 1981).
Note: the so-called dynamic turning, i.e. fast contralateral head turns, is not dose-dependent (p values: Mann-Whitney U-test).

opened by GABA), this output station of the caudate nucleus can be studied on condition that adequate control studies for the specificity and selectivity of the behavioral consequences elicited by these tools are included. Since the striato-nigral GABA-ergic fibers are known to increase or decrease nigral GABA release during activation or inhibition respectively of the nigro-striatal dopaminergic fibres (for review: Scheel-Kruger, 1983), it seems reasonable to use picrotoxin (supresssing GABA activity) within the SNR to block the arrival via the striato-nigral GABA-ergic fibers of reference signals for SNR (section 4). When such a zero-reference condition is produced by picrotoxin, the behavioral consequences should be marked by a characteristic loss of degree of freedom in the programming of an animal's motor behavior. Although picrotoxin is chosen to prevent the transmission of information carried by the striato-nigral fibers, there is no direct proof of this because of the fact that the SNR is loaded with GABA synapses of which only a part may belong to the striato-nigral fibers. Accordingly one needs additional information to specify whether the picrotoxin effects are indeed due to a direct interference with the incoming reference signals. As illustrated below, such information can be collected in studies using muscimol, which produces pharmacological effects diametrically opposite to those of picrotoxin.

Picrotoxin-induced effects

Cats receiving unilateral injections of picrotoxin into the SNR show highly characteristic deficits (Cools et al., 1983; Sontag et al., this volume; Wolfarth et al., 1981). First, they freeze on the spot, often in an unusual asymmetric position: they cannot use proprioceptive stimuli to adjust their body position. Second, they are unable to lift their hindlimbs when their forelimbs are put on a bar more than 2 m above the floor (the so-called bar test): they cannot use proprioceptive information from their hindlimbs to retract them. Third, they are unable to erect the pinna when its upper part is retroflexed: they cannot use proprioceptive stimuli from the pinna (Kolasiewicz et al., unpublished data). These effects are SNR-specific (because they are absent when the SNR is lesioned), dose-dependent, and inhibited by muscimol (Figs. 5-6; Table II). That the SNR-specific effects are bilateral despite unilateral injection is due to the fact that GABA-ergic inhibition on one side of the brain can produce analogous inhibition on the opposite side via crossed nigro-thalamo-caudato-nigral fibers (Chesselet et al., 1983).

It is evident that picrotoxin-treated cats are unable to switch to movements directed by input signals that are ultimately derived from proprioceptive stimuli. Thus, they are unable to execute propriotopic movements (section 3). In other words, reduced

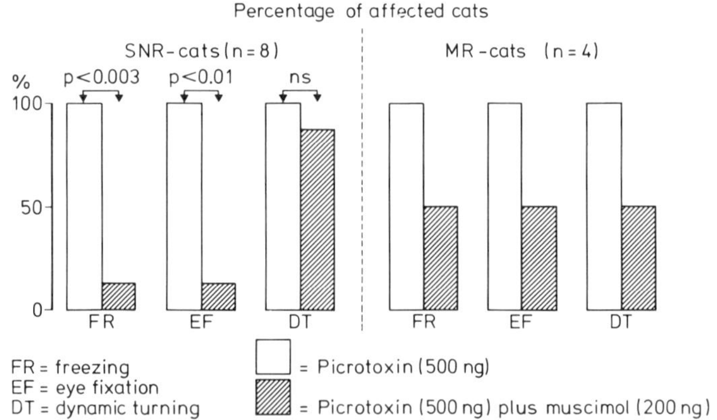

Fig. 6. Attenuating effects of muscimol injections (1 µl) into the SNR and adjacent structures in the mesencephalic (MR) region upon behavior elicited by picrotoxin injections (0.5 µl) into the same areas.
Note: the so-called dynamic turning is not attenuated, suggesting that this effect is not SNR-specific.

Table II. Locus specificity of GABA-ergic effects.

	percentage of affected cats		
	SNR	MR	SNR-lesion
freezing			
picrotoxin (0.5-2.0 μg; 0.5 μl)	100 (35)	100 (9)	0 (24)
dynamic turning			
picrotoxin (0.5-2.0 μg; 0.5 μl)	100 (35)	100 (9)	100 (8)
muscimol (0.4 μg; 1.0 μl)	100 (13)	100 (5)	100 (8)
propriotopic movements			
muscimol (0.4 μg; 1.0 μl)	100 (13)	0 (5)	0 (8)

the number of tested cats is given in brackets

SNR = substantia nigra pars reticulata injections
MR = injections into structures adjacent to the SNR
SNR-lesion = injection into SNR, which is mechanically lesioned by rotating the injection needle

Note: muscimol is always injected in a volume of 1.0 μl because of its limited diffusion in comparison with that of picrotoxin (0.5 μl; cf. Figs. 6, 7, 9 and 10)

GABA-ergic activity within the SNR prevents the serial ordering of propriotopic movements. It appears that such cats show a normal capacity to switch arbitrarily to exteroceptively directed movements: picrotoxin-treated cats trained to walk on a treadmill freeze as long as the belt stands still, but start walking as soon as the belt moves (Sontag et al., this volume). This indicates that stimuli provided by the treadmill apparatus can be still used for exteroceptively directed movements.

Muscimol-induced effects

Cats receiving unilateral injections of muscimol in the SNR show a highly complex series of movements (cf. Cools et al., 1983; Wolfarth et al., 1981). The ones discussed below are specific to the brain region under discussion: they disappear when the SNR is lesioned, they are dose-dependent, and they are antagonized by picrotoxin (Jaspers et al., unpublished data).

All the movements are restricted to spatio-temporal alternation between two given points, each of them characterized by fixed egocentric coordinates (cf. section 3). In practice the fully symmetric posture serves as point of departure for the drug-induced movements. As mentioned earlier, the egocentric coordinates of this point are always fixed. The drug-induced movements are directed towards another point of which the egocentric coordinates are also fixed. The latter movements are regularly interrupted by smooth movements directed back towards the point of departure (Fig. 7). Thus, muscimol elicits forced movements directed towards points whose coordinates show fixed deviations from the axes of the egocentric coordinate system. So the muscimol-treated cats continuously display movements between one naturally given point, i.e. the point of departure, and a point marked by drug-induced, fixed egocentric coordinates. As time progresses the part of the body forming the center of the egocentric coordinate system, i.e. the point of reference for describing the coordinates of the position to be reached, moved from the ears and eyes to the midline of the head and then in a cephalocaudal direction from the head to the shoulders and to the tail. Thus, the muscimol-treated cats initially move their ears, add eye movements, and progress by including head movements and movements involving neck, shoulders and forelimbs, and ultimately terminate with movements involving all parts of the body. Apart from the ear and eye movements which are not yet assessed in a quantitative way, all remaining movements are restricted to spatial alternation described above (Fig. 7).

Function of the Substantia Nigra Pars Reticulata

Before discussing the drug-induced effects it is useful to note that the SNR region under discussion is the area described in our previously published article (see Fig. 2a in Cools et al., 1983).

As mentioned above, muscimol alters the SNR system in such a manner that the animal persistently attempts to bridge the spatio-temporal gap between the point of departure, characterized by the axes of the egocentric coordinate system, and a point marked by drug-induced, fixed egocentric coordinates. This information allows us to draw two conclusions. First, the muscimol-treated cats show propriotopic movements (cf. section 3). Thus, the SNR system of muscimol-treated cats produces a code forcing the animal to switch to movements that are normally directed by input signals derived from proprioceptive stimuli. In muscimol-treated cats, however, the proprioceptive stimuli are not offered; accordingly, they are created by the treatment itself. Below we will return to this notion. The second conclusion deals with the question of specifying the actual target site of the chosen GABA-ergic drugs. Since muscimol-treated cats neither freeze nor let their hindlimbs hang when their forelimb are put on a bar (the previously mentioned bar test: Cools et al., 1983; Wolfarth et al., 1981), it appears that they are still able to

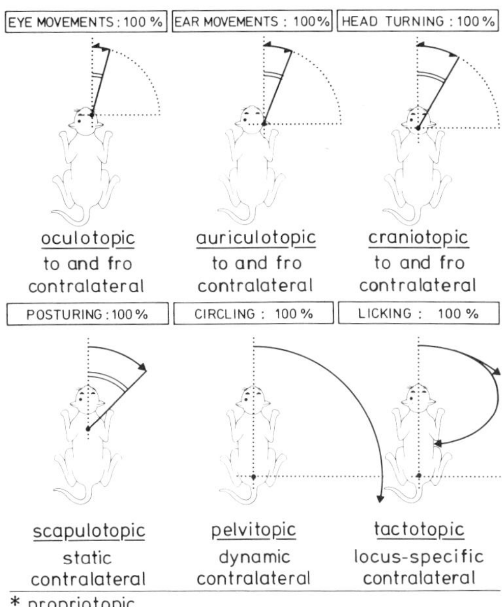

Fig. 7. Propriotopic movements elicited by unilateral SNR muscimol injections (400 ng/1.0 µl) in freely moving cats (n=20); the movements are observed in a familiar and static environment (for details of the cage: Cools et al., 1976). The percentages of cats showing the different movements are indicated. Characteristic features of the muscimol-induced movements are indicated by an asterisk (see also: Figs. 2 and 11).

switch to movements directed by information available from proprioceptive stimuli. Thus, they are still able to register changes in input signals derived from proprioceptive stimuli, to reduce the difference between input signals and reference signals, and to program correct output signals. Given this notion, the muscimol-induced effects have to be ascribed to interference with the reference signals, an outcome predicted on the basis of the known striato-nigral, GABA-ergic input. This implies that the muscimol-induced propriotopic movements result from muscimol's ability to fix the magnitude of the reference signals in the SNR system: consequently, muscimol creates a propriotopic code without using input signals derived from proprioceptive stimuli. This in turn implies that picrotoxin, by preventing the use of proprioceptive stimuli (see above), simply prevents the use of propriotopic codes.

Before summarizing this discussion it is relevant to recall that output signals of hierarchically superior systems direct programs of hierarchically inferior systems, and that behavioral consequences characteristic of disturbances in higher-order systems are successively replaced by behavioral consequences characteristic of disturbances in lower-order systems (Cools and van den Bercken, 1977). Given the fact that muscimol fixes the magnitude of the reference signals in the nigral system and, accordingly, produces fixed output signals of the nigral SNR system under certain circumstances, the animals should have behavioral deficits characteristic of fixed output signals in systems at successively lower-order levels within the hierarchy. This is what appears to happen in muscimol-treated cats: as time progresses, the part of the body forming the point of reference for describing the actual coordinates of the positions to be reached moves from the ears to the eyes and to the midline of the head and then to the shoulders and tail. Both the recovery from hypothalamic lesions and the ontogeny of movement patterning in rats are marked by analogous shifts as far as concerns the activation and re-integration of systems responsible for movement (for review: Teitelbaum et al., 1983). In view of these findings it appears justified to postulate that the successive shifts of the center of egocentric coordinate systems in muscimol-treated cats reflects the order in which systems inferior to the SNR system are successively affected.

Summarizing the above-mentioned considerations it appears that the SNR system receives the striatal code for programming the serial ordering of non-exteroceptively directed movements via the striato-nigral, GABA-ergic pathways. The SNR system itself apparently transforms this code into a new code for programming the serial ordering of propriotopic movements, i.e. movements whose egocentric coordinates are prescribed by a propriotopic code. In other words, the SNR system reduces the degree of freedom in programming behavior by adding information about the propriotopic coding of movements to be executed. In this manner the SNR system forms a next step in the process of transforming motor programming signals into motor commands.

6. SUPERIOR COLLICULUS AND SWITCHING MOTOR PROGRAMS

Given the SNR system as a main output station of the caudate nucleus, we have to trace the efferent pathways of this brain structure to delineate the hierarchical system immediately inferior to it. As there is no doubt about the existence of a three-neurone striato-nigro-collicular pathway (for review: Scheel-Kruger, 1983), we have to consider the superior colliculus, especially its deeper layers (CS-dl), as a station intercalated between the caudate nucleus and the lowest-order system, i.e. the spinal cord. As the nigro-collicular fibers contain GABA as neurotransmitter, muscimol and picrotoxin are the tools of choice to produce changes in the

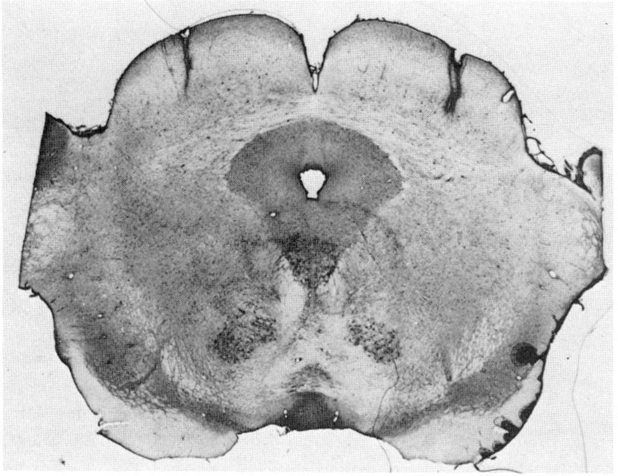

Fig. 8. Representative picture of the cat brain with "effective" injection sites of muscimol and picrotoxin respectively within the deeper layers of the colliculus superior (CS-dl).

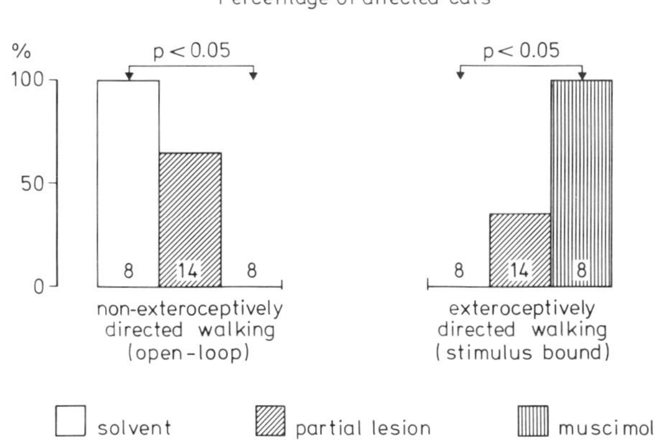

Fig. 9. Suppressant effect of unilateral muscimol injections (50 ng/1 µl) into the CS-dl of freely moving cats in the milk-test (see section 6): the non-exteroceptively directed movements allowing the cats to drink milk at the end of the bar are partly or fully replaced by exteroceptively directed stimulus-bound movements in lesioned cats (obliquely hatched column) and muscimol-treated cats (vertically hatched column) in comparison with solvent-treated cats (open column).

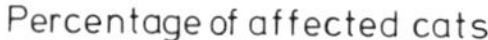

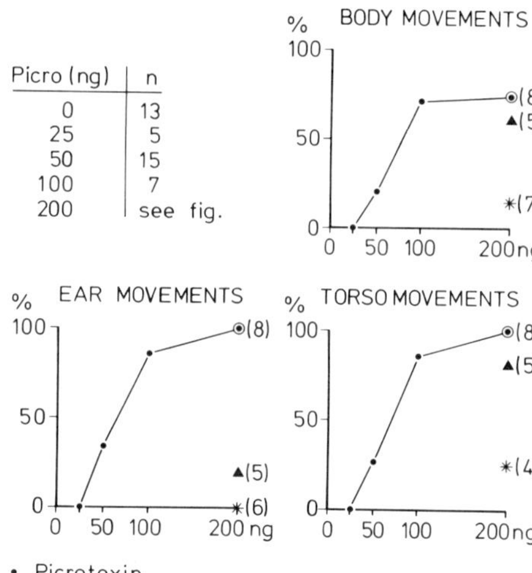

Fig. 10. Picrotoxin-induced exterotopic responses in CS-dl cats (picrotoxin is unilaterally injected). The number of tested cats per picrotoxin dose is given in the upper left part. The numbers of picrotoxin tested cats in the combined experiments (muscimol plus picrotoxin or bandage plus picrotoxin) are put in brackets (see also: Fig. 11).

function of the CS-dl system. Starting from the fact that nigro-collicular fibers increase and decrease their GABA release during inhibition and activation respectively of the striato-nigral fibers (for review: Scheel-Kruger, 1983), intracollicularly administered muscimol can be expected to block the arrival of reference signals at the CS-dl level in the nigro-collicular fibers carrying the output signals of the SNR system. Accordingly, CS-dl muscimol-treated cats should show behavioral consequences marked by loss of degree of freedom in programming behavior. To specify whether the muscimol effects are indeed due to a direct interference with incoming signals, again one needs additional studies of the effects of picrotoxin.

Muscimol-induced Effects

Cats receiving unilateral injections of muscimol in the CS-dl (Fig. 8) show characteristic deficits. When put on a narrow bar 5 cm wide, 2 m long and about 2 m above the floor, they are unable to bridge the gap between the point of departure at one end of the bar and the other end where they can collect milk from a small cup, unless their movements are continuously directed by visual and/or tactile stimuli. In contrast to solvent-treated cats, which walk straight towards the end of the bar without any visual fixation or wrong placing responses, CS-dl muscimol-treated cats either do not move at all or move forward slowly, fixating the bar visually just 30 to 40 cm ahead, and producing a large number of misplacements which they immediately correct (Fig. 9). These effects are specific for the region under discussion, since they are absent when the deeper layers are lesioned (Fig. 9); furthermore, they are dose dependent and antagonized by picrotoxin (Jaspers et al., unpublished data).

Although the muscimol-treated cats are apparently able to switch to movements directed by input signals that are ultimately derived from proprioceptive and/or exteroceptive stimuli, they differ from normal cats in the sense that the exteroceptive information intrinsic to the object to be reached (milk) is insufficient for programming the movements that allow them to reach that object: they need additional exteroceptive stimuli for executing the required program. Thus, intracollicularly administered muscimol produces a highly characteristic deficit; as suggested, this may be the direct consequence of a zero-reference condition of the CS-dl system. In order to verify the latter suggestion we need additional information on the effects of chemical interference producing pharmacological actions in the opposite direction. As illustrated below such studies also allow us to specify the actual deficit underlying the muscimol effects.

Picrotoxin-induced Effects

When picrotoxin instead of muscimol is unilaterally administered into the deep layers of the CS, a complex series of movements appears. The effects described below are specific for the brain region under study: they disappear when the CS-dl is lesioned, they are dose-dependent and are attenuated by muscimol (Fig. 10). In contrast to movements elicited by nigral muscimol, the movements elicited by collicular picrotoxin are not restricted to spatio-temporal alternations between points with fixed, egocentric coordinates. Instead, the movements are spatio-temporal alternations directed towards a point whose egocentric coordinates drift away from the axes of the egocentric coordinate system according to a particular rule. According to the terminology given in section 3, the cats are thus performing exterotopic movements, i.e. movements directed towards a spatial point whose egocentric coordinates are prescribed by an exterotopic code that is derived from exteroceptive stimuli and may vary according to the allocentric coordinate system involved. Globally seen, the cats move to and fro between two points in space. The fully symmetric posture again serves as point of departure for the drug-induced movements; as mentioned earlier this point has fixed egocentric coordinates. The movements are directed towards a spatial point whose coordinates are not fixed. The movements are regularly terminated when the body has reached a position marked by fixed deviations from the axes of the egocentric coordinate system. These movements are then replaced by others directed either towards the point of departure or towards a spatial point with coordinates determined by multiplying the initial degree of deviation from the axes of the egocentric coordinate system. Thus, picrotoxin elicits forced movements towards a spatial point whose coordinates deviate variably from the axes of the egocentric coordinate system, but does not prevent the execution of movements directed towards the axes of the egocentric coordinate system. As time progresses, the point of reference for describing the egocentric coordinates of the spatial point to be reached moves from the ears, to the eyes, to the midline of the head, and then cephalocaudally through shoulders to tail. Thus, the CS-dl picrotoxin-treated cats initially move their ears, add eye movements, progress to including head movements and movements involving head, neck, shoulders and forelimbs, and ultimately terminate with movements involving all parts of the body (Fig. 11). For instance, a bilaterally CS-dl picrotoxin-treated cat anteroflexes its head 10°, keeps its head position fixed for a while and then either turns its head to the point of departure or anteroflexes its head another 10°, and so on. Having reached the maximum degree of head bending it starts to bend its torso in that direction according to the same process of increasing the original deviation. Ultimately, the whole body becomes involved in the movements: the cat displays a great series of distinct movement patterns varying from bending its head between its limbs, bending its head along the outerpart of one of its limbs, to jumping backwards and then bending its head.

Apart from the ear and eye movements which are not yet assessed in a quantitative manner, all drug-induced movements are directed towards a point whose egocentric coordinates drift away from the axes of the egocentric coordinate system.

Function of the Superior Colliculus Deeper Layers

In the terminology of section 3, it appears that picrotoxin alters the CS-dl system so that the animal constantly attempts to bridge the gap between its instantaneously generated point of departure (i.e. the position determined by the axes of the allocentric coordinate system) and a point characteristized by drug-induced, allocentric coordinates, thereby forcing exterotopic movements. In other words, the CS-dl system somehow carries information about the exterotopic coding of the movements to be executed. Since picrotoxin-treated cats with their eyes bandaged still execute these movements (Fig. 10), exteroceptive stimuli cannot be used to generate this exterotopic code. Hence the treatment itself has created the exterotopic code. On the other hand, the

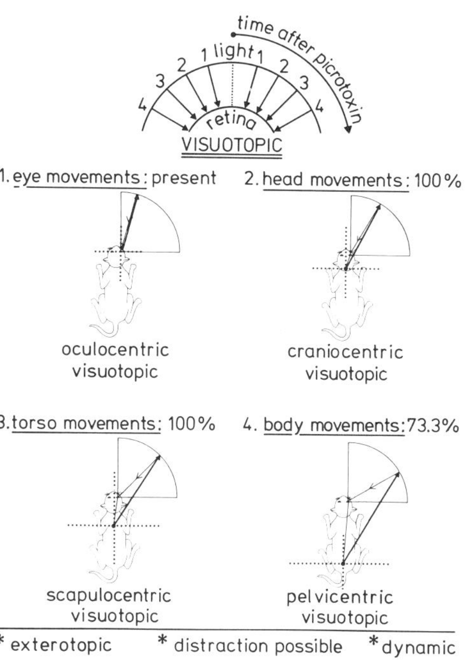

Fig. 11. Exterotopic movements elicited by ulilateral CS-dl picrotoxin injections (200 ng/0.5 µl) in freely moving cats (n=15); the movements are observed in a familiar and static environment (for details of the cage: Cools et al., 1976). The percentages of cats showing the different movements are indicated. Characteristic features of the picrotoxin-induced movements are indicated by an asterisk (see also: Figs. 3 and 7).

occurrence of movements directed towards the axes of the egocentric coordinate system implies that picrotoxin-treated cats are still able to perform propriotopic movements.

Given the finding that the picrotoxin-treated cats are quite capable of walking straight towards the end of a bar without visual fixation or any wrong placing responses (the test in which muscimol-treated CS-dl cats show characteristic deficits; see above), it appears that they are still able to switch to movements directed by information from exteroceptive stimuli. Thus they are still able to register changes in input signals derived from exteroceptive stimuli, to reduce the difference between input signals and reference signals, and to produce correct output signals. Given this, the picrotoxin-induced effects have to be ascribed to interference with the reference signals, an outcome predicted on the basis of the known nigro-collicular, GABA-ergic input. This implies that the picrotoxin-induced exterotopic movements result from picrotoxin's ability to fix the magnitude of the reference signals in the collicular CS-dl system: consequently, picrotoxin creates an exterotopic code without using input signals derived from exteroceptive stimuli. This in turn implies that muscimol, which prevents the use of exteroceptive stimuli intrinsic to the object to be reached (see above), simply prevents the use of exterotopic codes.

Summarizing, it has become evident that the CS-dl system receives the SNR code for programming the serial ordering of propriotopic movements via the nigro-collicular, GABA-ergic pathways. The CS-dl system itself transforms this SNR code into a new code for programming the serial ordering of exterotopic movements. In other words, the CS-dl system once more reduces the degree of freedom in programming motor behavior by adding information about the exterotopic coding of the movements to be executed. In this manner the CS-dl system represents a further step in the process of transforming programming signals into motor commands.

Before discussing the overall impact of the available data it is worthwhile considering the following remarks about the CS-dl system. First, the successively appearing shifts in the centre of the egocentric coordinate system of the spatial point to be reached by picrotoxin-treated CS-dl cats are identical to those seen in muscimol-treated SNR cats. This supports the validity of the earlier notion that these shifts may reflect the order in which systems inferior to the affected system are successively affected (see section 5). From this point of view it appears likely that the reticular formation plays a crucial role in this respect (cf. Sirkin et al., 1980). Second, CS-dl muscimol-treated cats are still able to switch to propriotopic movements, i.e. movements requiring correct output signals of the SNR system, as has been noted above. This finding may simply imply that the SNR system, which is superior to the CS-dl system, is still able to send output signals to systems inferior to the CS-dl system, bypassing the latter. Finally, both the SNR and the CS-dl systems are known to direct eye movements, especially saccades (Hikosaka and Wurtz, 1983a,b,c,d; Wurtz, 1978). In view of the above-discussed functions of these structures in directing propriotopic and exterotopic movements respectively (sections 5 and 6), it will be evident that there should exist two types of saccades: "propriotopic saccades", i.e. eye movements directed by a propriotopic code, and "exterotopic saccades", i.e. eye movements directed by an exterotopic code. In practice this appears to be true. Propriotopic saccades, i.e. saccades for which the point of departure is defined by reference to the axes of the involved egocentric system (the oculotopic system; Fig. 2), are found to occur in cats (Guitton et al., 1980; Roucoux et al., 1980). Exterotopic saccades, i.e. saccades for which the point of departure is defined by reference to the axes of the involved allocentric coordinate system (the visuotopic system; Fig. 3), are found to occur in monkeys

(Mohler and Wortz, 1976). Since both types of saccades could be elicited from the superior colliculus, it becomes worthwhile to consider the possibility that 1) propriotopic saccades are elicited from cells receiving reference signals directly from the SNR system, and 2) exterotopic saccades are elicited from cells sending output signals to systems hierarchically inferior to the colliculus superior. Anyhow, it now appears possible to reconcile the data about saccades in monkeys (cf. Mohler and Wurtz, 1976) and those from studies on cats (cf. Guitton et al., 1980).

7. BASAL GANGLIA AND DISORDERS IN SWITCHING MOTOR PROGRAMS

Animals

Having discussed how information sent to the caudate nucleus is successively transformed at the level of the caudate nucleus, substantia nigra pars reticulata, and superior colliculus (deeper layers) by adding more and more details about the motor program to be executed (Fig. 12) it becomes possible to consider the second question put forward in section 4: Does a dysfunction of the caudate nucleus have consequences for the functioning of the brain structures such as the substantia nigra pars reticulata (SNR system), and the superior colliculus (CS-dl system), i.e. brain structures intercalated between the caudate nucleus and the spinal cord?

Here it is useful to consider the behavioral consequences of striatal interventions that differ in degree of pathology. Let us analyse the behavioral consequences of increasing doses of haloperidol, a drug which ultimately reduces the output signals of the caudate nucleus to zero through its capacity to block the transmission of information from the nigro-striatal dopaminergic fibers to their corresponding postsynaptic receptors (section 4).

As mentioned earlier, low doses of haloperidol prevent the organism from switching arbitrarily to non-exteroceptively directed movements, because of the ultimately resultant zero-output condition of the caudate nucleus (section 4).

Intermediate doses of haloperidol produce artificial postures and positions; this phenomenon is classically described as catalepsy. According to the terminology discussed in sections 2 and 5, this deficit implies a reduced capacity to switch to propriotopic movements (cf. De Rijck et al., 1980). This phenomenon becomes understandable from the fact that haloperidol can produce a zero-output condition of the caudate nucleus. Since the output signals of the caudate nucleus are reference signals for the SNR system, haloperidol can also produce a zero-reference condition in the latter (cf. Fig. 12). As mentioned in section 5, such a reduction, produced by nigral injections of picrotoxin for instance,

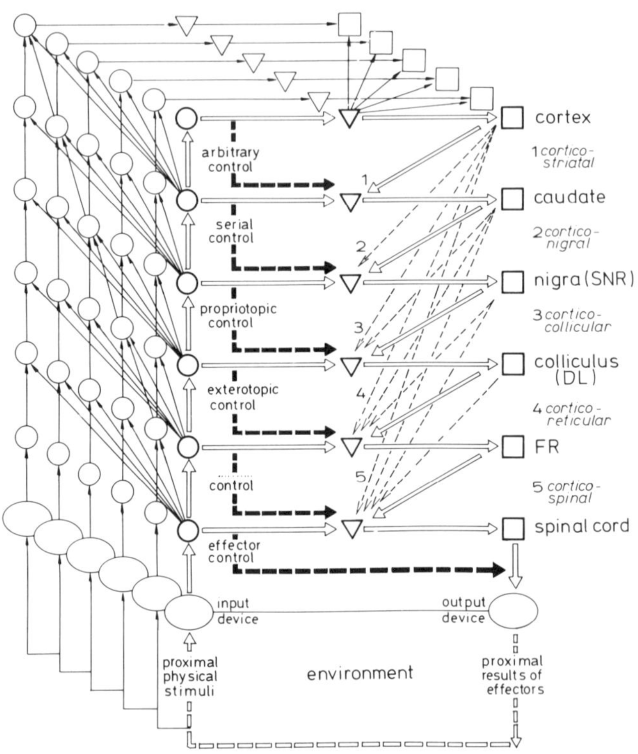

Fig.12. Schematic diagram of the non-linear, overlapping hierarchy of negative feedback systems controlling the input signals of the organism (see Sections 2 and 7). Stimuli delivered by 1) the ongoing behavior itself, 2) immediate surroundings, and 3) proprioceptive and interoceptive receptors are transformed into abstract, invariant input signals whose degree of abstraction increases at each higher level within the hierarchy. The latter input signals are compared with so-called reference signals, i.e. output signals of hierarchically superior systems, resulting thereby in error signals which, in turn, are transformed into output signals. Only the output signals of the lowest-order system are transformed into behavior.
Note: for the sake of simplicity the output signals leaving the caudate nucleus, substantia nigra pars reticulata, colliculus superior (deeper layers), and reticular formation respectively, and bypassing one or more hierarchically lower-order systems, are omitted.

can prevent the organism from using propriotopic codes for directing movements. The validity of this reasoning is underlined by the finding that haloperidol can reduce the release of GABA from striato-nigral, GABA-ergic fibers (for review: Scheel-Kruger, 1983), reducing thereby the magnitude of the reference signals in the SNR system.

Higher doses of haloperidol produce animals that are bound to tactile and pressure stimuli (Schallert and Teitelbaum, 1981). According to the terminology discussed in sections 2 and 6, this deficit implies a reduced capacity to switch arbitrarily to exterotopic movements. This phenomenon is also explained by the fact that haloperidol can ultimately produce a zero-output condition of the SNR system as a result of its ability to produce a zero-reference condition in this system. Since the output signals of the SNR system are reference signals of the CS-dl system (cf. Fig. 12), haloperidol can indirectly produce a zero-reference condition in the latter. As mentioned in section 6, such a reduction, produced by collicular injections of muscimol for instance, can prevent the animals from using exterotopic codes in directing movements. The validity of this reasoning is supported by the finding that chronically lowering striatal dopaminergic activity can increase the release of GABA from nigro-collicular, GABA-ergic fibers (for review: Scheel-Kruger, 1983), thereby reducing the magnitude of the reference signals for the CS-dl system.

Still higher doses of haloperidol are known to produce pure motor deficits. According to the terminology discussed in section 3, this deficit implies a reduced capacity to produce correct motor commands at the level of the spinal cord. Again, this becomes understandable if the whole process of sending information downstream in the hierarchy is marked by reducing the magnitude of reference signals at successively lower-order levels, as long as the zero-reference condition at the level of the caudate nucleus is kept invariant (cf. Fig. 12). The ultimate consequence of such a process will be a zero-output condition of the lowest-order system, i.e. the spinal cord, producing thereby pure motor disturbances. The validity of this reasoning is supported by the finding that manipulations of the SNR system, for instance, alter the function of the spinal cord (Sontag et al., this volume).

What is relevant in considering the above-mentioned effects is the recognition that a highly selective and specific interference with the magnitude of the output signals of a single hierarchical system, i.e. the caudate nucleus, produces a great variety of totally different movements; the degree to which systems inferior to the caudate nucleus are affected as a functional consequence determines the nature of the resulting movements (Fig. 12). The latter conclusion is actually inherent in any hierarchy of negative feedback control systems: systems superior to the lowest-order system direct "programs" but not "responses" (cf. Polit and Bizzi, 1979; Terzuolo and Viviani, 1980). Thus, movements are not simply represented in the brain at levels superior to the lowest-order system.

As a final remark in this context, it is interesting to consider the possibility that the increasing degree of pathology produced by increasing doses of haloperidol is related not to the number of dopamine receptors affected by haloperidol, but to the duration of the haloperidol-induced inhibition of dopamine receptors. In a hierarchy of feedback control systems the lowest-order system has the fastest response, and the higher in the hierarchy the output of these systems is affected, the more time it takes for sending the information downstream. Consequently, the degree to which fixed output signals of the caudate nucleus are able to produce fixed output signals of systems at successively lower-order systems is solely determined by the duration of experimentally induced or spontaneously occurring fixation of the striatal output signals.

Man

Having discussed the disorders pertaining to increasing degrees of dysfunction of the caudate nucleus in animals, it becomes possible to consider the third question put forward in section 4: Does knowledge about the function of the caudate nucleus offer perspectives for increasing our insight into Parkinson's disease?

The primary impact of the present data concerns the delineation of symptoms characteristic of a dysfunctioning caudate nucleus in man. As mentioned earlier it has recently been proven that patients suffering from Parkinson's disease, in whom at least the caudate nucleus suffers dysfunction, show a "shifting aptitude" disorder that manifests itself at the level of cognition and movement: such patients not only suffer from a reduced capacity to shift to non-exteroceptively directed movements, but also suffer from a reduced capacity to shift to non-exteroceptively directed figural and verbal concepts (Cools et al., 1984; cf. Barbeau, 1973; cf. Flowers, 1976, 1978a,b). Since the investigated patients did not have any symptoms of cortical dysfunction (depression, prefrontal syndrome, presenile or senile dementia) or of cerebrovascular disease (EEG and CT-scan abnormalities) (Cools et al., 1984), the deficits found in man with dysfunctional basal ganglia appear to be fully analogous to those found in animals with dysfunction of the caudate nucleus. This notion has several far-reaching implications. First, it underlines the validity of the animal models described for developing adequate frames of reference for detecting deficits in human beings; in fact, the clinical study was conceptualized by analogy to the animal experiments discussed in the present article. From this point of view it is relevant to emphasize that, in contrast to classical animal models, the one discussed here analyzes brain dysfunction in terms of deficits in programming behavior rather than in terms of deficits in symptomatology. Second, this analogy gives strong support to the hypothesis that the basal ganglia are, indeed, concerned with programming motor and cognitive behavior in man.

The second impact of the present data concerns the delineation of symptoms characteristic of different stages of Parkinson's disease (cf. Hoehn-Yahr rating scale: I-IV). Recalling the behavioral consequences of a hypofunctioning SNR system in animals, i.e. a reduced capacity to alter the serial ordering of propriotopic movements, it is reasonable to postulate that patients in whom the pathology of the caudate nucleus has indirectly led to a dysfunctioning SNR system will also suffer from a reduced capacity to shift the serial ordering of propriotopic movements. From this point of view it becomes possible to understand symptoms such as "kinesia paradoxa" and "freezing", i.e. phenomena characteristic of so-called on-off periods. Thus insight into the function of a disturbed SNR system in animals opens clear perspectives for developing anamnestic tools for detecting the symptoms of a dysfunctioning SNR system in human beings. The same holds true for the putative involvement of a dysfunctioning collicular system. Recalling the behavioral consequences of a hypofunction of the CS-dl system in animals, i.e. a reduced capacity to shift to the serial ordering of exterotopic movements, it is reasonable to postulate that patients in whom the pathology of the caudate nucleus has indirectly led to a dysfunctioning CS-dl system will also suffer from such a reduced capacity. From this point of view it becomes possible to understand the occurrence of abnormal eye movements in certain patients (White et al., 1983). Again, insight into the function of a disturbed CS-dl system in animals opens perspectives for developing methods of detecting symptoms of a dysfunctioning CS-dl system in human beings.

The final impact of the above data concerns the pharmacotherapy of Parkinson's disease. Given the possibility of diagnosing the degree of pathology in patients suffering from Parkinson's disease, it will be clear that agents ameliorating dopaminergic transmission within the caudate nucleus should be prescribed only for patients in whom the pathology has not yet extended to the SNR system. From this point of view it becomes understandable why only a limited number of Parkinsonian patients benefit from L-DOPA and related drugs. It remains to be investigated whether the so-called nonresponders to L-DOPA should be treated with agents improving GABA-ergic transmission within the SNR system.

Summarizing, the chosen approach opens up a completely new perspective for developing tools for diagnosis and treatment of patients with extrapyramidal disorders. In this light these diseases seem set for some exciting and potentially very important advances in the next few years.

REFERENCES

Barbeau, A., 1973, Biology of the striatum, in: "Biology of Brain Dysfunction," Vol. 2, G. Gaull, ed., Plenum Press, New York.

van den Bercken, J. H. L., and Cools, A. R., 1982, Evidence for a role of the caudate nucleus in the sequential organization of behavior, Behav. Brain Res., 4:319.

Chesselet, M. F., Chéramy, A., Reisine, T. D., Lubetzki, C., Desban, M., and Glowinski, J., 1983, Local and distal effects induced by unilateral striatal application of opiates in the absence or in the presence of nalaxone on the release of dopamine in both caudate nuclei and substantia nigrae of the cat, Brain Res., 258:229.

Cools, A. R., 1971, The function of dopamine and its antagonism in the caudate nucleus of cats in relation to the stereotyped behavior, Arch. int. Pharmacodyn., 194:259.

Cools, A. R., 1981a, Aspects and prospects of the concept of neurochemical and cerebral oganization of aggression: introduction of new research strategies in "Brain and Behavior" studies. in: "The Biology of Aggression," NATO Advanced Study Institutes Series D: Behavioral and Social Sciences No. 8, P. F. Brain and D. Benton, eds., Sijthoff and Noordhoff Int. Publishers, Rockville.

Cools, A. R., 1981b, Physiological significance of the striatal system: new light on an old concept, Adv. Physiol. Sci., 2:227.

Cools, A. R., and van den Bercken, J. H. L., 1977, Cerebral oganizatiion of behaviour and the neostriatal function, in: "Psychobiology of the Striatum," A. R. Cools, A. H. M. Lohman and J. H. L. van den Bercken, eds., Elsevier/North Holland Biomedical Press, Amsterdam.

Cools, A. R., Struyker Boudier, H. A. J., and van Rossum, J. M., 1976, Dopamine receptors: selective agonists and antagonists of functionally distinct types within the feline brain, Europ. J. Pharmacol., 37:283.

Cools, A. R., Jaspers, R., Kolasiewicz, W., Sontag, K. -H., and Wolfarth, S., 1983, Substantia nigra as a station that not only transmits, but also transforms incoming signals for its behavioural expression: striatal dopamine and GABA-mediated responses of pars reticulata neurons, Beh. Brain Res., 7:39.

Cools, A. R., van den Bercken, J. H. L., Horstink, M. W. I., van Spaendonck, K. P. M., and Bergerm H. J. C., 1984, Cognitive and motor shifting aptitude disorder in Parkinson's disease, J. Neurol. Neurosurg. Psychiat., in press.

Eccles, J. C., 1982, The initiation of voluntary movements by the supplementary motor area, Arch. Psychiatr. Nervenkr., 231:423.

Flowers, K. A., 1976, Visual 'closed-loop' and 'open-loop' characteristics of voluntary movements with Parkinsonism and intention tremor, Brain, 99:269.

Flowers, K., 1978a, Some frequency response characteristics of Parkinsonism on pursuit tracking, Brain, 101:19.

Flowers, K., 1978b, Lack of prediction in the motor behaviour of Parkinsonism, Brain, 101:35.

Guitton, D., Crommelinck, M., and Roucoux, A., 1980, Stimulation of the superior colliculus in the alert cat. I. Eye movements and neck EMG activity evoked when the head is restrained, Exp. Brain Res., 39:63.

Hikosaka, O., and Wurtz, R. H., 1983a, Visual and oculomotor functions of monkey substantia nigra pars reticulata. I. Relation of visual and auditory responses to saccades, J. Neurophysiol., 49:1230.

Hikosaka, O., and Wurtz, R. H., 1983b, Visual and oculomotor functions of monkey substantia nigra pars reticulata. II. Visual responses related to fixation of gaze, J. Neurophysiol., 49:1254.

Hikosaka, O., and Wurtz, R. H., 1983c, Visual and oculomotor functions of monkey substantia nigra pars reticulata. III. Memory-contingent visual and saccade responses, J. Neurophysiol., 49:1268.

Hikosaka, O., and Wurtz, R. H., 1983d, Visual and oculomotor functions of monkey substantia nigra pars reticulata. IV. Relation of substantia nigra to superior colliculus, J. Neurophysiol., 49:1285.

Jaspers, R., van Hoof, J., Sontag, K. -H., and Cools, A. R., 1983, Dopaminergic agents alter the caudate nucleus function in switching motor programmes, Neurosci. Letters, 14:S182.

Kuypers, H., 1978, The organization of the motor system in primates, in: "Recent Advances in Primatology," Vol. I, D. J. Chivers and J. Herbert, eds., Academic Press, London.

Mohler, C. W., and Wurtz, R. H., 1976, Organization of monkey superior colliculus: intermediate layers cells discharging before eye movements, J. Neurophysiol., 39:722.

Polit, A., and Bizzi, E., 1979, Characteristics of motor programs underlying arm movements in monkeys, J. Neurophysiol., 42:183.

Powers, W., 1973, "Behavior: The Control of Perception," Aldine Publishing Co., Chicago.

Roland, P. E., Larsen, B., Lassen, N. A., and Skinhøj, E., 1980, Supplementary motor area and other cortical areas in organization in voluntary movements in man, J. Neurophysiol., 43:118.

Roucoux, A., Guitton, D., and Crommelinck, M., 1980, Stimulation of the superior colliculus in the alert cat. II. Eye and head movements evoked when the head is unrestrained, Exp. Brain Res., 39:75.

Ryck, M. de Schallert, T., and Teitelbaum, P., 1980, Morphine versus haloperidol catalepsy in the rat: a behavioral analysis of postural support mechanisms, Brain Res., 201:143.

Schallert, T., and Teitelbaum, P., Haloperidol, catalepsy, and equilibrating functions in the rat: antagonistic interaction of clinging and labyrinthine righting reactions, Physiol. Behav., 27:1077.

Scheel-Kruger, J., 1983, The GABA receptor and animal behaviour, in: "GABA Receptors," S. Enna, ed., Humana Press, New Jersey, in press.

Sirkin, D. W., Schallert, T., and Teitelbaum, P., 1980, Involvement of the pontine reticular formation in head movements and labyrinthine righting in the rat, Exp. Neurol., 69:435.

Teitelbaum, P., Schallert, T., and Whishaw, I. Q., 1983, Sources of spontaneity in motivated behaviour, in: "Handbook of Behavioural Neurobiology," Vol. 6, E. Satinoff and P. Teitelbaum, eds., Plenum Publishing Corporation, New York.

Terzuolo, C. A., and Viviani, P., 1980, Determinants and characteristics of motor patterns used for typing, Neuroscience, 5:1085.

Vrijmoed-de Vries, M., and Cools, A. R., 1983, Disturbances in both social communication and motor behaviour can be elicited in the same region within the caudate nucleus of Java monkeys, Neurosci. Letters, suppl. 14:S395.

White, O. B., Saint-Cyr, J. A., and Sharpe, J. A., 1983, Oculomotor deficits in Parkinson's disease, Brain, 106:555.

Wolfarth, S., Kolasiewicz, W., and Sontag, K. -H., 1981, The effects of muscimol and picrotoxin injections into the cat substantia nigra, Naunyn-Schmiedeberg's Arch. Pharmacol., 317:54.

Wurtz, R. H., 1978, The primate superior colliculus and visually guided eye movements, in: "Recent Advances in Primatology," Vol. 1, D. J. Chivers and J. Herbert, eds., Academic Press, London.

STRUCTURE AND FUNCTION OF THE BASAL GANGLIA: A POINT OF VIEW

J. S. McKenzie

Department of Physiology
University of Melbourne
Parkville 3052 Australia

ANATOMICAL IDENTITY AND SPECIAL CHARACTERISTICS

Definition

Percheron, summarizing the terminological history of the "basal ganglia" (this volume), attempted to formulate an adequate, non-circular definition of them. One is reminded of old arguments about the term "pain" - everyone knows what it is, no-one can define it without tautology.

Percheron settled for a "core" of the basal ganglia being the "strio-pallidonigral system" (without specifying the subthalamic nucleus as a component). The deep masses of grey matter in the mammalian forebrain include nuclei which most of us would not want to count as proper members of the basal ganglia: thalamus, hypothalamus, septal nuclei; and the deep nuclei of the basal forebrain would have been excluded until Heimer (see Heimer and Van Hoesen, 1979) introduced the concept of "ventral striatum" to include the olfactory tubercle. The amygdala, although termed "archistriatum", so far remains stubbornly "limbic" rather than "striatal" in the eyes of brain scientists. The nucleus accumbens septi, which like the dorsal striatum receives dopaminergic afferents from the midbrain, and is similar in its cytoarchitectonics, is recognized as a principal component of the ventral striatum, its strong connections with the hippocampal formation giving it the status of an "interface" between the limbic system and other parts of the basal ganglia concerned with movement (Mogenson et al., 1980).

Because of its obvious connections and functional relations with striatum no-one wants to exclude the substantia nigra even though it is found in the midbrain not the forebrain. Analogous grounds are not usually accepted for including the tectal, ventral thalamic, and pedunculo-pontine nuclei which are closely related, as projection targets, to the system as conventionally understood. Of course, extended connectionist criteria would soon result in the inclusion of virtually all the brain in a meaningless term; but the question is where most logically to draw the line. One way would be to include those basal structures affected pathologically in one or other of the "extrapyramidal" motor syndromes, which would associate the definition with pragmatic origins, even if the clinical pathology is often not confined to what we all know as the basal ganglia.

Extrapyramidal system

The concept of an "extrapyramidal motor system", though disputed because it is claimed no such "system" operates independently of the motor cortex (Brodal, 1981), has traditionally been associated with the concept of basal ganglia. Perhaps some degree of independent status is due to be restored to it, since the thalamic targets of the basal ganglia project not to Area 4 motor cortex (in distinction from the cerebellar outflow), but to premotor, supplementary motor, and prefrontal cortex, i.e. "association areas" (DeLong and Georgopoulos, 1981), and thus engage the pyramidal tract system only by quite indirect paths. Furthermore there are now established a number of important, more direct outlets, to tecto-spinal neurones (e.g. Chevalier et al., this volume), and to lower brain stem nuclei (e.g. Jackson and Crossman, 1983) that, like the superior colliculus, may engage spinal motor structures, in pathways distinguishable anatomically and functionally from the specialized pyramidal cortico-spinal system.

Similarly, an extrapyramidal system may be broadly distinguished according to a set of characteristic abnormal movements and motor deficits (Bradley, this volume) that are associated with disease or damage to different parts of the basal ganglia, and are usually distinct from the real paralysis of a stroke affecting the cortico-spinal system.

Interconnected System of Basal Ganglia.

The anatomic interconnections of the striatum, globus pallidus, subthalamic nucleus (corpus Luysii), and substantia nigra appear to confer enough cohesion to justify the term "system". This system has some characteristic anatomic features, as reviewed by several in this volume:

Convergence. The most important distinctive feature is probably the extreme degree of convergence shown in the funnelling of pathways from the entire cortex to the output nuclei of the basal ganglia (Percheron, this volume). The inputs already converge through the striatal nuclei which comprise relatively numerous cells, down to the pallidonigral output neurones, which have dendritic fields wide enough to catch input from large bundles of axons derived from striatal neurones of many diverse functional relations as defined by the cortical territories projecting to them. Percheron (statement made in symposium discussion) considers this funnelling convergence to be so severe that only a remnant of the cortico-striatal topography will be left evident in pallidonigral output functions. The striatum thus transforms highly differentiated input from the cortex into a recoded output, using a code that is largely unknown since the interesting neural correlates of motor and complex behaviour studied by Anderson, DeLong, Rolls and others refer to cells not identified as output neurons. The recoded striatal output appears to operate on the pallidonigral targets not with reference to somatic or other sensory-motor topographic signatures, but on other principles. Such different coding may be needed, for example, to move a hand in a straight line in a particular direction with a predetermined speed (Rolls, symposium discussion).

Topography. Despite the marked convergence inferred from the three-dimensional reconstructions of Golgi material by Percheron, the results with small deposits of axon transport markers (Carpenter, this volume) indicate a systematic topographic representation in the striato-pallidonigral system, described by Faull (symposium discussion) as being reminiscent of the precise corticothalamic representation. There is then both topography and convergence, the details of each requiring further investigation. The scale of these opposing arrangements may be disputed according to the techniques used - impregnation with Golgi stain or intracellular horse-radish peroxidase, axon transport connections (Faull, Parent, this volume), or neurophysiological correlates of movement of body parts (Anderson, 1977; DeLong et al., 1983) - but the coexistence of the two contrary organizational tendencies challenges explanation.

The manner of connection appears to allow relatively strict topography within a geometry of extreme convergence. There is produced a form of neurophysiological local sign (without which no organized movement relationship would be possible) by some combination of information derived from sensorimotor and association areas. By obscure mechanisms, this results in, for example, putamen and pallidal cells firing in relation to direction and amplitude of limb movements, and determining "specific parameters of movement, independent of the muscular pattern" (De Long, 1982).

Of course, the convergence may be less intense in synaptic terms than it seems, for the statistically rare axo-dendritic synapses made by strio-pallidal axon bundles, for example, may not be random connections as is assumed in Percheron's model. It would be important to find out whether the different axons in the bundle traversing orthogonally a set of stacked pallidal dendritic discs, and converging from different striatal regions (see Percheron, Fig. 7, this volume) differ systematically in their mode of dendritic synaptic contact, whether according to local morphology, proximo-distal location on dendrites, or in selectivity for particular discs in the stack traversed by a bundle. The further possibility of chemical specificity will be mentioned below.

Characteristic outputs. The direct outputs of the basal ganglia via the nigro-tectal fibres to the cells of origin of the tecto-spinal tracts (Chevalier et al., this volume), provide a means to influence the postural bases and other antecedents of movements, particularly those related to attention, change of on-going motor pattern, and spatial orientation of action. The entrainment of spinal motor mechanisms at all levels, by neck muscle activities primarily influenced from tecto spinal fibres and by tecto-reticulo-spinal pathways, illustrates that the basal ganglia may thus have motor functions independent of the cortex. At the same time, motor cortical fibres to the subthalamic nucleus (Carpenter, this volume) provide the possibility of cortical intervention via subthalamic efferents in formulating pallidonigral outputs and influencing the pedunculo-pontine nucleus, whose relation to the midbrain "locomotor area" (Grillner, 1981, p.1210) requires more detailed investigation.

The basal ganglia outputs to ventral thalamic nuclei that in turn project to prefrontal, premotor and supplementary motor areas (Carpenter, this volume; DeLong and Georgopoulos, 1981) in addition imply that although it receives a disproportionate volume of sensorimotor cortical input, to the putamen, (Carpenter; Percheron, this volume), the striato-pallidonigral system influences pyramidal outflow almost exclusively through its contribution to the planning of action and programming of movement, a contribution synthesized using input derived from the whole cortex.

Recurrent pathways. A final characteristic of basal ganglia connections is the presence of many internal "feed-back" links. To the well-known loops involving the subthalamic nucleus, pallidum, substantia nigra, and pedunculopontine nucleus; the nigro-striato-nigral loop; and the pallidofugal loop through posterior intralaminar thalamus back to striatum, there should now be added striato-cortical (Jayaraman, 1980) and some pallido-striatal (Arbuthnott et al., this volume; Jayaraman, 1983) connections. Their neurophysiological and behavioural functions remain to be elaborated.

Neurochemical Features

The basal ganglia are notorious for holding an ever-expanding number of putative synaptic transmitters and modulators: acetyl-choline, dopamine, glutamate, GABA, serotonin, and the protean neuropeptides; but in this they are no longer to be distinguished from cortex, hypothalamus, or other CNS regions. Nonetheless, with its microscopically featureless large domain, the neostriatum appears like a neurochemical factory. The recently described "modules", compartments specifiable according to an increasing variety of chemical indices (Goldman-Rakic, 1983; Graybiel, 1983; Goedert et al., 1983, 1984) and more-or-less in register, suggest possibilities of microregional organization according to chemically-specific principles in both internal striatal data-processing and extrinsic input and output connections. This could provide a further mechanism for extracting specificity of output from the grossly convergent projections to pallidum and striatum. For the possibility arises that the biochemical "signature" of a module is preserved in some manner through to the pallidonigral outputs to thalamus, tectum, and brain stem tegmentum. Such functionally specific possibilities for the striatal modules remain entirely to be investigated.

CHARACTERISTICS OF NEURONAL PROCESSES

The neural mechanisms employed in the basal ganglia system have some distinctive properties.

Inhibitory cascade

The most familiar peculiarity is the sequence of GABAergic synapses, including projections from striatal output cells to pallidum and substantia nigra, from these nuclei to thalamus, tectum and tegmentum, from external pallidum to subthalamic nucleus, and from subthalamic nucleus to entopeduncular nucleus, nigra, and pedunculopontine nucleus (Rouzaire-Dubois et al., this volume). A cascading series of inhibitory connections implies the passage of information by disinhibition. This in turn requires parallel excitatory drive either by spontaneous intrinsic mechanisms, or from extrinsic excitatory inputs such as may be furnished in part by substance P axons of striatal Spiny II cells (Carpenter, this volume). The possible existence of other parallel excitatory pathways in the system requires systematic investigation, while the capacity for endogenous and truly spontaneous repetitive firing needs neurophysiological study in tissue-slices and tissue-cultures of the region.

Multiple modes of communication

The inhibitory GABAergic and excitatory substance P outputs of the striatum, though more or less parallel influences on pallidum and nigra, are not in total register at least in their regional striatal sources (Fonnum and Walaas, 1979). To them must be added enkephalinergic outputs (e.g. Haber and Nauta, 1983) and possibly other pathways leaving the striatum, using others of the peptides contained in striatal cells, such as neurotensin or angiotensin (Simonnet et al, 1979), as modulators of the dominant inhibition in mechanisms yet to be established. It is perhaps relevant that presynaptic modulation of transmission seems widespread in the system, judging from the evidence for presynaptic receptors on transmitter-characterized nerve terminals (Arbuthnott et al.; Beart, this volume). But since the EM evidence points to a scarcity of axo-axonal synapses, the influence of such receptors may depend on more diffuse chemical changes in the neuronal microenvironment, as now suggested for the significance of the dopaminergic input to striatum (Schulz et al., 1983).

Greenfield's suggestion (this volume) that enzymes transported along nigrostriatal axons and released from active terminals have the capacity to influence post-synaptic neural activity (e.g. AChEsterase), further blurs an already wavering distinction between transmitters and modulators at sites of interneuronal communication in the basal ganglia. This may well exemplify a more widespread form of "trophic" or "microenvironmental" interaction among neurons, that contributes to mechanisms for growth, differentiation, repair, code specification, plasticity and so on, all important areas not dealt with in this symposium.

It is only about 70 years since Ramon y Cajal so brilliantly modelled neural logic in many brain regions on the basis of Golgi-stained elements, with circuits which allowed for no inhibitory modulation of an excitatory network. Several decades later in a major monograph on the nervous system, central inhibition was interpreted in terms of the axonal subnormal period: "Inhibition.... means nothing more than failure to respond when stimulated" (Fulton, 1943, p.375) Humoral mediators of transmission were alright for the peripheral parasympathetic system, but unproven and unnecessary hypotheses for brain networks.

Nearly all of the multiform chemically and functionally specific interneuronal signalling that we begin to perceive has been identified in about the last decade of that time and the pace of discoveries demanding new concepts continues to grow. It is curious that a large a proportion of such recent discoveries having general significance have been made by students of the basal ganglia.

SPECIALIZED FUNCTIONS IN BRAIN AND BEHAVIOUR

Motor or Cognitive Relations

What is the special contribution of the basal ganglia to the control of behaviour? The many suggestions may be lumped into two classes: a traditional neurological view based on the prominent motor disorders characteristic of diseases affecting the basal ganglia (Bradley, this volume; Marsden 1980); and recent neuro-behavioural concepts derived from the more subtle disturbances of complex performance to be observed, without significant motor abnormality, in experimental animals following restricted interference in the system (Cools, this volume; Oberg and Divac, 1979). The two views may not be altogether distinct.

If interference with the basal ganglia in animals is sufficiently drastic in effect on strategic targets, as in destruction of the dopaminergic nigrostriatal pathway, then profound motor abnormalities such as akinesia or severe postural disturbances are produced in rats and monkeys (Ohye et al. this volume; Viallet et al., this volume) and are quite comparable to some of the clinical disorders.

The conventional view holds that the basal ganglia are specialized to form the cerebral end of a motor system, parallel to the pyramidal pathway controlling precise voluntary movements. This "extrapyramidal" system provides control of supporting postural activity, direction and "energy" of movement, (Anderson, this volume; Hallet and Khoskin, 1980; Viallet et al., this volume), and perhaps planning of movements (Marsden, 1980).

At a more complex level, the neostriatum would appear to control changes in the direction of attention or in motor program (Cools, this volume), translating cortical cognitive computations into action patterns (Rolls, this volume).

It is possible that the neostriatum can control the access of polysensory information to the cortex (Feltz et al., 1967; McKenzie and Rogers, 1973; Krauthamer, 1979). The evidence for this depends to a considerable extent on non-physiological pathway activation by focal electrical stimuli (Kunze et al. 1979). However, the existence of striato-cortical projections (Jayaraman, 1980) means that caudate stimulation may also succeed in engaging both orthodromic projections to sensory cortex, as well as polysynaptic pathways influencing medial thalamic and brain stem nuclei (Bendrups and McKenzie, 1981) known to participate in polysensory afferent conduction and possibly concerned with attention.

Involvement of the basal ganglia in more complex ("cognitive") processes intercalated between sensory input and motor execution has been supported with a variety of techniques. Lesions or drug injections in different definable striatal zones cause deficits selectively in delayed response or reversal types of behaviour (Oberg and Divac, 1979); electrical stimulation may have marked effects on various types of conditioning and memory behaviour (reviewed by Oberg and Divac, 1979) but is not free of confounding physiological artefacts as a result of antidromic and collateral axon activation (Albe-Fessard et al., 1983).

Potentially the most powerful method is to record neuronal activities correlated with well-specified actions. However, the correlations observed are necessarily limited by the range of behaviour systematically studied in the experimental animal. Not suprisingly, the investigation of simple movement correlates has contributed mainly to knowledge of the topographic properties of sensory motor representation, and of the types of motor act encoded in the basal ganglia (DeLong et al. 1983; Iansek and Porter, 1980). But Rolls (1979, and this volume), using a more comprehensive behavioural paradigm, succeeds in going further in the subdivision of response-categories, particularly in the caudate nucleus, where neurone firing is related to such things as use of cues and switching of motor programs.

New Concepts of Basal Ganglia Function

No salient reason appears for choosing, from the diverse evidence, particular overt activities as being the prime responsibility of the basal ganglia as a whole. There are two obvious responses to this situation.

On the one hand, it may be possible to parcellate the various functions associated with the basal ganglia into discrete categories, each more or less homogeneous and related to its own special striatal region; there is a precedent for this in the way functions are fractionated among cortical areas. For example, dorsal and ventral striatal regions clearly relate to different aspects of motor (DiChiara and Morelli, this volume) and "cognitive" behaviour (Divac, 1977; Rolls, this volume). It could well be that at finer grain the subdivisions of the striatum are concerned with quite limited parameters of behaviour, in dependence on their cortical connections as suggested by Divac (1977) some years ago. If so, such subdivision would probably be "orthogonal" to the microheterogeneous organization into modular, rostro-caudal "columnar" networks described by Goldman-Rakic (1983) and Graybiel (1983), in which the same neurochemical patchy mosaic seems to be re-represented throughout the striatum, and would have information-processing significance transcending particular behaviour-patterns.

On the other hand, the fact of convergent funnelling, even allowing possible chemically specific connections within its broad pattern (see above), provokes the attempt to combine conceptually what are superficially different functions of the system. Cools et al. (this volume) suggest a general function of the caudate nucleus in the arbitrary switching from one behaviour (in particular, a motor program) to another independently of exteroceptive cues, and provide a model for the involvement of the basal ganglia components in the programming of action. Their schema includes a detailed breakdown of the hierarchical contributions to the system's flow of information by descending tiers of nuclei, using feed-back for error-correction at each level, to give successively greater specification of the parameters of action (with a parallel loss of degree of freedom).

One might go further, to ask if there is possibly some fundamental function of the basal ganglia, a sort of "deep structural" function, that cannot be expressed directly in terms of overt behavioural or movement variables. This would imply some "abstract", information-processing function of the network, and generates subsidiary questions. For example, could the "cognitive" relations of the basal ganglia and their neural activity-patterns depend upon an underlying "fictive motor" function of switching among "internal models" of attentional or other actions, only then to be expressed in overt movement correlates? Could a failure to make use of environmental cues in preparations to act, which was suggested by Rolls to follow a loss of dopamine from the striatum, account for both the akinesia of Parkinsonism and the sort of subjective "explanation" of it by Bradley's patient? (see this volume p. 336).

Most of such questions are not answerable, or even capable of clear formulation. They and, it is to be hoped, more definite ones will be considered with greater confidence (or ignored as irrelevant) as new discoveries in the structure, processes, and function of the basal ganglia continue to be made.

REFERENCES

Albe-Fessard, D., Cesaro, P., and Hamon, B., 1983, Effect of striatal stimulation on cellular activities of medial thalamic neurons studied in rats, Exp. Brain Res., 50: 34

Anderson, M.E., 1977, Discharge patterns of basal ganglia neurons during active maintenance of postural stability and adjustment to chair tilt, Brain Res., 143: 325.

Brodal, A., 1981, "Neurological Anatomy in Relation to Clinical Medicine," 3rd Ed., Oxford University Press, New York.

De Long, M. R., 1982, Cortico-basal ganglia loops, communication to satellite symposium of 1st World Congress of IBRO: "Neural Coding of Motor Performance", Marseille,1982.

De Long, M.R., and Georgopoulos, A.P., 1981, Motor functions of the basal ganglia, in: "Handbook of Physiology", Section 1, Volume 2, Motor Control Part 2, J.M. Brookhart and V.B. Mountcastle, eds., American Physiological Society, Bethesda.

De Long, M.R., Crutcher, M.D., and Georgopoulos, A.P., 1983, Relations between movement and single cell discharge in the substantia nigra of the behaving monkey, J. Neuroscience, 3: 1599.

Divac, I., 1977, Does the neostriatum operate as a functional entity?, in: "Psychobiology of the Striatum," A.R. Cools, A.H.M. Lohman, J.H.L. Van Den Bercken, eds., North Holland Amsterdam.

Fonnum, F., and Walaas, I., 1979, Localization of neurotransmitter candidates in neostriatum, in: "The Neostriatum", I.Divac and R.G.E. Oberg, eds., Pergamon, Oxford.

Georgopoulos, A.P., De Long, M.R., and Crutcher, M.D, 1983, Relations between parameters of step-tracking movements and single cell discharge in the globus pallidus and subthalamic nucleus of the behaving monkey, J. Neuroscience, 3: 1586.

Goldman-Rakic, P.S., 1983, The corticostriatal fiber system in the rhesus monkey: organization and development, Progress in Brain Research, 58: 405.

Graybiel, A.M., 1983, Compartmental organization of the mammalian striatum, Progress in Brain Research, 58: 247.

Goedert, M., Mantyh, P.W., Emson, P.C., and Hunt, S.P., 1984, Inverse relationship between neurotensin receptors and neurotensin-like immunoreactivity in cat striatum, Nature, 307: 543.

Goedert, M., Mantyh, P.W., Hunt, S.P., and Emson, P.C., 1983, Mosaic distribution of neurotensin-like immunoreactivity in the cat striatum, Brain Res. 274: 176.

Grillner, S., 1981, Control of locomotion, in: "Handbook of Physiology", Section 1, Vol.2, Motor Control, Part 2, J.M. Brookhart and V.B. Mountcastle, eds., American Physiol. Soc., Bethesda.

Haber, S.N., and Nauta, W.J.H., 1983, Ramifications of the globus pallidus in the rat as indicated by patterns of immunohistochemistry, Neuroscience, 9: 245.

Heimer, L., and Van Hoesen, G., 1979, Ventral striatum, in: "The Neostriatum", I. Divac and R.G.E. Oberg, eds., Pergamon, Oxford.

Iansek, R., and Porter, R., 1980, The monkey globus pallidus: neuronal discharge properties in relation to movement, J. Physiol., 301: 439.

Jackson, A., and Crossman, A.R., 1983, Nucleus tegmenti pedunculopontinus: efferent connections with special reference to the basal ganglia, studied in the rat by anterograde and retrograde transport of horseradish peroxidase, Neuroscience, 10: 725

Jayaraman, A., 1980, Anatomical evidence for cortical projections from the striatum in the cat, Brain Res., 195: 29.

Jayaraman, A., 1983, Topographic organization and morphology of peripallidal and pallidal cells projecting to the striatum in cats, Brain Res., 275: 279.

Krauthamer, G.M., 1979, Sensory functions of the neostriatum, in: "The Neostriatum", I. Divac and R.G.E. Oberg, eds., Pergamon, Oxford.

Kunze, W., McKenzie, J.S., and Bendrups, A.P., 1979, An electrophysiological study of thalamo-caudate neurones in the cat, Exp. Brain Res., 36: 233.

Marsden,C.E., 1980, The enigma of the basal ganglia and movement, Trends in Neurosciences, 3: 284.

Mogenson, G.J., Jones, D.L., and Yim, C.Y., 1980, From motivation to action: functional interface between the limbic system and the motor system, Progress in Neurobiology, 14: 69.

Oberg, R.G.E., and Divac, I., 1979, "Cognitive" functions of the neostriatum, in: "The Neostriatum", I. Divac and R.G.E. Oberg, eds., Pergamon, Oxford.

Rolls, E.T., Thorpe, S.J., Maddison, S., Roper-Hall, A., Puerto, A., and Perret, D., 1979, Activity of neurones in the neostriatum and related structures in the alert animal, in: "The Neostriatum", I. Divac and R.G.E. Oberg, eds., Pergamon, Oxford.

Schultz, W., Ruffieux, A., and Aebischer, P., 1983, The activity of pars compacta neurons of the monkey substantia nigra in relation to motor activation, Exp. Brain Res., 51: 377.

Simonnet, G., Giorguieff, M.F., Carayon, A., Bioulac, B., Cesselin, F., Glowinski, J., and Vincent, J.D., 1981, Angiotensine II et système nigro-néostriatal, J. Physiol. (Paris), 77: 71.

CONTRIBUTORS

D. Albe-Fessard
Laboratoire de Physiologie des Centres Nerveux
Université Pierre et Marie Curie
4 Place Jussieu
75230 Paris 05, France

M.E. Anderson
Department of Physiology and Biophysics
University of Washington
Seattle, WA 98195 USA

G.W. Arbuthnott
M.R.C. Brain Metabolism Unit
University Department of Pharmacology
1 George Square
Edinburgh EHB 9J2 Scotland

P.J. Bate
Lincoln Institute of Health Sciences
625 Swanston St
Carlton 3053 Australia

P.M. Beart
Clinical Pharmacology
Austin Hospital
Heidelberg 3084 Australia

A.P. Bendrups
Lincoln Institute of Health Sciences
625 Swanston St
Carlton 3053 Australia

G. Bernardi
Cattedra de Clinica Neurologica
Universitá di Roma
Viale Dell'Universita 30
00185 Roma Italy

B. Bioulac
Université de Bordeaux 11
146 Rue Leo-Saignat
33076 Bordeaux France

K.C. Bradley
Emeritus Professor of Anatomy University of Melbourne
St. Andrews Hospital
East Melbourne 3002
Australia

L.L. Brown
The Saul R. Korey Department of Neurology
Albert Einstein College of Medicine of Yeshiva University
1300 Morris Park Avenue
Bronx, NY 10461 USA

M.B. Carpenter
Department of Anatomy
USUHS, 4301 Jones Bridge Rd
Bethesda MD 20014 USA

A.R. Cools
Department of Pharmacology
Katholieke Universiteit Nijmegen
PO box 9101
6500HB Nijmegen Netherlands

G. Di Chiara,
Institute of Experimental Pharmacology and Toxicology
University of Cagliari
09100 Cagliari Italy

I. Divac
Neurofysiologisk Institut
Panum Instituttet, Kovenhavns Universitet
Blegdamsvej 3 c
2200 Copenhagen N, Denmark

P.W. Everett
Physiology Department
University of Melbourne
Parkville 3052 Australia

R.L.M. Faull
Department of Anatomy
Auckland University
Auckland New Zealand

J. Féger
Laboratorie de Physiologie
Universite de Paris VI
97 Bld de L'Hôpital
75634 Paris 13 France

S.M. Feldman
see L.L. Brown, USA

P.G. Giacomini
see G. Bernardi, Italy

S.A. Greenfield
University Laboratory of Physiology
University of Oxford
Parks Rd
Oxford OX1 3PT England

I. Grofova
Department of Anatomy
Michigan State University
East Lansing, MI 48824 USA

R.G. Hall
Physiology Department
University of Melbourne
Parkville 3052 Australia

P.F. Harris,
School of Biological Sciences
Flinders University
Bedford Park 5042 Australia

C. Heim
Max-Planck-Institut für
Experimentelle Medizin
Hermann-Rein Strasse 3
D-3400 Gottingen West Germany

A. Jayaraman
1542 Tulane Ave
New Orleans LA 70112 USA

M. Joschko
School of Biological Sciences
Flinders University
Bedford Park 5042 Australia

R.E. Kemm
Physiology Department
University of Melbourne
Parkville 3052 Australia

W. Kunze
Physiology Department
University of Melbourne
Parkville 3052 Australia

J.M. McMeeken
Lincoln Institute of Health
Sciences
625 Swanston Street
Carlton 3053 Australia

J.S. McKenzie
Physiology Department
University of Melbourne
Parkville 3052 Australia

F.J. Noth
Medizinische Einrichtungen der
Universitat Dusseldorf,
Moorenstr 5
4000 Dusseldorf 1 West Germany

R.G.E. Oberg
Department of Neurology &
Neurosurgery
University Hospital
Copenhagen, Denmark

C. Ohye
Department of Neurosurgery, Gunma
University
3-39-22 Showa-Machi
Maebishi Gunma Japan

D.H. Overstreet
Department of Biological Sciences
Flinders University
Bedford Park 5042 Australia

A. Parent
Laboratoire de Neurobiologie
Universite Laval
Quebec Canada

P. Pasik
Department of Neurology
The Mount Sinai School of Medicine
of the City University of New York
New York NY 10029 USA

G. Percheron
INSERM Unité de Recherches
Neurophysiologiques - U3
Hôpital de la Salpêtrière
47 Bd de L'Hôpital
75013 Paris France

A.G. Phillips
Department of Physiology
University of British Columbia
Vancouver BC V6T 1W5 Canada

S. Polgar
Lincoln Institute of Health
Sciences
625 Swanston St
Carlton 3053 Australia

E.T. Rolls
Department of Experimental
Psychology
University of Oxford
South Parks Road
Oxford OX1 3UD England

G.J. Royce
Department of Anatomy
University of Wisconsin
Madison WI 53706 USA

M. Schwartz
Max-Planck-Institut für
Experimentelle Medizin
Hermann-Rein Strasse 3
D-3400 Gottingen West Germany

T. Shafton
Department of Physiology
Melbourne University
Parkville 3052 Australia

K.H. Sontag
Max-Planck-Instut für
Experimentelle Medizin
Hermann-Rein Strasse 3
D-3400 Gottingen West Germany

P. Stanzione
Cattedra di Clinica Neurologica
Universitá di Roma
Viale Dell'Universitá 30
00185 Roma Italy

Sir Sydney Sunderland
Emeritus Professor of
Experimental Neurology
University of Melbourne
Parkville 3052, Australia

E. Trouche
Département de Neurophysiologie
Générale
Institut de Neurophysiologie et
de Psychophysiologie
CNRS - INPS - BP 71
13277 Marseille 9 France

A. Ueki
Department of Neurology, Jichi
Medical School
Minami Kawachi-Machi
Tochigi-Ken Japan

L.N. Wilcock
Department of Physiology
University of Melbourne
Parkville 3052 Australia

G. Willis
Monash University
Department of Psychological
Medicine
Prince Henry's Hospital
St Kilda Rd
Melbourne 3004 Australia

INDEX